Lecture Notes in Physics

Springer
Berlin
Heidelberg
New York
Barcelona
Hong Kong
London
Milan
Paris
Singapore
Tokyo

Physics and Astronomy

ONLINE LIBRARY

http://www.springer.de/phys/

Editorial Policy

The series *Lecture Notes in Physics* (LNP), founded in 1969, reports new developments in physics research and teaching -- quickly, informally but with a high quality. Manuscripts to be considered for publication are topical volumes consisting of a limited number of contributions, carefully edited and closely related to each other. Each contribution should contain at least partly original and previously unpublished material, be written in a clear, pedagogical style and aimed at a broader readership, especially graduate students and nonspecialist researchers wishing to familiarize themselves with the topic concerned. For this reason, traditional proceedings cannot be considered for this series though volumes to appear in this series are often based on material presented at conferences, workshops and schools (in exceptional cases the original papers and/or those not included in the printed book may be added on an accompanying CD ROM, together with the abstracts of posters and other material suitable for publication, e.g. large tables, colour pictures, program codes, etc.).

Acceptance

A project can only be accepted tentatively for publication, by both the editorial board and the publisher, following thorough examination of the material submitted. The book proposal sent to the publisher should consist at least of a preliminary table of contents outlining the structure of the book together with abstracts of all contributions to be included.

Final acceptance is issued by the series editor in charge, in consultation with the publisher, only after receiving the complete manuscript. Final acceptance, possibly requiring minor corrections, usually follows the tentative acceptance unless the final manuscript differs significantly from expectations (project outline). In particular, the series editors are entitled to reject individual contributions if they do not meet the high quality standards of this series. The final manuscript must be camera-ready, and should include both an informative introduction and a sufficiently detailed subject index.

Contractual Aspects

Publication in LNP is free of charge. There is no formal contract, no royalties are paid, and no bulk orders are required, although special discounts are offered in this case. The volume editors receive jointly 30 free copies for their personal use and are entitled, as are the contributing authors, to purchase Springer books at a reduced rate. The publisher secures the copyright for each volume. As a rule, no reprints of individual contributions can be supplied.

Manuscript Submission

The manuscript in its final and approved version must be submitted in camera-ready form. The corresponding electronic source files are also required for the production process, in particular the online version. Technical assistance in compiling the final manuscript can be provided by the publisher's production editor(s), especially with regard to the publisher's own Latex macro package which has been specially designed for this series.

Online Version/ LNP Homepage

LNP homepage (list of available titles, aims and scope, editorial contacts etc.):
http://www.springer.de/phys/books/lnpp/
LNP online (abstracts, full-texts, subscriptions etc.):
http://link.springer.de/series/lnpp/

C. F. Barenghi R. J. Donnelly W. F. Vinen (Eds.)

Quantized Vortex Dynamics and Superfluid Turbulence

Springer

Editors

C.F. Barenghi
University of Newcastle
Mathematics Department
Newcastle NE1 7RU, United Kingdom

R.J. Donnelly
University of Oregon
Physics Department
Eugene, OR 97403, USA

W.F. Vinen
University of Birmingham
Physics Department
Birmingham B15 2TT, United Kingdom

Cover picture: Tangle of quantized vortex filaments computed in a periodic box by D. Kivotides, D. Samuels and C.F. Barenghi.

Library of Congress Cataloging-in-Publication Data applied for.

Die Deutsche Bibliothek - CIP-Einheitsaufnahme

Quantized vortex dynamics and superfluid turbulence / C. F. Barenghi ...
(ed.). - Berlin ; Heidelberg ; New York ; Barcelona ; Hong Kong ; London ;
Milan ; Paris ; Singapore ; Tokyo : Springer, 2001
  (Lecture notes in physics ; 571)
  (Physics and astronomy online library)
  ISBN 3-540-42226-9

ISSN 0075-8450
ISBN 3-540-42226-9 Springer-Verlag Berlin Heidelberg New York

Typesetting: Data conversion by Steingraeber GmbH, Heidelberg
Cover design: *design & production*, Heidelberg

Printed on acid-free paper
SPIN: 10792065   55/3141/du - 5 4 3 2 1 0

Preface

This book springs from the programme *Quantized Vortex Dynamics and Superfluid Turbulence* held at the Isaac Newton Institute for Mathematical Sciences (University of Cambridge) in August 2000. What motivated the programme was the recognition that two recent developments have moved the study of quantized vorticity, traditionally carried out within the low-temperature physics and condensed-matter physics communities, into a new era.

The first development is the increasing contact with *classical fluid dynamics* and its ideas and methods. For example, some current experiments with helium II now deal with very classical issues, such as the measurement of velocity spectra and turbulence decay rates. The evidence from these experiments and many others is that superfluid turbulence and classical turbulence share many features. The challenge is now to explain these similarities and explore the time scales and length scales over which they hold true. The observed classical aspects have also attracted attention to the role played by the flow of the normal fluid, which was somewhat neglected in the past because of the lack of direct flow visualization. Increased computing power is also making it possible to study the coupled motion of superfluid vortices and normal fluids. Another contact with classical physics arises through the interest in the study of superfluid vortex reconnections. Reconnections have been studied for some time in the contexts of classical fluid dynamics and magneto-hydrodynamics (MHD), and it is useful to learn from the experience acquired in other fields.

The second development arises from *atomic physics* and is the discovery of Bose–Einstein condensation in confined clouds of alkali atoms. The study of superfluidity and quantized vorticity is now possible in a wide range of other systems besides helium II. The rapid progress in this area has given momentum to the use of the Gross–Pitaevskii Equation or Nonlinear Schroedinger Equation (NLSE). Researchers have become more aware of the approximations and limitations involved in the NLSE model, but also of its range of validity and great power of prediction. The use of the NLSE has become more established, and the NLSE is proving to be a powerful tool for modeling problems such as vortex nucleation, reconnections and even turbulence.

A further development arises from the results of preliminary theory and experiments in turbulent Helium 3 which suggest that there are significant differences with turbulence in Helium 4 and these are likely to be explored in the future.

It is apparent from this background that the contributions to this book come from investigators with a wide range of backgrounds and expertise: condensed-matter physics and low-temperature physics, classical fluid dynamics and applied mathematics, MHD, atomic physics, and engineering (for the applications of helium II as a cryogenic coolant).

The book is divided into topical chapters. Each chapter begins with one or two introductory review articles, which are suitable for students and new investigators interested in entering the field. The introductory articles are followed by shorter, more specialized papers.

Chapter 1 introduces us to the problem of quantized vorticity and superfluid turbulence, and it summarizes the key aspects and problems which are currently studied. Chapter 2 is devoted to turbulence experiments. Chapter 3 considers the fundamental problem of friction and vortex dynamics. The theory of superfluid turbulence and the interpretation of the experimental results is the subject of Chap. 4. Chapter 5 is devoted to the application of the NLSE model to superfluidity and vortices. Chapter 6 moves away from helium and considers Bose–Einstein Condensation and vortices in the context of alkali atoms. Chapter 7 is concerned with some aspects of classical turbulence and MHD which are relevant in the study of superfluid turbulence. Finally, Chap. 8 deals with Helium 3 and other systems.

We are grateful for the support and encouragement of Professor Keith Moffatt, Director of the Newton Institute, and we would like to thank Tracey Andrew who helped in the preparation of the manuscripts for publication.

Newcastle, Eugene and Birmingham,
June 2001

Carlo Barenghi
Russ Donnelly
Joe Vinen

Contents

Part V   The NLSE and Superfluidity

Part VII   Vortex Reconnections and Classical Aspects

List of Contributors

C.S. Adams
Physics Department
University of Durham
Durham DH1 3LE, UK
c.adams@dur.ac.uk

C.F. Barenghi
University of Newcastle
Mathematics Department
Newcastle upon Tyne, NE1 7RU
UK
c.f.barenghi@ncl.ac.uk

N. Berloff
Mathematics Department
University of California
Los Angeles, CA 90095-1555, USA
nberloff@math.ucla.edu

R. Blossey
Department of Physics
University of Essen
45117 Essen, Germany
blossey@theo-phys.uni-essen.de

A. Brandenburg
NORDITA
Blegdamsvej 17
2100 Copenaghen, Denmark
brandenb@nordita.dk

H.P. Buechler
Theoretical Physics
ETH
8093 Zuerich, Switzerland
buechler@itp.phys.ethz.ch

R.J. Donnelly
Physics Department
University of Oregon
Eugene, OR 97403, USA
russ@vortex.uoregon.edu

L. Eaves
School of Physics and Astronomy
University of Nottingham
Nottingham NG7 2RD, UK
Laurence.Eaves@nottingham.ac.uk

A. Fetter
Geballe laboratory for Advanced
Materials
Stanford University
Stanford, CA 94305-4045, USA
fetter@stanford.edu

K.L. Henderson
Faculty of Computer Studies and
Mathematics
University of the West of England
Bristol BS16 1QY, UK
karen.henderson@uwe.ac.uk

D.D. Holm
Theoretical Division
Mail Stop B284
Los Alamos National Laboratory
Los Alamos NM 87545, USA
holm@lanl.gov

G. Hornig
Dept. Theoretical Physics IV
Ruhr-Universitaet Bochum
44780 Bochum, Germany
gh@egal.tp4.ruhr-uni-bochum.de

C. Huepe
James Frank Institute
University of Chicago
5640 S. Ellis Avenue
Chicago, IL 60637, USA
cristian@eclectise.uchicago.edu

O. Idowu
Center for Turbulence Research
Stanford University
Stanford, CA 94305-3030, USA
idowu@nas.nasa.gov

D. Kivotides
Mathematics Department
University of Newcastle
Newcastle NE1 7RU, UK
demosthenes.kivotides@ncl.ac.uk

J. Koplik
Levich Institute, T-1M
City College of New York
New York, NY 10031, USA
koplik@sci.ccny.cuny.edu
koplik@lid3a0.engr.ccny.cuny.edu

M. Krusius
Low Temperature Laboratory
Helsinki University of Technology
02015 HUT
Finland
krusius@neuro.hut.fi

H. Kuratsuji
Department of Physics
Ritsumeikan University
Kusatsu City 525-8577, Japan
kra@se.ritsumei.ac.jp

M. Leadbeater
Physics Department
University of Durham
Durham DH1 3LE, UK
Mark.Leadbeater@durham.ac.uk

T. Lipniacki
Institute of Fundamental Technological Research
Świętokrzyska St. 21
00-049 Warsaw, Poland
tlipnia@ippt.gov.pl

P.V.E. McClintock
University of Lancaster
Physics Department
Lancaster LA1 4YB, UK
p.v.e.mcclintock@lancaster.ac.uk

S. Nazarenko
Mathematics Institute
University of Warwick
Coventry CV4 7AL, UK
snazar@math.warwick.ac.uk

S.K. Nemirovskii
Institute of Thermosphysics
630090 Novosibirsk, Russia
nem@nsu.ru

J.J. Niemela
Physics Department
University of Oregon
Eugene, OR 97403, USA
joe@vortex.uoregon.edu

M. Niemetz
Institut für Experimentelle
und Angewandte Physik
Universität Regensburg
93040 Regensburg, Germany
michael.niemetz@
physik.uni-regensburg.de

V. Penna
Dipartimento di Fisica
and INFM
Politecnico di Torino
C.so Duca degli Abruzzi 24
10129 Torino, Italy
penna@athena.polito.it

L.M. Pismen
Department of Chemical Engineering
and Minerva Center for Nonlinear
Physics of Complex Systems
Technion - Israel Institute
of Technology
32000 Haifa, Israel

S. Rica
CMM CNRS UCHILE
Av Blanco Encalada 2120
Santiago, Chile
rica@ens.fr
rica@dim.uchile.cl

L.R. Ricca
Mathematics Department
University College London
Gower Street
London WC1E 6BT, UK
ricca@math.ucl.ac.uk

P.H. Roberts
Mathematics Department
University of California
Los Angeles, CA 90095-1555, USA
roberts@math.ucla.edu

W. Schoepe
Institut für Experimentelle und
Angewandte Physik
Universität Regensburg
93040 Regensburg, Germany
wilfried.schoepe@
physik.uni-regensburg.de

B.K. Shivamoggi
University of Central Florida
Orlando, FL 32816, USA
ijjmms@pegasus.cc.ucf.edu

E. Sonin
Racah Institute of Physics
Hebrew University of Jerusalem
Givat Ram,
Jerusalem 91904, Israel
sonin@cc.huji.ac.il

L. Skrbek
KFNT MFF UK
Charles University
V Holesovickach 2
180 00 Prague 8, Czech Republic
skrbek@fzu.cz

B. Svistunov
Russian Research Center
Kurchatov Institute
123182 Moscow, Russia
svist@kurm.polyn.kiae.su

A. Tsinober
Faculty of Engineering
Tel Aviv University
Tel Aviv, Israel
tsinober@eng.tau.ac.il

M. Tsubota
Department of Physics
Osaka City University
Osaka 558-8585, Japan
tsubota@sci.osaka-cu.ac.jp

S.W. Van Sciver
National High Magnetic Field
Laboratory
Florida State University
Tallahassee FL 32310, USA
vnsciver@magnet.fsu.edu

É. Varoquaux
CNRS-Université Paris-Sud
Laboratoire de Physique des Solides
Bâtiment 510
F-91405 Orsay Cedex, France
varoquaux@drecam.saclay.cea.fr

W.F. Vinen
Physics Department

University of Birmingham
Birmingham B15 2TT, UK
w.f.vinen@bham.ac.uk

G. Williams
Department of Physics and Astronomy
University of California
Los Angeles, CA 90095, USA
gaw@ucla.edu

Part I

Introduction

Introduction to Superfluid Vortices and Turbulence

Carlo F. Barenghi

Mathematics Department, University of Newcastle, Newcastle NE1 7RU, UK

1 The Two-Fluid Model

My aim in this article is to introduce the basic properties of quantized vortex lines in Helium II and summarize the main experimental observations of superfluid turbulence. Then I shall discuss a selection of the theoretical methods used to study quantized vorticity and turbulence and the results obtained using these methods.

The liquid state of ^{4}He exists in two phases: a high temperature phase called Helium I, and a low temperature phase, called Helium II. The two phases are separated by a transition called the *lambda transition*, which occurs at the critical temperature $T = T_\lambda = 2.172\ K$ at saturated vapour pressure and marks the onset of *Bose Einstein condensation* (BEC) and quantum order. The phenomenon of BEC is described in the article of Stringari. Helium I is a classical fluid which obeys the ordinary Navier - Stokes equations. Hereafter the focus of attention is only Helium II.

A simple, phenomenological model which explains the motion of Helium II is the *two - fluid theory* of Tisza and Landau [1]. In this model Helium II is described as the intimate mixture of two fluid components which penetrate each others, the *normal fluid* and the *superfluid*. Each fluid component has its own density and velocity field, ρ_n and $\mathbf{v}_n$ for the normal fluid and ρ_s and $\mathbf{v}_s$ for the superfluid. The total density of Helium II is $\rho = \rho_n + \rho_s$. The superfluid component is irrotational, and, since it carries nor entropy nor viscosity, is similar to a classical, inviscid Euler fluid. The normal fluid component is a gas of thermal excitations called *phonons* and *rotons* depending on the wavenumber. The normal fluid carries the entire entropy and viscosity of Helium II and is similar to a classical, viscous Navier - Stokes fluid.

The relative proportion of normal fluid and superfluid is determined by the absolute temperature T. At $T = 0$ Helium II is entirely superfluid: $\rho_s/\rho = 1$ and $\rho_n/\rho = 0$. If the temperature is increased the superfluid fraction decreases and the normal fluid fraction increases, until, at $T = T_\lambda$, Helium II becomes entirely normal: $\rho_s/\rho = 0$ and $\rho_n/\rho = 1$. The temperature dependence of the two fluid's fractions is nonlinear: for example ρ_n/ρ drops from 100 percent at $T = T_\lambda$ to 55 percent at $2.0\ K$ and to 7.5 percent at $1.4\ K$, and is effectively negligible at temperatures below $1\ K$.

The two - fluid model explains many observed phenomena. Among them it is worth mentioning *second sound* and *thermal counterflow* because they are important in the study of turbulence. Ordinary sound in Helium II is called *first sound*. A first sound wave is an oscillation of density ρ and pressure P in which temperature T and entropy S remain almost constant and $\mathbf{v}_n$ and $\mathbf{v}_s$ move in phase with each others. A second sound wave, on the contrary, is an oscillation of T and S in which ρ and P remain almost constant and $\mathbf{v}_n$ and $\mathbf{v}_s$ move in antiphase. Second sound is used to detect quantized vorticity.

Thermal counterflow is Helium II's special way to transfer heat. Consider a channel which is closed at one end and open to the helium bath at the other end. At the closed end a resistor dissipates a known heat flux W. In an ordinary fluid, such as water or Helium I, heat is transferred away from the resistor by conduction, provided that one is careful to prevent convective motion, so the heat flux W is proportional to the temperature gradient ∇T and there is a well defined thermal conductivity at small W. In Helium II the heat is carried by the normal fluid away from the resistor, $W = \rho S T v_n$. Because of the presence of the closed end, however, the mass flux is zero, $\mathbf{j} = \rho_s \mathbf{v}_s + \rho_n \mathbf{v}_n = 0$, so some superfluid must flow toward the resistor to conserve mass, $\mathbf{v}_s = -(\rho_n/\rho_s)\mathbf{v}_n$. In this way a *counterflow* $\mathbf{v}_{ns} = \mathbf{v}_n - \mathbf{v}_s$ is generated which is proportional to the applied heat flux, $v_{ns} = W/(\rho_s S T)$. If W exceeds a critical value, then superfluid turbulence is generated. Turbulence limits the heat transfer properties of Helium II, so it is relevant in the engineering applications, as described in the article of Van Sciver.

2 Quantized Vortex Lines

The quantization of the circulation of the superfluid makes the hydrodynamics of helium II particularly interesting [2]. Superfluid vortex lines appear when helium II rotates or moves faster than a critical velocity. This *vortex nucleation* process has been the subject of many investigations and is described in the articles of Adams and Rica. Superfluid vortex lines can be spatially organized (laminar vortex flows) or disorganized (turbulent vortex tangles). The key property of a superfluid vortex line, discovered by Onsager and developed by Feynman, is that the circulation is quantised, that is to say

$$\int_C \mathbf{v}_s \cdot \mathbf{dl} = \frac{h}{m},\tag{1}$$

where C is a circular path around the axis of the vortex. The ratio $\Gamma = h/m$ of Plank's constant and helium's mass is called the *quantum of circulation* and has value $\Gamma = 9.97 \times 10^{-4}\ cm^2/sec$.

The simplest way to create superfluid vortex lines is to rotate a cylinder filled with Helium II at constant angular velocity Ω. Provided that Ω is large enough, superfluid vortex line appear and form on ordered array of areal density $n = 2\Omega/\Gamma$, all vortex lines being aligned along the axis of rotation. In this way the superfluid mimics the vorticity 2Ω of the solid body rotating normal fluid, each vortex line contributing one quantum to the total circulation.

3   Modelling the Vortex Lines

3.1   Microscopic Model

Using cylindrical coordinates r, ϕ, z it follows from equation (1) that the superfluid velocity around the axis of a vortex line is $v_{s,\phi} = \Gamma/2\pi r$, hence it diverges as $r \to 0$. To understand what happens as $r \to 0$ it is instructive to use the nonlinear Schroedinger equation (NLSE), which is discussed in the article of Roberts. The NLSE models a gas of bosons which interact with each others via a delta function repulsive potential of strength V_0:

$$i\hbar\frac{\partial\psi}{\partial t} = -\frac{\hbar^2}{2m}\nabla^2\psi - m\mathcal{E}\psi + V_0\psi|\psi|^2, \tag{8}$$

Here $\mathcal{E}$ is the energy per unit mass and $\hbar = h/2\pi$. The wavefunction ψ can be written as $\psi = Ae^{i\Phi}$ in terms of an amplitude A and a phase Φ. In this way one can define the condensate's density $\rho_{BEC} = mA^2$ and velocity $\mathbf{v}_{BEC} = (\hbar/m)\nabla\Phi$; the last relation confirms that the superfluid is irrotational as envisaged by Landau. This transformation establishes the hydrodynamics of the model: equation (8) is equivalent to a continuity equation and an Euler equation (modified by the so called quantum pressure term).

The NLSE has a vortex solution: if Φ is the azimuthal angle ϕ then we have $v_{BEC,\phi} = \Gamma/(2\pi r)$ which is the Onsager - Feynman vortex. Substitution into the NLSE yields a differential equation for ρ_{BEC}. One finds that ρ_{BEC} tends to the bulk value $m^2\mathcal{E}/V_0$ for $r \to \infty$, and that $\rho_{BEC} \to 0$ for $r \to 0$. The characteristic distance over which ρ_{BEC} changes from its bulk value to zero is $a_0 \approx 10^{-8}$ cm. This distance is called the *vortex core parameter*. We conclude that the superfluid vortex line is hollow at the core. Geometrically, a vortex line transforms the volume occupied by the superfluid into a multiply connected region.

Hereafter we identify ρ_{BEC} with ρ_s at absolute zero and $\mathbf{v}_{BEC}$ with $\mathbf{v}_s$. It must be noted that this identification is convenient from the point of view of having a simple hydrodynamics model but is not entirely correct. The reason is that Helium II is a dense fluid, not the weakly interacting Bose gas described by the NLSE, so the condensate is not the same as the superfluid component. One should compare the case of Helium II (in which the NLSE is a rather approximate model) with the case of BEC in clouds of trapped alkali atoms (in which the NLSE is a better model because the bosons' interaction is weaker). Another drawback of the NLSE is that it fails to describe the observed dispersion relation of Helium II at high momenta. If one studies small oscillations of the uniform solution of the NLSE and interprets them as thermal excitations, one finds a dispersion relation $E = E(p)$ in which the energy E is proportional to the momentum p at small p (phonons) and then becomes quadratic in p at high p (free particles). The spectrum of excitations observed in Helium II is different, because the phonon part is followed by the rotons' minimum. Despite these shortcomings, of which one must be aware, the NLSE is much used as a convenient hydrodynamical model of Helium II at $T = 0$.

3.2   Mesoscopic Model

An important feature of the NLSE model is that it makes visible what happens on the microscopic scale ($\approx 10^{-8}$ *cm*) of the vortex core parameter. This makes the NLSE a useful tool to investigate phenomena such as vortex nucleation and vortex reconnections. There are however other problems in which one is concerned with the behaviour of a large number of vortex lines. In these problems the resolution of what happens on the microscopic scale is not necessary and can even be a waste of computer resources. It is more appropriate to use the classical vortex dynamics model.

The vortex dynamics model was pioneered by Schwarz [3] and consists in representing a vortex line as a curve $\mathbf{s} = \mathbf{s}(\xi, \mathbf{t})$ in three-dimenensional space, where ξ is arclength and t is time. We call $\mathbf{s}' = d\mathbf{s}/d\xi$ and note that the vectors $\mathbf{s}$, $\mathbf{s}'$ and $\mathbf{s}' \times \mathbf{s}''$ are perpendicular to each others and point along the tangent, normal and binormal respectively. To determine the equation of motion of $\mathbf{s}$ we must identify the forces acting upon the line: the Magnus force $\mathbf{f}_M$ and, at nonzero temperature, the drag force $\mathbf{f}_D$.

The Magnus force arises when a body with circulation about it moves in a flow: the circulation creates an increased total velocity of fluid on one side, which results in excess pressure from the other side. Since the key ingredient is the circulation rather than the details of the body, we apply the concept to a vortex line and write

$$\mathbf{f}_M = \rho_s \Gamma \mathbf{s}' \times (\mathbf{v}_L - \mathbf{v}_{s,tot}), \tag{2}$$

where $\mathbf{v}_L = d\mathbf{s}/dt$ is the velocity of the line in the laboratory frame and $\mathbf{v}_{s,tot}$ is the total velocity of the surrounding superfluid, also in the laboratory frame. The velocity $\mathbf{v}_{s,tot}$ consists of two parts: any superfluid velocity applied externally and the self - induced velocity of the vortex line.

$$\mathbf{v}_{s,tot} = \mathbf{v}_s + \mathbf{v}_i, \tag{3}$$

The self induced velocity $\mathbf{v}_i$ describes the motion which a vortex line induces onto itself because of its own curvature and is determined by the Biot - Savart (BS) law

$$\mathbf{v}_i(\mathbf{s}) = \frac{\Gamma}{4\pi} \int \frac{(\mathbf{z} - \mathbf{s}) \times d\mathbf{z}}{|\mathbf{z} - \mathbf{s}|^3}, \tag{4}$$

The Biot - Savart law is sometimes replaced by the Local Induction Approximation (LIA), which is

$$\mathbf{v}_i \approx \beta \mathbf{s}' \times \mathbf{s}'', \tag{5}$$

where $\beta = \Gamma/(4\pi)log(1/(|\mathbf{s}''|a_0))$. Since $|\mathbf{s}''| = 1/R$ where R is the local radius of curvature, we have

$$\mathbf{v}_i \approx \frac{\Gamma}{4\pi} \ln(\frac{R}{a_0})\hat{\mathbf{b}}, \tag{6}$$

where $\hat{\mathbf{b}}$ is the binormal.

The drag force $\mathbf{f}_D$ arises from the *mutual friction* between the superfluid vortex lines and the normal fluid [4]. Normal fluid flowing with velocity $\mathbf{v}_n$ past a vortex core exerts a frictional force $\mathbf{f}_D$ per unit length on the superfluid in the neighborhood of the core given by

$$\mathbf{f}_D = -\alpha\rho_s\Gamma\mathbf{s}' \times [\mathbf{s}' \times (\mathbf{v}_n - \mathbf{v}_{s,tot})] - \alpha'\rho_s\Gamma\mathbf{s}' \times (\mathbf{v}_n - \mathbf{v}_{s,tot}), \qquad (7)$$

The dimensionless parameters α and α' are temperature dependent and are often written in terms of *mutual friction coefficients* B and B' defined by $\alpha = \rho_n B/(2\rho)$ and $\alpha' = \rho_n B'/(2\rho)$. The values of B and B' are known from experiments. Samuels and Donnelly [5] showed that, at least in the high temperature range in which most experiments are performed, the friction arises from the scattering of rotons from the velocity field of a vortex line. The calculation of the mutual friction parameters over the entire temperature range is still an open question and has subtle aspects, as explained in the article by Sonin. An important effect of the mutual friction is that it modifies the propagation of second sound. By measuring the second sound attenuation one can determine the superfluid vortex line density L_0, defined as the length of vortex line per unit volume.

Now that $\mathbf{f}_M$ and $\mathbf{f}_D$ are identified we can make use of the fact that the sum of all forces is zero as the line's inertia is negligible:

$$\mathbf{f}_D + \mathbf{f}_M = 0, \qquad (8)$$

Hence, solving for $d\mathbf{s}/dt$, we obtain Schwarz's equation

$$\frac{d\mathbf{s}}{dt} = \mathbf{v}_s + \mathbf{v}_i + \alpha\mathbf{s}' \times (\mathbf{v}_n - \mathbf{v}_s - \mathbf{v}_i) + \alpha'(\mathbf{v}_n - \mathbf{v}_s - \mathbf{v}_i), \qquad (9)$$

An algorithm to numerically simulate the time evolution of any arbitrary configuration of vortex lines can be developed on the basis of Schwarz's equation and is described in the article of Samuels. Here it suffices to say that an initial vortex configuration is discretized into N points. The time evolution of each point is calculated using (9), given externally applied fields $\mathbf{v}_s$ and $\mathbf{v}_n$ and given the temperature T, which determines the friction coefficients α and α'. The transverse part of the mutual friction, proportional to α', is smaller and is sometimes neglected. The number of points N and the time step must be allowed to vary during the evolution to take into account the appearance of regions of high or low curvature. Note that, if one uses the BS law, the computational time is proportional to N^2, while, if one uses the LIA, this time is only proportional to N. Numerical simulations of vortex tangles based on the BS are therefore computationally expensive. However the use of the LIA can give misleading results [6]. Finally the numerical simulation must be able to perform vortex reconnections when two vortex lines come sufficiently close to each others. This process is an arbitrary assumption in the context of the dynamics of vortex filaments, but is justified by a microscopic calculation [7] performed using the NLSE, as explained in the article by Koplik.

3.3   Macroscopic Model

Besides the NLSE model (in which the vortex core is visible) and the vortex dynamics model (in which the core is not visible but the vortex line is) there is a third *macroscopic* model in which the individual vortex lines are not visible and Helium II is considered as a continuous vortex flow. The third model, called the HVBK model[8] [9], is useful to describe laminar flows in which the vortex line are spatially organized. Examples are solid body rotation, flows in an rotating annulus or cavity, and Taylor - Couette flow. The HVBK model is a generalization of Landau's equations to include the presence of vortices. The fluid particles of the model are assumed to be large enough to be threaded by many vortex lines which are aligned in the same direction. In this way the individual vortex lines are not visible, the superfluid is treated as a continuum and we can define a macroscopic, nonzero superfluid vorticity $\boldsymbol{\omega}_s$, despite the fact that, microscopically, the superfluid velocity field obeys $\nabla \times \mathbf{v}_s = 0$. Clearly the HVBK equations are valid only if the vortex lines are organized spatially and not randomly oriented, and if the length scales of the flow under consideration are much bigger than the average separation between the vortex lines. An example is the simple case of Helium II inside a rotating cylinder, for which $\boldsymbol{\omega}_s = 2\Omega\hat{z}$.

The incompressible HVBK equations are

$$\frac{\partial \mathbf{v}_n}{\partial t} + (\mathbf{v}_n \cdot \nabla)\mathbf{v}_n = -\frac{1}{\rho}\nabla P - \frac{\rho_s}{\rho_n}S\nabla T + \nu_n\nabla^2\mathbf{v}_n + \frac{\rho_s}{\rho}\mathbf{F}, \qquad (10)$$

$$\frac{\partial \mathbf{v}_s}{\partial t} + (\mathbf{v}_s \cdot \nabla)\mathbf{v}_s = -\frac{1}{\rho}\nabla P + S\nabla T + \mathbf{T} - \frac{\rho_n}{\rho}\mathbf{F}, \qquad (11)$$

where we have defined

$$\boldsymbol{\omega}_s = \nabla \times \mathbf{v}_s, \qquad (12)$$

$$\mathbf{F} = \frac{B}{2}\widehat{\boldsymbol{\omega}}_s \times [\boldsymbol{\omega}_s \times (\mathbf{v}_n - \mathbf{v}_s - \nu_s\nabla \times \widehat{\boldsymbol{\omega}}_s)] + \frac{B'}{2}\boldsymbol{\omega}_s \times (\mathbf{v}_n - \mathbf{v}_s - \nu_s\nabla \times \widehat{\boldsymbol{\omega}}_s), \qquad (13)$$

$$\widehat{\boldsymbol{\omega}}_s = \boldsymbol{\omega}_s/|\boldsymbol{\omega}_s|, \qquad (14)$$

$$\mathbf{T} = -\nu_s\boldsymbol{\omega}_s \times (\nabla \times \widehat{\boldsymbol{\omega}}_s), \qquad (15)$$

$$\nu_s = (\Gamma/4\pi)\log(b_0/a_0), \qquad (16)$$

The quantities $\mathbf{F}$, $\mathbf{T}$ and ν_s are respectively the friction force, the tension force and the vortex tension parameter, and $b_0 = (2\omega_s/\Gamma)^{-1/2}$ is the intervortex spacing. Note that ν_s has the same dimension of a kinematic viscosity, but physically it is very different: it is related to the ability of a superfluid fluid particle to oscillate because of the vortex waves which can be excited along the vortex lines threading the fluid particle itself. Note that without $\mathbf{F}$ and $\mathbf{T}$ the HVBK equations are formally the same as the original two - fluid equations of Landau. The HVBK equations have interesting limits. If $T \to T_\lambda$ then $\rho_s \to 0$ so the normal fluid equation (10) becomes the classical Navier - Stokes equation. If $T \to 0$ then $\rho_n \to 0$ so the superfluid equation (11) describes a pure superflow; by setting Plank's constant equal to zero we have then $\nu_s = 0$ and the pure superflow equation becomes the classical Euler equation.

The HVBK model has been used with success to study the transition from Couette flow to Taylor vortex flow [10]: Barenghi's predictions [11] of the critical

Reynolds number of the transition and its temperature dependence were confirmed by the experiments [12]. Taylor - Couette flow has also been studied in the nonlinear Taylor vortex flow regime [13] [14] and the results are in agreement with the observations, providing a further test of the theory. These results are described in the article by Henderson.

4 Turbulence

Superfluid turbulence manifests itself as a tangle of vortex lines and can be generated in many ways. Turbulent thermal counterflow was the first turbulent flow which was studied in detail in a series of pioneering papers by Vinen [15]. Since this flow has no classical analogy it deserves a separate discussion. Other ways to generate turbulence are more classical in character, and we refer to them as *turbulent coflows*.

4.1 Turbulent Counterflows

As said in the first section, laminar counterflow breaks down if the heat flux W exceeds a critical value W_c. Corresponding to W_c there is a critical counterflow velocity $v_{ns} = v_{c1} = W_c/(\rho_s ST)$. At this critical velocity a vortex tangle appears. The simplest way to characterize the tangle is to measure or compute its vortex line density L_0. A great number of measurements of L_0 were performed in pipes and channels and showed that $L_0 = \gamma v_{ns}^2$ for $v_{ns} > v_{c1}$ where γ is some temperature dependent parameter. Further measurements in circular pipes indicated that there exists a second critical velocity v_{c2} at which the vortex line density L_0 becomes suddenly larger. The region of weak turbulence $v_{c1} < v_{ns} < v_{c2}$ and the region of strong turbulence $v > v_{c2}$ are called the T-1 and the T-2 *turbulent states* respectively [16]. On the theoretical side, the numerical simulations of Schwarz [3] based on the vortex dyanmics approach confirmed the existence of a self - sustaining vortex tangle driven by a constant, spatially uniform v_{ns}, and gave values of L_0 consistent with the observations in the T-2 state. The nature of the first turbulent states and the physical meaning of the transition at $v_{ns} = v_{c2}$ were a puzzle until Melotte and Barenghi [17] showed that the transition from the T-1 to the T-2 state is related to the onset of normal fluid turbulence (see section 6).

4.2 Turbulent Coflows

Turbulence can also be induced in more traditional ways by driving a mass flow, spinning discs or propellers, towing a grid or a sphere, using shocks, ultrasound, jets and rotating cylinders (Taylor - Couette flow). Considering all these results together, the general trend is that the slow, laminar flow of Helium II, with or without vortices, tends to be rather different from the flow of a classical fluid, but when Helium II moves fast and is driven turbulent it seems to behave like a classical turbulent flow.

To characterize the turbulence we use the Reynolds number $Re = UL/\nu$ where L is the length scale, U the velocity scale and ν the kinematic viscosity. Examples of the observed classical features of Helium II turbulence are the following. Mass flow rates and pressure drops at $Re \approx 10^6$ can be well described by using classical relations for high Reynolds number classical flows [18]. Experiments on large scale turbulent vortex rings at $Re \approx 4 \times 10^4$ detect normal fluid vorticity and superfluid vorticity moving together as a single structure [19] Experiments on turbulent Taylor - Couette flow at $Re \approx 4 \times 10^3$ show the typical structures of classical turbulent Taylor - Couette flow [20]. Experiments on the decay of superfluid vorticity created by towing a grid show that the decay in time obeys the same laws as of the decay of classical turbulence [21]. Moreover the decay appears to be independent of temperatures in the explored range, from the lambda region down to 1.4 K, where the normal fluid fraction is only 7.5 percent. Experiments on turbulence created by rotating blades [22] show the classical Kolmogorov $-5/3$ power spectrum in the temperature range explored, from the lambda region down to $T = 1.4$ K again. Finally experiments on the drag on a moving sphere ($Re \approx 10^5$) show the same drag crisis observed in a classical fluid [23].

The temperature independence of these observations is interesting. The normal fluid must be responsible for these classical aspects, but the dynamical importance of the normal fluid should be related to the fraction ρ_n/ρ, so it should be negligible at temperatures as low as 1.4 K. Since a large number of quantized vortex lines must be present in these turbulent flow, it is speculated that they are able to lock together the two fluid components of Helium II into a single fluid which behaves somewhat like a classical turbulent fluid. This is a topic of much current interest'[24] [25] and is discussed in the article of Vinen.

5   Motion of Superfluid Vortices for a Given Normal Fluid

In the next two sections we discuss some selected examples about the interaction of superfluid vortices and the normal fluid. The numerical calculations done by Schwarz using the vortex dynamics model (section 3.2) were concernd with thermal counterflow. The calculations were successful; above all, they gave great insight into the nature of turbulence, confirming the existence of a vortex tangle, and brought into attention the issue of vortex reconnections. However Schwarz's approach had an important limitation: it assumed that the quantity $\mathbf{v}_n - \mathbf{v}_s$ is constant in time and space. The work of Schwarz was followed up by other calculations which made different assumptions about the normal fluid: uniform flow [26], Poiseuille flow [26] [27], a Gaussian vortex [28], ABC flows [29]. All these calculations were *kinematic* in character, because the driving field ($\mathbf{v}_n$ for mass flow and $\mathbf{v}_n - \mathbf{v}_s$ for counterflow) were imposed at the beginning of the calculation and never changed, neglecting the back reaction of the superfluid vortices. Not surprising, the vortex tangles calculatedusing different driving fields looked different from each others. The success of the original calculation of Schwarz in reproducing a vortex line density L_0 consistent with the experiments in the T-2 state was probably due to the fact that the uniform $\mathbf{v}_n - \mathbf{v}_s$ profile used by Schwarz modelled well the average, flattened turbulence profile in the channel flows under consideration.

Despite this limitation, kinematic calculations are clearly useful and can shed light on important physical mechanisms. A particularly interesting mechanism which is relevant to turbulence is the *Ostermeier- Glaberson instability*. This is an instability of Kelvin vortex waves which takes place if the component of the normal fluid velocity in the direction parallel to the vortex lines exceeds a critical value. The instability was first observed by Cheng, Cromar and Donnelly [30], but is was Ostermeier and Glaberson [31] who explained it and it was Samuels [28] who realized its importance in turbulence. Another physical mechanism, which is apparent in the kinematic numerical calculations [28] [29] is vorticity matching: once the Ostermeier - Glaberson instability has generated superfluid vortex lines by extracting energy from the normal fluid, then the vortex lines become attracted to the regions of concentrated normal fluid vorticity. Therefore, although the local superfluid velocity pattern in the bundles is very complicated, the averaged vorticity $\boldsymbol{\omega}_s$ is similar to the vorticity of the driving normal fluid. A further interesting application of the Ostermeier - Glaberson instability is that it creates a damping length scale ℓ for superfluid turbulence: superfluid structures at length scale smaller than ℓ will lose energy to the normal fluid and be dissipated [25].

6   Motion of the Normal Fluid at Given Superfluid Vortices

The kinematic approach can also be used to study the behaviour of the normal fluid given the superfluid vortices rather than viceversa. An example is the calculation of Melotte and Barenghi [17] who studied the stability of normal fluid motion induced by heat transfer in the presence of a uniform tangle of superfluid vorticity of density L_0. The governing equation is a Navier - Stokes equation modified by the introduction of a friction term

$$\frac{\partial \mathbf{v}_n}{\partial t} + (\mathbf{v}_n \cdot \nabla)\mathbf{v}_n = -\frac{1}{\rho}\nabla P - \frac{\rho_s}{\rho_n}S\nabla T + \nu_n \nabla^2 \mathbf{v}_n + \frac{1}{\rho_n}\mathbf{F}, \qquad (20)$$

where

$$\mathbf{F} = \frac{B\rho_s\rho_n}{2\rho}g\omega_s(\mathbf{v}_s - \mathbf{v}_n), \qquad (21)$$

$\omega_s = \Gamma L_0$, and, assuming an isotropic tangle, $g \approx 2/3$. Melotte and Barenghi's stability calculation determined the critical vortex line density at which the normal fluid's profile becomes turbulent. The results are in agreement with the measurements of the transition from the weak T-1 state (in which the superfluid is turbulent but the normal fluid is not) to the strong T-2 state (in which both fluids are turbulent).

The same kinematic approach has been used to study how the superfluid vortices affect the stability of normal fluid in channel flows. This modified Orr - Sommerfeld problem[32] is described in the article by Godfrey.

7 Fully Coupled Motion of Superfluid Vortices and Normal Fluid

The limitations of the kinematic approach can be overcome if one allows the normal fluid and the superfluid vortices to determine each others self - consistently during the evolution. Essentially one has to combine the (Lagrangian) calculation of the superfluid vortices based on vortex dynamics (section 2) with with the (Eulerian) calculation of the normal fluid based on a modified Navier - Stokes equation (section 5). However, since the velocity field $\mathbf{v}_n$ which is computed is local, the equation of Schwarz (9) requires a modification[33]. At the place of (9) one has

$$\frac{d\mathbf{s}}{dt} = \mathbf{v}_s + h_1\mathbf{v}_{si} + h_2\mathbf{s}' \times (\mathbf{v}_n - \mathbf{v}_i) - h_3\mathbf{s}' \times \mathbf{s}' \times \mathbf{v}_n, \qquad (22)$$

where $\mathbf{v}_i$ is given by (4). In the absence of friction we have $h_2 = h_3 = 0$ and $h_1 = 1$. The normal fluid is determined by a modified Navier - Stokes equation like (20), but now $\mathbf{F}$ is obtained numerically by considering the friction force on the normal fluid per unit length of superfluid vortex line

$$\mathbf{f} = D\mathbf{s}' \times \mathbf{s}' \times (\mathbf{v}_n - \mathbf{v}_L) + D_t\mathbf{s}' \times (\mathbf{v}_n - \mathbf{v}_L), \qquad (23)$$

where D and D_t are mutual friction coefficients, and summing the contribution of each segment of $\mathbf{s}$ that falls within the computational grid cell of the normal fluid.

The drawback of the fully - coupled approch is the computational cost. A useful compromise is to implement it in two dimensions x, y neglecting the z dependence: in this way the vortex lines becomes vortex points (the intersection of vortex lines with the plane $z = 0$) and the normal fluid can be obtained easily using the stream function - vorticity formulation. It is found that a single superfluid vortex point induces an elongates normal fluid jet [34] whose intensity depends on the temperature, hence creating a dipolar vorticity structure in the normal fluid. A similar calculation in three dimensions performed by Kivotides, Barenghi and Samuels [35] showed that a superfluid vortex ring creates around itself a normal fluid structure which consists of two coaxial vortex vortex rings of opposite polarity.

Since the normal fluid is accelerated by the superfluid vortices by friction which depends on the difference between $\mathbf{v}_n$ and $\mathbf{v}_L$, the normal fluid speed cannot exceed the vortex lines' speed. If one considers the typical vortex line density L_0 of experiments and estimates $v_L \approx \Gamma/2\pi\delta$ using the average intervortex spacing $\delta \approx L_0^{-1/2}$, one finds an upper bound for v_n, using which one estimates the normal fluid Reynolds number to be in the range from 1 to 10, too small for turbulence[36].

This conclusion refers only to the ability of vortex line to make the normal fluid turbulent by directly stirring it. A second mechanism in which the vortex line can make the normal fluid turbulent is by causing instabilities in a mean flow, as showed in section 6. Finally normal fluid turbulence can also be induced directly by the boundaries.

8   Discussion

I have mentioned only a selection of topics of current interest in the study of quantized vorticity. Much work is in progress. It is clear that the NLSE will be used more and more to understand the details of fundamental processes: an example is vortex reconnections[37]. Similarly, the use of vortex dynamics simulations in the absence of friction is proving useful to understand processes such as the turbulence cascades of vortex waves [38]. At finite temperatures, fully coupled numerical simulations are proving useful to understand the effects of the reconnections on the normal fluid[39] and hopefully will be able in the near future achieve contact between theory and experiment.

References

1. L.D. Landau and E.M. Lifshitz: *Fluid Mechanics*, 2nd edn. (Pergamon Press, London 1987)
2. R.J. Donnelly:*Quantized Vortices in Helium II*, Cambridge University Press, Cambridge (1991).
3. K.W. Schwarz: Phys. Rev. Lett. **49**, 283 (1982); Phys. Rev. B **31** 5782 (1985); Phys. Rev. B **38** 2398 (1988)
4. C.F. Barenghi, R,J, Donnelly and W.F. Vinen: J. Low Temp. Phys. **52** 189 (1983)
5. D.C. Samuels and R.J. Donnelly: Phys. Rev. Lett. **65** 187 (1990).
6. R.L. Ricca, D.C. Samuels and C.F. Barenghi: J. Fluid Mech. **391** 29 (1999)
7. J. Koplik and H. Levine: Phys. Rev. Lett. **71**, 1375 (1993).
8. H.E. Hall and W.F. Vinen: Proc. Roy. Soc. London A **238**, 215 (1956).
9. R.N. Hills and P.H. Roberts: Arch. Rat. Mech. Anal. **66**, 43 (1977).
10. C.F. Barenghi and C.A. Jones: J. Fluid Mech. **197** 551 (1988)
11. C.F. Barenghi: Phys. Rev. B **45**, 2290 (1992)
12. C.J. Swanson and R.J. Donnelly: Phys. Rev. Lett. **67** 1578 (1991)
13. K.L. Henderson, C.F. Barenghi and C.A. Jones: J. Fluid Mech. **283** 329 (1995).
14. K.L. Henderson and C.F. Barenghi: Phys. Lett. A **191** 438 (1994).
15. W.F. Vinen: Proc. Roy. Soc. A **240**, 114 (1957); ibidem **240**, 128 (1957); ibidem **242**, 493 (1957); ibidem **243** 400 (1957).
16. J.T. Tough: 'Superfluid turbulence'. In: *Progress in Low Temperature Physics, vol. VIII*, ed. by D.F. Brewer (North Holland, Amsterdam 1987) p. 133.
17. D.J. Melotte and C.F. Barenghi: Phys. Rev. Lett. **80** 4181 (1998).
18. P.L. Walstrom, J.G. Weisend, J.R. Maddocks and S.V. VanSciver: Cryogenics **28** 101 (1988).
19. H. Borner, T. Schmeling and D.W. Schmidt: Phys. Fluids **26** 1410 (1983)
20. F. Bielert and G. Stamm: Cryogenics **33** 938 (1993)
21. M.R. Smith, R.J. Donnelly, N. Goldenfeld and W.F. Vinen: Phys. Rev. Lett. **71** 2583 (1993)
22. J. Maurer and P. Tabeling: Europhysics Lett. **43** 29 (1998)
23. M.R. Smith, D.K. Hilton and S. V. VanSciver: Phys. Fluids. **11** 751 (1999)
24. W.F. Vinen: Phys. Rev. B **61** 1410 (2000)
25. D.C. Samuels and D. Kivotides: Phys. Rev. Lett. **83** 5306 (1999)
26. R.G.K.M. Aarts and A.T.A.M. deWaele: Phys. Rev. B **50** 10069 (1994)
27. D.C. Samuels: Phys. Rev. B **46** 11714 (1992)

28. D.C. Samuels: Phys. Rev. B **47** 1107 (1993)
29. C.F. Barenghi, G. Bauer, D.C. Samuels and R.J. Donnelly: Phys. Fluids **9** 2631 (1997)
30. D.K. Cheng, M.W. Cromar and R.J. Donnelly: Phys. Rev. Lett. **31** 433 (1973)
31. R.M. Ostermeier and W.I. Glaberson: J. Low Temp. Phys. **21** 191 (1975).
32. S.P. Godfrey, C.F. Barenghi and D.C. Samuels: Physica B **284** 66 (2000); Phys. Fluids **13** 983 (2001).
33. O.C. Idowu, D. Kivotides, C.F. Barenghi and D.C. Samuels: J. Low Temp. Physics, **120** 269 (2000).
34. O.C. Idowu, A. Willis, D.C. Samuels and C.F. Barenghi: Phys. Rev. B **62** 3409 (2000)
35. D. Kivotides, C.F. Barenghi and D.C. Samuels: Science **290** 777 (2000).
36. C.F. Barenghi, D. Kivotides, O. Idowu and D.C. Samuels: J. Low Temp. Physics, **121** 377 (2000).
37. M. Leadbeater, T. Winiecki, D.C. Samuels, C.F. barenghi and C.S. Adams: Phys. Rev. Lett. **86** 1410 (2001).
38. D. Kivotides, C. Vassilicos, C.F. Barenghi and D.C. Samuels: Phys. Rev. Lett. **86** 3080 (2001).
39. D. Kivotides, C.F. Barenghi and D.C. Samuels: to be published in Europhys. Lett. (2001).

Part II

Turbulence Experiments

An Introduction to Experiments on Superfluid Turbulence

Russell J. Donnelly

Cryogenic Helium Turbulence Laboratory, Department of Physics,
University of Oregon, Eugene, Oregon 97403

1 Introduction

A description of the experimental background of superfluid turbulence was assigned to me for this lecture. Superfluid turbulence, or as some call it, quantum turbulence, has been an active field of physics since the 1950's. The field was pioneered experimentally and theoretically by Joe Vinen and as such is approaching a half century in age. It is safe to say that with few exceptions the results are unknown to those investigators who are interested in classical turbulence, that is the kind of investigation which has been pioneered by Taylor, Landau, Kolmogorov and others.

It has only recently been realized that liquid helium I, liquid helium II and cryogenic (critical) helium gas are attractive candidates for investigating classical turbulence problems, and in the process many have decided to look at the kinds of challenges encountered in using helium II, that phase of liquid helium which exhibits superfluidity.

In preparing this talk I had hoped to cover, however briefly, the entire corpus of experimental work on the subject. It soon became evident that there was too much material by far than could be covered in a single one-hour lecture. Fortunately, I had recently written a review article in honor of Joe Vinen's retirement from the University of Birmingham [1] and Skrbek, Niemela and myself had written a second paper for the same occasion [2]. These papers contain virtually all the known results on cryogenic fluid mechanics. This talk will instead concentrate mostly on future directions, and results obtained since the articles cited above were written.

2 Update on Pipe Flow

Cryogenic helium is of significant value in generating, and studying the highest possible Reynolds and Rayleigh number flows under controlled laboratory conditions, primarily due to its extremely low value of kinematic viscosity. Pipe flow is an example where the highest Reynolds numbers can be achieved by using helium I. Our apparatus is shown in Fig.1. The pipe is a polished stainless steel tube 4.67 mm in diameter and 25 cm long. The working fluid is drawn into the bellows by moving the push rod at the top upward. At a preset time the rod is pushed down at constant velocity and discharges a known mass of liquid per

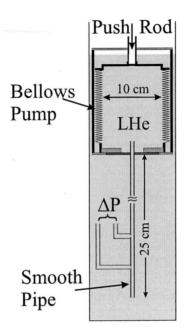

Fig. 1. Bellows driven flow apparatus for measuring the friction factor in a smooth pipe. The advantage of cryogenics can be appreciated by comparing the size of this apparatus with the famous apparatus of Nikuradze built in the early 1930's and weighing many tons. The highest Reynolds numbers achieved are about the same in both Nikuradze's and our apparatus.

unit time. Flow velocity is limited by the power of the drive motor. A sensitive capacitance manometer has been developed to measure the pressure gradient in all working fluids and is located 38 diameters down stream of the inlet. The in-situ differential pressure gauge was specifically designed to avoid mechanical strains at low temperature, which can overwhelm the 0.05 Pascal resolution of the device [3].

This experiment, primarily the effort of Chris Swanson of the University of Oregon, with help from Gary Ihas, visiting from the University of Florida, has revealed several subtleties in such measurements, including the need for high purity of helium, fine stability of temperature and pressure, and care to maintain a time-independent state. Agreement with data at room temperature using air and water as a fluid is quite good.

The primary advantage of this apparatus is that it accommodates any non-reactive fluid at a temperature between 1 K and 300 K, allowing unprecedented *ranges* of Reynolds numbers to be obtained. For example, we have used room temperature SF_6, air, and helium gas, as well as cryogenic helium I at 4.2 K to measure the friction factor from Reynolds number 10 to 5,000,000, nearly six decades of Reynolds numbers. The fact that the results agree with standard engineering data from many sources attests to the utility of this new approach.

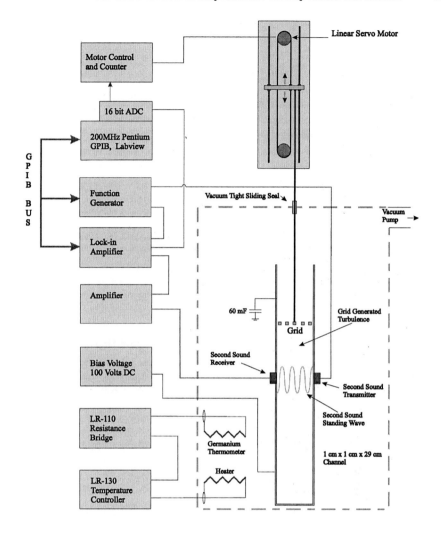

Fig. 2. Schematic diagram of the University of Oregon towed grid apparatus in helium II. Turbulence is generated by sweeping the grid upward, and is observed by measuring the attenuation of second sound in the presence of the turbulence which can be interpreted to give the vorticity in the superfluid.

It would be of interest to extend these results to helium II, where other evidence suggests the results will be about the same [4]. However this is yet to be accomplished in a satisfactory way.

3   Update on Towed Grid Experiments

Superfluid turbulence in counterflow turbulence takes the form of a tangle of quantized vortex lines, a frictional drag between the moving normal fluid and the cores of the lines (mutual friction) serving to maintain the turbulence. Counterflow turbulence has no classical analogue. Other types of superfluid turbulence have been studied over the years, but only recently has there been an experimental study of superfluid turbulence behind a steadily moving grid. This turbulence is analogous to the much studied, and fundamentally important, case of homogeneous isotropic turbulence created by flow through a grid in a wind tunnel.

The towed grid apparatus, shown in Fig.2, consists primarily of a 1 cm x 1 cm square tube fitted with a sliding grid with about 65 % porosity. The grid is swept upward at a preset time and creates a slug of turbulence of quite high vorticity filling the channel. The probe is a pair of second sound transducers which measure the attenuation of second sound owing to quantized vorticity in the channel.

The results obtained with this apparatus have been fascinating and a series of publications have come from this apparatus which show, among other things, that the decay of the turbulence can most easily be understood treating the decay as if it were from a classical fluid having the viscosity of the normal fluid, but the density of the total fluid [5–8]. The physics underlying this phenomenon has been carefully discussed by Vinen [9]. His point of view is also further developed in a lecture at this conference [10].

Two new experiments have been performed which had not appeared in print at the time of this conference. A brief description of both follows.

3.1   The Nature of Grid Turbulence in Helium II

A detailed experimental and theoretical study of superfluid grid turbulence was carried out by Stalp et al [8]. As an observational technique they used the attenuation of second sound, which serves to measure the length of vortex line per unit volume, L, in the superfluid component. They showed that a classical model could explain the observed time decay of line length. According to this model superfluid helium behaves as a single fluid with a kinematic viscosity ν'. The rate of dissipation of turbulent energy per unit mass has the classical form

$$\varepsilon = \nu'\langle\omega^2\rangle = \nu'\kappa^2 L^2 \tag{1}$$

where $\langle\omega^2\rangle$ is the mean square vorticity in the coupled fluids, and the identification of $\langle\omega^2\rangle$ with $\kappa^2 L^2$ is discussed in [9]. Therefore according to the model second sound measures $\langle\omega^2\rangle$, as a function of time. After an initial time interval the turbulent energy spectrum has the classical Kolmogorov form

$$E(k) = C\epsilon^{2/3}k^{-5/3} \tag{2}$$

within a range of wave numbers $k_e < k < k_\eta$. Energy is being continually transferred in a cascade from lower to higher wave numbers by inertial forces, dissipation by viscosity taking place eventually at $k \geq k_\eta = (\varepsilon/\nu'^3)^{1/4}$ (the Kolmogorov wave number). C is the Kolmogorov constant and is equal to about 1.5 for a classical fluid, and k_e is the wave number of the energy-containing eddies. $k_e(t)$ decreases until it saturates at a value equal to approximately the inverse of the channel width.

The model made predictions that were in agreement with the experiments reported in [5–7] in the temperature range studied, from the λ-temperature down to about 1.3K; however, although ν' was shown to be of order μ/ρ, where μ is the viscosity of the normal fluid and ρ is the total density of the helium, no careful analysis of the temperature dependence of ν' was carried out.

It turns out that within the inertial range of wave numbers ($k_e < k < k_\eta$) the relevant length scales are significantly larger than the expected spacing of quantized vortex lines, which turns out to be close to k_η^{-1}. Mutual friction is then sufficient to keep the two fluids locked together on these length scales, the locked fluids behaving as a single turbulent fluid in which there is negligible dissipation. Therefore, a Kolmogorov spectrum forms, as in a classical fluid, provided of course that energy is dissipated by some means at or beyond the Kolmogorov wave number k_η. The model requires that this dissipation be described by equation (1). The fact that the vortex line spacing is of order k_η^{-1} implies that in reality the two fluids cannot remain locked together for wave numbers of order or greater than k_η. It follows that dissipation must be due to a combination of viscous dissipation in the normal fluid and mutual friction, and it is therefore not obvious that an equation of the form (1) should hold [9]. It was predicted in [9] that the resulting rate of dissipation would be given by a formula similar to (1), with a value of ν' that is weakly dependent on temperature and of order μ/ρ as measured above 1 K.

We conclude that energy dissipation in grid turbulence is likely to be described correctly by (1), in spite of the fact that the mechanism of energy dissipation in the superfluid is quite different from that in a classical fluid. The precise value of ν' depends on the dissipative mechanisms operating in the turbulence. These mechanisms operate on a length scale of order the vortex line spacing, and are therefore strongly influenced by quantum effects in the superfluid.

We see that the parameter ν', which is analogous to the kinematic viscosity in a classical fluid, is fundamentally important in superfluid turbulence. The experimental technique was described in [5–7]. Data analysis is carried out for that period of the decay during which the vortex line density falls as $t^{-3/2}$, when the theory of the classical model gives

$$L(t) = \frac{d}{2\pi\kappa\sqrt{\nu'}}(3C)^{3/2}t^{-3/2} \tag{3}$$

where d is the width of the channel through which the grid is towed. At times earlier than this period the scale of the energy-containing eddies has not yet saturated at the channel width (see section 3.2 below) and the decay is slower. At later times the inertial range of wave numbers becomes narrow and the

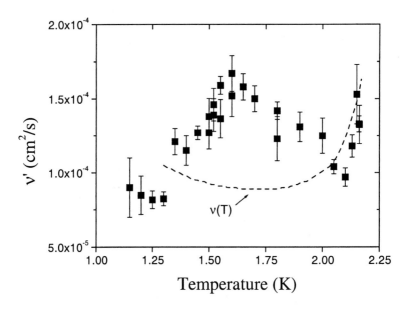

Fig. 3. The effective kinematic viscosity as a function of temperature, as measured from experiment.

decay rate increases. The length of line is obtained from second sound resonance measurements using the relation

$$L(t) \approx \frac{16\Delta_0}{\kappa B}(\frac{A_0}{A(t)} - 1) \qquad (4)$$

where Δ_0 is the full-width at half maximum for the second-sound (power) resonance curve in the absence of vortex lines (at the 50th harmonic), B is the mutual friction constant obtained from measurements on uniformly rotating helium [10], and A(t) and A_0 are respectively the peak second sound amplitude with and without vortex lines present. Using (4) and (5) we see that the experiment measures $\sqrt{C^3/\nu'}$. Note that Maurer and Tabeling [11] performed an experiment with rotating disks in liquid helium in which they were able to measure the Kolmogorov spectrum by means of pressure fluctuations using a probe inserted at a selected location in the flow and connected to a quartz pressure transducer. Their results showed that the Kolmogorov spectrum could be observed in both helium I and helium II (2.3K, 2.0K and 1.4K) and reached the important conclusion that the Kolmogorov constant C is the same above and below the lambda transition. With that information, we see that our experiment determines ν' providing we know C. The Kolmogorov constant is taken to be 1.5 at all temperatures, which is the accepted classical value [12]. The resulting values of ν' are shown as a function of temperature in Fig. 3. The error bars

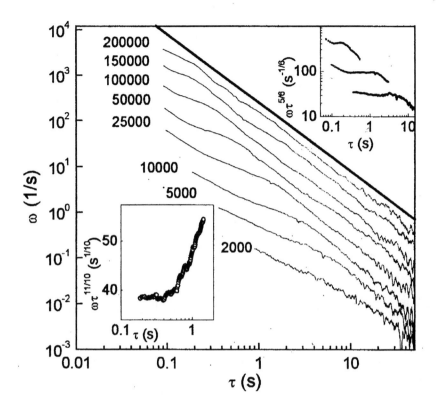

Fig. 4. The decaying helium II vorticity measured at T=1.3 K for the indicated Re_M.

include contributions due to uncertainties in the values of B, and the frequency-dependence of B was taken into account [13].

The experimental results reported here relate to temperatures greater than 1.1 K. There is a clear need to extend the temperature range downwards, especially because the theory then makes interesting predictions. The only experiments so far reported on superfluid turbulence at these low temperatures are those of Davis, Hendry and McClintock [14].

3.2   Four Regimes of Decaying Grid Turbulence in Helium II [15]

Attenuation of second sound in helium II has been used to observe up to six orders of magnitude of decaying vorticity displaying four distinctly different regimes of decaying grid turbulence in a finite channel. A purely classical spectral model for homogeneous and isotropic turbulence describes most of the decay of helium II vorticity in the temperature range $1.2K < T < 2K$. The four regimes switch successively as the energy containing and dissipative Kolmogorov length scales gradually grow during the decay, finally both being saturated by the size of the channel. In Fig.4 each curve represents an average of three individual decays. As the decay curves tend to collapse on the universal curve, we shifted them for clarity by a factor of two downwards, the uppermost remaining unchanged. The early part of the vorticity decay displays a power law with exponent -11/10 (see left inset, showing normalized data for $Re_M = 10^4$) and later $-5/6$ (see right inset, showing normalized data for $Re_M = 1.5 \times 10^5, 2.5 \times 10^4$ and 5×10^3). After saturation, typically several orders of magnitude of decaying vorticity closely follow the power law with exponent $-3/2$, represented by the thick solid line.

For the first time we report the fourth and last regime - a late exponential decay - not shown separately in Fig.4. The nature of this last decay is not understood at present.

This experiment explores over 8 orders of magnitude of decaying turbulent energy, an impossible task for a wind tunnel, which would need to have a 1000 km test section to observe the same thing.

4   Agenda for the Future

4.1   The University of Oregon 6 cm Wind Tunnel

So far we have not had a continuous flow facility to work on. About two years ago we decided to build a small wind tunnel using critical helium gas as the working fluid. The advantage of critical helium gas is the enormous range of properties which can be reached by adjusting the pressure and temperature. Flow velocities available depend on the density. At low densities flow velocities up to $1m/s$ are possible, at high densities velocities to $30cm/s$ are available. One of the optimal operating points for high Re flow will be at 4 bar pressure and 6 K at which the kinematic viscosity is $3.1 \times 10^{-8} m^2/s$. Mesh Reynolds numbers of 30,000 to 100,000 will be available, and corresponding microscale Reynolds numbers will range from 150 to 280.

We have recently successfully operated this tunnel at 6.5 K and mesh Reynolds numbers around 1500. The grid generated turbulence is probed with 10 micron diameter cryogenic hot wire anemometer also developed in our laboratory. These hot wires are observed to obey King's law relating velocity and voltage, and pre-liminary velocity time series show standard statistical features. Note that this is the first cryogenic tunnel to operate below liquid nitrogen temperature.

4.2   Wind Tunnels for Model Testing

We have given considerable attention to the conceptual design of larger wind tunnels designed to be useful for model testing. These have been discussed recently in [2] and do not need repeating in this article. The importance for superfluid turbulence of these devices is the possibility of operating them in helium II.

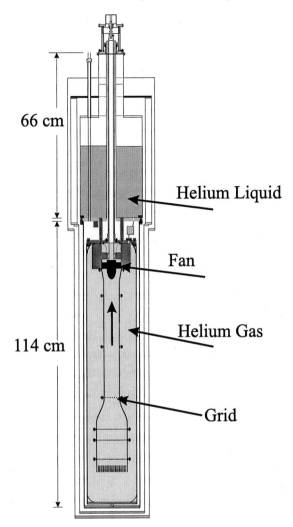

Fig. 5. Sketch of the University of Oregon 6 cm wind tunnel.

The acoustic characteristics of models being tested in such a tunnel is very important, and unexplored at the present time. Acoustics is a very large field having to do with the generation and interaction of sound with mechanical structures. It is of particular importance for submarine testing.

4.3   Tow Tanks

Liquid helium offers much promise for tow tank design, a subject also reported in some detail in [2]. Again the importance for superfluid turbulence is the possibility of operating with helium II as the working fluid.

5   Challenges for the Future

5.1   The Challenge of Instrumentation

Although helium I is a Navier-Stokes fluid, the possibility of generating ever higher Reynolds numbers carries with it ever decreasing Kolmogorov lengths. Of primary concern is the smallness of the flow structures such as turbulent eddies and boundary layers. Consider flow through a smooth pipe at high Reynolds number. To be definite let us take a pipe 10 cm in diameter using liquid helium having a kinematic viscosity $\nu = 2 \times 10^{-4} (cm)^2$ flowing at a mean velocity $\bar{u}$, and density $\rho = 0.146 gm/(cm)^3$. For turbulent flow a viscous sublayer is generated at the walls which scales with the friction velocity . The friction velocity is given by $u^* = \sqrt{\tau/\rho}$, where τ, the wall stress, is $\tau = \lambda \rho \bar{u}^2/8$ and λ is the empirical friction factor [16]. The corresponding length scale is

$$y = \frac{\nu}{u^*} \tag{5}$$

and the viscous sublayer has a thickness of a few times y. This y can be thought of as the scale of the smallest eddies in the turbulent flow. Table 1 outlines some typical values for helium I. Note that in helium I we can in principle use velocities of flow up to the velocity of sound before encountering shock waves. We cannot use such large velocities in helium II because second sound shock waves will intrude at about an order of magnitude lower velocity. The speed of sound in the gas is lower than in the liquid, hence the highest Reynolds number generated with helium should be generated with helium I.

It can be seen that the length scales can range from microns to Angstroms. For comparison with theories it is important to be able to probe the flow at the smallest scales. The last entry for y in table 1 is 42 Angstroms. This should be compared with the cube root of the atomic volume of liquid helium which is 3.6 Angstroms. Kolmogorov lengths in the Angstrom ranges, therefore, can approach conditions where the continuum hypothesis for the Navier Stokes equation is going to be in question. Probing such flows will be a major instrumentation challenge. We must learn to build not only micron sized, but even nanometer sized transducers in order to take advantage of the new range of high Reynolds numbers afforded by cryogenic helium.

Another experimental challenge is to visualize the normal fluid. This problem has always been with research with helium II, and not a great deal of progress has been made except in specialized cases [18]. Perhaps the most promising avenue for visualization will be PIV (mentioned below).

We now summarize briefly the diagnostic tools available for cryogenic fluid mechanics. These are discussed in a recent review [17].

- The measurement of average flow velocity can be measured in many instances by means of acoustic flowmeters or venturi tubes.
- Flow velocity at a point can be measured in convecting helium gas, for example, by means of bolometer fluctuation correlations.

- Hot wires, Laser Doppler Velocimeters (LDV) and Particle Image Velocimeters (PIV) should all work in principle in helium gas, helium I and helium II.
- RMS vorticity can be measured in helium II by second sound attention in both open flows and counterflows. There is some speculation that chemical potential probes would give local information on vorticity fluctuations, but this has yet to be implemented [17].
- Temperature gradients can be measured by means of standard germanium thermometry and pressure gradients, for example, by means of capacitance manometers [3].
- Lift and drag on models can best be measured using Magnetic Suspension and Balance Systems (MSBS), discussed in [18] by Britcher. Of course in a cryogenic environment the magnets can be superconducting.
- Wall stress gages can be fabricated for work in helium gas and helium I.
- Ion trapping can be used to measure vorticity in helium II [25].

Table 1. Viscous Sub Layer Thickness at Various Reynolds Numbers

$\overline{u}$ (cm/sec)	Re	λ	τ erg/cm^3	u^* cm/sec	y (cm)
10	5×10^5	0.0132	0.0240	0.405	4.93×10^{-4}
10^2	5×10^6	0.00898	1.64	3.35	5.97×10^{-5}
10^3	5×10^7	0.00649	118.4	28.48	7.02×10^{-6}
10^4	5×10^8	0.00489	8,940	247	8.08×10^{-7}
2×10^4	1×10^9	0.00453	33,100	476	4.20×10^{-7}

Hot wire anemometers are resistive self-heating devices which balance heat lost to flow, which depends on the fluid velocity. Calibration provides the correlation between fluid velocity and electrical power supplied. Standard hot wires have $d \sim 5$ microns and $L \sim 1000$ microns. Standard materials (e.g. platinum) used at room temperature are insensitive at low temperature. Our cryogenic hot wires are made on a quartz fiber of diameter 10 microns. They consist of an evaporated Au-Ge film of thickness of order 3000Å. A small sensitive region in middle of the fiber is defined by masking the fiber and evaporating a metallic film over it as shown in Fig.7. Masking is achieved by laying a small diameter fiber perpendicular to the first fiber. The metal film provides electrical contact to the sensitive region.

Figure 8 shows how the small sensors are mounted. Sensor support dimensions follow the "rule of 10": they are placed ten times (more or less) their characteristic dimension away from the sensitive region. Electrical contact to the fiber is made through stainless steel wires, which are isolated from the brass ring by epoxy. The fiber is epoxied to the wires using an electrically conducting epoxy.

We have recently tested a few prototype 10 micron sensors in the wind tunnel of Fig.5. A helium gas flow at 7 K was generated and we applied a relatively large

Hot Wire Anemometry

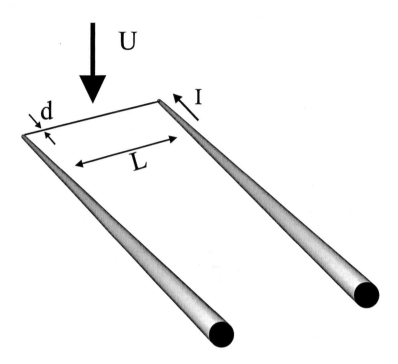

Fig. 6. A hot wire anemometer.

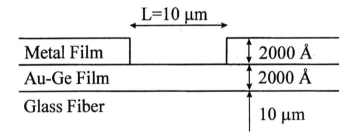

Fig. 7. Principle of the cryogenic hot wire anemometer developed at the University of Oregon.

sinusoidal voltage to overheat one of the sensors and measured the temperature of a nearby sensor. The characteristic heat signature at twice the input frequency was clearly seen on the monitoring sensor. Furthermore, we found that the signal

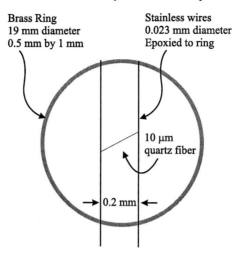

Brass Ring
19 mm diameter
0.5 mm by 1 mm

Stainless wires
0.023 mm diameter
Epoxied to ring

10 µm
quartz fiber

0.2 mm

Fig. 8. Mounting the 10 micron hot wire anemometer

to noise ratio to be the same as that predicted by our noise analysis. Thus we feel confident that the sensors are working according to our expectations.

We have made substantial strides in detector development. Our 10 micron detectors are to be contrasted with the bolometers we use in our large convection apparatus, which are 250 microns on side. But as we see from Table 1, we need to push our detectors down in size by orders of magnitude. Such advances are likely to depend on optical techniques.

5.2   Challenges for Understanding Counterflow Turbulence

Counterflow experiments have a history of many decades, but much remains to be accomplished [1]. The brief outline below suggests some directions needing exploration.

Comparing the decay of towed grid turbulence and counterflow turbulence. We need to understand the differences between the decay of towed grid turbulence and the decay of counterflow turbulence after the counterflow is turned off. Fig.9 below shows decay of both types of turbulence in the same apparatus, beginning at about the same vorticity and demonstrating nearly identical decay at long times. The experimental channel was identical in both cases except the grid was removed for the counterflow experiment. The form of the upper curve has been discussed by Schwarz and Rozen [19] and by Smith [5]. The decay curves coincide at long times and correspond exactly to the decay of classical turbulence (see Section 2.2 above).

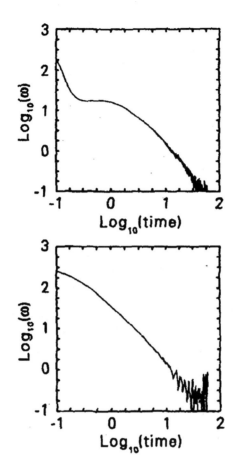

Fig. 9. An illustration of the dramatic differences in decay of two turbulent flows of the same initial vorticity produced by a counterflow (upper curve) and towed grid (lower curve).

The first critical velocity. What accounts for the first critical velocity in counterflow and its temperature dependence? This phenomenon is likely a case of extrinsic nucleation, but lacks any quantitative explanation.

The TI-TII transition. The TI-TII transition needs further attention. We have Melotte and Barenghi's insight which suggests that TI corresponds to turbulence in the superfluid, and TII marks the onset of disturbances in the normal fluid. Their linear stability analysis gives the right order of magnitude for the transition, but does not capture the temperature dependence.

The Tough classification. Why do large and small aspect geometries in counterflow exhibit such different behavior?

Combined rotation and heat flow. Combined rotation and heat flow is a relatively new area of investigation. Prior to the investigation discussed in this section it was assumed (from earlier experiments) that the ordered array of vortex lines produced by steady rotation and the disordered tangle produced by counterflow preserved their identities in a combined experiment. Measurements at Oregon with improved sensitivity by Barenghi, Swanson and Donnelly [20] showed that the picture just described is far from true. The observations consisted of measuring the amount of vortex line present owing to counterflow or rotation alone using second sound attenuation, and comparing the observed line density with what would be expected if the two sources of vorticity simply added. The results are complicated, but appear to be relatively simple in two limits:

(i) Limit of large line density L_H due to heat, slow rotation. Here the effect of rotation is not simply to add line density $L_R = 2\Omega/\kappa$. Instead the tangle appears to be polarized to accomplish the rotation. The effective polarization increases with rotation Ω by analogy to a gas of magnetic dipoles in a magnetic field. The results scale with L_R/L_H by analogy to $\mu H/kT$. Thus rotation appears to produce alignment in the tangle, as does a magnetic field for dipoles, and L_H appears to play the role of disordering heat bath in the statistical mechanics of superfluid turbulence. Indeed, it takes very little polarization of a dense tangle to produce rotation at the relatively small angular velocities of the apparatus.

(ii) Limit of fast rotation and small axial heat flux. Any rotation eliminates the critical velocity v_c . In this limit two critical counterflow velocities appear, v_{c1} and v_{c2}, which scale as $\Omega^{\frac{1}{2}}$.

We might speculate here that the first critical velocity appears to correspond to the Donnelly-Glaberson instability, excitation of helical waves by the counterflow on the vortex lines induced by rotation [1]. The second appears to be a transition to turbulence, with the rotation-induced array becoming a vortex tangle. A more formal investigation of these effects is likely to be rewarding.

5.3   Challenges for Understanding Periodic Boundary Layer Experiments

The decay of torsional oscillations of various pendulums was a veritable "cottage industry" in low temperature physics in the 1950's. The overall situation was reviewed by Donnelly and Hollis Hallett in 1958 [21]. The situation can be appreciated by looking at the damping of oscillations of liquid helium in a U-Tube, as reported by Donnelly and Penrose [22] and reviewed in [21].

Figure 10 summarizes the data. In helium I the damping is independent of amplitude up to a critical amplitude (or velocity) called h_t. In helium II, however, the results show damping accounted for by the Landau two fluid model below a temperature-dependent critical amplitude called h_c, a steady increase in damping up to an amplitude h_n, and a further critical amplitude, also named

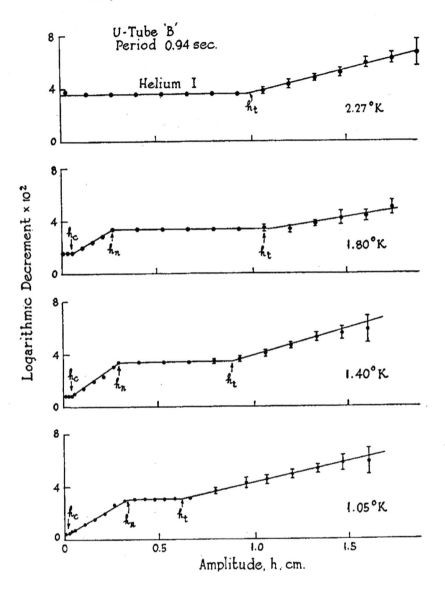

Fig. 10. The variation with amplitude of the damping of gravity oscillations of liquid helium in a U-tube at a period of 0.94 sec. After Donnelly and Penrose [22] and Donnelly and Hollis Hallett [21].

h_t. The region of damping between h_n and h_t seems to be accounted for by the Navier-Stokes equation. The authors assumed that the region between h_n and h_t was characterized by an increasing coupling between the two fluids. The details of these observations are unclear even today.

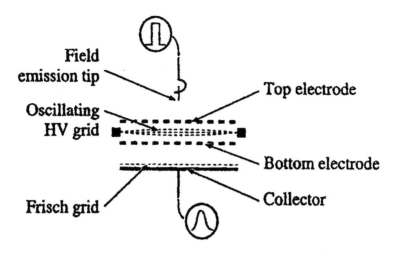

Fig. 11. Apparatus constructed by Davis, Hendry and McClintock [14] to measure the decay of superfluid turbulence at 70 mK.

What causes the two fluid model to break down at a critical amplitude h_c? What happens at h_t? What would happen if experiments (including the pendulums) were ramped up in amplitude instead of starting suddenly at high amplitude and allowed to decay. Oscillating cylinders and spheres are likely best for such novel experiments.

5.4 Instrumentation to Detect Vortices Below 1 K

Apparatus constructed by Davis, Hendry and McClintock [14] is the only one to address the question of the nature of superfluid turbulence in the absence of normal fluid experimentally. They arranged to generate turbulence by means of an oscillating grid structure, and to examine it by trapping of negative ions coming from a field emission tip on the vortex tangle. Their results indicate a temperature independent decay process. The absolute line density of the vortices remains unknown.

Second sound will not work when the temperature is too low because the normal fluid has disappeared. Ion capture may be difficult because of lack of knowledge of the dynamics of capture at very low temperatures, and experimental evidence that the capture cross section is decreasing at lower temperatures [23]. Samuels and Barenghi [24] have suggested that there is enough heat energy in a vortex tangle to be measured calorimetrically when the tangle has decayed.

Another possibility might be to design some sort of phonon (or roton)detector fast enough to follow the decay of turbulence from a towed or oscillating grid.

5.5   The Normal Fluid and the Vortex Tangle

Understanding the relationship between the normal fluid and the vortex tangle was an important topic of discussion at this conference. Theoretical progress on the self-consistent interaction between the normal fluid and quantized vortices will likely lead to the need to develop measurement techniques which will give more information than just the RMS line density L.

5.6   Flow over Blunt Objects, Testing Models such as Submarines

Turbulence can easily be generated by high Reynolds number flows over simple objects like cylinders and spheres. Here the flows are shear flows and are not homogeneous and isotropic. Indeed the whole subject of the flow over experimental models is mostly in the future. However Van Sciver has reported preliminary measurements of flow over a sphere in a companion article in this volume [24]. We look forward to more complete measurements and interpretation as this work progresses.

Acknowledgements

First of all I am grateful to Keith Moffatt, Director of the Newton Institute, for the opportunity to have this remarkable gathering here. I am indebted to Joe Vinen, Steve Stalp, Ladislav Skrbek, Carlo Barenghi, David Samuels, Renzo Ricca and Peter McClintock for many useful discussions. My research is supported by the National Science Foundation under grant DMR-9529609.

References

1. R. J. Donnelly, J. Phys Condensed Matter, **11**, 7783-7834 (1999).
2. L. Skrbek, J. J. Niemela and R. J. Donnelly, J. Phys Condensed Matter, **11**, 7761-7782 (1999).
3. Chris J. Swanson, Kris Johnson and Russell J. Donnelly, Cryogenics **38**, 673-677 (1998).
4. P. L. Walstrom, J. G. Weisend II, J. R. Maddocks and S. W. Van Sciver, Cryogenics **28**, 101 (1988).
5. M.R. Smith, R. J. Donnelly, N. Goldenfeld and W.F. Vinen, Physical Review Letters **71**, 2583 (1993).
6. M.R. Smith "Evolution and Propagation of Turbulence in Helium II". PhD Dissertation, University of Oregon (1992).
7. S. Stalp, "Decay of Grid Turbulence in Superfluid Helium". PhD Dissertation, Physics, University of Oregon (1998).
8. S.R. Stalp, L. Skrbek, R.J. Donnelly, Phys. Rev. Lett., **82**, 4831 (1999)
9. W. F. Vinen, Phys. Rev. **B61**, 1410 (2000)

10. W. F. Vinen "An Introduction to the Theory of Superfluid Turbulence" Paper in this volume
11. J. Maurer and P. Tabeling, Europhys. Lett., **43**, 29 (1998).
12. K.R. Sreenivasan, Phys. Fluids, **7**, 2778 (1995).
13. C.F. Barenghi, R.J. Donnelly, W.F. Vinen, J. Low Temp. Phys., **52**, 189 (1983).
14. S. I. Davis, P. C. Hendry and P. V. E. McClintock, Physica B , **280**, 43 (2000).
15. L. Skrbek, J. J. Niemela and R. J. Donnelly, Phys. Rev. Lett., **85**, 2973 (2000).
16. H. Schlichting, *Boundary Layer Theory*, 7th ed, McGraw-Hill (1979).
17. C. F. Barenghi C. E., Swanson, and R. J. Donnelly, J. Low Temp. Physics, **100**, 385 (1995).
18. R. J. Donnelly (Editor), Liquid and Gaseous Helium as Test FluidsSpringer-Verlag, 1991). In *High Reynolds Number Flows Using Liquid and Gaseous Helium*(Springer-Verlag, 1991). Also R. J. Donnelly and K. R. Sreenivasan (Editors) *Flow at Ultra-High Reynolds and Rayleigh Numbers*(Springer-Verlag, 1998)
19. K. W. Schwarz, and J. R. Rozen, Phys. Rev. B , **44**, 7563 (1991).
20. C. F. Barenghi C. E., Swanson, and R. J. Donnelly, Phys. Rev. Lett. , **50**, 190 (1983).
21. R. J. Donnelly and A. C. Hollis Hallett, Annals of Physics, **3**,320 (1958).
22. R. J. Donnelly and O. Penrose, Phys. Rev. , **103**, 1137 (1956).
23. R. M. Ostermeier and W. I. Glaberson, Phys. Lett. **49A**, 223 (1974).
24. D. Samuels and C. Barenghi, Phys. Rev. Lett. **81**, 4381,1998.
25. R. J. Donnelly,*Quantized Vortices in Helium II*,(Cambridge University Press, 1991).
26. S. W. Van Sciver, "The Drag Crisis on a Sphere in Flowing HeII" Paper in this volume.

The Experimental Evidence for Vortex Nucleation in ^{4}He

Éric Varoquaux[1], Olivier Avenel[2], Yury Mukharsky[2], and Pertti Hakonen[3]

[1] CNRS–Université Paris-Sud, Laboratoire de Physique des Solides,
   Bâtiment 510, F-91405 Orsay Cedex, France
[2] Commissariat à l'Énergie Atomique, Service de Physique de l'État Condensé,
   Bâtiment 772, Centre de Saclay, F-91191 Gif-sur-Yvette Cedex, France
[3] Low Temperature Laboratory, Helsinki University of Technology,
   02150 Espoo, Finland

Abstract. This update on the problem of vortex nucleation in superflows through micro-apertures follows the recent reviews of the subject by Zimmermann [1] and one of the authors [2]. Recent developments of the model of vortex nucleation involving a vortex half-ring are assessed with an emphasis on the statistical properties of the critical velocity transition. The occurrence of collapses and multiple slips is then discussed in relation with the problem of vortex nucleation.

1   Single Vortex Nucleation

Thermally assisted vortex nucleation. When superflow is forced through a micro-aperture and the critical velocity is reached, energy dissipation occurs by phase slippage. Phase slips by 2π take place according to a scenario in which $\frac{1}{2}$–rings nucleate at the wall of the aperture, at a site where the critical velocity v_c is exceeded [3–6]. A remarkable feature of the critical velocity for phase slips is its dependence on temperature, which is very nearly a straight line from about 2 K down to ~ 0.15 K:

$$v_c = v_0(1 - T/T_0) .\tag{1}$$

The value of T_0 varies somewhat with the micro-aperture but lies in the vicinity of 2.5 K. A sample of the experimental data obtained in various laboratories is shown in Fig. 1.

Since superfluid ^{4}He is nearly fully in its ground state below 1 K, such a large temperature dependence gives a strong indication that an Arrhenius-type process must be taking place. The end products of this nucleation process are vortices because phase slips by 2π are observed [7]. The point of view taken here is that these vortices are nucleated (i.e., created ex nihilo). However, it has been suggested that solitons [8] or bubbles [9] can appear as intermediate states in the nucleation process. Other mechanisms based on pure fluid-mechanical motion such as vortex mills [10,11], which will be discussed in Sect.2, are difficult to reconcile with the large temperature dependence of v_c below 1 K.

Thermal nucleation implies the existence of an energy barrier to the spontaneous formation of vortices which is overcome with the help of thermal fluctuations. As it can be surmised that a critical velocity would be reached even

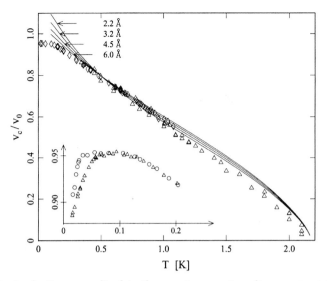

Fig. 1. Critical velocity, normalised to the zero temperature *linear* extrapolation value v_0, versus T, in Kelvin: ($\triangle$) [5], ($\diamond$) [1], for ultra-pure ^{4}He. The curves are computed from the half-ring model for $a_0 = 2.2, 3.2, 4.5, 6.0$ Å and are normalised to match the experimental value at 0.5 K. The inset shows the influence of ^{3}He impurities on v_c, ($\circ$) 3 ppb, ($\triangle$) 45 ppb.

in the absence of thermal (or quantum, as discussed below) fluctuations, this energy barrier vanishes for a finite value v_{c0} of the critical velocity. The hydrodynamic instability threshold v_{c0} at which vortices appear spontaneously has been shown to occur in numerical simulations of flows past an obstacle using the Gross-Pitaevskii equation by Frisch et al. first [12] and by others [13,14]. Close to the point where it vanishes, i.e. for $v \simeq v_{c0}$, a well-behaved energy barrier takes the form [5]:

$$E_a = \frac{2}{3} E_J \left[1 - \frac{v^2}{v_{c0}^2} \right]^{\frac{3}{2}} , \qquad (2)$$

where E_J depends on temperature and pressure.

The nucleation rate for thermally activated process is given in terms of the activation energy by Arrhenius' law:

$$\Gamma = \frac{\omega_0}{2\pi} \exp \left\{ -\frac{E_a}{k_B T} \right\} , \qquad (3)$$

$\omega_0/2\pi$ being the attempt frequency. In experiments performed in a Helmholtz resonator, such as those shown in Fig. 1, the velocity varies periodically at the resonance frequency as $v_p \cos(\omega t)$ and the probability that a phase slip takes place during the half-period $\omega t_i = -\pi/2$, $\omega t_f = \pi/2$ is

$$p = 1 - \exp\left\{ - \int_{t_i}^{t_f} \Gamma(P,T,v_p \cos(\omega t')) \, dt' \right\}$$

$$= 1 - \exp\left\{ -\frac{\omega_0}{2\pi\omega} \sqrt{\frac{-2\pi k_B T}{v_p \, \partial E_a / \partial v|_{t=0}}} \exp\left\{ -\frac{E_a}{k_B T} \right\} \right\}. \tag{4}$$

Eq.(4) stems from an asymptotic evaluation of the integral at the saddle point $t = 0$. The accuracy of the asymptotic evaluation (4) becomes questionable as $T \to 0$ where the energy barrier vanishes. It has been checked by direct numerical integration for typical cases and found to be quite satisfactory.

The critical velocity v_c is defined as the velocity for which $p = 1/2$. This definition is independent of the experimental setup, except for the occurrence in (4) of the natural frequency of the Helmholtz resonator ω. The implicit equation for v_c reads:

$$\frac{\omega_0}{2\pi\omega} \sqrt{\frac{-2\pi k_B T}{v_c \, \partial E_a / \partial v|_{v_c}}} \exp\left\{ -\frac{E_a(P,T,v_c)}{k_B T} \right\} = \ln 2 . \tag{5}$$

If E_a is given by (2), (5) can be solved analytically [5]:

$$v_c = v_{c0} \left\{ 1 - \left[\frac{3}{2} \frac{k_B T}{E_J} \gamma \right]^{\frac{2}{3}} \right\}^{\frac{1}{2}}, \tag{6}$$

where, with logarithmic accuracy, $\gamma \simeq \ln(0.1\omega_0/\omega)$. It is immediately apparent that the empirical linear dependence of v_c on T, (1), is not well reproduced by (6); this dependence could be satisfied by the following functional form for E_a,

$$E_a = E_e(1 - v/v_{c0}) , \tag{7}$$

which is quite different from (2).

Thus, the temperature dependence of v_c, (1), strongly suggests the existence of a thermally activated nucleation process, but a well-behaved energy barrier such as (2) leads to too fast an increase of v_c as $T \to 0$.

Statistical width of the critical transition. The velocity v_c of each critical event is a stochastic quantity. Its statistical spread can be characterised by the 'width' of the probability distribution defined [5,15] as the inverse of the slope of the distribution at v_c, $(\partial p/\partial v|_{v_c})^{-1}$. This critical width is found to be expressed by:

$$\Delta v_c = -\frac{2}{\ln 2} \left[\frac{1}{2} \left\{ \frac{1}{v_c} + \frac{\partial^2 E_a}{\partial v^2}\bigg|_{v_c} \bigg/ \frac{\partial E_a}{\partial v}\bigg|_{v_c} \right\} + \frac{1}{k_B T} \frac{\partial E_a}{\partial v}\bigg|_{v_c} \right]^{-1} . \tag{8}$$

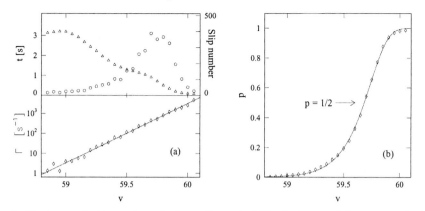

Fig. 2. Vortex nucleation rate and phase slip probability in terms of the mean velocity in the micro-aperture expressed as phase winding numbers [16] in ^{4}He at 12 mK, 0.6 bar, with 100 ppb of ^{3}He impurities: frame (a) top, time spent, in seconds ($\triangle$, left scale), and number of phase slips ($\circ$, right scale) per velocity bin of size 0.1; frame (a) bottom, slip rate in s^{-1} as a function of flow velocity, in units of 2π; frame (b), cumulative probability obtained from the histogram of the slip velocities. The plain curve is a least square fit to the data with the functional form corresponding to (4). The value $p = 1/2$, shown by the arrow, defines the critical velocity (here, $v_c = 59.68$).

At low temperatures and large critical velocities, the quantity in curly brackets in the right hand side of (8) is small with respect to the last term so that the width is simply expressed as $\Delta v_c = -(2/\ln 2)\, k_B T \left(\partial E_a/\partial v|_{v_c} \right)^{-1}$. Thus, the statistical width is an approximate measure of the inverse of the slope of E_a in terms of v.

This quantity is derived from p, itself obtained by integrating the histogram of the number of nucleation events ordered in velocity bins. This procedure is illustrated in Fig. 2: p shows an asymmetric-S shape characteristic of the double exponential dependence of p on v, (4), a consequence of Arrhenius'law, (3), being plugged into a Poisson probability distribution. The observation of this asymmetric-S probability distribution constitutes another experimental clue for the existence of a nucleation process.

The critical transition of v_c displays a measurable width, shown in Fig. 3, which implies (as does the T-dependence of v_c) that the energy barrier is neither very high compared to $k_B T$ nor very steep in terms of v. Fig. 3 contains data from several groups. These data are somewhat scattered, especially above 0.5 K, but they do show overall agreement within this scatter. They provide experimental input on $\partial E_a/\partial v|_{v_c}$ and put tight limits on the theoretical models discussed below.

It is also possible to measure the nucleation rate Γ directly as a function of v. This quantity is the ratio, for a given velocity bin, of the number of slips which have occurred at that velocity to the total time spent by the system at the same velocity. The procedure is illustrated in Fig. 2. The slope of $\Gamma(v)$ yields $\partial E_a/\partial v|_{v_c}$; the value of Γ at v_c gives a combination of ω_0 and $E_a(v_c)$. Non-linear

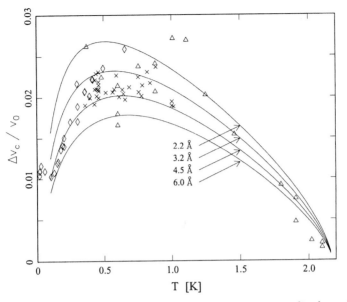

Fig. 3. Statistical width of the critical velocity transition, normalised to the linear extrapolation limit at $T = 0$, v_0, in terms of temperature: ($\diamond$) [5], ($\triangle$) [1], ($\times$) [17] .

fits of p to (4) give estimates of ω_0, E_J and v_c/v_{c0} which are independent of the specific features of a given model. These estimates compare well with those of the more precise analysis described below.

Quantum tunnelling. Below 0.15 K, v_c ceases abruptly to vary with T, as seen in Fig. 1. For ultra-pure ^{4}He (less than 1 part in 10^9 of ^{3}He impurities), $v_c(T)$ remains flat down to the lowest temperatures (~ 10 mK). The crossover from one régime to the other is very sharp. At the same crossover temperature T_q, Δv_c also levels off sharply. It is believed on experimental grounds that this saturation is intrinsic and is not due to parasitic interferences [5].

If the nucleation barrier were undergoing an abrupt change at T_q, for instance because of a bifurcation toward a vortex instability of a different nature [8], in all likelihood Δv_c would jump to a different value characteristic of the new process (presumably small since v_c reaches a plateau). Such a jump is not observed in Fig. 3. Furthermore, v_c levels off below T_q, which would imply through (5) that E_a becomes a very steep function of v, but Δv_c also levels off, which, through (8), would imply the contrary. We are led to conclude [18] that, below T_q, thermally-assisted escape over the barrier gives way to quantum tunnelling under the barrier. In which case the attempt frequency is related to the crossover temperature by

$$\hbar \omega_0 = 2\pi k_B T_q . \tag{9}$$

Relation (9) holds for barriers of the form (2) even in the presence of moderate damping [19].

In the quantum régime, the tunnelling rate for a cubic-plus-parabolic potential well [20] can be used in place of (3) to compute v_c and Δv_c. The corresponding formulae are obtained by the following substitution in (3), (5) and (8):

$$\text{a)}\ \frac{\omega_0}{2\pi} \implies \frac{\omega_0}{2\pi} \left(864\pi \frac{E_a}{\hbar\omega_0} \right)^{\frac{1}{2}}, \qquad \text{b)}\ \frac{E_a}{k_B T} \implies \frac{36\, E_a}{5\,\hbar\omega_0}.$$

This procedure yields the smooth, if rapid, transition in v_c and Δv_c observed at T_q. The value of ω_0 given by (9) ($\omega_0/2\pi = 2 \times 10^{10}$ Hz for $T_q = 0.147$ K) is consistent with the attempt frequency appropriate to the thermally-activated régime [21] and that found directly from the fits to the probability p as shown in Fig. 2. Furthermore, it agrees (for $a_0 = 5$ Å) with the eigenfrequency of the highest Kelvin mode that a vortex filament in ^{4}He can sustain, $\omega^+ = \kappa/\pi a_0^2 = \omega_0$ where κ is the quantum of circulation. Thus, assuming vortex nucleation by quantum tunnelling below T_q is fully consistent with the thermally-assisted nucleation régime that prevails above T_q.

The $\frac{1}{2}$–ring model. Described in full in [5], this model has a long history, following the work of Volovik [22], itself based on the theory of homogeneous nucleation of vortices by Iordanski, and by Langer and Fischer [23]. It has been very successful in accounting for vortex nucleation by ions [24]. In essence, the model is based on the hydrodynamics of an Eulerian fluid assumed valid to scales of the order of the vortex core diameter a_0, that is down to atomic size in ^{4}He. Nucleation is assumed to occur on an asperity on the walls of the micro-aperture where i) the flow velocity is largest, ii) the superfluid density is depressed. The perturbed volume over which nucleation takes place is, on heuristic grounds, of the order of a_0^3. The radius of the nucleated half-ring when it escapes to the bulk fluid turns out of the order of 15 Å. As a consequence, the size and shape of the asperity can be neglected in a first approximation; the asperity only serves the purpose of breaking translational invariance. The bases of the $\frac{1}{2}$–ring model have been re-examined critically by Sonin [25] and by Fischer [26]. It can also be mentioned at this point that Nore et al. [27] have shown by numerical simulations of 3D flows past an obstacle in the Gross-Pitaevskii model that vortex filaments do evolve spontaneously into one, or possibly a few, half-rings which thus constitute the preferred configuration of small vortices close to walls.

The $\frac{1}{2}$–ring model is amenable to an expansion of the form (2) but the full expression of E_a can straightforwardly be evaluated numerically and substituted into the expressions of v_c and Δv_c, (5) and (8) [5]. The energy barrier impeding vortex nucleation is found to vanish at

$$v_{c0} = 0.432\, \frac{\kappa}{2\pi a_0}. \tag{10}$$

This value, obtained for an Eulerian fluid, is to be compared with that obtained by Rica for the Gross–Pitaevskii model, $v_{c0} = 0.262\kappa/2\pi a_0$ [28], comparison which points up quantitative inadequacies in the $\frac{1}{2}$–ring model outlined above.

Comparison with experiment and discussion. The outcome of the numerical evaluation of $v_c(T)$ and $\Delta v_c(T)$ is plotted in Figs. 1 and 3 for various values of the core parameter a_0 which is the only free parameter in the problem. The temperature dependence of a_0 and ρ_s is taken from refs.[29] and [30] respectively. In the bulk, the low temperature value of $a_0 \simeq 1$Å . At a wall boundary, it is at least of the order of the static penetration depth (2.5 Å) [29], and probably rather higher because of the large applied flow velocity [31].

The temperature dependence of the computed velocity shown in Fig. 1 matches that of the data above 0.5 K but markedly bends upward below that temperature; the larger a_0, the more pronounced the bending. As noted before, this behaviour comes from the form (2) of the energy barrier and is therefore built into the model.

It is apparent in Fig. 3 that the width at high temperature is in reasonable agreement with the smaller values of a_0 (i.e. $\sim$ 3 Å), while the low temperature end calls for larger values ($\sim$ 4.5 Å). This trend is the same as for the critical velocity in Fig. 1, and, as noted above, it can certainly be expected that a_0 would increase close to the wall, hence with increasing velocities and decreasing nucleated vortex size [3].

The critical velocities scale as the inverse of a_0 (This scaling does not appear in Fig. 1). An estimate of the absolute, *local*, value of v_c, i.e., the value at the nucleation site, has been made using ^{3}He impurities as local probes [32]. The presence of a tiny concentration of ^{3}He quasiparticles strongly affects the critical velocity, as shown in the insert of Fig. 1, because quasiparticles condense on the vortex core and lower its energy. The local concentration of ^{3}He depends on the local pressure, itself governed by the velocity according to the Bernoulli equation. Detailed measurements of this effect have led to a value of $\sim$ 22 m/s for the *local* nucleation velocity on the quantum plateau. This value is to be compared with the *mean* flow velocity in the aperture which is measured in the experiment and which is of the order of 5 m/s. Using the curves in Fig. 1 and (10), the corresponding value of a_0 is found to be $\sim$ 2.65 Å.

Thus the values of the vortex core parameter a_0 which account for the various experimentally measured quantities, the magnitude and the temperature dependence of v_c and of Δv_c, agree with one another within a factor of about two in spite of the blatant oversimplifications of the $^1/_2$–ring model. Hence, taking account of the details of the nucleation site geometry and chemical environment, of the exact nature of the boundary layer at the wall and of the properties of the small nascent vortex cannot be expected to vastly improve the description of vortex nucleation given above. On another hand, refinements are needed, especially for the conceptual grounds underlying the model since it is known that very small vortices do not even exist [33].

Discussion of the $^1/_2$–ring model. There have been several attempts to reformulate the $^1/_2$–ring model, both to put it on firmer theoretical grounds and to possibly obtain a better description of the experimental results [1,3,25,26,34]. The $^1/_2$–ring model main inadequacies lie at the high velocity (small vortex) end. Burkhart et al. [3] have introduced corrections to the vortex energy and momentum due to the proximity of a boundary as computed with the Gross-Pitaevskii equation. These corrections go in the right direction but cannot be considered as a full reformulation of the small-vortex-at-a-wall problem.

A reassessment of the model, due to Sonin [25] and aiming in particular at a quasiclassical reformulation of vortex quantum tunnelling, yields the following elegant analytical form for the energy barrier

$$E_a = \frac{\kappa^3}{32\,\pi} \frac{\rho_s}{v} \left[\ln\left(\frac{\kappa}{4\pi\,va_0}\right) \right]^2$$

in the thermally activated régime and a corresponding expression for quantum tunnelling. This energy barrier does vanish for $v^* = \kappa/4\pi a_0$ but quadratically in $1 - v/v^*$. This functional dependence is further away from the empirical relation (7) than the low temperature limit of the $\frac{1}{2}$–ring model (2). Consequently, and as also shown by direct numerical computation, the curvature of $v_c(T)$ at low T is even more pronounced. An even more serious discrepancy is found for the width Δv_c: the computed value is much too large for any reasonable value of a_0. These results may look surprising since the work of [25] starts from the same premises as the conventional $\frac{1}{2}$–ring model. The difference has been checked numerically to reflect the cost of the approximations made to obtain the analytical form of E_a [35].

Sonin also discusses the influence of the geometry of the asperity but there is no direct evidence from experiment that this geometry plays an important rôle. On the contrary, the fact that the slope of $v_c(T)$ with temperature, that is T_0 in (1), is found to be the same in quite different apertures shows that vortex nucleation is not very sensitive to the details of the geometry. It is a little more puzzling that the crossover temperature T_q from quantum tunnelling to thermal activation has been found independent of cooldowns for two different apertures at Saclay, although a somewhat different value of T_q has been reported by the Berkeley group [36]. The overall agreement between various experiments basically indicates that finer specific details are not very relevant and that the simplifications of the $\frac{1}{2}$–ring model are reasonably well founded. It nonetheless remains that the nature and geometry of a typical nucleation site are quite undetermined and that the enhancement factor between the mean aperture velocity and the velocity at the nucleation site is not under control, as shown in particular by Shifflett and Hess [37].

Theoretical fits comparable to those in Figs. 1 and 3 have been obtained by Zimmermann et al. [1] with a variant of the $\frac{1}{2}$–ring model in which a_0 is kept fixed to the Hills-Roberts value but additional parameters are introduced to force the vortex energy and momentum to go to zero for vanishing vortex sizes faster than the classical vortex expressions. Also, the attempt frequency $\omega_0/2\pi$ and v_{c0} are treated as fitting parameters. These authors find that $\ln(\omega_0/2\pi)$ should be comprised between 15 and 25 (3.3×10^6 and 7.2×10^{10} Hz for $\omega_0/2\pi$). The latter value is compatible with $v_{c0} = 26.5$ m/s, which is in fair agreement with the value of 22 m/s found in [32]. Thus, different approaches give converging results and the $\frac{1}{2}$-ring model can be said to give a semi-quantitative description of vortex nucleation when single phase slips are involved.

2   Multiple Slips and Collapses

Single phase slips are observed in experimental situations which may be loosely characterised as 'clean', that is, for uncontaminated apertures of relatively small sizes (a few micrometres at the most), with low background of mechanical and acoustical interferences, etc ..., and with probing techniques which do not man-handle the superfluid, namely, with low frequency Helmholtz resonators. When these conditions are not met, flow dissipation occurs in a more or less erratic manner by large events - giant slips or 'collapses' of the superflow. This last situation is quite commonly met in practice, as discussed in [38].

Collapses constitute an apparent disruption of the vortex nucleation mechanism described in the previous section. Their properties have been studied in phase slippage experiments and are reviewed in this section, together with possible mechanisms for their formation. It is likely that these events provide a bridge between the 'clean' single phase slip case and the usual situation of critical velocities which are temperature-independent below 1 K and which depend on the channel size d according to $v_F \sim (1/d)\ln(d/a_0)$ [39]. It is also possible that they take part in the build-up of vortex tangles forming superfluid turbulence [40]. Neither problems are fully resolved at present.

The two types of large slips. Examples of multiple slips can be seen in Fig. 4 which shows the peak amplitude chart of a two-aperture resonator at 12.5 mK, 24 bars, in a 100 ppb ^{3}He in ^{4}He sample. The very large amplitude drop shown in Fig. 4 and in the insert is rare (one in 10^4 to 10^5 slips) under the conditions of this particular experiment. This type of events, called in [38] 'singular' collapses and discussed further below, may occur at velocities much below the vortex nucleation threshold (down to at least a third of v_c).

Besides the usual single slip pattern, there appears in Fig. 4 occasional double slips (i.e. involving phase changes by 4π) and infrequent triple slips. Raising the temperature to 80 mK, again for this particular cooldown, causes these multiple slips to occur much more frequently and to involve more circulation quanta on the mean. Lowering the pressure to 0 bar results in an almost complete disappearance of multiple slips at all temperatures. These features are described in detail in [38].

Some degree of understanding of the formation of multiple slips can be gained by plotting the mean value of the phase slip sizes, expressed in number of quanta, against the flow velocity at which the slips take place [41]. This flow velocity is close to the critical velocity for single phase slips, i.e. the nucleation velocity; it is varied by changing the temperature, the pressure, the resonator drive level. Such a plot is shown in Fig. 5 for $\langle n_+ \rangle$, i.e. in flow direction conventionally chosen as the + direction. Slips in the opposite (−) direction behave qualitatively in the same manner but the phenomenon displays a clear quantitative asymmetry. As

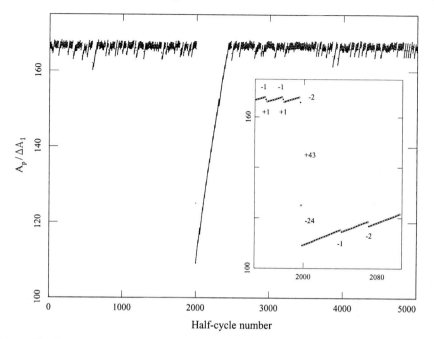

Fig. 4. Absolute peak amplitudes at successive half-cycles of the resonance, normalised to the amplitude drop of a single slip versus time expressed in half-cycle number. The half-period is 31.8 milliseconds. Individual measurements cannot all be resolved on the scale of the figure. The top trace shows a succession of amplitude drops which correspond, for the largest part, to succession of phase slips by 2π of opposite sign, with occasional larger slips. When no slip occurs during the half-cycle, the resonance amplitude grows under the action of the electrostatic drive continuously applied to the resonator. The large feature around the 2000th half-cycle is a 'singular' collapse. The insert shows the details of this particular collapse, ● for positive peaks, ○ for negative peaks.

can be seen in Fig. 5, the mean slip size decreases on either side of the quantum plateau, as does the nucleation velocity, but increases with pressure, contrarily to the nucleation velocity which decreases with increasing pressure. It appears clearly that the magnitude of the superflow velocity does not directly control, by itself, the occurrence of multiple slips. This implies, as will be discussed further below, that the phenomenon under study is not purely hydrodynamical in the bulk of the fluid but involves some complex interplay with the boundaries. As shown in Fig. 5, the velocity threshold for the appearance of multiple slips depends on hydrostatic pressure; in fact, the P-dependence of the upturn of $\langle n_{+} \rangle$ vs v exactly tracks that of the critical velocity for single phase slip nucleation. This indicates that multiple slips appear because of an alteration of the nucleation process itself.

The pattern of formation of multiple slips changes from cooldown of the cell from room temperature to cooldown but remains stable for each given cooldown.

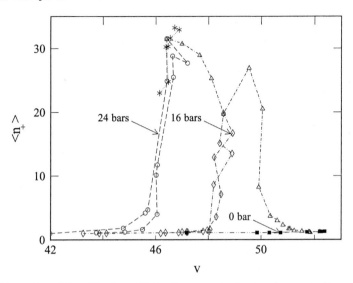

Fig. 5. Mean size of (positive) multiple slips vs velocity in phase winding number in nominal purity ^{4}He (100 ppb ^{3}He): ($\triangle$) pressure sweep from 0.4 to 24 bars at 81.5 mK (all even values of P, and 0.4, 1, 3, 5, 7 bars) - ($\diamond$) temperature sweep at 16 bars - ($\circ$) temperature sweep at 24 bars - ($*$) drive level sweep at 24 bars, 81.5 mK - ($\blacksquare$) temperature sweep at 0 bar. For the temperature sweeps, from 14 to 200 mK approximately, v first increases, reaches the quantum plateau and then decreases, as shown in the insert of Fig. 1. Lines connect successive data points in the temperature and pressure sweeps.

It seems to depend on the degree of contamination of the cell, degree which cannot easily be controlled experimentally. The detailed microscopic configuration of the aperture wall where nucleation takes place probably plays an major rôle in multiple slip formation. Multiple slips are different from 'singular' collapses and the underlying mechanisms responsible for both phenomena are bound to be different as will be discussed below.

Remanent vorticity and vortex mills. Remanent vorticity in ^{4}He, which had been long assumed, has been shown directly to exist by Awschalom and Schwarz [42]. This trapped vorticity, according to Adams et al [43], either is quite loosely bound to the substrate and disappears rapidly, or is strongly pinned. To account for these observations, Schwarz has proposed the following formula, based on numerical experiments, for the velocity at which vortices unpin,

$$v_u \lesssim \frac{\kappa}{2\pi D} \ln(\frac{b}{a_0}) ,$$

D being the size of the pinned vortex and b being a characteristic size of the pinning asperity. Thus, vortices pinned on microscopic defects at the cell walls can in principle exist under a wide range of superflow velocities.

It has been suggested by Glaberson and Donnelly [10], in connection with the critical velocity problem in an aperture, that imposing a flow to a vortex pinned between opposite lips of the aperture would induce deformations such that the vortex would twist on itself, undergo self-reconnections, and mill out free vortex loops. As shown by numerical simulations of 3D flows involving few vortices only [44], vortex loops and filaments are stable even against large deformations. Vortices are not prone to twist on themselves and foster loops. It takes the complex flow fields associated with fully developed vortex tangles to produce small rings, as discussed at the Workshop [45,46]. And it takes some quite special vortex pinning geometry to set up a vortex mill that actually works; Schwarz has demonstrated by numerical simulations that a vortex pinned at one end and floating along the flow streamlines with its other end free moving on the wall develops a helical motion, a sort of driven Kelvin wave, and reconnects sporadically to the wall when the amplitude of the helical motion grows large enough [47]. This helical mill does churn out fresh vortices.

The above remarks on the stability of vortex loops or half-loops in their course make it unlikely that the multiple slips be due to the production of small rings by a vortex after having left the vicinity of the nucleation centre. Furthermore, such a purely hydrodynamical process would depend on the velocity of the flow only, contrary to the results shown in Fig. 5. What appears more likely is that multiple slips are produced by a transient vortex mill of the helical type suggested by Schwarz operating very close to the nucleation site. The pinning of the mill vortex and its subsequent release would take place immediately after nucleation when the velocity of the vortex relative to the boundary is still small and the capture by a pinning site easy. This process depends on the precise details of the pinning site configuration and of the primordial vortex trajectory, factors which allow for the variableness of multiple slips on contamination and pressure (i.e. single phase slip nucleation velocity) [48]. That pinning does take place close to sites where vortices are nucleated is shown below.

In-situ **contamination by atomic clusters: pinning and collapses.** In a series of experiments conducted at Saclay [49,50], it was found that heavy contamination of the cell by atomic clusters of air or H_2 caused numerous collapses of the 'singular' type to occur. The peak amplitude charts of the resonator became mostly impossible to interpret, except in a few instances where two apparent critical velocities for single slips were observed. The higher critical velocity corresponded to the one observed in the absence of contamination. The lower critical velocity is thought to reveal the influence of a vortex pinned in the immediate vicinity of the nucleation site. This vortex induces a local velocity which adds to that of the applied flow and causes an apparent decrease in the critical velocity for phase slips. Because of this change, the presence of the pinned vortex could be monitored, the lifetime in the pinned state and the unpinning velocity could be measured, yielding precious information on the pinning process.

A observation about vortex nucleation which comes from these experiments is rather straightforward: existing vortices, either pinned or moving, do contribute to the nucleation of new vortices at the walls of the experimental cell.

Another observation is that a large number of unpinning events were taking place at an 'anomalously low' unpinning velocity. A parallel can be made [50] with singular collapses, which may also occur at 'subcritical' velocities and which were also quite numerous, suggesting that the two effects have a common cause. Noting furthermore that pinning and unpinning processes were also quite frequent in these experiments, releasing a fair amount of vagrant vorticity, it appears quite plausible that both singular collapses and low velocity unpinning events are caused by vagrant vortices hopping from pinning sites to pinning sites, eventually passing by close to the pinning centre or the nucleation site, and giving a transient boost to the local velocity which pushes a pinned vortex off its perch or causes a burst of vortices to be shed.

As shown in this section, the capability to observe the behaviour of single vortices in resonator experiments, interpreted with the hindsight provided by finely tuned numerical simulations, leads to a partial understanding of the phenomena of pinning, multiple slips and collapses which are taking place on a nanometric scale at the rims of the micro-aperture in strong superflows. Vagrant vortices reentering the region where the vortex nucleation site lies, nanoscopic vortex mills à la Schwarz, offer explanations for singular collapses and for multiple slips respectively which account for the detailed signatures of these effects. Obtaining a clearer and better grounded picture would require a knowledge of the microstructure of the nucleation site and its environment which is not available at present.

A large number of results on topics relevant to this work were presented at the QVD workshop to which the authors feel unable to refer appropriately here and to which they hope that other contributions in this volume will give fuller coverage. One of them (E.V.) wishes to acknowledge the hospitality of the Isaac Newton Institute where part of the manuscript was prepared.

References

1. W. Zimmermann Jr., C.A. Lindensmith, J.A. Flaten, J. Low Temp. Phys. **110**, 497 (1998), and references therein
2. E. Varoquaux, in: *Topological Defects and the Non-Equilibrium Dynamics of Symmetry Breaking Phase Transitions*, ed. by Y.M Bunkov, H. Godfrin (Kluwer Academic, The Netherlands 2000) p. 303 and references therein
3. S. Burkhart, M. Bernard, O. Avenel, E. Varoquaux: Phys. Rev. Lett. **72**, 380 (1994)
4. W. Zimmermann, Jr: J. Low Temp. Phys. **93**, 1003 (1993)
5. O. Avenel, G.G. Ihas, E. Varoquaux: J. Low Temp. Phys. **93** 1031 (1993)
6. E. Varoquaux, O. Avenel: Physica B **197**, 306 (1994) and references therein
7. W. Zimmermann, Jr.: Contemporary Phys. **37**, 219 (1996)
8. C. Josserand, Y. Pomeau: Europhys. Lett. **30**, 43 (1995)
9. C. Josserand, Y. Pomeau, S. Rica: Phys. Rev. Lett. **75**, 3150 (1995)
10. W.I. Glaberson, R.J. Donnelly: Phys. Rev. **141**, 208 (1966)

11. K.W. Schwarz: Phys. Rev. Lett. **64**, 1130 (1990)
12. T. Frisch, Y. Pomeau, S. Rica: Phys. Rev. Lett., **69**,1644 (1992) C. Josserand, Y. Pomeau, S. Rica: Physica D **134**, 111 (1999) S. Rica: this Workshop
13. B. Jackson, T. Winiecki, M. Leadbeater, J.F. McCann, C.S. Adams: this Workshop, and references therein
14. P.H. Roberts, N. Berloff: this Workshop, and references therein
15. W. Zimmermann Jr., O. Avenel, E. Varoquaux: Physica B, **165&166**, 749 (1990)
16. Phase winding numbers are obtained from velocities in cm/s by multiplication by l_h/κ, the 'hydraulic' length l_h characterising the geometry of the aperture. For a phase slip by 2π, the phase winding number changes by one unit and the trapped circulation in the resonator loop by one quantum.
17. J. Steinhauer, K. Schwab, Yu. Mukharsky, J.C. Davis, R.E. Packard: Phys. Rev. Lett. **74**, 5056 (1995) J. Low Temp. Phys. **100**, 281 (1995). The criticism of the $\frac{1}{2}$–ring model made in these references has been refuted by E. Varoquaux and O. Avenel: Phys. Rev. Lett. **76**, 1180 (1996)
18. G.G. Ihas, O.Avenel, R. Aarts, R. Salmelin, E. Varoquaux: Phys. Rev. Lett. **69**, 327 (1992)
19. V.I. Mel'nikov: Phys. Reports **209**, 1 (1991)
20. A.O. Caldeira: PhD Thesis, quoted in A.O. Caldeira, A.J. Leggett: Ann. Phys. (N.Y.) **149**, 374 (1983)
21. E. Varoquaux, M.W. Meisel, O. Avenel: Phys. Rev. Lett. **57**, 2291 (1986)
22. G.E. Volovik: Sov. Phys. JETP Lett. **15**, 81 (1972)
23. J.S. Langer, J.D. Reppy: Prog. Low Temp. Phys., Vol. 6, ed. C. J. Gorter (North-Holland , Amsterdam 1970)
24. C.M. Muirhead, W.F. Vinen, R.J. Donnelly: Phil. Trans. Roy. Soc. A **311**, 433 (1984) Proc. R. Soc. London A **402**, 225 (1985)
25. E.B. Sonin: Physica B **210**, 234 (1995)
26. U.R. Fischer: Phys. Rev. **B58**, 105 (1998) Physica B **255**, 41 (1998)
27. C. Nore, C. Huepe, M.E. Brachet: Phys. Rev. Lett. **84**, 2191 (2000)
28. S. Rica: Physica D (to be published) and this Workshop
29. R.N. Hills and P.H. Roberts: J. Phys. C **11**, 4485 (1978) P.H. Roberts, R.N. Hills and R.J. Donnelly: Phys. Lett. **70**A, 437 (1979)
30. J. Maynard: Phys. Rev. B **14**, 3868 (1976)
31. Numerical simulations in the framework of the Gross-Pitaevskii equation presented at the Workshop by C. Adams and N. Berloff illustrate the deformation of the boundary layer with velocity quite vividly. It has also been shown in the same framework that the vortex core radius diverges at a wall [3].
32. E. Varoquaux, G.G. Ihas. O. Avenel, R. Aarts: Phys. Rev. Lett. **70**, 2114 (1993)
33. C.A. Jones, P.H. Roberts: J. Phys. A: Math. Gen. **15**, 2599 (1982)
34. F.V. Kusmartsev: Phys. Rev. Lett. **76** 1880 (1996)
35. The form of the logarithmic term in the expression for the energy barrier cannot be considered as accurate (E. Sonin, private communication)
36. J.C. Davis, J. Steinhauer, K. Schwab, Yu. Mukharsky, A. Amar, Y. Sasaki, R.E. Packard: Phys. Rev. Lett. **69**, 323 (1992)
37. G. M. Shifflett, G. B. Hess: J. Low Temp. Phys. **98**, 591 (1995)
38. O. Avenel, M. Bernard, S. Burkhart, E. Varoquaux: Physica B **210**, 215 (1995)
39. E.Varoquaux, W. Zimmermann Jr., O. Avenel: in Proc. NATO workshop on Excitations in 2D and 3D Quantum Fluids, ed. by A.F.G. Wyatt, H.J. Lauter (Plenum Press, NY 1991) p. 343 and references therein
40. K.W. Schwarz: Phys. Rev. Lett. **50**, 364 (1983)

41. E. Varoquaux, O. Avenel, M. Bernard, S. Burkhart: J. Low Temp. Phys. **101**, 821 (1995)
42. D.D. Awschalom, K.W. Schwarz: Phys. Rev. Lett. **52**, 49 (1984)
43. P.W. Adams, M. Cieplak, W.J. Glaberson: Phys. Rev. Lett. **78**, 3602 (1985)
44. K.W. Schwarz: private communication to E.V.; the 3D simulations presented at the QVD Workshop by C. Adams also show that vortex loops can undergo severe deformations and not break apart (to appear in Euro. Phys. Lett.)
45. B.V. Svistunov: Phys. Rev. **52**, 3647 (1995)
46. M. Tsubota, T. Araki, S.K. Nemirovskii: Phys. Rev. B (in press) and this Workshop
47. K.W. Schwarz: Phys. Rev. Lett. **64**, 1130 (1990)
48. W. Zimmermann: Jr., J. Low T. Phys. **91**, 219 (1993) J. Flaten: PhD Thesis, Univ. of Minnesota (1997, unpublished). The large dissipation events reported by these authors apparently fall in yet another category than the multiple slips and the singular collapses discussed here as they persist over a number of resonator periods.
49. P. Hakonen, O. Avenel, E. Varoquaux: Phys. Rev. Lett. **81**, 3451 (1998)
50. E. Varoquaux, O. Avenel, P. Hakonen,Yu. Mukharsky: Physica B **255**, 55 (1998)

Applications of Superfluid Helium in Large-Scale Superconducting Systems

Steven W. Van Sciver

National High Magnetic Field Laboratory, Florida State University, USA

Abstract. The application of superfluid helium (He II) in large-scale superconducting systems in reviewed. For most of these systems, He II is the coolant of choice, compared to alternative low-temperature helium cooling methods, for its combination of excellent thermal transport properties at reduced temperature. These advantages come at some cost to system designers as the hardware associated with delivering He II is more complex. The paper provides overviews of several major superconducting systems that utilize He II cooling in terms of the relevant thermal fluid properties. Included are issues of transient and steady-state heat transport, high Reynolds number flows, and two-phase phenomena. An effort is made to show the connection between the fundamental superfluid properties and the need of the application. Areas where further understanding of the superfluid state could benefit applications are also discussed.

1 Introduction

Liquid helium has a wide range of uses in low temperature technology today. The properties of liquid helium that make it particularly valuable are its low temperature, persistence of the liquid state to the lowest achievable temperatures and the existence of the superfluid state with its associate unique transport phenomena. These properties combined with an ever-improving cryogenic engineering infrastructure have allowed the development of a number of large-scale systems that use liquid helium cooling.

There are primarily two classes of large-scale applications for liquid helium:

1. Large superconducting systems such as magnets or RF cavities for high-energy physics accelerators, fusion systems and other magnet facilities.
2. Large space-based instruments for infrared astronomy and other fundamental studies.

Each of these applications has a unique set of requirements that in turn place requirement on the coolant, which determines the preferred state for the liquid helium. In many cases, the best choice for the coolant is superfluid helium (or He II) typically in the temperature range 1.4 K to 2.1 K. The decision to select He II as a coolant is usually driven by a combination of the lower temperature and beneficial heat transport characteristics.

The present paper begins with an overview of the basic design features and operating characteristics employed in several large-scale He II cooled applications. The motivation for He II cooling, the unique design features, and the required engineering database are reviewed in the context of these applications. In the subsequent section, the current understanding of He II as an engineering fluid is presented followed by comments about areas where further work is needed.

By far the largest and most varied application for He II cooling is in the area of large-scale superconducting magnet systems. These include magnet systems for high energy physics, fusion experiments as well as other specialized magnet facilities. In addition, there are several large-scale electron accelerators that utilize superconducting RF cavities and are also cooled by He II. Space does not permit a detailed discussion of all these applications, so the present discussion will concentrate on three such systems, which represent the range of technical demands placed on He II cryogenics:

1. The dipole magnet system for the Large Hadron Collider (LHC) which is under construction at the European Organization for Nuclear Research (CERN) laboratory in Geneva, Switzerland.
2. The recently completed outsert superconducting magnet of the 45-T Hybrid Magnet at the National High Magnetic Field Laboratory in Tallahassee, Florida.
3. The RF cavity system for the Teravolt Electron Synchrotron Linear Accelerator (TESLA) under design at the Deutsches Elektronen-Synchrotron (DESY) laboratory in Hamburg, Germany.

There are a number of advantages and some disadvantages to the use of He II for superconducting systems. The first and most obvious advantage is simply that the lower temperature provides improved superconductor properties. For example, NbTi superconductors, which are used in more than 95% of the operating superconducting magnets today, have substantially improved current densities with operation near 1.8 K as compared to the normal 4.2 K operation in atmospheric He I. This improved performance allows the achievement of higher magnetic fields. For example, a standard 8 T NbTi laboratory magnet will often achieve 10 T when operated at 1.8 K. Similarly, niobium RF cavity used in electron accelerators have losses which decrease significantly with temperature justifying operation near 2 K.

The other, somewhat more subtle, advantage to the use of He II cooling in superconducting systems is in the uniquely high effective thermal conductivity of the fluid; a property which provides improved thermal stability to the superconducting system. In application, superconductors dissipate negligible amounts of heat during steady current operation. However, there are number of special conditions that can lead to power dissipation in the superconductor. To minimize this effect, the superconductor is co-processed with a normal metal (usually copper) that acts as a shunt for current should conditions prevent the superconductor from carrying the full current. The heat necessary to bring about the momentary normalization of the superconductor can come from a variety of sources (dB/dt, mechanical motion) and must be guarded against. Efficient heat removal from the conductor is therefore a critical aspect of good reliable design.

Since He II can transport large heat fluxes (compared to that of He I), its stabilizing effect on magnet systems can be very beneficial. The understanding of the relevant thermal processes in He II has been a subject of numerous recent experimental investigations.

The principal disadvantage to the use of He II cooling in large superconducting systems originates from the increased complexity and cost of the cryogenic system. In very simple terms, the cost of refrigeration scales with the Carnot factor, which at low temperatures goes as T^{-1}. Therefore, a unit of power deposited in a He II cryogenic system operating at 1.8 K will require a refrigeration system 2.3 times larger than the equivalent system operating at 4.2 K. Since the cost of refrigeration systems scales monotonically with power, the advantages of superconductor performance and stability must outweigh this cost issue. Thus, the selection of a He II cooling system is only warranted in cases where the overall system design is optimized by this approach. What follows are three such examples.

2 Superconducting Systems That Use He II Cooling

2.1 Accelerator Magnet System for LHC

The LHC accelerator currently under construction at CERN will, when complete, operate the world's largest He II cryogenic system [1]. In this case, all the main ring magnets (dipole and quadrupoles) are made with NbTi superconductor and cooled to 1.9 K in a He II bath pressurized to atmospheric pressure. The dipole magnets, shown in cross section in Fig. 1, operate at 8.3 T, which is higher than previously built accelerator dipoles such as for the SSC and therefore require the lower temperature and improved thermal stability that He II operation affords. There are eight refrigeration plants around the 26.7 km circumference of the accelerator, each cooling one 45° sector. The 1.8 K refrigeration capacity for each sector is between 2.1 and 2.4 kW.

The system uses pressurized He II to cool the magnets, which reduces the possibility of trapped vapor within the windings of the coil. The use of pressurized He II to cool magnets was first demonstrated on a large scale in a development program for the Tore Supra tokamak plasma experiment in France [2]. The approach requires the use of a low pressure of He II heat exchanger as part of the installation. In the case of LHC, the heat exchanger is a corrugated pipe within the magnet vessel and immersed in the He II reservoir. A schematic of this system is shown in Fig. 2. The heat exchanger is partially filled with saturated He II. The two phase He II heat exchanger provides a much more efficient heat removal method than can be obtained through counterflow heat transport in the bulk fluid. Although heat transport is very efficient in bulk He II, the long distances between refrigeration stations would demand very large He II cross sections to minimize the temperature gradient.

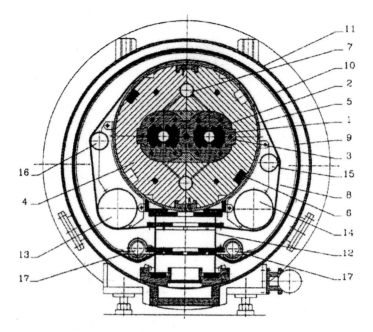

Fig. 1. Schematic of LHC dipole magnet

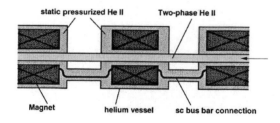

Fig. 2. Cooling system for LHC dipole magnets with two-phase heat exchanger

2.2 High Field Solenoid for the NHMFL 45-T Hybrid

The 45-T hybrid is a large combined superconducting and resistive magnet recently put in service at the NHMFL [3]. The superconducting portion of this magnet is a 14 T solenoid with a 710 mm cold bore operating at 10 kA. The outer coil of this magnet is made with NbTi conductor, while the high field inner sections use Nb_3Sn for its superior high field characteristics. The conductor in this magnet is a Cable-in-Conduit type (CICC), cf. Fig. 3, consisting of a superconducting stranded cable in a rectangular steel jacket. This is an advanced conductor design developed for leading edge magnet systems. The liquid helium coolant, which is contained within the jacketed conductor, absorbs the transient heat generation from the conductor and conducts it to the surrounding path. The NHMFL hybrid outsert is the first CICC magnet that employs He II cooling. As in the case of LHC, He II cooling is applied to achieve the highest current densities in the conductor, while providing a good thermal stabilizing environment.

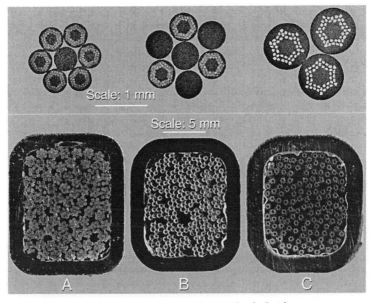

Fig. 3. Cable in Conduit conductor for hybrid magnet

A schematic cross section of the 45-T hybrid magnet system is shown in Fig. 4. The coils are located in the magnet vessel, which is separate from the He II refrigeration system and cooled by counterflow through a horizontal duct. Heat extraction occurs at a set of saturated heat exchangers located in the supply cryostat. These heat exchangers are supplied from the upper reservoir of the cryostat with the vapor flow going through room temperature vacuum pumps. The total steady state heat load on this system is less than 10 W at 1.7 K [4]. Thus, the scale of the refrigeration system is modest compared to the LHC.

2.3   RF Cavity Systems for the TESLA Electron Collider

The TESLA electron collider is one of the proposed major accelerator systems that will follow the completion of the LHC [5]. This system is a linear electron collider as compared to a circular LHC machine, which accelerates protons. Linear machines do not require the large number of dipole bending magnets that are the signature of the hadron colliders. Rather the electron colliders depend on high gradient RF cavities for accelerating the beam and use a relatively small number of magnets for beam steering and focusing. The RF cavities can be either

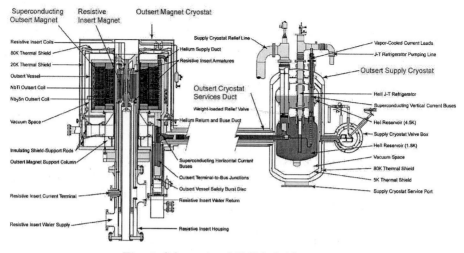

Fig. 4. Schematic of 45-T hybrid magnet

made from copper and operate at room temperature or from superconducting niobium operating near 2 K. This latter approach has been demonstrated in the Continuous Beam Accelerating Facility (CEBAF) at the Thomas Jefferson Laboratory in Virginia. The use of niobium cavities provides higher accelerating voltages and lower overall losses compared to operation at higher temperature (e.g. 4.2 K).

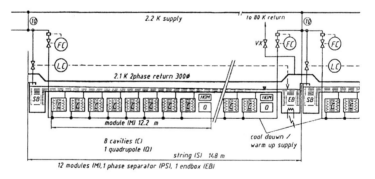

Fig. 5. Schematic of TESLA cooling system

A schematic of the cryogenic system for TESLA [6] is shown in Fig. 5. It consists of a long cryostat, which contains the RF cavities, filled with saturated He II typically at 2 K. Unlike superconducting magnets, large local disturbances are not expected to occur in RF cavities, so the complexity of a pressurized He II system is not required. The RF cavity cryostats, cross section shown in Fig. 6, are maintained at constant temperature over a long length by linking them to a common vacuum line, which may be partially filled with liquid. Thus, the principal technical issue associated with the design of the He II system pertains to the behavior of two-phase He II/vapor in near horizontal channels of relatively large diameter.

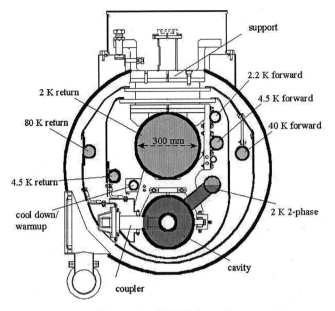

Fig. 6. Schematic of TESLA cavity cryostat

3   Application Relevant He II Properties

The application of He II cooling to large systems requires an understanding of the fundamental properties of this unique fluid. Of greatest interest are those properties that affect engineering design of the system. In some cases, the strong demands are placed on the He II coolant and it is only through practical investigation combined with basic understanding that one can apply He II cooling in an optimal way. The present section reviews current practical knowledge of He II as it affects applications. Some discussion of the need for further work is also given.

3.1   Second Sound Pulse Transport

The transport of heat pulses in He II is primarily of relevance to the stability of superconducting systems (mostly conductors) in He II. This is a very complex problem, the solution of which continues to elude scientists and engineers. The primary reason for this problem comes from an incomplete understanding of the source and characteristics of the heat deposition spectrum that originates from a working superconducting magnet. Such disturbances come from a variety of sources (AC losses in the conductor or mechanical motion of the winding) and can vary considerably from magnet to magnet. Thus, most of the effort is placed on understanding the transient thermal response of the He II coolant. One would like to understand the development of the fully turbulent state in He II and the limits of heat transfer. Nemirovskii and Tsoi [7] have presented this problem in the form of a regime map, which qualitatively defines the transition of He II from the ideal superfluid state through the development of the fully developed turbulent state to the limit of heat transfer film boiling. For short times, the heat transport is by a burst of second sound pulse, which leads to the development of turbulence. This second sound pulse can be of benefit to the operation of very high current density magnets, which sometimes experience very short duration intense disturbances.

Several experiments have reported on intense heat pulse propagation in He II [7,8]. Shimazaki et al. [8] measured the shape and heat content of thermal shock pulses in a counterflow He II channel. Fluxes as high as 40 W/cm^2 for up to 1 ms duration were applied and shown to propagate at second sound velocities. Turbulence generated in the He II followed the pulse an effect that limited the maximum energy transported to about 10 mJ/cm^2. This is a large energy deposition for high field compact superconducting magnets and thus could affect design and performance. However, one would like to better understand how this process transforms into the fully developed turbulent state, which is discussed in the next section.

3.2   Transient and Steady Transport in the Mutual Friction Regime

In many large-scale superconducting systems, the He II can be assumed to be fully turbulent and the heat transport governed by the mutual friction regime. There are several features of these systems that support this assumption. First, most of these systems have relatively large channels and He II reservoirs, so that the He II already contains a considerable amount of turbulence. Also, large magnets have slower thermal response times, typically 10 ms to 1 s, so that propagation of second sound pulses can only be a small portion of the total flux. This assumption is the design basis of most large-scale superconducting systems that are cooled with He II.

With the assumption of fully developed turbulence, one can then apply the mutual friction form of the two fluid equations to the general problem of heat transport in the He II [9]. Thus, the temperature gradient in the fluid is given by the relationship,

$$q = -\left[f^{-1}\frac{dT}{dx}\right]^{1/3} \tag{1}$$

where $f = \frac{A\rho_n}{\rho_s^3 s^4 T^3}$ and A is the usual Gorter Mellink parameter. The interesting development related to this process is the realization that (1) can be applied much like Fourier's Law for conduction in materials, although the form is proportional to the cube root of the temperature gradient. Thus, slowly varying heat transport processes should have a diffusion-like character and be governed by the He II diffusion equation,

$$\rho C_p \frac{\partial T}{\partial t} = \frac{\partial}{\partial x}\left(f^{-1}\frac{\partial T}{\partial x}\right)^{1/3}. \tag{2}$$

This formulation has been successful at treating a wide variety of transient heat transport problems related to superconducting magnet systems such as the thermal stability of a composite superconductor.

In a superconducting composite conductor, the power generated in the conductor is a function of time. For short times, the power usually peaks as the current distributes within the conductor and the initial heat deposited diffuses away. After the peak, the power usually levels off and corresponds to the steady-state joule heat in the copper stabilizer. A simplified form for the heat generated in the conductor is shown by the cross-hatched portion in Fig. 7. To ensure that the conductor returns to the superconducting state, it is necessary to have the heat transfer to the He II exceed the heat generated. The highest level of reliability is obtained by requiring that the Joule heat never exceed the maximum steady-state heat flux in the He II. However, to optimize the design, it is really only necessary to have the integrated average heat transfer exceed the generation. This condition, which was first proposed by P. Seyfert et al. [10], is shown in Fig. 7. The curve represents the maximum heat transport by the He II governed by the diffusion process, see (2). As long as the excess heat generation, area A, does not exceed the excess heat tranfer, area B, the magnet should be stable.

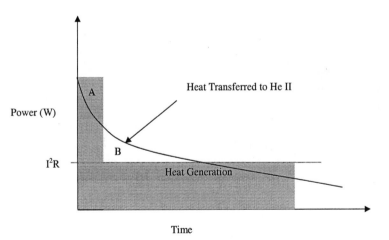

Fig. 7. Stability criterion for He II cooled conductor

3.3 The He II Energy Equation

Some applications of He II require forced circulation through a closed loop. This approach adds complexity to the system, but can provide significantly enhanced heat removal particularly in distributed or remote systems. For example, such a forced flow system was considered for the 45-T hybrid outsert magnet, but not accepted in the final design due to the additional complexity and cost. Adding a net fluid velocity to He II provides some advantage in design as well as introducing a new variable into the analysis of the fluid behavior.

As in the case of static He II applications, if one assumes that the He II is governed by the fully developed turbulent mutual friction process, then forced flow He II heat transport should obey a modified energy equation. This equation has the form [11],

$$\frac{\partial}{\partial x}\left[\left(f^{-1}\frac{\partial T}{\partial x} - \frac{1}{\rho s}\frac{\partial p}{\partial x}\right)\right]^{1/3} - \rho u C\frac{\partial T}{\partial x} - u\frac{\partial \rho}{\partial x} = \rho C\frac{\partial T}{\partial t} \qquad (3)$$

where the pressure gradient terms take into account the frictional losses and change of internatl energy with pressure. As before, this relationship has successfully modeled the behavior of heat transport in He II for a variety of experimental systems. An interesting observation about this formulation is the fact that He II will display the Joule Thomson effect when experiencing a pressure drop through an insulated tube [12]. The result will be an increase in temperature, which will generally have a negative impact on applications.

An example of heat transport in forced flow He II is displayed in Fig. 8. The flow is from left to right and a heat pulse is deposited at $x = 0$. The plot displays calculated and experimental time-dependent temperature profiles at different locations along a channel containing He II. Note that the peak in the temperature profile broadens due to thermal diffusion as time progresses and the pulse propagates. Also, the location of the peak moves at approximately the speed of the fluid, suggesting a rough method of measuring fluid velocity in such systems. At these high velocities, the Joule Thomson effect contributes by producing a gradual increase of the fluid temperature as seen by the increase in the baseline temperature as the fluid moves through the tube.

From the application viewpoint, the steady-state and transient heat transport characteristics a fully turbulent He II can be understood in terms of the He II diffusion and energy equations. There are limits to this representation and these limits need to be explored. Work is continuing on the development of turbulence and heat transport at very high velocities. This seems to be an area where some fundamental work would be able to contribute.

3.4   Fluid Dynamics of Forced Flow He II

Technical interest in forced flow He II has grown out of a number of applications. As mentioned previously, the LHC accelerator magnet system and the NHMFL hybrid magnet both had early designs that utilized forced flow He II cooling. In addition, several space-based infrared telescopes, including the currently underway Space Infrared Telescope Facility (SIRTF), have evaluated ways of circulating He II. The fundamental question that arose out of these studies

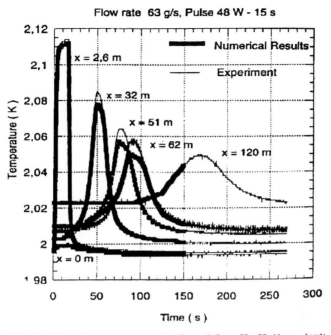

Fig. 8. Pulsed heat transport in forced flow He II (from [17])

was whether the He II pressure drop and associated fluid dynamic properties would be unique. Since that time, a considerable body of research has come to support the notion that the dynamic behavior of forced flow He II is essentially similar to that of classical fluids. Thus, He II when flowing through tubing displays a pressure drop that can be described by classical correlation [13]. This result has been supported by theoretical investigations that suggest the two fluids are coupled through turbulent interactions and thus flow together. Figure 9 displays recent measurements of the friction factor for He II flowing in a 10 mm ID tube at Re$>10^7$ (u$>$ 10 m/s). These data continue to support the previous observations.

Forced flow He II at high Reynolds number has more recently become interesting as a test fluid for basic fluid dynamic investigations [14]. The fluid has a very small kinematic viscosity allowing high Reynolds numbers to be achieved in sub-sonic flows. A question that continues to be raised is to what extent can one ignore superfluid effects. This is still an open issue and a subject for further investigation. In a companion paper at this conference, we report on drag coefficient measurements for a sphere in flowing He II [15]. The measurements suggest a temperature dependence to C_D not seen in normal fluids and this dependence appears to correlate with the normal fluid density, ρ_n. Clearly, this topic needs further experiment supported by theoretical analysis.

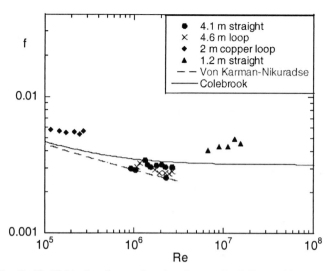

Fig. 9. He II friction factor for pipe flow at high Reynolds number

3.5   He II/Vapor Two Phase Flow

An area that has received recent investigation driven by the needs of large accelerators such as LHC or TESLA is two phase He II/vapor flow. In many of these large systems, there are tubing sections partially filled with He II and in near horizontal configuration. This operating condition leads to some interesting phenomena due to the existence of the free surface. As is in the case of single phase He II systems, the heat transfer and fluid dynamics behavior are of interest to designers. Further, because of the free surface and relative velocity between the liquid and vapor phases, there are a variety of flow stability issues that need to be addressed. These are topics of current investigation [16–18]. Modeling these systems is complex due to the interaction between the two phases [19,20]. To date, most studies have treated the He II as a classical fluid with heat transport character governed by the He II energy equation. Clearly, this fluid system is complex and further experimental and theoretical work is needed.

3.6   Fountain Effect (Fluid Management)

The thermomechanical fountain effect in He II has been considered during the design of a variety large scale applications. The most successful application of this effect occurred with the Superfluid Helium On Orbit Transfer (SHOOT) experiment which flew on the Space shuttle in 1993 [21]. This experiment demonstrated the ability of a large-scale fountain pump to tranfer He II in micro-gravity conditions. Also, this experiment as well as most other space-based He II systems used a porous plug phase separator to contain the liquid helium within the dewar and vent the vapor. The porous plug phase separator uses the heat of evaporation to remove the small internal heat generated in the dewar. The temperature difference between the He II bath and exciting vapor provides the thermomechanical pressure head to hold back the liquid.

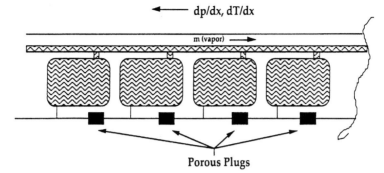

Fig. 10. Schematic of method to use porous plugs to manage He II fluid level

The use of the thermomechanical fountain effect is much more limited in large-scale superconducting systems. It has been considered in the design of space-based magnet systems for similar reasons to its use in infrared telescope technology, i.e. mostly as a phase separator. The fountain effect may also be useful as part of the fluid management system for ground-based accelerator systems containing horizontal He II/vapor two phase flows. One concern in these systems is "dry out" of the piping due to the pressure and temperature gradient along the channel. Higher temperature means higher vapor pressure, which will depress the liquid level compared to that down stream. Thus, the liquid in the channel will slope upward in the downstream direction if there is no net liquid flow. One possible method to overcome this concern is to use the fountain effect to return and circulate He II to the upstream side of the channel. A schematic of how this process can be achieved is shown in Fig. 10. The porous plugs are part of a parallel fluid handling system. The increase in temperature on the upstream side would have the effect to force liquid through the porous plug thus raising the level and preventing dry-out. There is no current application for this concept, but some technical sub-components demonstrated the approach [22].

4   Conclusions

Superfluid helium has become an engineering fluid for a number of technical applications in superconductivity and space-based instrumentation. A considerable volume of practical data has been accumulated through the development of these systems. Steady-state and slowly varying transient thermal processes can be described to be a diffusive process much like conduction. Forced flow at relatively high velocities appears to obey clssical correlations. There are a number of areas where further study is needed. We do not have adequate understanding of the development of turbulence particularly at high heat fluxes. Also, two-phase flow is an entire subject that has only recently been under investigation. Future applications will no doubt require additional investigation.

Acknowledgements

The National High Magnetic Field Laboratory is jointly funded by the National Science Foundation and the State of Florida.

References

1. P. Lebrun, Cryogenics for the Large Hadron Collider. IEEE Trans. On Applied Super., Vol 10, 1500 (2000)
2. G. Claudet and R. Aymar, Tore Supra and He II Cooling of Large High Field Magnets, Adv. Cryog. Engn. 35A, 55 (1990)
3. J.R. Miller, et al., An Overview of the 45-T Hybrid Magnet System for the NHMFL, IEEE Trans. On Magnetics, Vol 30, 1563 (1994)
4. S.W. Van Sciver, et al., Design, Development and Testing of the Cryogenic System for the 45-T Hybrid, Adv. Cryog. Engn. Vol 41, 1273 (1996)
5. R. Brinkmann, et al., Conceptual Design of a 500 Gev e+e- Linear Collider with Integrated X-Ray Laser Facility, DESY 1997-048 ECFA 1997-182, May 1997
6. G. Horlitz, et al., The TESLA 500 Cryogenic System and He II Two Phase Flow, Cryogenics Vol 37, 719 (1997)
7. S.K. Nemirovskii and A.N. Tsoi, Transient Thermal and Hydrodynamic Processes in Superfluid Helium, Cryogenics Vol 29, 985 (1989)
8. T. Shimazaki, M, Murakami and T. Iida, Temperature measurement in Transient Heat Transport Phenomena Though a Thermal Boundary Layer in High Vortex Density, Adv. Cryog. Engn. Vol 41, 265 (1996)
9. S.W. Van Sciver, Chap. 10: Helium II (Superfluid Helium), in: Handbook of Cryogenic Engineering, J.G. Weisend II (ed.). Taylor & Francis (1998)
10. P. Seyfert, Practical Results on Heat Transfer in Superfluid Helium, in: Stability of Superconductors in He I and He II, IRR Commission A 1/2 (1981), pp. 53-62
11. B. Rousset, Pressure Drop and Transient Heat Transport in Forced Flow Single Phase He II at High Reynolds Number, Cryogenics Vol 34 supplement, 317 (1994)
12. P.L. Walstrom, Joule Thomson Effect and Internal convection Heat Transfer in Turbulent He II Flow, Cryogenic Vol 28, 151 (1988)
13. P.L. Walstrom, et al., Turbulent Flow Pressure Drop in Various He II Transfer System Components, Cryogenics Vol 28, 101 (1988)
14. R.J. Donnelly, Ultra High Reynolds Number Flows Using Cryogenic Helium: An Overview, in: FLow at High Reynolds and Rayleigh Numbers, R.J. Donnelly and K. Sreenivasan (eds.), Springer (1998)
15. M.R. Smith, Y.S. Choi and S.W. Van Sciver, The Temperature Dependent Drag Crisis on a Sphere in Flowing He II, (this publication)
16. P. Lebrun, et al., Cooling Strings of Superconducting Devices Below 2 K: The He II Bayonet Heat Exchanger, Adv. Cryog. Engn., Vol 43, 419 (1998)
17. B. Rousset, et al. Behavior of He II in Stratified Counter-Current Two Phase FLow, in: Proceedings of the ICEC17, Institute of Physics Publishing (1998) pp. 671-674
18. J. Panek and S.W. Van Sciver, Heat Transfer in a Horizontal Channel Containing Two Phase He II, Cryogenics Vol 39 (1999)

19. L. Grimaud, et al., Stratified Two Phase Superfluid Helium Flow, Cryogenics Vol 37 (1997)
20. Y. Xiang, et al. Numerical Study of Two Phase He II Stratified Channel with Inclination, IEEE Trans on Applied Super. Vol 10, 1530 (2000)
21. M. DiPirro and P.J. Shirron, The SHOOT Orbital Operations, Cryogenics Vol 32 (1992)
22. J. Panek, Y Zhao and S.W. Van Sciver, Liquid Level Control Using a Porous Plug in a Two Phase He II System, Adv. Cryog. Engn., Vol 43, 1401 (1998)

The Temperature Dependent Drag Crisis on a Sphere in Flowing Helium II

Yeon Suk Choi[2], Michael R. Smith[1], and Steven W. Van Sciver[1,2]

[1] National High Magnetic Field Laboratory, Florida State University, Tallahassee, FL 32310, USA
[2] Mechanical Engineering Department, FAMU-FSU College of Engineering, Tallahassee, FL 32310, USA

Abstract. In a previous paper, we reported observing a drag crisis on a sphere in flowing He I and He II. Data in He II suggested a possible temperature dependence to the critical Reynolds number, as well as the magnitude of the crisis. In this paper, we explore temperature dependence more completely. Dynamical similarity arguments, which lead to Reynolds number scaling in the case of the Navier–Stokes equations, are applied to the two-fluid equations. The result is a modified Reynolds number involving the factor 1-δ, where $\delta \equiv \rho_s/\rho$. The ramifications of this argument, together with other possible scaling relationships, are discussed. Data and critical Reynolds numbers are plotted for several temperatures between 1.6 K and 2.0 K. Results appear to agree well with the proposed scaling for He II.

1 Introduction

The low kinematic viscosity ($\nu = \eta/\rho$, where η and ρ are dynamic viscosity and total density, respectively) of liquid helium makes it an attractive fluid for modern dynamical similarity studies, where one wishes high Reynolds numbers (Re=Ud/ν, where U and d are the characteristic velocity and dimension of the flow field) without transonic effects. While research suggests that helium above 2.176 K (He I) behaves as a classical fluid, He II (the liquid phase below 2.176 K) is a quantum fluid with a wide range of non-classical macroscopic properties. Still, studies have found that classically generated turbulence in He II may behave classically in certain experiments [1–3]. Uncertainties about the microscopic character of classically generated turbulence in He I and He II motivated the previous work [4,5] in which the form drag on a sphere in flowing He I and He II was calculated from the observed pressure distribution over the surface.

If the critical Reynolds number for the drag crisis in He II is temperature dependent, then the dimensionless equations of motion for the two fluid system must scale with other dimensionless parameters, in addition to or instead of the Reynolds number. In the classical fluid dynamics of ordinary fluids (including He I), dynamical similarity and scaling arguments for expressing experimental data in terms of Reynolds number, coefficients of drag, lift and so on, spring rigorously from non-dimensionalizing the Navier–Stokes equations. Expressing He II data in terms of an effective Reynolds number however, has been more of an empirical convenience. Perhaps one reason for this is the empirical nature of the two-fluid equations themselves, and the sometimes unspoken question of whether the current set of equations accurately represent all aspects of the physics. In the spirit of investigation however, we present one such analysis as a framework for discussing our data. [6]

The generally accepted form for the two-fluid equations in one dimension [7] is shown in (1) and (2), where we have restricted the problem to one dimension for simplicity. The entropy terms describe the exceptionally high effective thermal conductivity of the system through a mechanism called counterflow. The equations for the superfluid and normal fluid components are coupled through the velocity difference $v_{ns} = v_n - v_s$, and the mutual friction term F_{ns}. One expression for the mutual friction is shown in (3), where α is a temperature dependent parameter. The local line length per unit volume of quantized vortex lines existing in the superfluid component is given by L, and K is the quantum of circulation around a vortex line.

$$\rho_s \frac{dv_s}{dt} = -\frac{\rho_s}{\rho}\frac{dP}{dx} + \rho_s S \frac{dT}{dx} + \frac{\rho_s \rho_n}{2\rho}\frac{dv_{ns}^2}{dx} - F_{ns} \tag{1}$$

$$\rho_n \frac{dv_n}{dt} = -\frac{\rho_n}{\rho}\frac{dP}{dx} - \rho_s S \frac{dT}{dx} - \frac{\rho_s \rho_n}{2\rho}\frac{dv_{ns}^2}{dx} + F_{ns} + \eta \frac{d^2 v_n}{dx^2} \tag{2}$$

$$F_{ns} = -\frac{2}{3}\alpha k \rho_s L v_{ns} \tag{3}$$

Now, we define non-dimensional quantities in terms of representative velocities, distances and temperatures as follows;

$$
\begin{aligned}
x_0 &= x/D & t_0 &= U \cdot t/D \\
L_0 &= L \cdot D^2 & T_0 &= (T_\lambda - T)/T_\lambda \\
v_{n0} &= v_n/U & S_0 &= T_\lambda S_0/U^2 \\
v_{s0} &= v_s/U & P_0 &= P/\rho U^2
\end{aligned}
$$

where the zero subscript refers to the dimensionless quantity and $T_\lambda = 2.176$ K is the lambda point. Upon substituting these quantities into (1-3), and rearranging terms a bit, we arrive at;

$$\frac{dv_{s0}}{dt_0} = -\frac{dP_0}{dx_0} - S_0 \frac{dT_0}{dx_0} + \frac{\frac{2}{3}\alpha k}{UD} \cdot L_0 v_{ns0} + \left(\frac{1-\delta}{2}\right)\frac{dv_{ns0}^2}{dx_0} \tag{4}$$

$$\frac{dv_{n0}}{dt_0} = -\frac{dP_0}{dx_0} + \left(\frac{\delta}{1-\delta}\right) \cdot \left(S_0 \frac{dT_0}{dx_0} - \frac{\frac{2}{3}\alpha k}{UD} \cdot L_0 v_{ns0}\right) - \left(\frac{\delta}{2}\right)\frac{dv_{ns0}^2}{dx_0} \tag{5}$$
$$+\frac{1}{1-\delta} \cdot \frac{1}{Re}\frac{d^2 v_{n0}}{dx_0^2}$$

The terms involving the temperature gradient drive counterflow, in which the two fluids flow in opposite directions, and heat is transported within the system. Note that the effect of a thermal gradient upon the normal fluid component

(Eq. 6) becomes larger than $1 - \delta \to 0$ where $\delta \equiv \rho_s/\rho$. This is comforting, since it agrees well with our expectations. Furthermore, the classical equations of motion are recovered in the limit $\delta \to 0$. From a more general perspective, Eqs. (4) and (6) describe a wide spectrum of physics, which goes beyond fully developed co-flowing turbulent fields. The driving terms for such non-classical phenomena as counterflow however (we include for consideration at this point terms involving v_{ns0}), must affect the dynamics of all flows at suitably small length scales. Since the co-flowing condition ($v_{ns0} = 0$) generally depends upon mutual friction, it must depend upon $v_{ns0} \neq 0$ across some range of length scales within the flow.

In addition to the many factors of $(1-\delta)$ in (4) and (6), which underscore the strong role of temperature in the dynamics of He II, there is one additional dimensionless quantity, $UD/(2/3)\alpha k$, which is tied into the mutual friction. Although we will be discussing co-flowing turbulence, where v_{ns0} may zero, L_0 may still be quite large. Thus, the term on the whole may fluctuate dramatically on a local scale.

For the sake of the argument at hand however, consider the case where v_{ns0} is strictly zero (fully coupled superfluid turbulence), with zero temperature gradient. Then at suitably large length scales, we are left with

$$\frac{dv_{s0}}{dt_0} = -\frac{dP_0}{dx_0} \tag{6}$$

$$\frac{dv_{s0}}{dt_0} = -\frac{dP_0}{dx_0} + \frac{1}{1-\delta} \cdot \frac{1}{Re} \frac{d^2 v_{n0}}{dx_0^2} \tag{7}$$

The most important point to notice is that we never truly recover classical equations of motion, except in the limit as $\delta \to 0$. The viscous term is driven by $(1-\delta) \cdot Re$, which provides a mechanism through which the effective critical Reynolds number associated with the drag crisis might scale with temperature. At first glance, this appears to pose a dilemma. A large body of data characterizing skin friction in pipes scales nicely with Reynolds number, alongside classical data, yet without any apparent temperature dependence. One possible explanation for this may lie in the fundamental difference between pipe flow, and flow over a bluff body, such as a sphere. Pipe flow is a fully developed turbulent field, where the two fluids largely flow together [1], except at very small length scales and within a very narrow region close to the wall. Here, the non-slip condition on the normal fluid provides the observed pipe friction and associated pressure drop. Flow over a sphere however, possesses a stagnation point on the leading edge, from which the two fluids accelerate under different boundary conditions, with a non-zero velocity difference. This brings into play the other terms in (4) and (6). Thus, we expect δ, and thereby temperature, to play a greater role in flows where the two fluids are allowed to have substantially different velocity fields ($v_{ns} \neq 0$).

As a final preliminary speculation, note that the viscous term in (7) becomes quite large as $1-\delta \to 0$, similar to inviscid flow (Re $\to 0$). Thus, it may be that the drag crisis itself changes character and magnitude as the normal component vanishes ($\rho_n \to 0$). Ultimately in this limit, one expects to recover some sort of potential flow.

2   Experimental Apparatus and Protocol

The apparatus was described in a previous publication [4]. A 10 mm diameter sphere is suspended upon a strut oriented perpendicular to the oncoming flow. A single pressure tap located on the surface of the sphere is used to map out the pressure distribution on the azimuth connecting the upstream and downstream points by rotating the sphere/strut. This pressure distribution is then integrated to yield the form drag.

The principal difference from the previous work is a matter of protocol. Drag versus Reynolds number were taken at several intermediate temperatures between 1.6 K and 2.0 K in order to determine the critical Reynolds number for the drag crisis as a function of temperature. Additionally, drag and pressure distribution were observed at different temperatures for fixed Reynolds number.

3   Results and Discussion

The measured pressure distribution in He II at 2.0 K, expressed in terms of the coefficient of pressure, $C_P = (P(\Theta) - P_0)/\frac{1}{2}\rho U^2$, is shown in Fig. 1. Points which comprise the curve are the result of averaging a series of individual measurements. Since many of the important dynamics are dependent upon the equatorial velocity ($\Theta = 90°$), the Reynolds number was calculated based upon the mean velocity in this smallest cross-section. In Fig. 1, the Reynolds number spans the range from 1.1×10^5 to 7.8×10^5. Profiles for Reynolds number at or above 1.8×10^5 show supercritical behavior with the boundary layer separation occurring at approximately 100 degrees. The data corresponding to the two lowest Reynolds numbers exhibit variations which may be due to transition, together with very low signal levels.

Assuming azimuthal symmetry of the pressure distribution with respect to the oncoming flow, together with the spherical shape of the surface, we may integrate the coefficient of pressure to calculate the coefficient of drag directly. For the discrete data here, this is easily performed by a summation. Figure 2 shows drag coefficients for our experiment, together with published results for the smooth sphere, shown as the solid line. Error bars are derived from the statistical scatter in the individual measurements which comprise the points on the curve in Fig. 1. The lower Reynolds number data have larger error bars since these measurements correspond to lower signal level. Although He II exhibits a drag crisis at approximately the same Reynolds number as He I, the variability in the He II data led us to speculate about a temperature dependence to the turbulent transition within the boundary layer.

The results displayed in Fig. 2 suggest that the drag coefficient in He II increases with decreasing temperature. For fixed Reynolds number, the coefficient is the largest at 1.6 K decreasing monotonically to the value in He I. Further,

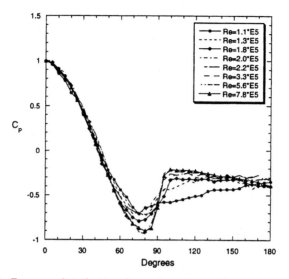

Fig. 1. Pressure distribution for various Reynolds number at 2.0 K

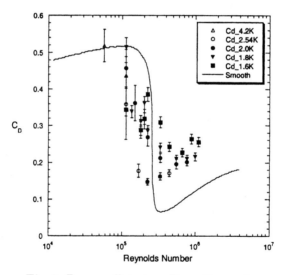

Fig. 2. Drag coefficient vs. Reynolds number

the minimum in the drag coefficient occurring just above the transition also increases with decreasing temperature. These two observations can be evaluated in more detail by careful study of the data.

Figure 3 is a plot of the drag coefficient versus temperature for fixed Reynolds number. Below the lambda transition, the coefficients increase approximately linearly although not with the same slope for each case. It is also interesting to note that the extrapolation of the measured He II drag coefficients to T_λ

appear to coincide with the approximately constant values measured in He I. Although these data need further confirmation, it certainly appears that the drag coefficient is temperature dependent in He II.

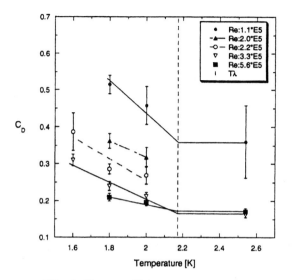

Fig. 3. Drag coefficient vs. temperature

To test the two fluid equation scaling arguments, we plot in Fig. 4 the drag coefficients versus Re $(1-\delta)$, as suggested by [5]. The open symbols correspond to He I data and the closed symbols are for He II.

Comparing Fig. 4 to Fig. 2, we note that the scaling shifts the minimum in the drag coefficient to approximately the same value of the modified Reynolds number for each temperature. That value, Re $(1-\delta) \approx 2.3 \times 10^5$, is in reasonable agreement with the minimum in the classical drag coefficient curve at approximately Re $= 3 \times 10^5$.

4   Conclusion

We have measured the pressure distribution over the surface of a sphere in flowing He II as a function of Reynolds number. The He II data shows clear evidence of a drag crisis at approximately the same Reynolds number. The coefficients of drag have a minimum value at same modified Reynolds number based on scaling the He II two fluid equations, Re $\times(1 - \delta) \approx 2.3 \times 10^5$. The variability in the He II data confirms a temperature dependence to the turbulent transition within boundary layer.

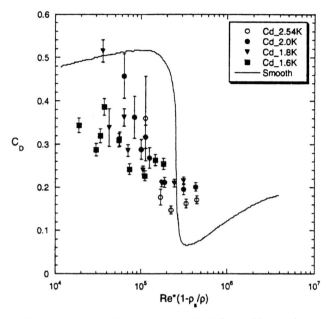

Fig. 4. Drag coefficient vs. modified Reynolds number

Acknowledgments

We wish to thank the National High Magnetic Field Laboratory and the Department of Energy, Division of High Energy Physics for their financial support. Thanks to David K. Hilton for useful conversation and Scott Maier for technical assistance.

References

1. P.L. Walstrom, J.G. Weisend II, J.R. Maddocks and S.W. Van Sciver, Turbulent pressure drop in various He II transfer system components, Cryogenics 28, 101, 1988.
2. D.C. Samuels, Velocity matching and Poiseuille pipe flow of superfluid helium, Phys. Rev. B 46, 11714, 1992.
3. C.F. Barenghi, D.C. Samuels, G.H. Bauer and R.J. Donnelly, Numerical evidence for vortex-coupled superfluidity: Quantized vortex lines in an ABC model of turbulence, Phys. Fluids 9, 2631, 1997.
4. M.R. Smith and S.W. Van Sciver, Measurement of the pressure distribution and drag on a sphere in flowing He I and He II, Advances in Cryogenic Engineering, Vol 43, 1473, 1998.
5. M.R. Smith, D.K. Hilton and S.W. Van Sciver, Observed drag crisis on a sphere in flowing He I and He II, Physics of Fluids, Vol. 11, No.4, 1999.
6. A similar analysis was originally presented in Evolution and Propagation of Turbulence in Helium II, Ph.D. Thesis, M.R. Smith, 1992, University of Oregon.
7. R.J. Donnelly, *Quantized Vortices in Helium II*, Cambridge, New York, 1991.

Experiments on Quantized Turbulence at mK Temperatures

S.I. Davis, P.C. Hendry, P.V.E. McClintock, and H. Nichol

Department of Physics, Lancaster University, Lancaster LA1 4YB, UK.

Abstract. An experiment to investigate the free decay of quantized turbulence in iso-topically pure superfluid ^{4}He at millikelvin temperatures is discussed. The vortices are created by a vibrating grid, and detected by their trapping of negative ions. Preliminary results suggest the existence of a temperature-independent vortex decay mechanism be-low $T \sim 70$ mK.

1 Background

The renaissance of interest in the turbulent hydrodynamics of HeII has led to the realisation that, in many respects, it exhibits unexpected similarities to anal-ogous flows in classical fluids at high Reynolds number [1–3]. Unlike a classical fluid, HeII is well described by a two-fluid model, with a normal (dissipative) component mutually interpenetrating with a superfluid (inviscid) component with quantized circulation

$$\kappa = \oint v_s.dl = n \left(\frac{h}{m_4} \right) \tag{1}$$

where the integral is taken around a loop enclosing the vortex, v_s is the superfluid velocity, m_4 is the ^{4}He atomic mass and the quantum number n is an integer.

The flow properties of HeII in an open geometry are dominated by singly-quantized vortex lines [4], linear singularities around which the superfluid flows at tangential velocity v_s. Unless velocities are kept extremely small, the liquid flowing through a tube becomes filled with a tangled mass of such vortex lines. Because of their quantization, they represent a particularly simple form of tur-bulence. In that the vortex cores can be considered as part of the normal fluid component, but the encircling superflow field in accordance with (1) is of super-fluid component, vortices provide a weak coupling (mutual friction) between the two components. So it is not at all clear, at first sight, why this complex liquid system should ever behave like a single-component classical fluid. The question has recently been discussed in considerable detail by Vinen [3]. One of the aims of the present project is to establish the properties of the turbulent liquid when it really does consist of just a single component, i.e. in the low temperature limit where the normal fluid density is negligible.

In the conventional scenario at higher temperatures $1 < T < T_\lambda$, a vortex tangle can be maintained by the work done by the driving force, which could be e.g. a pressure or temperature gradient for bulk flow, or for thermal counterflow (in which the normal and superfluid components move in opposite directions), respectively. On removal of the driving force, the tangle decays according to the Vinen [5] equation

$$\frac{dL}{dt} = -\chi_2 \frac{\hbar}{m_4} L^2 \tag{2}$$

where L is the length of vortex line per unit volume and χ_2 is a (weakly temperature-dependent) dimensionless constant. The physical mechanism driving the decay has been discussed by Schwarz [6], who concluded that it involves crossings and consequent reconnections of lines. The rapid self-induced motion of the resultant sharp cusps through the normal fluid is dissipative, and causes the rounding off of cusps with consequent line shrinkage. The presence of the normal fluid component is thus a key component of the decay mechanism.

Details of the decay process, e.g. the existence of two distinct decay rates [5,7], are not at all well understood, partly because it sometimes seems to involve turbulence in the normal fluid component as well as in the superfluid. The experiment that we describe below avoids this complication. Vortex decay is investigated at sufficiently low temperatures (in the mK range) that normal fluid component as such is absent. Under these conditions, it is far from obvious how the tangle will decay, but it appears that there are two possibilities –

1. Essentially the same decay process occurs as above 1K, but now driven by the dilute phonon gas rather than by normal fluid, or
2. A new decay mechanism comes into play.

If the former applied, decay times would be expected to become exceedingly long as T decreased, given that the vortex-phonon cross section is tiny [8] and that the phonon density is falling as T^3. The question is difficult to settle on theoretical grounds, and best resolved by experiment.

2   Creation and Detection of Vortices

Above 1K vortices can conveniently be created [4] by thermal counterflow in which a thermal gradient causes the two fluids to flow in opposite directions above a critical velocity. Vortices are usually detected [4] by their attenuation of second-sound, an entropy-temperature wave in which the two fluids move in antiphase. Neither of these techniques can be used at mK temperatures, because there is no normal fluid component. New techniques have therefore been required.

To create vortices we have employed a resonantly-excited circular grid [9]. The expectation was that, in the absence of normal fluid, the amplitude of vibration would grow until a critical velocity was attained, after which the energy drawn from the exciting field would be converted to vorticity.

To detect the vortices we have observed the attenuation of ion signals caused by trapping of the ions on vortex cores as the ion clouds pass through the vortex tangle. An essential requirement of the experimental design is the necessity of preventing the ions from themselves creating vortex rings – converting to

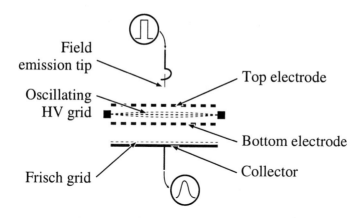

Fig. 1. The experimental arrangement (schematic). Some of the ions created by the field emission tip may get trapped on vortices created previously by the oscillating grid, thereby reducing the signal arriving at the collector. The perforated top and bottom electrodes complete the double capacitor needed to excite oscillations of the grid, and the Frisch grid screens the collector from the approaching charge.

large slowly-moving charged-vortex-ring complexes, and thus being lost from the signal. An ion in HeII at mK temperatures, in the absence of normal fluid component, will accelerate steadily under the influence of any electric fluid, however small, until it attains a critical velocity. It is necessary to ensure that this is the Landau critical velocity v_L, and not the critical velocity for vortex creation. This can be accomplished by choosing negative (rather than positive) ions [10], by applying pressures $P > 11$ bar [10], by using isotopically pure ^{4}He [11], and by ensuring that the electric field is not too large [12]. Under these conditions the ions do not create vortex rings but, rather, travel freely through the liquid with an average speed slightly in excess of v_L [13].

The mechanism by which a rapidly moving ion can get trapped on a vortex under these conditions is unclear. The vortex presents an effective potential [14] and, above 1K, normal fluid dissipation enables the ions to lose enough energy to get trapped on the vortex core. In the absence of normal fluid, and with the ion moving almost ballistically, like a free particle in a vacuum, we hoped that the ions would lose enough energy to be trapped by exciting vortex waves.

3   The Experiment

The electrode structure used for the experiment is shown schematically in Fig. 1. The operating procedure was performed in two stages. First, a high constant voltage (usually 500 V) was applied to the vortex-generating grid and a periodic driving voltage of ±270 V was applied to one of the adjacent plates. The drive was maintained for several seconds, to build up a tangle of vorticity. Secondly, the potentials on the electrodes were adjusted so that the electric field would draw ions from the field-emitter to the collector. The field-emitter was then pulsed to create an ion cloud, which travelled down the cell. As it passed the Frisch screen-grid it induced a signal in the collector which was amplified and recorded using a Nicolet NIC-80 data processor. The sequence was then repeated, ensemble-averaging the collector signals to enhance the signal/noise ratio.

4   Preliminary Results

Figure 2 shows a typical sequence of ion signals. The first signal is a reference, recorded before the grid had been vibrated, and the others show how the signal gradually recovered after the grid vibration had been completed. It is evident that there is significant attenuation as a result of the grid oscillation – demonstating immediately that the technique described in the preceding sections enables us both to create and to detect vorticity in the mK temperature range.

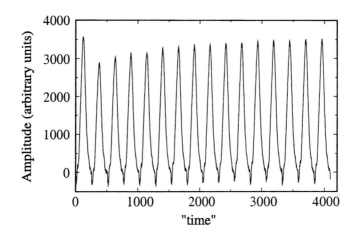

Fig. 2. A set of collector signals. The duration of each of them is ~ 200 μs, and a period of 1.5 s in real time separates each signal from its neighbour. The first signal is for reference, recorded before the grid was vibrated.

In Fig. 3, the signal heights are plotted as a function of time for several temperatures T. It appears that for $T < 70$ mK nothing changes, within experimental error: the decay mechanism is apparently temperature-independent within this range.

5   Discussion

Our preliminary data are too scattered for us to be able to draw definite conclusions about the form of the decay and there is, in any case, no theoretical

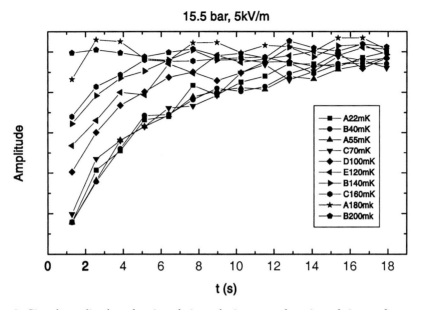

15.5 bar, 5kV/m

Fig. 3. Signal amplitudes, showing their evolutions as a function of time t, for several temperatures.

form with which to compare. Nonetheless a plot of $\ln(S_0 - S)$ against time t, where S_0 and S are repectively the unattenuated and attenuated signals, produces what is definitely a curve, showing that the decay is non-exponential. It is also interesting to compare with the form of decay that occurs above 1 K in the presence of normal fluid component. By integration of the Vinen equation (2), it is straightforward to demonstate that

$$\left[\ln\left(\frac{S_0}{S}\right)\right]^{-1} = \frac{A}{\kappa d}\left(\chi_2\frac{\kappa}{2\pi}t + L_i^{-1}\right) \tag{3}$$

where A is a constant containing the ion-vortex trapping cross-sections, L_i is the initial vortex line density at $t = 0$, and d is the length of the vorticity-filled region. Thus a plot of $[(\ln(S_0/S)]^{-1}$ against t would be expected to yield a straight line. It does so, within experimental error, as shown in Fig. 4.

It is interesting to speculate on the physical nature of the low temperature T–independent decay mechanism inferred from the results of Fig. 3. The fact that the decay is non-exponential would appear to rule out spontaneous decay processes, e.g. where there was a constant probability per unit time that any given element of vortex might emit a phonon and become incrementally shorter. But the T–independence strongly suggests that the phonon gas plays no role, given that the phonon density is falling as T^3 within the range of interest. One possibility, perhaps, is that phonons are emitted during reconnections, leading to line-shrinkage. Simulations by Tsubota et al [15] suggest that reconnections are indeed the key to the problem, but that the resultant decay arises because

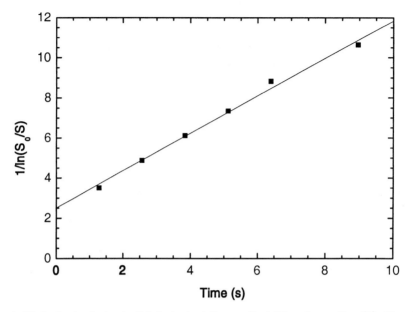

Fig. 4. Plot of a typical set of data to test the applicability of equation (3). S and S_0 are respectively the amplitudes of a signal and of the reference signal.

of the strongly kinked lines that are produced, resulting [3] in Kelvin waves and a cascade of energy towards smaller and smaller length scales until it is radiated as sound.

A difficulty in interpreting the results is that the absolute vortex line densities are unknown, because the ion-vortex trapping cross-section is unknown under the conditions of the experiment. We can obtain a very approximate estimate from the measured linewith of the grid resonance which, for typical electrode potentials and driving amplitudes implies that the energy dissipation of the grid is $(0.4\pm0.2)\mu W$; insertion of this value in (2), on the assumptions that all of the dissipation goes into vortex creation, χ takes the same value as above 1K and that the vortex tangle remains mostly between the electrodes, yields a steady state line density of $\sim 10^{10}$ m^{-2}. Any or all of the assumptions could be in error, however, and the only unambiguous way to clarify the situation will be through direct measurement of the ion-line trapping cross-section in a rotating cryostat where the line density is known precisely.

6   Conclusions

In conclusion, we would emphasize the preliminary character of these results, and the large number of unknowns. As pointed out above, we know nothing about the spatial distribution of the vorticity, although we imagine that it stays mostly between the top and bottom electrodes of the triple capacitor in Fig. 1.

Nor do we know anything about the nature of (possible) temperature dependences of the vortex-generation and ion-trapping mechanisms. Our ignorance of vortex/ion trapping cross-section represents an even more serious lacuna, because it means that we cannot calculate absolute vortex line densities from the ion signal attenuation.

Nonetheless, we can tentatively conclude, first, that it is possible to generate and detect vorticity in HeII at mK temperatures and, secondly, that the decay of quantized vorticity becomes temperature-independent below $T \sim 70$ mK.

Acknowledgements

It is a pleasure to acknowledge helpful discussions with C F Barenghi, R J Donnelly, L Skrbek and W F Vinen. The work was supported by the Engineering and Physical Sciences Research Council (UK).

References

1. S.R. Stalp, L. Skrbek, R.J. Donnelly: Phys. Rev. Lett. **82**, 4831 (1999)
2. C.F. Barenghi: J. Phys.: Condens. Matter **11**, 7751 (1999)
3. W.F. Vinen: Phys. Rev. B **61**, 1410 (2000)
4. R.J. Donnelly: *Quantized Vortices in He II* (Cambridge University Press, Cambridge 1991)
5. W.F. Vinen: Proc. R. Soc. A **242**, 493 (1957)
6. K.W. Schwarz: Phys. Rev. B **18**, 245 (1978); **31**, 5782 (1985); **38**, 2398 (1988)
7. K.W. Schwarz, J.R. Rozen: Phys. Rev. Lett. **66**, 1898 (1991)
8. A.L. Fetter: Phys. Rev. A **136**, 1488 (1964)
9. M.I. Morell, M. Sahraoui-Tahar, P.V.E. McClintock: J. Phys. E: Sci. Instrum. **13**, 350 (1980)
10. L. Meyer, F. Reif: Phys. Rev. **123**, 727 (1961)
11. R.M. Bowley, P.V.E. McClintock, F.E. Moss, P.C.E. Stamp: Phys. Rev. Lett. **44** 161 (1980)
12. R.M. Bowley, P.V.E. McClintock, F.E. Moss, G.G. Nancolas, P.C.E. Stamp: Phil. Trans. R. Soc. Lond. A **307**, 201 (1982)
13. D.R. Allum, P.V.E. McClintock, A. Phillips, R.M. Bowley: Phil. Trans. R. Soc. Lond. A **284**, 179 (1977)
14. R.J. Donnelly, P.H. Roberts: Proc. R. Soc. Lond. A **312**, 519 (1969)
15. M. Tsubota, T. Araki, S.K. Nemirovskii, J. Low Temperature Phys. **119**, 337 (2000)

Grid-Generated He II Turbulence in a Finite Channel – Experiment

J.J. Niemela, L. Skrbek, and S.R. Stalp

Cryogenic Helium Turbulence Laboratory, Department of Physics,
University of Oregon, Eugene, OR 97403, USA

Abstract. We present experimental data on decaying turbulence, generated by towing a grid through a stationary sample of He II. We describe in detail the experimental apparatus and physical principles that allow observation of up to six orders of magnitude of decaying vortex line density over three orders of magnitude in time using the second sound attenuation technique.

1 Introduction

Superfluid turbulence has long been an area of study, with an emphasis largely on flows created by applying a heat current in He II; i.e. on thermal counterflow[1]. It is an advantage that, experimentally, counterflow turbulence requires no moving parts for its generation. The connection of this type of flow with classical turbulence, however, is not obvious. More recently, it has been of some interest to explore this connection by generating turbulent flows in He II in a similar manner as for classical fluids. In particular, turbulence created in the wake of a grid can create nearly homogeneous and isotropic turbulence (HIT)[2], and application of this procedure to He II has led to new insights and a few surprises[3–5]. In particular, a deep similarity appears to exist between grid turbulence in classical fluids and in He II, a quantum fluid.

In this article, we focus on the experimental apparatus and techniques used to generate grid turbulence in He II, and discuss the observed decay of the vortex line density behind a towed grid. Measurements of second sound attenuation allow detection of up to six orders of magnitude of decaying vortex line density L, which can be converted into roughly eight orders of magnitude of decaying turbulent energy[6] - at present hardly a feasible goal for any laboratory experiment on classical turbulence. In a companion article in this book, we will further interpret the data in terms of classical hydrodynamics. The underlying quantum nature of this experiment has been recently discussed by Vinen [7].

2 Experimental Setup

The schematic of the experimental apparatus is shown in Fig.1. The turbulence is generated by towing a grid through a stationary sample of He II. We use a 65% open brass monoplanar grid of rectangular tines, 1.5 mm thick, with a mesh size, M (tine spacing) of 0.167 cm[3,4]. The grid is attached to a central stainless

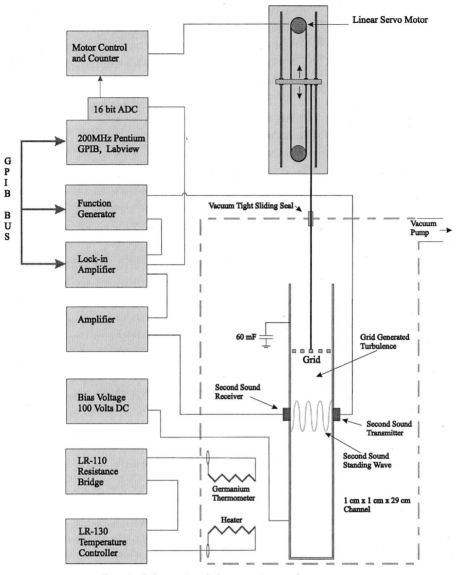

Fig. 1. Schematic of the experimental apparatus

steel pulling rod of diameter 0.24 cm which exits the cryostat via a pair of tight sliding seals. The space between the seals is continually evacuated to prevent the introduction of impurities inside the cryostat during the experiment. Above the cryostat, the rod is attached to a computer controlled linear servo motor that positions the grid with 0.01 cm accuracy and provides the towing velocity, v_g, up to 2.5 m/s. This enables the exploration of a wide range of mesh Reynolds numbers $2 \times 10^3 \leq Re_M = v_g M \rho / \mu \leq 2 \times 10^5$, where μ is the dynamic viscosity of the normal fluid and ρ the total density. It is worth noting that the linear servo motors, developed at the University of Oregon, accomplish this precise positioning without the problems associated with electrical switching noise and mechanical resonances that are characteristic of the more commonly used stepper motors.

The channel of square cross section is 29 cm long and has a width $d = 1$ cm. It was manufactured by an electroforming process with a tolerance of 25 μm and a surface roughness less than 0.5 μm. The channel is suspended vertically in the helium cryostat and during the measurement is totally submersed in superfluid helium, which enters the channel via 16 holes, 500 μm in diameter, placed near the top end. Except for these, the channel is leak-tight. There is no free surface of the liquid inside the channel during the experiment, and second sound coupling to the free surface of the bath is minimized by the use of the small diameter holes for mass transfer between the channel and the bath.

The temperature is measured and controlled via a germanium resistance thermometer and heater placed in the He II bath. The thermometer is calibrated against the saturated vapor pressure with a temperature accuracy better than 1 mK and a resolution within 10 μK[4]. Details of possible temperature fluctuations inside the channel that might occur during and just after pulling the grid are unknown, as the thermometer is located outside the channel and its response time is about 100 ms. For maximum cooling, the main bath is pumped via a 20 cm diameter line connected to a 300 CFM vacuum pump and a 1400 CFM roots blower. This enables bath temperatures down to 1 K to be reached easily. It is difficult, however, to assure temperature stability below about 1.1 K, so this marks the lowest reliable temperature for the experiment. Experiments close to the lambda temperature (above 2 K) require special care (see below) mainly due to the strong temperature dependence of the second sound velocity, u_2.

To probe the quantized vortex line density resulting from pulling the grid, we excite and detect second sound using vibrating nuclepore membranes 9 mm in diameter mounted flush on opposing walls of the channel. These 6 μm thick polycarbonate membranes have a dense distribution of 0.1 μm holes and on one side is evaporated an approximately 95 Å thick layer of gold, which is then pressed against the channel wall. This gold layer forms one electrode of a capacitor transducer, with the other being a brass electrode pressed towards the opposite side of the membrane from outside the channel. Applying an ac signal ($\sim 0.3 - 1$ V$_{PP}$) in addition to a dc bias (~ 100 V) results in an oscillatory motion of the membrane. In He II, the normal fluid is clamped by viscosity inside the small holes of the membrane, while the superfluid component passes freely, thereby exciting second sound, an entropy wave in He II. Directly across the channel the second sound wave produces a corresponding oscillation of the other membrane (i.e., the receiver) and the induced signal is input to a lock-in amplifier referenced to the transmitter frequency. The channel acts as a second sound resonator. The excitation amplitude is adjusted to be the upper half of the linear response range and the n-th harmonic- with n about 50- of the fundamental frequency is used: typically $30 - 40$ kHz. A Lorentzian resonance peak is obtained, having a full width at half maximum that is temperature dependent and typically reaches values of $\Delta_0 = 20 - 500$ Hz without quantized vortices in the channel.

To understand the use of second sound in detecting the vortex line density and its relation to averaged *rms* vorticity, we consider the seminal work of Hall and Vinen[8]. In experiments with a rotating container of He II, they observed an excess attenuation of second sound in a direction perpendicular to the rotation axis due to the presence of quantized vortices, $\alpha_L = B\Omega/2u_2$. This extra attenuation resulted from the scattering of the elementary excitations - normal fluid - by the vortex lines and was absent for second sound propagating parallel to the rotation axis. Here B is the dimensionless mutual friction parameter (B generally depends both on temperature and frequency[9]) and Ω denotes the angular velocity of rotation. It is now well known that the rotating bucket of He II displays, on average, the same shape of the surface meniscus as any classical fluid, since the superfluid mimics solid body rotation by creation of a lattice of rectilinear quantized vortices aligned in the direction of the rotation axis. In this case, the vorticity $\omega = 2\Omega = \kappa L$, where κ is the circulation quantum ($\kappa = h/m_4$, where h is Planck's constant and m_4 the mass of the helium atom) and L is the total length of the vortex line per unit volume. It is often assumed that this relation between ω and L holds in general. By considering a second sound resonance as an infinite series of reflected waves in a rotating cavity, the extra attenuation (in a limit of small attenuation) due to quantized vortices becomes [4]

$$\alpha_L = \frac{B\kappa L}{4u_2} = \frac{\pi \Delta_0}{u_2}\left(\frac{A_0}{A} - 1\right) \tag{1}$$

where A and A_0 are the amplitudes of the second sound standing wave resonance with and without vortices present, respectively. We can extend this formula to the case of a homogeneous vortex tangle, taking into account that vortices oriented parallel to the second sound propagation do not contribute to the excess attenuation. Then we have[3]

$$L = \frac{16\Delta_0}{B\kappa}\left(\frac{A_0}{A} - 1\right) \tag{2}$$

It can be shown[4] that for arbitrary attenuation one has to use the more general formula

$$L = \frac{8u_2}{\pi B\kappa d}\ln\left[\frac{1 + p^2 P + \sqrt{2p^2 P + p^4 P^2}}{1 + P + \sqrt{2P + P^2}}\right] \tag{3}$$

where $p = A_0/A$ and $P = 1 - \cos(2\pi d\Delta_0/u_2)$ that for small $d\Delta_0/u_2$ reduces to (2). It is essential to use formula (3), as using the approximate formula (2) at some experimental conditions leads to results that are more than an order of magnitude off!

It is important to consider the time response of the measuring system with regard to the finite velocity of second sound. Our calculations show that in most

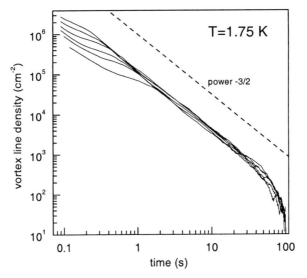

Fig. 2. The log-log plot of the decaying vortex line density versus time after grid passes 2 mm above the measuring volume. Each decay curve represents an average of three identical pulls. The decay curves, in order, correspond to $Re_M = 2 \times 10^5$ (the uppermost one), 1.5×10^5, 10^5, 5×10^4, 2.5×10^4, 10^4, 5×10^3 and 2×10^3. For each Re_M, the decaying vortex density displays an inertial range with power law exponent -3/2.

cases (except close to the lambda temperature) there is negligible error introduced into the deduced vortex line density, for the following reasons. Immediately after the grid is towed through the measuring volume the quality factor is very low, of order unity, and the detecting system can be described rather as a second sound pulse technique with the characteristic time response given by the time of flight $d/u_2 \sim 10^{-3}$s, where $u_2 \sim 20$ m/s [10]. As the turbulence decays, the characteristic time constant increases with the (temperature dependent) quality factor. Without the vortex tangle, the typical linewidth of the second sound resonance is 20-500 Hz, the typical frequency used is 30-40 kHz, so the quality factor reaches 60-2000 and the time response gradually rises to about 0.1-1 sec at the very end of the decay, where it constitutes an error of less than 1%.

There are also other time restrictions than the finite velocity of second sound. As it takes a time $\tau_g \cong d/v_g$ to tow the grid through measuring volume, we use only that data obtained on the time scale longer than τ_g and also exceeding $8\tau_{LI}$, where τ_{LI} is the time constant of the lock-in amplifier used for detection of the amplitude of the second sound signal. Another time scale restriction involves consideration of how soon the flow can be assumed as nearly HIT [6]. It was discussed in[11] that a physical criterion to estimate this time leads to about 1-2 turnover times of the largest eddies present in the flow, i.e., those of the size of the channel. It follows that the minimum time needed to assume nearly HIT conditions is about 1-2 widths downstream from the grid, or using the Taylor

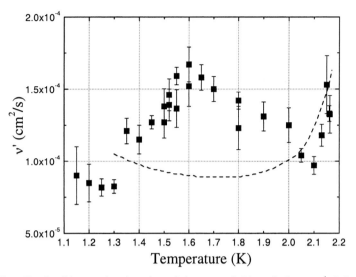

Fig. 3. The effective kinematic viscosity of the superfluid turbulence ν' deduced from vorticity decay data at various temperatures, and assuming a value of the Kolmogorov constant consistent with classical experiments. The dashed line represents $\nu = \eta/\rho$ which has a dissimilar temperature dependence indicative of quantization effects, where η is the normal fluid viscosity and ρ is the total density of helium II.

frozen hypothesis, the time it takes the towed grid to pass 1-2 widths of the channel.

The data acquisition process can be briefly described as follows. The cryostat is filled with liquid helium, the bath being pumped and the temperature controlled to a desired value. With the grid "parked" towards the top of the channel, the second sound resonance curve is measured and fitted to a Lorentzian, giving the value of Δ_0. The grid is then lowered to the bottom and after a necessary waiting time (typically 2 minutes) pulled through the channel such that its velocity is constant and equal v_g for at least 5 cm below and above the experimental volume. It is then slowed down and smoothly "parked" against the top again. The data acquisition is triggered when the grid passes a predetermined position 2 mm above the measuring volume, as determined by a signal from a photodiode. Typically 100 s of data are recorded at a rate of 100 Hz from the output signal of the lock-in amplifier, representing the recovering second sound standing wave amplitude, $A(t)$. After another waiting period to ensure that the turbulence has decayed down to a negligibly low level, the reference amplitude A_0 is read. The decaying vortex line density is then calculated using formula (3). It implicitly assumes that there is no additional attenuation of the second sound that might be caused by a turbulence in the normal fluid.

Since the second sound is transmitted and detected via membranes on opposing sides of the channel, we obtain information from a measuring volume of order $d^3 \cong 1$ cm^3. It is this natural integration that bypasses tedious statistical analysis involved in any local velocity measurements in conventional turbulence, provides enormous sensitivity and unprecedented dynamical range of the method, making it very useful and complementary to classical turbulence studies.

Examples of decay data and the effective kinematic viscosity of the superfluid turbulence ν', are shown in Figs. 2 and 3 respectively. The decay curves represent up to six orders of magnitude of decaying vortex line density over three decades of time. The overall form of the decay does not change with temperature. At any temperature, after some initial period depending on Re_M, the decay curves tend to collapse and display the power law decay $L \propto t^{-3/2}$[5,12]. We note if vorticity and L are assumed to be related in a similar manner as discussed above for the rotating bucket experiments, then this feature and others are understandable in terms of a classical spectral decay model[6,11,12], further discussed in a companion article in this book. Quantum effects show up in the measured temperature dependence of the effective kinematic viscosity (see Fig. 3). Finally, we have had many useful discussions at this workshop concerning complementary measurements to probe the normal fluid turbulence and/or energy dissipation directly, and also extending the protocol to near zero temperature to complement the work of McClintock's group reported here. These ideas are presently receiving attention.

Acknowledgements

The towed grid He II experiment has been developed at the University of Oregon over many years. We would like to acknowledge R.J. Donnelly, W.F. Vinen and M.R. Smith for their valuable contributions to the conception and design of this experiment. This research was supported by NSF under grant DMR-9529609.

References

1. R.J. Donnelly: *Quantized vortices in helium II.* Cambridge University Press (1991)
2. G. Comte-Bellot, S. Corrsin: J. Fluid Mech. **25**, 657 (1966); **48**, 273 (1971)
3. M.R. Smith: Evolution and propagation of turbulence in helium II. PhD Thesis, University of Oregon, Eugene (1992)
4. S.R. Stalp: Decay of grid turbulence in superfluid helium. PhD Thesis, University of Oregon, Eugene (1998)
5. M.R. Smith, R.J. Donnelly, N. Goldenfeld, W.F. Vinen: Phys. Rev. Lett. **71**, 2583 (1993)
6. L. Skrbek, J.J. Niemela, R.J. Donnelly: Phys. Rev. Lett. **85**, 2973, (2000)
7. W.F. Vinen: Phys. Rev. B **61**, 1410 (2000)
8. H.E.Hall, W.F. Vinen: Proc. Roy. Soc. London A**238**, 204 (1954); **238**,215(1954)
9. C.F. Barenghi, R.J. Donnelly, W.F. Vinen: J. Low Temp. Phys. **52**, 189 (1983)
10. R. J. Donnelly, C. F. Barenghi: J.Phys. Chem. Data **27**, 1217 (1998).
11. L. Skrbek, S.R. Stalp: Phys. Fluids **12**, 1997 (2000)
12. S.R. Stalp,L. Skrbek, R.J. Donnelly: Phys. Rev. Lett. **82**, 4831(1999)

Intermittent Switching Between Turbulent and Potential Flow Around a Sphere in He II at mK Temperatures

Michael Niemetz, Hubert Kerscher, and Wilfried Schoepe

Institut für Experimentelle und Angewandte Physik, Universität Regensburg, D-93040 Regensburg, Germany

Abstract. Intermittent switching between potential flow and turbulence is observed with an oscillating sphere in HeII below 0.5 K, where there is no normal fluid component and no viscosity. The remaining dilute phonon gas is in the ballistic regime and therefore turbulence in the pure superfluid can be investigated. The amplitude of the driven oscillations is a measure of the damping which in case of potential flow is due to residual ballistic phonon scattering or, when the flow is turbulent, is due to a large nonlinear turbulent drag. In an intermediate range of driving forces the flow is observed to be unstable, intermittently switching between both patterns. We have investigated this phenomenon down to 25 mK and have made a statistical analysis of the time series measured at various constant driving forces and temperatures. We obtain a temperature independent probability density for switching, different for both directions. We find a regime of metastable laminar flow whose lifetime is limited by natural radioactivity and cosmic rays.

1   Experiment

The experimental setup used in our investigations consists of a ferromagnetic microsphere (radius $r = 124$ mm, $m = 27$ mg) suspended between two niobium electrodes by superconducting levitation. While cooling the horizontally arranged electrodes forming a parallel plate capacitor (distance $d = 1$ mm, diameter 2 mm) a dc-voltage of several hundred volts is applied to the capacitor giving rise to an electric charge at the surface of the electrodes and the sphere. As the electrodes become superconducting, the magnetic sphere is repelled and levitates at an equilibrium position between the two electrodes. Horizontal stability is provided by trapped flux in the electrodes. As the sphere carries an electric charge q vertical oscillations ($f \approx 150$ Hz) can be excited by applying a resonant ac electric field to the capacitor corresponding to a force $F = q \cdot U_{ac}/d$ on the sphere. The oscillations can be detected by measuring the current $I = v \cdot q/d$ induced in the electrodes of the capacitor by the moving charge. The space between the superconducting electrodes is filled with pure ^{4}He (^{3}He concentration ≤ 1 ppb). This setup provides a simple geometry without disturbance by mechanical suspension elements and a low background dissipation (Q-factors above 10^6 are achieved when the cell is evacuated). We measure the velocity amplitude of the oscillating sphere for different driving forces at different temperatures. For a more detailed description of the experimental technique and its applications, see [1–4].

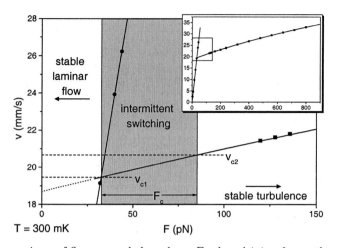

Fig. 1. Three regimes of flow around the sphere: For low driving forces there is stable laminar flow (•), resulting in a linear dependence $v(F) = F/\lambda$. For large driving forces there is stable turbulent flow (■) resulting in large dissipation and low oscillation amplitudes. (The solid line is a fit of the turbulent $v(F)$ dependence derived in the text.) Between these two regions there is a regime where both flow patterns are unstable and which extends from v_{c1} over an interval F_c of driving forces to v_{c2} where turbulence becomes stable. In this regime the system switches between laminar and turbulent flow intermittently. The inset shows a larger range of driving forces and velocities. The enlarged region is marked by a box

2   Results

The inset of Fig. 1 shows a typical result of the experiment. For small driving forces the flow around the sphere remains laminar corresponding to a linear drag force λv, which is given by ballistic scattering of phonons (left regime in Fig. 1). For large driving forces there is stable turbulent flow instead, accompanied by large nonlinear dissipation. Between these two regimes we find an interval of driving forces, where neither laminar nor turbulent flow is stable but where instead the fluid switches between these states intermittently.

2.1   Stable Turbulent Flow

Fully developed turbulence in a classical fluid causes a drag force

$$|F_d| = \gamma v^2, \quad \gamma = C_d \varrho \pi r^2 / 2 \tag{1}$$

($C_d \approx 0.4$) on a sphere moving through the fluid. As we perform an ac experiment where $v(t) = v \sin(2\pi f t)$ we employ the energy balance between the energy input per half period $T/2$ at resonance and the dissipation

$$\int_0^{T/2} F \sin(\omega t)\, v \sin(\omega t)\, dt = \int_0^{T/2} F_d(v(t))\, v \sin(\omega t)\, dt \tag{2}$$

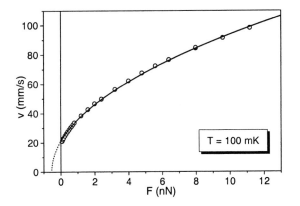

Fig. 2. Turbulent flow around the sphere: Data points taken up to one hundred times larger driving forces than in Fig. 1 are well described by the turbulent drag force introduced in the text. The shift of the apex of the parabola towards negative F values can clearly be observed

to find the corresponding $F(v)$ dependence $F(v) = 8\gamma v^2/3\pi$. In Fig. 2 a larger range of driving forces is shown, providing a more complete view of the turbulent regime. The solid line is a fit, using a similar drag force as in (1) but reduced by a constant value:

$$|F_\mathrm{d}| = \gamma(v^2 - v_0^2) = \gamma v^2 - F_0 \ . \tag{3}$$

By applying the energy balance (2) this drag force corresponds to

$$F(v) = \frac{8\gamma}{3\pi}\left(v^2 - \frac{3}{2}v_0^2\right) = \gamma'(v^2 - v_{c1}^2) \tag{4}$$

which perfectly fits the data points obtained in the superfluid (Fig. 2). A similar drag force has been calculated for turbulent flow of a two dimensional dilute Bose-Einstein condensate around a cylinder [5]. Even if these calculations are not directly applicable to liquid helium, there might be a similar mechanism in both cases for the reduced drag force. In our case turbulent drag exists only above a critical velocity $v_{c1} = 19.4\,\mathrm{mm/s}$ (s. Fig. 1).

2.2   Intermittent Switching

In contrast to classical fluids there is no smooth transition from laminar flow to fully developed turbulence in the superfluid, as reported earlier [2]. Instead, the fluid switches between laminar and turbulent flow intermittently. We recorded the velocity amplitude of the oscillations as a function of time while holding temperature and driving force constant. Figure 3 shows three sections of 13 minutes out of 4 h time series obtained for different driving forces at 300 mK. At the lowest driving force (series (a)) we observe quite long laminar phases, where the velocity amplitude exponentially approaches the equilibrium value given by phonon drag and driving force. The laminar phases are interrupted by turbulent phases intermittently and the velocity amplitude drops to the lower value v_t given by the large nonlinear turbulent drag. When using a larger driving force (series (b)) we find a steeper increase of the velocity amplitude corresponding to a higher equilibrium value that is not reached any more, as the lifetimes of

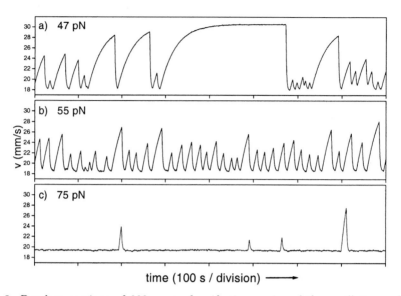

Fig. 3. Random sections of 800 s out of a 4 h time series of the oscillation velocity amplitude for three different drives at 300 mK. Series (**a**) was taken at the lowest driving force, showing the velocity increase during laminar phases and its exponential saturation at a level given by phonon drag and driving force. Laminar phases are interrupted intermittently by turbulent phases accompanied by a sharp drop of the oscillation velocity amplitude to an equilibrium value given by the large turbulent drag and driving force. Series (**b**) was taken using a larger driving force, resulting in a higher saturation value (that is not reached any more) and slightly longer turbulent phases. At the largest drive (series (**c**)) there is turbulent flow most of the time, interrupted by laminar phases

the laminar phases are shorter than in series a. Taking a closer look at the lifetimes of the turbulent phases shows that they in turn are longer than in series a. Increasing the driving force further results in turbulent flow most of the time, interrupted by short laminar phases. As the switching between both flow patterns is intermittent, the results are discussed in terms of a statistical analysis of the velocity amplitudes reached during laminar phases and their lifetimes as well as the lifetimes of the turbulent phases. We apply reliability theory [6] for the evaluation of the time series.

2.3   Turbulent Phases

A typical distribution of lifetimes of turbulent phases is shown in Fig. 4a, where the number P of turbulent phases exceeding a certain lifetime t is plotted versus t. The data points are described very well by a straight line in the semilogarithmic plot, corresponding to exponentially distributed lifetimes $\exp(-t/\mu)$. The slope of the straight line gives the mean lifetime μ. We have performed this analysis for many time series at different temperatures ranging from 28 mK up to 400 mK and

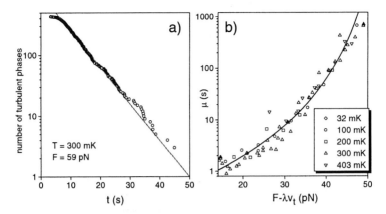

Fig. 4. Analysis of the lifetimes of the turbulent phases. (a) The number of turbulent phases living longer than a certain time t shows an exponential decay (*solid line*) corresponding to exponentially distributed lifetimes. The reciprocal slope of the distribution in the plot gives the mean lifetime μ. (b) Mean lifetimes of turbulent phases for different driving forces and temperatures. There is no temperature dependence, but a strong dependence on the driving force. The solid line is a fit of a fourth-power law divergence of the mean lifetime μ at a critical driving force value

in a wide range of driving forces and have always found exponentially distributed lifetimes. A comparison of the mean lifetimes obtained is shown in Fig. 4b. The values of μ are independent of temperature and collapse to an universal drive dependence if the strongly temperature dependent laminar drag λv (where $v = v_t$ is the velocity amplitude of the turbulent phase) is subtracted from the external driving force. The mean lifetime of turbulent phases increases with the driving force and diverges at a critical value $F_c = 54\,\mathrm{pN}$ approximately with a fourth-power law and stays infinite at larger drives corresponding to turbulent velocities above v_{c2}, see Fig. 1. The power dissipated at F_c is $0.6\,\mathrm{pW}$, corresponding to the production of $\approx 1\,\mathrm{mm}$ vortex lines per half period. This is equivalent to producing a vortex ring with a diameter of ≈ 1.4 times the diameter of the sphere.

2.4   Laminar Phases

A similar analysis can be performed for the laminar phases, but now it is convenient to analyze first the amplitudes reached during laminar phases. Figure 5a shows the number of laminar phases which exceed a given amplitude $Dv = v - v_t$. The data are perfectly described by a parabola in the semilogarithmic plot, $P(Dv) = P(0) \exp{-(Dv/v_w)^2}$, which means that the probability density function of the amplitudes corresponds to a Weibull distribution [6]. This result holds for all driving forces and temperatures and the fitting parameter v_w is constant (Fig. 5b). The results for a given experiment fit well into a 10% bandwidth around a mean value. In order to extract the probability of breakdown of the metastable laminar flow it is necessary to analyze the lifetimes t of the laminar phases, i.e. $P(t) = P(Dv(t))$. The failure rate $\Lambda(t)$ (i.e. the probability

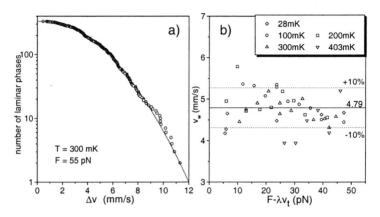

Fig. 5. Analysis of velocity amplitudes reached during laminar phases. (a) The number of laminar phases having an amplitude larger than Dv shows a quadratic dependence on Dv in this semilogarithmic plot (*solid line*), corresponding to a Weibull probability density of velocities. (b) Fitting parameter v_{w} for different driving forces and temperatures. We find no systematic dependence on these parameters. The values vary in a 10% bandwidth around a mean value of 4.8 mm/s.

of breakdown after the laminar phase has survived for a time t) is given by [6]

$$\Lambda(t) = -\frac{d}{dt}\ln P(\mathrm{D}v(t)) = \frac{d}{dt}\left(\frac{\mathrm{D}v(t)}{v_{\mathrm{w}}}\right)^2 = \frac{2}{v_{\mathrm{w}}^2}\,\mathrm{D}v\,\mathrm{D}\dot{v}\,. \tag{5}$$

Because the velocity amplitude is exponentially approaching an equilibrium value with a time constant $\tau = 2m/\lambda$, this result has several implications: First, as Dv is constant for small values of Dv (or $t \ll \tau$), the failure rate is increasing with the velocity amplitude of the sphere. Second, with the velocity reaching the equilibrium value (or $t \gg \tau$), Dv goes to zero, and so does $\Lambda(t)$. This means that the laminar phase, having survived for many τ and having reached its stationary velocity amplitude, will live forever although the velocity is clearly above v_{c1}. Experimentally, however, this is not exactly true. This can be seen in Fig. 6, where the probability distribution for large lifetimes is shown. The decrease slows considerably down for lifetimes greater than approximately 100 s (corresponding to 3τ at 300 mK), but the distribution does not approach a constant value. Instead, a constant failure rate is found, leading to a mean lifetime of 25 min (indicated by the straight line in Fig. 6). Obviously, there must be another mechanism causing the breakdown of those long lived laminar phases. We can exclude mechanical vibrations or acoustic noise to be the origin, as we tried to destroy such long laminar phases by slamming the door, jumping on the floor, or even refilling helium into the cryostat. But placing a small radioactive source (^{60}Co, 74 kBq) outside the dewar, had a dramatic effect on the lifetimes. As can be seen from Fig. 6, the mean lifetime changes by a factor 8.3 from 25 minutes to 3.0 minutes. We have measured the dose rate of the source at the position of the measuring cell inside the cryostat (taking into account a measured 20% loss in the dewar walls) to be $440\,\mathrm{nGy/h}$ ($\pm 5\%$). Comparing this value with a measured dose rate due to natural background radiation in our laboratory of $50\,\mathrm{nGy/h}$ ($\pm 10\%$), which is typical for our area, we obtain an increase of the dose rate due to the source by a factor of $(440 + 50)/50 = 9.8$ This compares well with the ratio of lifetimes of metastable laminar phases obatined in the experiment. Therefore, it is obvious that natural background radioactivity limits the lifetime of metastable laminar phases above the critical velocity v_{c1}. This effect may be attributed to local vorticity generated by ions produced by radiation, inducing the breakdown of the metastable laminar flow around the sphere.

3   Conclusion

The turbulent flow of the superfluid causes a drag force on the sphere which is very similar to turbulence in classical fluids except for a constant offset in analogy with a dilute Bose-Einstein condensate. Furthermore, we have found that the transition from potential flow to turbulent flow around a sphere in superfluid ^{4}He at mK temperatures occurs by intermittent switching between both flow patterns instead of the gradual transition observed in viscous fluids. A statistical analysis of this switching phenomenon has shown that the lifetimes of turbulent phases diverge at a critical driving force. The probablity for breakdown of the laminar phases has been obtained. Finally, there exist metastable laminar phases above the critical velocity v_{c1}. Their lifetime is limited only by natural background radioactivity [7].

We have repeated our experiment several times by heating the measuring cell above T_c of niobium in order to prepare a new levitating state of the sphere. All our observations were reproducible. The three quantities v_{c1}, v_w and F_c which we expect to be affected by the properties of the sphere have standard deviations of 8%, 13% and 15%, respectively, probably due to asymmetries of the surface of the sphere.

We understand now why above $0.5\,\mathrm{K}$ the intermittent switching changes into the hysteretic behavior observed earlier [1,2]: the velocity increases $\mathrm{D}v$ become

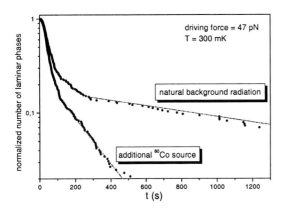

Fig. 6. Influence of radioactivity on the lifetime of laminar phases. By adding a radioactive source the mean lifetime for long living laminar phases ($t \gg \tau = 31\,\mathrm{s}$ at $300\,\mathrm{mK}$) is reduced from 25 minutes to 3.0 minutes. The mean lifetimes were obtained by analyzing the approximately exponential tail at large lifetimes (*straight lines*)

very small at higher phonon drag which implies a very low failure rate of the laminar phase. But if it fails (i.e. when the driving force F is largely increased) the following turbulent phase is stable because the critical drive is exceeded.

With our present results in superfluid ^{4}He it appears extremely promising to extend these experiments to superfluid ^{3}He in order to investigate the transition to turbulence in this very different quantum fluid.

References

1. J. Jäger, B. Schuderer, and W. Schoepe: Physical Review Letters **74**, 566 (1995).
2. J. Jäger, B. Schuderer, and W. Schoepe: Physica B **210**, 201 (1995)
3. P. Eizinger, W. Schoepe, K. Gloos, J.T. Simola, and J.T. Tuoriniemi: Physica B **178**, 340 (1992)
4. M. Niemetz, W. Schoepe, J.T. Simola, and J.T. Tuoriniemi: Physica B **280**, 559 (2000)
5. T. Winiecki, J.F. McCann, and C.S. Adams: Physical Review Letters **82**, 5186 (1999)
6. B.V. Gnedenko, Yu.K. Belayev, and A.D. Solovyev: *Mathematical Methods of Reliability Theory* (Academic Press, New York, 1969)
7. M. Niemetz, H. Kerscher, and W. Schoepe: http://arXiv.org/abs/cond-mat/0009299 and to be published

Part III

Vortex Dynamics

Vortex Filament Methods for Superfluids

David C. Samuels

Dept. of Mathematics, Univ. of Newcastle, Newcastle upon Tyne, NE1 7RU, UK

1   Introduction

Vortex filaments are an idealized form of rotational flow where the vorticity is confined to a small core region, of radius a, around a one dimensional line embedded in the three dimensional flow. Outside of this core region the flow is potential. When the dynamics of the core size are not important these objects may also be referred to as vortex lines.

In classical fluid mechanics, by which I mean solutions of the Navier-Stokes or Euler equations, vortex filaments are a useful tool for understanding the geometry and dynamics of a flow. But after an initial popularity in the early 1980's [1] [2] [3] [4] [5], the use of vortex filament methods fell out of favor in classical fluid mechanics. Though there has been some slight resurgence in this method recently [6] [7] due to the rapidly increasing computational power available and the development of new computational algorithms, direct numerical simulations and large eddy simulations have become the methods of choice for calculating the motion of fluids. One reason for the decreased use of vortex filament methods is that while they give a clear and intuitive understanding of a flow through the easy visualization of the vortex filaments, this representation was often just a rough cartoon of the true flow. Vortex filaments are only a convenient idealization in a classical flow. The vorticity in a realistic classical flow rarely takes the form of clearly discrete vorticity filaments.

But in superfluids like helium II vortex filaments are real [8]. Due to the quantization of circulation, vorticity in a helium II flow can only exist within vortex filaments with a core size of a. Since this core size is very small in helium II, about 1 Angstrom, the thin-core vortex filament idealization is actually a very accurate description of the true superfluid flow.

There are a few special qualities of the superfluid vortex filament [8] that make it even simpler than the standard model of a vortex filament in a classical fluid. The circulation around each superfluid vortex filament in helium II is set by quantum mechanics to be an integral multiple of $\kappa = 9.97\mathrm{x}10^{-4}\mathrm{cm}^2/\mathrm{sec}$. Since an n quantum vortex filament contains more energy than n single quantum vortices, it is generally assumed that only single quantum vortex filaments will be commonly observed. Thus all helium II superfluid vortices have identical circulation. The core size a is also determined by quantum mechanics and is closely related to the concept of a *healing length*, the length scale required for a wave function to change from its bulk value to zero. In the classical fluid vortex filament the core size is an important variable and much of the complications, and the interesting dynamics, of classical vortex filaments are due to changes in the core size along a filament [9]. Without a variable core size, the dynamics of a quantized vortex filament are much simpler than those of its classical counterpart. Of particular importance is the effect of the constant core size on vortex stretching. When a classical vortex filament is stretched, the core size shrinks and the local vorticity rises along the filament. When a quantum vortex filament is stretched, it can only elongate with no change in core size. Due to this difference, many prefer to reserve the phrase "vortex stretching" for the classical process, and will say that vortex stretching does not occur for individual quantized vortex lines. One should note, however, that a bundle of parallel quantized vortex filaments can undergo vortex stretching, when the core size is interpreted as the radial size of the bundle of filaments.

Many of the complications of classical vortex filament calculations have to do with vortex stretching and with other core dynamics which are assumed negligible in helium II quantized vortex filaments. We thus have a much simpler system to deal with, both in terms of physical behaviour and in terms of the computation of the motion of the filaments. My discussion in this paper will be concerned directly with the superfluid vortex filaments of helium II, and when necessary I will take the values of constants to be those of helium II.

2   Vortex Filament Motion

We represent the superfluid vortex filament by a series of mesh points distributed along the centerline of the vortex filament. The motion of the vortex filaments is calculated as the motion of each of these mesh points.

Since the vortex core size is on the atomic scale, it is usually assumed that the effective mass of the vortex is negligible and the motion of the vortex can be found by setting the sum of the forces at each point on the vortex filament to be zero and solving the resulting equation for the velocity of the vortex line [8]. From these procedure we get the basic equation of motion for the superfluid vortex filament,

$$d\boldsymbol{S}/dt = \boldsymbol{V}_S + \boldsymbol{V}_I + \alpha \boldsymbol{S}' \otimes (\boldsymbol{V}_N - \boldsymbol{V}_S - \boldsymbol{V}_I) - \alpha' \boldsymbol{S}' \otimes [\boldsymbol{S}' \otimes (\boldsymbol{V}_N - \boldsymbol{V}_S - \boldsymbol{V}_I)] \quad (1)$$

where $\boldsymbol{S}$ is a position of a point on the vortex line, $\boldsymbol{V}_S$ is the local superfluid flow due to all non-vorticity sources, $\boldsymbol{V}_I$ is the superfluid velocity due to the presence of the vortex filaments (the induced velocity). $\boldsymbol{S}'$ is the derivative of $\boldsymbol{S}$ with respect to arclength along the vortex line. $\boldsymbol{V}_N$ is the local normal fluid velocity and α and α' are temperature dependent mutual friction parameters. It is common to claim that $\alpha > \alpha'$ and then neglect the last term in the equation of motion. However, this inequality is not true for temperatures very near the lambda transition and there should also be the more general concern that since these α and α' terms in (1) are at right angles to each other it does not strictly matter that one term may be of smaller magnitude than the other. The α' term produces a qualitatively different type of motion of the vortex line than does the α term, and it should therefore not be discarded in an offhand manner.

The two fluid model presents the motion of helium II as a superposition of two fluids, the normal fluid and the superfluid. The vortex filaments that concern us here are vorticity structures of the superfluid. They interact with the normal fluid through the mutual friction force, and the motion of these vortex filaments can be strongly affected by the flow in the normal fluid. The reverse is also true; the presence and motion of the superfluid vortex filaments can strongly affect the flow of the normal fluid. This coupled interaction between the two fluids is the subject of many of the articles in this volume. Calculating the fully coupled motion of the entire two-fluid system is a very complicated problem (which is discussed more fully in the article by Idowu et al). While work on the fully coupled calculations is just beginning, most superfluid vortex calculations up to this time have only been concerned with the motion of the superfluid vortices without considering the response of the normal fluid flow to the mutual friction force. This kinematic approach requires us to define a normal fluid flow, with spatial and possibly even time dependence, and use (1) to calculate the motion of the vortex line due to its self induced motion, advection in the superfluid potential flow V_S and the response to mutual friction with the normal fluid flow V_N. This is the problem that we will consider in this article. The vortex filament techniques discussed here are also needed for the fully coupled calculations described elsewhere in this volume.

2.1 The Biot–Savart Law and the Local Induction Approximation

The first difficulties in the vortex filament method come from the definition of the term V_I in the equation of motion. This term represents the advection of the point on the vortex filament by the velocity field due to all the superfluid vortex filaments present in the flow. This advection velocity at the point r is given by the Biot-Savart law.

$$V_I(r) = \frac{\kappa}{4\pi} \int \frac{(S - r) \otimes dS}{|S - r|^3} \tag{2}$$

The line integral is taken over the entire length of vortex filament present in the fluid. The Biot-Savart law can be used to find the flow at any point r in the fluid, but the immediate difficulty with this equation comes from its use in calculating the advection velocity in (1). In this case, the point r lies on the vortex line and the Biot-Savart law contains a singularity at that point. To heal this singularity we must include some aspect of the core structure of the vortex filament. Many different methods of de-singularizing the Biot-Savart law have been used in classical fluid dynamics [10] [11] [12] [13], where the core dynamics of the filaments can often be quite complicated and important, but in superfluid vortex methods I have only seen one method used [14]. This method breaks the line integral into two parts; an integral over a local neighborhood around the point r (this contains the singularity), and the rest of the Biot-Savart integral (this part is non-singular). Then the local integral must be replaced by an algebraic approximation that takes into account the core size of the filament.

It is convenient to split up the Biot-Savart integral using the local meshing of the vortex line. In calculating the motion of the J'th mesh point on the filament, the local section of the integral is taken to be section containing the mesh points $J-1$, J, and $J+1$. By calculating the Biot-Savart integral over a circular vortex ring excluding the section of line between the mesh point $J-1$ and $J+1$, and then subtracting this value from an analytic expression for the velocity of an ideal vortex ring, the remainder can be identified as the Biot-Savart integral over the section of the vortex line from $J-1$ to $J+1$. Thus this section of the Biot-Savart integral (containing the singularity) is replaced by the term

$$V_{local} = \frac{\kappa}{4\pi}(S' \otimes S'') \ln\left(\frac{2\sqrt{l_+ l_-}}{a_{eff}}\right). \tag{3}$$

S' is the unit tangent vector to the vortex line at the J'th mesh point and S'', the second derivative of the line with respect to arclength, is the local curvature vector. The length scales l_+ and l_- are distances between the J'th mesh point and the $J+1$ and $J-1$ points respectively. The parameter a_{eff} is an effective core size since numerical constants of order one have been absorbed into this parameter. This choice for the local part of the Biot-Savart integral gives the correct velocity for a planar vortex ring. Defining the velocity of a planar vortex ring requires that you specify some details about the core structure of the vortex (hollow core vs solid body rotation, for example). Different choices of the core structure will alter the log term in (3), but unless you are calculating flows on very small length scale, on the order of the core size, then these details make negligible changes to the value of (3).

Once you prevent the singularity by splitting the Biot-Savart law into two sections you are still left with the difficulty of calculating the integral over the rest of the vortex line. In a simulation, we only know the positions of the discrete mesh points along the line, so the mesh points must be joined in some numerically convenient manner to calculate the integral between the mesh points. Any interpolation method through the mesh points will do the job. I prefer to use the simplest method of integrating over piece-wise linear vortex line segments between the mesh points. The velocity at a point r (which may be a mesh point on the vortex line or may be any point in the volume of the fluid) due to the Biot-Savart integral over the segment of vortex line between the J and $J+1$ mesh points is

$$V_{segment}(r) = \frac{\kappa}{4\pi} \int_{S_J}^{S_{J+1}} \frac{(S-r) \otimes dS}{|S-r|^3}. \tag{4}$$

Taking this integral over a straight line between the two mesh points gives

$$V_{segment}(r) = \frac{\kappa}{4\pi} \frac{(|R_J| + |R_{J+1}|)(R_J \otimes R_{J+1})}{|R_J||R_{J+1}|(|R_J||R_{J+1}| + R_J \cdot R_{J+1})} \tag{5}$$

where $R_J = S_J - r$ and $R_{J+1} = S_{J+1} - r$. Other interpolation methods will have more complicated results for the Biot-Savart integral.

One of the reasons that vortex filament methods have fallen out of favor in classical fluid dynamics is the fact that the calculation of the Biot-Savart integral requires order N^2 operations over a mesh of N points. The calculation of this integral is computational expensive whenever a large number of mesh points are needed to represent the vortex lines. This calculation is often not practical, even with the speed of today's computers, if the amount of vortex line is large, or if the fine detail of the vortex line shape needs to be captured using a very fine meshing. And these are often exactly the cases that are of the most interest to us.

One way around this difficulty is the *Local Induction Approximation*, (LIA) [15]. In this method we keep the local term in the induced motion (3) and neglect completely the non-local Biot-Savart integral. This is typically done with some minor adjustments to the log term so that we have

$$V_{I,LIA} = \frac{\kappa}{4\pi}(S' \otimes S'')\ln\Big(\frac{2R}{a_{eff}}\Big), \tag{6}$$

where R is a length scale of the filament. R may be taken as a constant, such as the length scale of the computational box, or it may be taken as the local radius of curvature of the vortex line. In the first case, the log term is a constant, and it is often absorbed into a non-dimensional timescale. In the second case, the log term will vary along the filament and with time, but unless you are capturing a very wide range of curvatures with your mesh, or the curvatures are only about a factor of 10 larger than the effective core size, the log term is nearly constant.

LIA is a very convenient approximation. It is simple and easy to calculate. The time required to calculate (6) increases only linearly with the number of mesh points. The interpretation of (6) is simple: the induced velocity is in the direction of the local binormal $S' \otimes S''$ and is inversely proportional to the local radius of curvature. It is commonly used in analytic investigations of the properties of vortex filaments. And LIA correctly describes the motion calculated by the Biot-Savart law of simple vortex line geometries such as the planar vortex ring and low amplitude vortex waves. But it is a very severe approximation to the true equation of motion!

LIA clearly works very well for calculating the motion of single vortex lines which do not loop around so that sections of the filament far apart in arclength are actually close to each other. Also it probably works well in a random vortex tangle, where the non-local part of the Biot-Savart law may tend to cancel out when integrated over the random vortex lines. LIA will not work well for vortex configurations which tend to have parallel, or antiparallel, sections of vortex filament near one another (for example, vortex knots [16]). LIA will not work well for flows which tend to develop any alignment of vorticity. Unfortunately, aligned superfluid vorticity does tend to be formed by flows at non-zero temperatures where the normal fluid has some local vorticity structure (the superfluid vortices tend to align with the normal fluid vorticity [17]). In the case of turbulent flows, I personally doubt that the LIA captures enough of the physics of the interactions between the vortex filaments to give much useful information on these flows, though others would debate this point.

Some of the qualitative differences between the motion of vortices under LIA and the full Biot-Savart law should be pointed out. A superfluid flow calculated by LIA can never develop any rotational flow, aside from the rotation around a single filament. In an LIA calculation, the energy of the flow is just an energy per unit length of the vortex lines and when there is no mutual friction (at zero temperature) the length of vortex line does not change with time (without further assumptions to the model, such as phonon emission). Under the Biot-Savart law, the kinetic energy in the flow can be quite difficult to calculate, and even at zero temperature the length of vortex line can change while the kinetic energy remains constant. In a LIA calculation, the physical effect of vortex stretching cannot occur while it can occur in a Biot-Savart calculation whenever there are even just two approximately parallel vortex lines.

With the vortex filaments represented by N mesh points, the vortex equation of motion (1) becomes N coupled, first order ordinary differential equations. This coupled set of ODEs can be solved by any general method. I prefer to use a fifth order Runge-Kutta method with an adaptive stepsize. An adaptive method like this calculates the result of one time step in two different ways (one fourth order method and one fifth order method) and compares the results at each of the N mesh points to estimate the error. An allowed error range is defined and if the estimated error lies below this allowed error range for all N points, the time step is increased, while if the estimated error is above the allowed range for any of the N mesh points, the time step is decreased and that timestep is recalculated. This is a sturdy method, capable of automatically handling the rapid changes in vortex line velocities which can occur, often as a result of vortex line reconnections. It is not a very efficient method however. Whenever any region of very high curvature (or a close approach of two vortex filaments under the Biot-Savart law) occurs then the timestep for the entire vortex tangle can drop significantly. Some type of 'multi time step' method, where the motion of fast moving sections of the vortex line could be calculated using more time steps than the slow moving sections of the vortex, would would be far more efficient, particularly when the equation of motion is as expensive to evaluate as is the Biot-Savart law.

2.2   Boundary Conditions

There are three basic types of boundary conditions used in vortex filament calculations. In order of increasing complexity these are: an infinite fluid, periodic boundary conditions, and solid boundaries (no penetration of the superfluid through the boundary). In an infinite fluid calculation no boundary conditions are required in the calculations, as long as you set up the initial conditions as closed vortex loops. Many flow quantities, such as the total kinetic energy, have simple line integral definitions that only work in the infinite fluid simulations, where the vorticity is zero at spatial infinity. This is a good choice of boundary conditions for studying simple closed loop vortex structures, such as vortex knots, but it is not a good choice for complicated vortex tangles.

Periodic boundary conditions work well for vortex tangles since they allow a homogeneous flow to develop and allow simple definitions of averaged quantities

such as the vortex line density. Periodic boundary conditions are easily pro-
grammed for LIA simulations, though care must be taken at the places where
neighboring mesh points along a filament lie on opposite sides of your periodic
boundary. But these problems are easily handled. The main difficulty with periodic boundary conditions comes from the Biot-Savart calculations. In principle,
periodic BCs make the Biot-Savart integral infinitely long. In practice, this integral must be cut off at some arbitrary point. The costs of periodic BC in Biot-
Savart calculations can be staggering. For a three dimensional periodic cube,
including the first layer of periodic vortices around the central cubic volume will
increase the number of vortex mesh points by a factor of 27. Including even this
level of periodicity is usually prohibitively costly. To deal with problem I will
sometimes define a buffer layer around the central cubic volumn, with a width
of 1/2 to 1/4 of the width of the computational volume, and I will only include
in the Biot-Savart calculation the periodic vortex filaments that fall within this
fairly thin buffer zone. This type of method actually only ensures that vortex
filaments moving through one side of the periodic volume and re-entering on the
other side, experience a more smoothly varying velocity field during the transition. Ideally, this type of problem could be removed by an analytic method to
calculate the velocity field due to this infinitely repeating vorticity distribution.
While this has been done in two dimensions, I know of no such method in 3D.

In LIA, solid boundary conditions are quite easy to implement [14]. In this
case, for any vortex filaments which end on the solid surface we must set the
normal vector at the ends of the vortex filament equal to the local normal vector
of the solid surface. This is the only condition needed no matter how complex the
boundary geometry. In Biot-Savart calculations we must use image vortices for all
the vortex filaments in the flow. For a single flat boundary, these image vortices
are simply vortices in the mirror image positions, and with their orientation
reversed. For a single flat surface this doubles the length of the Biot-Savart
calculations, which is not a terribly onerous increase. But for multiple solid
surfaces, you must then include the images of images, ad infinitum (just as
you must do in electrostatic calculations). This naturally introduces the same
computational difficulties as in the periodic BC case, and it must be solved in
the same manner, either by cutting off the image calculations at some arbitrary
number of images or by developing an analytic summation method to collapse
the infinite images to a reasonable calculation. I know of no one who has done
the latter for a 3D vortex flow, though the tools may exist in the electrostatics
literature for us to use.

2.3   Meshing of the Filaments

Since the CPU time cost of a vortex simulation rises rapidly as the number of
mesh points increases it is important to manage this number carefully. As the
vortex filaments are locally compressed or elongated the spacing between the
mesh points will change with time. New mesh points will need to be inserted on
some sections on the filament to keep a certain mesh resolution, and to keep the
number of mesh points at a minimum some mesh points will need to be removed

from other sections of the filament as they move too close to one another (close in terms of arclength). The criteria that you choose for remeshing depends on the behaviour that you are trying to capture with a specific simulation.

For this discussion, let the arclength distance between two neighboring mesh points, J and $J+1$ be represented by $\delta_{J,J+1}$. The simplest remeshing criteria is to attempt to keep an approximately uniform distance between the mesh points. In this method, the mesh point J would be removed if $\delta_{J-1,J} + \delta_{J,J+1} < \delta$ and a mesh point would be added between the J and $J+1$ points if $\delta_{J,J+1} > 2\delta$. This will keep all the distances along the mesh roughly in the range $\delta/2$ to 2δ. This is the most straightforward meshing method, but it suffers from the limitation that it can only resolve structures on the vortex filament of scale δ or larger. This method is best if you are only interested in the vortex behaviour at one particular length scale.

If you want to calculate the development of vortex filament structures over a wide range of length scales then you must use a more complicated remeshing method. We would like to have more mesh points in the parts of the vortex filament with high curvature and fewer in regions of low curvature. This can be done by interpreting the meshing length scale δ as a variable, proportional to the local radius of curvature. Typically, I will take $\delta = 2\pi R/32$, where R is the local radius of curvature of the filament. This will mesh a planar ring, at any length scale, with approxamately 32 mesh points. Of course, there must be a limit on the range of this meshing. This is done by setting a range of R over which this remeshing will be used, with minimum and maximum mesh lengths to be used outside this range. With this method it is possible for the mesh to represent a range of two or three orders of magnitude in radius of curvature without having an unreasonably large number of mesh points. It may be possible to extend this range by another order of magnitude by setting the meshing length δ to decrease slower than linearly with R. Techniques like this are used in 2D vortex simulations in the meshing of the surface of vorticity blobs [18].

However you choose to remesh, removing mesh points poses no problems but in adding mesh points you must make a choice on the position of the new mesh point. Once again, the simplest choice is to use a piece-wise linear interpolation, and simply place the new point at the center of the straight line between the J and $J+1$ point. But if you do this, then you are likely to change the first and second derivatives, $\boldsymbol{S}'$ and $\boldsymbol{S}''$ at the points J and $J+1$. Worse than this, the new point would be introduced with zero local curvature, almost certainly causing a rapid (and artificial) change in curvature along the filament. For adding mesh points a better interpolation must be used, and almost any interpolation will perform well as long as it is not piece-wise linear. I use a method that inserts new mesh points at a position where the curvature vector, $\boldsymbol{S}''$, at the new mesh point is the average of the curvature vectors at the two neighboring points, J and $J+1$. This prevents the intruduction of sudden jumps in the vortex line curvature, errors which would radiate vortex waves and affect more than just the local segment of the vortex line.

For ease of explanation in this paper I have been labelling neighboring mesh points with consecutive number, J and $J+1$ for example. Any remeshing naturally will destroy this order, and consecutive numbering is not practical. Instead, a data structure should be used that has for each mesh point, the labels of the two neighboring mesh points. Then a remeshing involves only the addition or deletion of a mesh point and the necessary changes in this data structure for the neighboring mesh points.

One further remeshing routine should be used. Occasionally, very small vortex rings may develop. These may be formed by the decay through mutual friction of larger vortex loops. Or they may be generated by the pinching off of a small loop in a reconnection event. However they are formed, these small loops should be removed from the calculation. By definition, the curvature of these loops will be very large, and thus their motion will be very fast. With an adaptive stepsize technique, the code will automatically drop the time step to a very small value to attempt to accurately calculate the rapid motion of these small loops. If you are not using an adaptive stepsize, then the motion of these small loops will not be accurately calculated, possibly leading to the sudden and artificial expansion of the small loop. In either method these small loops must be removed from the calculation. It is a good idea to keep a record at least of the length of vortex line removed from the simulation by this routine.

3   Reconnections of Filaments

When two classical Navier-Stokes vortex filaments cross, viscous effects will reconnect the vortex filaments. Lacking these viscous effects, the crossing of two classical Euler vortex filaments is believed to lead to a singularity and the breakdown of the Euler equations [19]. This singularity formation is not observed in calculations of the crossing of superfluid vortex filaments, where vortex reconnection occurs without the need for viscosity [20]. In this way, superfluid vortices are clearly not Euler vortex filaments. The difference lies in the core structure of the superfluid vortices. At the centerline of each vortex filament, quantum mechanics requires the density of the superfluid (given by the amplitude of the ground state wave function) to go to zero and the vorticity of the flow is singular along this centerline.

Studies of superfluid vortex filament reconnections must be done with a quantum theory calculating the evolution of the wavefunction of the superfluid ground state. A more detailed discussion of these calculations can be found in the articles by Roberts and by Adams in this volume. These studies have traditionally been done using a Non-Linear Schrodinger Equation (NLSE). It is recognized that the NLSE is probably a poor representation of helium II superfluid because the dispersion curve calculated from the NLSE contains only phonons and no rotons. The NLSE does, however, seem to be a very good representation of the new Bose condensed alkali atom superfluids. Without a fundamental microscopic theory of helium II superfluidity, we must consider the results of NLSE calculations, keeping in mind the possible errors of these calculations. The NLSE calculations clearly show that when two superfluid vortex lines cross, a reconnection event occurs and the topology of the vortex lines changes [20]. This behavior must be included as a basic assumption in the superfluid vortex filament model since it is not a consequence of the equation of motion. This discussion will be made in terms of the crossing to vortex filaments in the bulk of the fluid. In simulations with boundaries, reconnections of filaments with the boundaries must also be considered, but these reconnections may be easily detected as vortex segments approach and eventually hit the boundaries so we need not discuss them further here.

The reconnection assumption leaves us with the problem of detecting the crossings of 1-D lines moving through 3-D space, when we only know the positions of a finite number of mesh points along these lines at discrete time intervals. Though the assumption can be quite simply stated, it has proved very difficult to develop any satisfactory algorithms to implement this. The first reconnection algorithm, due to Schwarz [21], was a simple, intuitive one: vortex lines are reconnected whenever two mesh points come within a pre-defined distance Δ of each other. While this algorithm is very simple to use, the objections to it are many and strong. This algorithm will obviously trigger a reconnection event early, since a true reconnection event does not occur until the distance between the lines is on the order of a core radius (practically zero in filament simulations). Sometimes, a large section of vortex filament, including many mesh points, will move within the reconnection distance triggering a series of reconnection events resulting in the formation of a number of small vortex loops where only one crossing event would actually have occured. Even worse, in some cases this algorithm will trigger a reconnection where no crossing of the vortex lines would actually occur. This error happens when vortex lines pass by each other, without intersection but within the distance Δ. This may at first seem to be a rare and thus negligible event, but it is not. We now know that vorticity concentrations in the normal fluid generate bundles of well-aligned superfluid vortex filaments at the center of the normal fluid rotation [17]. These superfluid vortex filaments may be closer to each other than the defined reconnection distance Δ but they cross only rarely. With the Δ reconnection algorithm, these vorticity bundles will reconnect wildly.

The Δ reconnection model also introduces a new and completely artificial length scale to the simulation; the reconnection length scale Δ. If you set Δ too large, you will have many spurious reconnections. If you set it too low, you will miss many real reconnections, as vortex lines jump past one another in the finite time step, plus you will still have some spurious reconnections of nearly aligned vortices. Some of the problems of this algorithm can be helped (though not eliminated) by making the reconnection length scale variable along the vortex filament and setting it equal to the local mesh size (a natural choice). This refinement to the algorithm still suffers from the separate problems of premature and also spurious reconnection.

A different reconnection detection algorithm, the one that I currently use, is based on the idea that a vortex filament reconnection event is a dissipative event, even in a superfluid. The reconnection is assumed to involve the emission of phonons and possibly rotons, thus converting superfluid energy to heat. With this in mind, we assume that reconnections can only occur if they are accompanied by a loss of energy in the superfluid. Identifying vortex line energy with vortex line length (an exact proportionality in LIA, but only approximate for the Biot-Savart law), we reconnect the vortex lines when the reconnection will directly lower the line length. The reconnections that occur in this algorithm still happen when the filaments are slightly less than a local mesh length apart, so it still suffers from the premature reconnection problem. But the generation of additional small vortex loops in a reconnection rarely occurs in this algorithm and spurious reconnections from near misses and between aligned vortices are almost eliminated.

It is worth taking care with the reconnection algorithm. Differences in the implementation of reconnections [22] [23] [24] were initially blamed for inconsistancies between the vortex tangle simulations of Schwarz and Buttke in the late 80's, though the more serious error of the introduction of an unphysical 'mixing step' by Schwarz was eventually determined to be the prime cause of the disagreement [25]. We must admit that the reconnection algorithms in use today still leave much to be desired and can be improved. A reconnection is a major event, drastically changing the vortex filament motion, curvature distribution, and topology. The close approach and eventual crossing of two vortex filaments create superfluid flow structures on the full range of length scales, down to the core size scale. And the motion of a large section of the vortex filament can be altered by a reconnection as the initially localized, but extreme, change in curvature radiates out from the reconnection point as vortex waves. It is still worthwhile to carry out detailed investigations of individual vortex reconnections [26] (see Lipniacki's contribution to this volume) in order to develop better reconnection models for use in both the filament simulations and in analytic theories of the averaged quantities in superfluid turbulence.

4   Analysis of the Superfluid Flow

Once you have developed and tested your vortex filament code you are faced with the analysis problem common to all 3-D fluid dynamics simulations: how to make sense of all that mass of data. Though the problem is the same, the solution must be different for vortex filaments due to the very different nature of the data. Many quantities that can be calculated easily from the 3-D velocity fields of a classical fluid dynamics simulation are actually very difficult to calculate with vortex filament methods. One example of this is kinetic energy. Conversely, there are at least a few things (such as flow topology) that are easier to calculate in a vortex filament simulation than in a standard fluid dynamics simulation.

The simplest measurable quantity in a vortex filament simulation is the total length of vortex line. In a finite volume V this is measurement gives you directly the vortex line density $L = (\int d\xi)/V$, the length of vortex line per unit volume where the integral is taken over the arclength ξ of the vortex filaments. This is the fundamental averaged quantity of a vortex tangle and many of the analytic model of superfluid vortex turbulence have been based on this quantity alone. These models are differential equations for the evolution of the line density as a function of the average normal fluid velocity and some temperature dependent parameters. The most basic of these models is the Vinen equation [27]

$$\frac{dL}{dt} = \chi_1 \alpha V_{ns} L^{3/2} - \frac{\chi_2 \kappa}{2\pi} L^2, \tag{7}$$

where χ_1 and χ_2 are temperature dependent parameters and V_{ns} is the average relative velocity between the normal fluid and the superfluid. The standard interpretation of the Vinen equation is that the first term represents the growth of L due to mutual friction and the second term represents the decay of the tangle due to reconnection events, though other interpretations of these terms are have been made [28].

Now let us define some measure of the isotropy of the tangle. Very little work has been done on measures of isotropy in vortex tangles. One set of measures defined by Schwarz [28] defines three quantities which measure the isotropy of a tangle relative to two perpendicular unit vectors, $\hat{r}_{\|}$ and $\hat{r}_{\perp}$.

$$I_{\|} = \frac{1}{VL} \int [1 - (\boldsymbol{S}' \cdot \hat{r}_{\|})^2] d\xi \tag{8}$$

$$I_{\perp} = \frac{1}{VL} \int [1 - (\boldsymbol{S}' \cdot \hat{r}_{\perp})^2] d\xi \tag{9}$$

$$I_{\ell} = \frac{1}{VL^{3/2}} \int (\boldsymbol{S}' \otimes \boldsymbol{S}'') \cdot \hat{r}_{\|} d\xi \tag{10}$$

In flows with a unidirectional normal fluid flow the unit vectors $\hat{r}_{\|}$ and $\hat{r}_{\perp}$ are usually taken to be parallel and perpendicular to the direction of the normal fluid flow. The interpretation of these three isotropy measures is discussed in [28]. Other isotropy measures can be defined, such as the length of line in a projection along a vector $\hat{r}$

$$J(\hat{r}) = \frac{1}{VL} \int \sqrt{1 - (\boldsymbol{S}' \cdot \hat{r})^2} d\xi. \tag{11}$$

It is possible that more refined versions of (7) could be defined using these isotropy measures as well as L.

The local rate of extension (or compression if negative) of the vortex line is defined by

$$r(\xi) = -\boldsymbol{S}'' \cdot \boldsymbol{V_L} \tag{12}$$

where $\boldsymbol{V_L}$ is the vortex line velocity at the arclength position ξ. Integrating $r(\xi)$ over the arclength gives the instantaneous rate of change of the vortex line density.

$$\frac{dL}{dt} = \frac{1}{V} \int r d\xi \tag{13}$$

Simulations show that even in a steady state vortex tangle the line length density has strong fluctuations, so one must be careful in interpreting instantaneous measures such as (13).

It is tempting to define a whole range of quantities averaged over the vortex tangle. One that must be mentioned is the averaged curvature

$$C = \frac{1}{\ell} \int |S''| d\xi, \tag{14}$$

where $\ell = \int d\xi$ is the total vortex filament length. This quantity can be used to define an average length scale of the tangle, $1/C$, and an average velocity scale κC. Following the evolution of this quantity with time can detect if the vortices are 'crinkling' (increasing C) or smoothing (decreasing C). But I urge against an over-reliance on averaged quantaties. It is easy enough to measure the full distribution of values of quantities such as the curvature in the tangle, thus measuring the full range of behavior of the vortices and not just some average value. Again, little has been done with this approach.

But what about the more traditional fluid mechanics measures such as kinetic energy and linear and angular impulse? While these are quite easily determined in a simulation where the velocity is defined on a grid extending throughout the fluid, they are actually quite difficult to measure in a vortex filament simulation. Purely for diagnostic purposes, one can define a 3D grid and calculating the velocity on the grid by the Biot-Savart law (2) and then such quantities can be easily measured. But it is important when doing this to check the convergence of your results with grid resolution. The $1/r$ velocity field of the vortex lines can make resolution of the energy and momentum difficult. It is possible to avoid this problem by using definitions of these quantities in terms of the line integrals over the vortex filaments [29]. Some useful integrals are: kinetic energy

$$T = \kappa \rho_s \oint \boldsymbol{V}_s \cdot (\boldsymbol{S} \otimes d\boldsymbol{S}), \tag{15}$$

linear impulse

$$\boldsymbol{I} = \frac{1}{2} \kappa \rho_s \oint \boldsymbol{S} \otimes d\boldsymbol{S}, \tag{16}$$

angular impulse

$$\boldsymbol{A} = \frac{1}{3} \kappa \rho_s \oint \boldsymbol{S} \otimes \boldsymbol{S} \otimes d\boldsymbol{S}, \tag{17}$$

and helicity

$$\boldsymbol{J} = \kappa \oint \boldsymbol{V}_s \cdot d\boldsymbol{S}. \tag{18}$$

These integrals are limited to flows with no vorticity at infinity, and the integrals must be taken over closed vortex loops.

While simple things such as kinetic energy can become quite difficult to measure in vortex filament methods, there are some qualities of the flow which are simpler to define and measure in filament methods. There has been much recent interest in the topology of vorticity in Euler flows, interest which is mainly generated by the possibility of producing singularities in these flows in finite time [19]. Remembering that the vorticity in a classical Euler fluid may be distributed thoughout space (not just as vortex filaments!), the topology of the vorticity field can be difficult to define, or picture, easily. But since the vorticity of the superfluid is confined to discrete vortex filaments, vortex topology is a natural and simple quality in a superfluid [16]. In an Euler flow, the topology of the vorticity is conserved as the flow develops because filament reconnections are not allowed, and the conservation of topology can lead to strong constraints on the development of the flow [30]. But since superfluid vortices can reconnect, the topology of a superfluid vortex is not conserved, and may change with every reconnection. Despite this complication, it is possible that by determining the topology (knotted and linked loops) of a vortex tangle we can develop a measure of the complexity of a tangle, a quality that is missing from all averaged measurements on the vortex tangle.

5 Alternative Approaches

The computational cost of calculating the full Biot-Savart law is high, due to its non-locality. The usefulness of local induction approximation calculations is suspect because of its complete locality. Is there not a middle way that contains some, but not all, of the non-locality of the Biot-Savart integral? The recent development of *fast multipole methods* seems to fill this need [31] [6]. Originally developed for simulations of large numbers of gravitationally interacting particles, these methods can lower the computational time of pairwise interactions between particles from order N^2 to order $N \log(N)$. In these methods the interacting particles are grouped into clusters, and then into clusters of clusters, et cetera, in a hierarchical tree structure. The interactions between the clusters are calculated by a finite multipole series expansion, greatly decreasing the number of calculations needed while retaining the primary non-local effects. The generalization from the original gravitational interaction to a vortex motion calculation is straightforward and these new methods have fueled much of the recent resurgence (however slight) in vortex methods for classical fluid dynamics. These methods work best in cases where the fluid vorticity has some strong spatial structure. The drawback to these methods is the high overhead cost associated with the determination of the hierarchical tree data structure, and the necessity of rebuilding this data structure as the vortices move, and groups of vortices break up or coalesce. For a moderate number of vortex mesh nodes, up to a few thousand in my experience, the Biot-Savart calculations are still faster than the fast multipole methods, though this limitation may become outdated soon as new implementation algorithms are being developed. For simulations af actual superfluid turbulence, where we must reproduce a wide range of spatial scales in the superfluid vorticity (requiring a large number of mesh points) and include the non-local vorticity interactions of turbulence (ruling out the LIA approach) we must use some non-local approximate method in our simulations, and the fast multipole methods are currently our best hope.

Yet another approach is to drop the idea of vortex filaments completely and use a locally averaged set of equations for the superfluid vorticity field, ω_s. These approaches use the Hall-Vinen-Bakharevich-Khalatnikov (HVBK) equations (or modifications of these equations) for the coupled evolution of continuous vorticity fields in the superfluid and the normal fluid. Since this approach does not use vortex filaments it lies somewhat out of the scope of this article, but is dealt with in the articles by Henderson and Holme in this volume.

6 Conclusions: What Needs to Be Done

The primary research challenge before us today is to develop a detailed understanding of the interaction between the normal fluid and the superfluid [32][33], particularly in turbulent flows. In terms of simulations, this presents us with the double task of calculating the turbulent flow of the normal fluid (including the forcing on the normal fluid due to mutual friction, see the paper by Idowu et al, this volume), and simultaneously calculating the motion of a large length of superfluid vortex lines. The simulation method must allow for the possibility of the formation of organized structures in the superfluid vorticity, so non-local vortex interactions must be included. The solution of the full superfluid turbulence problem will be difficult and will take some time, both CPU time and time to develop an understanding of the problem. But there is much to be learned from simpler studies, short of the full turbulence solution. In kinematic simulations (simulations where the normal fluid velocity is given and only the superfluid motion is calculated) the most complex normal fluid flow that has been used so far is an ABC flow [34]. We could learn many things from kinematic simulations with more complicated normal fluid flow structures. We should take a hint from the closely related field of magnetohydrodynamics and investigate the development of superfluid vorticity driven by a turbulence model to represent the normal fluid velocity. These types of studies would give us valuable information on the reaction of the superfluid to complex normal fluid flows. Does the superfluid vortex tangle lack structure (as it does under a uniform V_N) or do structures develop in the superfluid vorticity? If so, are there any relationships between the flow structures in the normal fluid flow and those developing in the superfluid? What are the scaling laws for kinetic energy in the turbulent superfluid? How do the velocity correlation functions in the superfluid compare to those in classical turbulence? We could have at least preliminary and approximate answers to these questions from kinematic simulations. In fully coupled simulations (solving for the motion of both V_N and the superfluid vorticity) we must investigate the possible changes in the normal fluid turbulence itself due to the mutual friction forcing from the superfluid. This problem bears a resemblance to turbulence with polymer additives, where the Navier-Stokes turbulence is clearly affected by the 1-D forcing from the polymers.

A not-entirely separate issue is the problem of superfluid turbulence at zero temperature [35] (see the articles by Tsubota and McClintock, this volume). In studies of this type we need a firmer understanding of the loss of energy by phonon and roton emission from reconnecting vortices [36] and possibly even from just the motion of vortices. Does this 'zero temperature dissipation' clearly affect the properties of the vortex tangle? Is it enough to produce an inertial range in the pure superfluid turbulence? Can we develop an understanding of the energy cascade in a pure superfluid vortex tangle? I say that this is a not entirely separate issue from that of the fully coupled helium II turbulence because we need to understand the behaviour of the fully coupled turbulence in both the high and low temperature limits. In the high temperature limit, helium turbulence must surely be classical Navier-Stokes turbulence after the lambda transition temperature is passed and some experiments show no detectable change in the measured turbulence quantities as this transition temperature is passed. Pure superfluid turbulence is the low temperature limit of fully coupled helium II turbulence and there is no reason not to expect this transition to also be very smooth. Helium II turbulence experiments show little, if any, temperature dependence down to approximately 1.4 Kelvin [37], where the normal fluid relative density and the mutual friction interaction both become quite small. An understanding of both classical Navier-Stokes turbulence and pure superfluid turbulence will be helpful in developing our understanding of coupled helium II turbulence.

References

1. A. J. Chorin: SIAM J. Sci, Stat. Comput. **1**, 1 (1980)
2. A. Leonard: J. Comput. Phys. **37** 289, (1980)
3. B. Couet, O. Buneman, A. Leonard: J. Comput. Phys. **39** 305, (1981)
4. A. Leonard: Ann. Rev. Fluid Mech. **17** 523 (1985)
5. W. T. Ashurst, E. Meiburg: J. Fluid Mech. **189** 87 (1988)
6. G-H. Cottet, P. D. Koumoutsakos: *Vortex Methods* (Cambridge University Press, Cambridge, 2000)
7. R. Cortez: J. Comput. Phys. **160** 385 (2000)
8. R. J. Donnelly: *Quantized Vortices in Helium II* (Cambridge University Press, Cambridge, 1991)
9. V. M. Fernandez, N. J. Zabusky, V. M. Gryanik: J. Fluid. Mech. **299** 289 (1995)
10. T. Y. Hou: Math. of Comp. **58** 103 (1992)
11. R. Klein, O. Knio: J. Fluid Mech. **284** 275 (1994)
12. M. F. Lough: Phys. Fluids **6** 1745 (1994)
13. R. Klein, L. Ting: Appl. Math. Lett. **8** 45 (1995)
14. K. W. Schwarz: Phys. Rev. B **31** 5782 (1985)
15. R. L. Ricca: Fluid Dyn. Res. **18** 245 (1996)
16. R. L. Ricca, D. C. Samuels, C. F. Barenghi: J. Fluid Mech. **391** 29 (1999)
17. D. C. Samuels: Phys. Rev. B **47** 1106 (1993)
18. D. G. Dritschel, N. J. Zabusky: Phys. of Fluids **8** 1252 (1996)
19. R. B. Pelz: Phys. Rev. E **55** 1617 (1997)
20. J. Koplick, H. Levine: Phys. Rev. Lett. **71** 1375 (1993)
21. K. W. Schwarz: Phys. Rev. Lett. **49** 283 (1982)
22. T. F. Buttke: Phys. Rev. Lett. **59** 2117 (1987)
23. K. W. Schwarz: Phys. Rev. Lett. **59** 2118 (1987)

24. T. F. Buttke: J. Comp. Phys. **76** 301 (1988)
25. K. W. Schwarz: J. Comp. Phys. **87** 237 (1990)
26. T. Lipniacki: Eur. J. Mech. B - Fluids **19** 361 (2000)
27. W. F. Vinen: Proc. Roy. Soc. Lond. A **242** 493 (1957)
28. K. W. Schwarz: Phys. Rev. B **38** 2398 (1988)
29. P. G. Saffman: *Vortex Dynamics* (Cambridge University Press, Cambridge, 1992)
30. H. K. Moffatt, A. Tsinobar: *Topological Fluid Mechanics* (Cambridge University Press, Cambridge, 1990)
31. L. Greengard, V. Rokhlin: J. Comp. Phys. **73** 325 (1987)
32. D. Kivotides, C. F. Barenghi, D. C. Samuels: Science **290** 777 (2000)
33. O. C. Idowu, A. Willis, C. F. Barenghi, D. C. Samuels: Phys. Rev. B **62** 3409 (2000)
34. C. F. Barenghi, D. C. Samuels, G. H. Bauer, R. J. Donnelly: Phys. Fluids **9** 2631 (1997)
35. B. V. Svistunov: Phys. Rev. B **52** 3647 (1995)
36. M. Leadbeater, T. Winiecki, D. C. Samuels, C. F. Barenghi, C. S. Adams: Phys. Rev. Lett. **86** 1410 (2001)
37. J. Maurer, P. Tabeling: Europhys. Lett. **43** 29 (1998)

Introduction to HVBK Dynamics

Darryl D. Holm

Theoretical Division and Center for Nonlinear Studies, Los Alamos National
Laboratory, MS B284, Los Alamos, NM 87545

Abstract. We review the Hall-Vinen-Bekarevich-Khalatnikov (HVBK) equations for
superfluid Helium turbulence and discuss their implications for recent measurements
of superfluid turbulence decay.

A new Hamiltonian formulation of these equations renormalizes the vortex line
velocity to incorporate finite temperature effects. These effects also renormalize the
coupling constant in the mutual friction force between the superfluid and normal fluid
components by a factor of ρ_s/ρ (the superfluid mass fraction) but they leave the vortex
line tension unaffected. Thus, the original HVBK form is recovered at zero temperature and its mutual friction coefficients are renormalized at nonzero temperature. The
HVBK equations keep their form and no new parameters are added. However, a temperature dependent trade-off does arise between the mutual friction coupling and the
vortex line tension.

The renormalized HVBK equations obtained via this new Hamiltonian approach
imply a dynamical equation for the space-integrated vortex tangle length, which is the
quantity measured by second sound attenuation experiments in superfluid turbulence.
A Taylor-Proudman theorem also emerges for the superfluid vortices that shows the
steady vortex line velocity becomes columnar under rapid rotation.

1 HVBK Equations

Recent experiments establish the Hall-Vinen-Bekarevich-Khalatnikov (HVBK)
equations as a leading model for describing superfluid Helium turbulence. See
Nemirovskii and Fiszdon [1995] and Donnelly [1999] for authoritative reviews.
See Henderson and Barenghi [2000] for a recent fluid mechanics study of steady
cylindrical Couette flow using computer simulations of the incompressible HVBK
equations.

In the Galilean frame of the normal fluid with velocity $\mathbf{v}_n$, the HVBK equations may be expressed as follows, upon ignoring thermal diffusivity and viscosity,

$$\partial_t \rho = -\operatorname{div}\mathbf{J}\,, \qquad \partial_t J_i = -\partial_k T_i^k\,, \qquad \partial_t S = -\operatorname{div}(S\mathbf{v}_n) + R/T\,,$$

$$\rho_s \partial_t \mathbf{v}_s + \rho_s(\mathbf{v}_s \cdot \nabla)\mathbf{v}_s = -\rho_s \nabla\left(\mu - \frac{1}{2}|\mathbf{v}_s - \mathbf{v}_n|^2\right) + \rho_s \mathbf{f}\,, \tag{1}$$

and summing over pair of upper and lower repeated indices. One may consult,
e.g., Bekarevich and Khalatnikov (BK) [1961] and Donnelly [1999] to compare
these equations with the form they take in the reference frame of the superfluid.[1]

---

[1] In making this comparison it is useful to recall the Galilean transformation of the
chemical potential, $\mu' = \mu - \frac{1}{2}|\mathbf{v}_s - \mathbf{v}_n|^2$, where μ' is evaluated in the superfluid

Notation. Here ρ and ρ_s denote the total and superfluid mass densities, respectively. The entropy density of the normal fluid is S, its temperature is denoted T, and $\rho_n = \rho - \rho_s$ denotes its mass density. The superfluid velocity is denoted $\mathbf{v}_s$ and

$$\mathbf{J} = \rho_s \mathbf{v}_s + \rho_n \mathbf{v}_n$$

is the total momentum density. In the entropy equation R is the rate that heat is produced by the phenomenological friction and reactive forces in $\mathbf{f}$, which must be specified to close the theory. We also denote

Stress tensor: $\quad T_i^k = \rho_s\, v_{s\,i}\, v_s^k + \rho_n\, v_{n\,i}\, v_n^k + (P + \lambda\cdot\omega)\,\delta_i^k - \lambda_i \omega^k\,,$

Euler's pressure law: $\quad P = -\varepsilon_0 + TS + \rho\mu\,,$

Superfluid First Law: $\quad d\varepsilon_0 = \mu\, d\rho + T dS + (\mathbf{J} - \rho\mathbf{v}_n)\cdot d(\mathbf{v}_s - \mathbf{v}_n) + \lambda\cdot d\omega\,.$ (2)

In the superfluid First Law, $\omega = \mathrm{curl}\,\mathbf{v}_s$ is the superfluid vorticity with magnitude $|\omega| = \hat\omega\cdot\omega$, where $\hat\omega = \omega/|\omega|$ is its unit vector. BK [1961] takes the energy density ε_0 to depend on the magnitude of the superfluid vorticity, $|\omega|$, as

$$\varepsilon_0 = \frac{\rho_s \kappa\, |\omega|}{4\pi}\ln\frac{R}{a}\,.$$

This is the energy per unit length of a superfluid vortex line, $\rho_s(\kappa^2/4\pi)\ln(R/a)$, with quantum of circulation $\kappa = h/m \simeq 10^{-3}(\mathrm{cm}^2/\mathrm{sec})$ and ratio R/a of mean distance between vortices R to effective vortex radius a, times the vortex length per unit volume, $|\omega|/\kappa$. Hence, we find

$$\lambda = \frac{\partial\varepsilon_0}{\partial|\omega|}\frac{\partial|\omega|}{\partial\omega} = |\lambda|\,\hat\omega \quad \text{with} \quad |\lambda|/\rho_s \approx \frac{\kappa}{4\pi}\ln\frac{R}{a} = \lambda_0\,,$$

where one ignores the derivative of $R \simeq \sqrt{\kappa/|\omega|}$ inside the logarithm. The appearance of λ in the stress tensor T_i^k shifts the pressure P, and $\mathrm{div}\,T$ introduces an additional force $-\omega\cdot\nabla\lambda \equiv -\rho_s\mathbf{T}$ into the motion equation. The quantity $\mathbf{T}$ is called the "vortex line tension."

BK [1961] assigned the following form to the phenomenological coupling force $\mathbf{f}$ appearing in the superfluid velocity equation in (1),

$$\mathbf{f} = (\mathbf{v}_L - \mathbf{v}_s) \times \omega\,, \qquad \mathbf{v}_L = \mathbf{v}_n - \rho_s\,(\alpha\, \mathbf{s}^0 + \beta\hat\omega \times \mathbf{s}^0)\,, \qquad (5)$$
$$\text{where} \qquad \mathbf{s}^0 = \mathbf{v}_\ell^0 - \mathbf{v}_n\,, \qquad \mathbf{v}_\ell^0 = \mathbf{v}_s + \rho_s^{-1}\mathrm{curl}\,\lambda\,, \qquad (6)$$

with "vortex velocity" $\mathbf{v}_L$ and "slip velocity" $\mathbf{s}^0$ introduced as auxiliary quantities. The HVBK equations written as (1) in the normal-fluid frame conserve the energy,

$$E = \int \left[\frac{1}{2}\rho|\mathbf{v}_n|^2 + (\mathbf{J} - \rho\mathbf{v}_n)\cdot\mathbf{v}_n + \varepsilon_0\right]d^3x\,. \qquad (7)$$

---

frame and μ in the normal-fluid frame. See Putterman [1974] for a clear discussion of the role of Galilean transformations in superfluid hydrodynamics.

The BK [1961] form of the phenomenological force $\mathbf{f}$ also implies the dissipative heating rate,

$$R = (\mathbf{J} - \rho\mathbf{v}_n + \operatorname{curl}\lambda) \cdot \omega \times (\mathbf{v}_L - \mathbf{v}_n),$$

which is Galilean invariant and positive. Substituting this form of $\mathbf{f}$ into the superfluid motion equation in (1) and taking its curl provides the following equation for the superfluid vorticity, $\omega = \operatorname{curl}\mathbf{v}_s$,

$$\partial_t \omega = \operatorname{curl}(\mathbf{v}_L \times \omega).$$

This vorticity equation implies the **HVBK superfluid Kelvin circulation theorem**,

$$\frac{d}{dt}\iint_S \omega \cdot d\mathbf{S} = \frac{d}{dt}\oint_{\partial S(\mathbf{v}_L)} \mathbf{v}_s \cdot d\mathbf{x} = 0.$$

The Kelvin formula (10) expresses conservation of the flux of superfluid vorticity through any surface S whose boundary ∂S moves with the velocity $\mathbf{v}_L$, so $\mathbf{v}_L$ may be regarded as the local velocity of a vortex line. Equivalently, the Kelvin formula expresses conservation of superfluid velocity circulation around any loop that moves with the vortex line velocity $\mathbf{v}_L$.

In a key phenomenological step that closed the theory, BK [1961] assigned the undetermined functions α and β in the force $\mathbf{f}$ and its auxiliary vortex line velocity $\mathbf{v}_L$ as

$$1 + \alpha\rho_s = \frac{B'\rho_n}{2\rho}, \quad \beta\rho_s = \frac{B\rho_n}{2\rho}. \tag{11}$$

The *dimensionless* coefficients B and B' were introduced earlier in Hall and Vinen (HV) [1956] to parameterize the Gorter-Mellink [1949] "mutual friction" force (B) and its reactive component (B'). Hence the name, **HVBK equations** for this closure.

The assignments in BK [1961] of the undetermined functions α, β in (11), as well as the vortex slip velocity $\mathbf{s}^0$ in (6) are designed to reproduce the phenomena observed in HV [1956] and yet still conserve mass, momentum and energy. This phenomenological approach used in BK [1961] is indeterminate, however, in the sense that some freedom still remains in making these assignments. The HVBK equations (1) that result from this approach do possess the desired conservation laws for mass, momentum and energy. And they also possess a Kelvin theorem for the circulation of superfluid velocity. However, because of the indeterminacy inherent in the phenomenological approach, the HVBK equations (1) are not unique in possessing these properties. An alternative assignment of the vortex slip velocity is

$$\mathbf{s} = \mathbf{v}_\ell - \mathbf{v}_n \quad \text{with} \quad \mathbf{v}_\ell = \bar{\mathbf{v}} + \rho^{-1}\operatorname{curl}\lambda, \quad \text{with} \quad \bar{\mathbf{v}} = \rho^{-1}(\rho_s\mathbf{v}_s + \rho_n\mathbf{v}_n). \tag{12}$$

The mean velocity $\bar{\mathbf{v}}$ also figures prominently in Hills and Roberts [1977] discussion of the HVBK equations. As we shall see, the alternative expression (12) for the auxiliary vortex slip velocity in terms of the mean velocity $\bar{\mathbf{v}}$ arises naturally in the Hamiltonian derivation of a slightly modified set of HVBK equations.

These equations possess the same formal conservation and circulation properties as HVBK, modulo redefining the vortex slip velocity as $\mathbf{s}$ rather than $\mathbf{s}^0$. The vortex slip velocity $\mathbf{s}$ in (12) is defined relative to the Galilean frame of the normal fluid, which is present only at finite temperature. The HVBK $\mathbf{s}^0$ in (6) is the limit of the vortex slip velocity $\mathbf{s}$ for zero temperature, at which no normal fluid remains. The vortex slip velocity $\mathbf{s}$ in (12) is a slight modification of $\mathbf{s}^0$ in (6) necessary to incorporate finite temperature effects, without changing the form of the HVBK theory, obtained from a Hamiltonian derivation of these equations in the normal fluid frame. In the Hamiltonian framework, the energy-momentum conservation laws and Kelvin circulation theorem are all natual consequences. Moreover, the velocities $\mathbf{v}_\ell$ and $\mathbf{v}_n$ are identified as being dual to the momenta given by $\rho \mathbf{v}_s$ and $\rho_n(\mathbf{v}_n - \mathbf{v}_s)$, respectively.

Outline. We shall use a Hamiltonian approach with Lie-Poisson brackets to derive the expression (12) for the vortex line velocity $\mathbf{v}_\ell$ at finite temperature from first principles by using the energy E in (7) as the Hamiltonian. The momenta conjugate to the velocities $\mathbf{v}_\ell$ and $\mathbf{v}_n$ shall be our basic dynamical variables. The finite temperature vortex line velocity $\mathbf{v}_\ell$ and slip velocity $\mathbf{s}$ determined this way turn out to be

$$\mathbf{v}_\ell = \rho^{-1}(\mathbf{J} + \operatorname{curl} \lambda) , \quad \text{and} \quad \mathbf{s} = \mathbf{v}_\ell - \mathbf{v}_n \simeq \left(\frac{\rho_s}{\rho}\right)(\mathbf{v}_\ell^0 - \mathbf{v}_n) . \qquad (13)$$

At zero temperature, $\rho \to \rho_s$ and these reduce to the BK [1961] phenomenological expressions with $\mathbf{v}_\ell^0$ given by (6). Thus, the finite temperature corrections found by using the Hamiltonian approach **renormalize** the HVBK slip velocity in the mutual friction force $\mathbf{f}$ by the factor ρ_s/ρ (the superfluid mass fraction). Aside from this renormalization, the vortex line tension is left unaffected by this renormalization, the superfluid vortex equation keeps its form and no new parameters are added.

Technical details of deriving this renormalized theory from its Hamiltonian and Lie-Poisson brackets are given in the Appendix.

Main results. We shall use the superfluid vortex dynamics for the renormalized HVBK equations obtained via this Hamiltonian approach to write a dynamical equation for the space-integrated total **vortex tangle length**, which is the quantity measured in the Oregon experiments on superfluid turbulence reported in Skrbek, Niemela and Donnelly [1999].

We shall also study the restriction of the renormalized HVBK equations for the **incompressible case**, in which ρ_n and ρ_s are constants and one takes $\nabla \cdot \mathbf{v}_n = 0$ and $\nabla \cdot \mathbf{v}_s = 0$. Finally, we shall demonstrate the invariance of the forms of these equations upon transforming into a rotating frame. The Coriolis force in such a rotating frame couples to the vortex line velocity $\mathbf{v}_L$, which of course differs from both the superfluid velocity and the normal velocity. We shall derive a **Taylor-Proudman theorem** for steady superfluid vortices under rapid rotation. According to this superfluid Taylor-Proudman theorem, the vortex line velocity becomes columnar under sufficiently rapid rotation. That is, the lateral vortex line velocity is nondivergent and independent of the axial coordinate, and the axial velocity decouples from the lateral motion. Therefore, under sufficiently rapid rotation, the superfluid vortex filaments will straighten and become parallel to the axis of rotation as they approach a steady state.

Numerical implications. This renormalization of the vortex line element slip velocity in the HVBK equations from $\mathbf{s}^0 \to \mathbf{s} \simeq \mathbf{s}^0 \rho_s/\rho$ is sensitive to temperature, but it does not affect the vortex line tension. Therefore, a temperature sensitive trade-off arises between mutual friction and vortex line tension that may be worth testing in numerical simulations such as those reported in Henderson and Barenghi [2000]. The HVBK equations are thought to break down in the presence of strong counterflow. However, as general conservation laws there is no mechanism in the equations that would signal this breakdown. A rotating Rayleigh-Besnard experiment might be useful in testing the range of validity of the HVBK equations (Barenghi, private communication). Such an experiment might also indicate how these equations should be modified in the presence of strong counterflow.

Experimental implications. Temperature sensitivity of the coupling between the superfluid vortices and the normal fluid component is an area of intense current investigation in superfluid Helium turbulence, see Donnelly [1999]. One would like to know whether the ρ_s/ρ renormalization of the mutual friction forces relative to the vortex line tension would matter significantly in comparisons of the predictions of the HVBK equations with modern experiments in Helium turbulence at low, but finite temperatures.

Superfluid vortex dynamics. To begin addressing this issue, we may use the superfluid vorticity equation for the renormalized HVBK equations obtained in the Appendix via the Hamiltonian approach to write an explicit equation for the dynamics of **Vinen's vortex length density** $L = |\omega|/\kappa$. In the superfluid turbulence decay experiments reported by Skrbek, Niemela and Donnelly [1999] the spatial integral of this quantity is measured as a function of time to decrease over *six decades* as $t^{-3/2}$. The integrated vortex length measured in these experiments is predicted by the renormalized HVBK equations to be governed by the superfluid vorticity dynamics alone.

Upon including mutual friction, the superfluid vortex dynamics for the renormalized HVBK equations is expressed as, cf. equation (9),

$$\partial_t \omega = \mathrm{curl}(\mathbf{v}_L \times \omega),$$

in which the renormalized total vortex line velocity given by, cf. equation (5),

$$\mathbf{v}_L = \mathbf{v}_\ell - \frac{B'\rho_n}{2\rho}\mathbf{s} - \frac{B\rho_n}{2\rho}\hat{\omega} \times \mathbf{s}, \quad \text{where} \quad \mathbf{s} = \mathbf{v}_\ell - \mathbf{v}_n, \tag{15}$$

and its Hamiltonian limit is found to be

$$\mathbf{v}_\ell = \bar{\mathbf{v}} + \rho^{-1}\mathrm{curl}\,\lambda, \quad \text{with} \quad \bar{\mathbf{v}} = \rho^{-1}\mathbf{J} \quad \text{and} \quad \lambda = \lambda\hat{\omega}. \tag{16}$$

Thus, relative to the Hamiltonian approach, the terms in B and B' are additional velocities introduced by phenomenology, while $\mathbf{v}_\ell$ is the vortex line velocity in the absence of mutual friction.

The HVBK superfluid vorticity equation implies the following dynamics for the integrated vortex length measured in the turbulence decay experiments,

$$\frac{d}{dt} \underbrace{\int L\, d^3x}_{\textbf{length}} = \underbrace{\int \hat{\omega} \cdot \partial_t \omega / \kappa \, d^3x}_{\textbf{vorticity dynamics}}$$

$$= \underbrace{\int L\, \tilde{\mathbf{v}} \cdot (\hat{\omega} \times \operatorname{curl} \hat{\omega}) \, d^3x}_{\textbf{transport} \cdot \textbf{curvature}}$$

$$- \underbrace{\int L\, \frac{B\rho_n}{2\rho^2} \, (\hat{\omega} \times \operatorname{curl} \lambda \hat{\omega}) \cdot (\hat{\omega} \times \operatorname{curl} \hat{\omega}) \, d^3x}_{\textbf{damping by curvature}}$$

$$+ \underbrace{\oint L\, (\hat{\mathbf{n}} \times \hat{\omega}) \cdot \left((\hat{\omega} \times \mathbf{v}_\ell) + \frac{B\rho_n}{2\rho} (\mathbf{v}_\ell - \mathbf{v}_s) \right) dS}_{\textbf{creation and destruction at the boundary}} . \quad (17)$$

Here $\hat{\omega}$ is the unit vector tangent to a superfluid vortex filament, so $\kappa = (\hat{\omega} \times \operatorname{curl} \hat{\omega})$ is its local curvature. The transport and damping of the vortex tangle length is proportional to this local curvature. The effective velocity $\tilde{\mathbf{v}}$ in the transport term is given by

$$\tilde{\mathbf{v}} = \overline{\mathbf{v}} - \frac{B'\rho_n}{2\rho} \left(\overline{\mathbf{v}} - \mathbf{v}_n \right) - \frac{B\rho_n}{2\rho} \hat{\omega} \times \left(\overline{\mathbf{v}} - \mathbf{v}_n \right). \quad (18)$$

According to the last term in (17), vortex length is created or destroyed at the boundary, unless the vortex filaments approach it in the normal direction, so that $\hat{\mathbf{n}} \times \hat{\omega} = 0$.

Formula (17) for the evolution of the total superfluid vortex length presents a trade-off between the mass-weighted velocity $\overline{\mathbf{v}}$ and the local induction velocity (or filament curvature) $\hat{\omega} \times \operatorname{curl} \hat{\omega}$, in the interior of the domain. This trade-off in the interior competes with the process of creation and destruction at the boundary. For example, in counterflow turbulence, the superfluid moves toward the heater at the boundary, so the term in $\overline{\mathbf{v}}$ would tend to be nonzero. In contrast, for grid turbulence, $\overline{\mathbf{v}}$ is small, so this term would tend to contribute less. This formula governs the dynamics of the experimentally measured quantity $\int L\, d^3x$. However, it does not yet show how to obtain the $t^{-3/2}$ decrease seen in this quantity by Skrbek, Niemela and Donnelly [1999] in their experiments on decay of turbulence.

Suppose the main source of decay were the term labeled "damping by curvature" in formula (17) and the flow were isothermal and incompressible. This would imply

$$\frac{1}{\langle L \rangle} \frac{d}{dt} \langle L \rangle = -c_0(T)\, \lambda_0 \, \frac{\langle L/R^2 \rangle}{\langle L \rangle} = -\frac{3}{2\,(t+t_0)}, \quad (19)$$

where $\lambda_0 = \lambda/\rho_s = (\kappa/4\pi)ln(b/a)$ is the quantum vortex constant, t_0 is a time shift in the experimental analysis, $c_0(T) \equiv B\rho_n\rho_s/(2\rho^2)$ and angle brackets $\langle \cdot \rangle$ denote spatial integral over the measurement domain. In particular,

$$\langle L \rangle \equiv \int L \, d^3x, \quad \langle L/R^2 \rangle \equiv \int L \, |\hat{\omega} \times \text{curl}\,\hat{\omega}|^2 \, d^3x. \tag{20}$$

The measured $t^{-3/2}$ decrease in $\langle L \rangle$ implies via formula (19) that the length-weighted mean curvature of the vortex tangle $\langle L/R^2 \rangle/\langle L \rangle$ decays due to mutual friction as t^{-1}. Thus, on the average **as the vortex length decays, the vortices tend to straighten**, under the effects of mutual friction damping.

Preservation of helicity versus preservation of vortex length. The helicity, or linkage number for the superfluid vorticity is defined as

$$\Lambda = \int (\mathbf{v}_s \cdot \omega) \, d^3x.$$

The helicity satisfies an evolution equation obtained from the superfluid vortex dynamics,

$$\frac{d\Lambda}{dt} = - \oint (\hat{\mathbf{n}} \cdot \omega)\left(\mu - \frac{1}{2}v_n^2 - \mathbf{v}_s \cdot (\mathbf{v}_L - \mathbf{v}_n)\right) dS - \oint (\hat{\mathbf{n}} \cdot \mathbf{v}_L)(\mathbf{v}_s \cdot \omega) \, dS.$$

Therefore, even with mutual friction, helicity is created and destroyed only on the boundary. Moreover, helicity will be preserved, provided both ω and $\mathbf{v}_L$ are *tangential at the boundary*. The former condition, however, is the opposite of that required for the creation and destruction of vortex length at the boundary to cease. Therefore, no equilibrium should be expected that preserves both the helicity and the vortex length in a superfluid.

Superfluid vortex equilibria are not ABC flows. The steady equilibrium solutions of the superfluid vorticity dynamics satisfy

$$\text{curl}\,(\mathbf{v}_L \times \omega) = 0. \tag{23}$$

For example, a steady equilibrium exists when ω and $\mathbf{v}_L$ are parallel. Note that these "super-Beltrami flows" are *not* eigenfunctions of the curl. Therefore, they are not Arnold-Beltrami-Childress (ABC) flows, as occur for the Euler equations.

2   Incompressible Renormalized HVBK Flows

To express the renormalized HVBK equations in the incompressible limit, we begin by recollecting the compressible equations and abbreviating $|\mathbf{v}_s - \mathbf{v}_n|^2 = v_{sn}^2$,

$$\partial_t S \qquad\qquad = - \text{div}(S\mathbf{v}_n) + R/T,$$
$$\partial_t \rho \qquad\qquad = - \text{div}(\rho_s\mathbf{v}_s + \rho_n\mathbf{v}_n),$$
$$\rho_s\big(\partial_t\mathbf{v}_s + (\mathbf{v}_s \cdot \nabla)\mathbf{v}_s\big) = - \rho_s\nabla\big(\mu - \frac{1}{2}v_{sn}^2\big) + \rho_s(\mathbf{v}_L - \mathbf{v}_s) \times \omega,$$
$$\partial_t\big(\rho_s v_{s\,i} + \rho_n v_{n\,i}\big) \quad = - \partial_k\big(\rho_n v_{n\,i}v_n^k + \rho_s v_{s\,i}v_s^k\big) - \partial_i P - \partial_k\tau_i^k,$$
$$\tau_i^k \qquad\qquad = \epsilon^{klm}\partial_l(v_{s\,i}\lambda_m) - \lambda_i \omega^k + \delta_i^k \lambda \cdot \omega.$$

As we have seen, finite temperature effects renormalize the total vortex line velocity as

$$\mathbf{v}_L = \mathbf{v}_\ell - \frac{\rho_n}{\rho}\left(\frac{B}{2}\,\hat{\omega} \times \mathbf{s} + \frac{B'}{2}\mathbf{s}\right), \quad \text{where} \quad \mathbf{s} = \mathbf{v}_\ell - \mathbf{v}_n\,, \qquad (24)$$

and the Hamiltonian part of the line velocity (with corrections for finite temperature) is defined as

$$\mathbf{v}_\ell = \bar{\mathbf{v}} + \rho^{-1}\mathrm{curl}\,\lambda\,, \quad \text{with} \quad \bar{\mathbf{v}} = \rho^{-1}(\rho_s\mathbf{v}_s + \rho_n\mathbf{v}_n) \quad \text{and} \quad \lambda = \lambda\hat{\omega}\,. \qquad (25)$$

To the extent that ρ, ρ_s, ρ_n and S all may be taken as constants for a given temperature and the heating rate R is negligible, then the velocities $\mathbf{v}_n$ and $\mathbf{v}_s$ are incompressible, i.e.,

$$\nabla \cdot \mathbf{v}_n = 0 \quad \text{and} \quad \nabla \cdot \mathbf{v}_s = 0\,.$$

In this situation, the pressure P may be obtained from the **Poisson equation**,

$$-\nabla^2(P + \lambda \cdot \omega) = \mathrm{div}\Big(\rho_s(\mathbf{v}_s \cdot \nabla)\,\mathbf{v}_s + \rho_n(\mathbf{v}_n \cdot \nabla)\,\mathbf{v}_n - \omega \cdot \nabla\lambda\Big)\,, \qquad (27)$$

found from the divergence of the total momentum equation. Combining the superfluid motion equation with total momentum conservation results in an equation for the normal fluid velocity in the incompressible case

$$\rho_n\partial_t\mathbf{v}_n + \rho_n(\mathbf{v}_n \cdot \nabla)\mathbf{v}_n = -\nabla\Big(P' - \rho_s\mu + \frac{\rho_s}{2}v_{sn}^2\Big) - \rho_s(\mathbf{v}_L - \mathbf{v}_s) \times \omega + \omega \cdot \nabla\lambda\,,$$

where $\mathbf{v}_L$ is given in equation (24). We set $P' \equiv P + \lambda \cdot \omega$ and take it as the total pressure. (One also could have absorbed $\lambda \cdot \omega$ into P earlier, by including it in Euler's pressure law.) Since $\lambda = |\lambda|\hat{\omega}$ and $\hat{\omega}$ is a unit vector, we find *for constant* ρ_s the standard relation for the **vortex line tension** denoted as **T**. Namely,

$$\omega \cdot \nabla\lambda = -\lambda_0\,\rho_s\,\omega \times \mathrm{curl}\,\hat{\omega} \equiv \rho_s\mathbf{T}\,,$$

where $\lambda_0 = \lambda/\rho_s = (\kappa/4\pi)\ln(b/a)$ is a constant.

Remark. We note that the quantity **T** known as the vortex line tension first appears in the *normal fluid equation*, as a reaction to the presence of the superfluid. The standard convention for introducing the mutual friction force has the effect of shifting **T** into the superfluid equation. By action and reaction, though, **T** could appear in either equation.

These equations of motion must be completed by providing an equation of state relation for the quantity $\mu - \frac{1}{2}v_{sn}^2$. BK [1961] assumes a law of partial pressures,

$$P_n = \frac{\rho_n}{\rho}P' = P' - P_s \quad \text{and} \quad P_s = \frac{\rho_s}{\rho}P' = \rho_s\mu - \frac{1}{2}\rho_s v_{sn}^2\,.$$

In this case, the renormalized HVBK motion equations for incompressible flow reduce to

$$\partial_t \mathbf{v}_s + (\mathbf{v}_s \cdot \nabla)\mathbf{v}_s = -\frac{1}{\rho}\nabla P' - \frac{\rho_n}{\rho}\mathbf{F}_{ns} + \mathbf{T}, \qquad (31)$$

$$\partial_t \mathbf{v}_n + (\mathbf{v}_n \cdot \nabla)\mathbf{v}_n = -\frac{1}{\rho}\nabla P' + \frac{\rho_s}{\rho}\mathbf{F}_{ns}. \qquad (32)$$

In these superfluid motion equations, the **renormalized mutual friction force** $\mathbf{F}_{ns}$ is defined as the sum (with $\omega = \operatorname{curl} \mathbf{v}_s$)

$$\mathbf{F}_{ns} = (\mathbf{s}^0 \times \omega) + \left(\frac{\rho_s}{\rho}\right)\mathbf{F}_{ns}^0, \quad \text{where} \quad \mathbf{F}_{ns}^0 = \left(\frac{B}{2}\hat{\omega} \times \mathbf{s}^0 + \frac{B'}{2}\mathbf{s}^0\right) \times \omega. \quad (33)$$

Here $\mathbf{F}_{ns}^0$ is the HVBK mutual friction force without any finite temperature corrections. To acquire these formulas, we used the relations for the incompressible case,

$$\mathbf{s} = \mathbf{v}_\ell - \mathbf{v}_n = \frac{\rho_s}{\rho}(\mathbf{v}_s + \lambda_0 \operatorname{curl}\hat{\omega} - \mathbf{v}_n) = \frac{\rho_s}{\rho}(\mathbf{v}_\ell^0 - \mathbf{v}_n) = \frac{\rho_s}{\rho}\mathbf{s}^0, \qquad (34)$$

with $\mathbf{s}^0 = \mathbf{v}_s + \lambda_0 \operatorname{curl}\hat{\omega} - \mathbf{v}_n$, and we eliminated $\mathbf{v}_L$ by using the relation

$$-\rho_s(\mathbf{v}_L - \mathbf{v}_s - \lambda_0 \operatorname{curl}\hat{\omega}) = \rho_n \mathbf{s} + \frac{\rho_s \rho_n}{\rho}\left(\frac{B}{2}\hat{\omega} \times \mathbf{s} + \frac{B'}{2}\mathbf{s}\right). \qquad (35)$$

In equation (33) for $\mathbf{F}_{ns}$, the quantity $(\mathbf{s}^0 \times \omega)$ is the **Hamiltonian reactive force** (which could be naturally absorbed into Vinen's B' parameter) and $\mathbf{F}_{ns}^0$ is the phenomenological mutual friction force defined according to the standard convention as in BK [1961] and Donnelly [1999]. The finite-temperature corrections contribute an **overall factor** of ρ_s/ρ to the standard zero-temperature expression $\mathbf{F}_{ns}^0$ for the phenomenological mutual friction force. No new parameters are added, but a **temperature dependent trade-off** is identified between the renormalized mutual friction coupling and the vortex line tension, since the vortex line tension remains *unaffected* by the finite-temperature corrections.

In the isothermal case, the motion equations are closed by the Poisson equation for P, since the other coefficients (B, B', ρ_n/ρ, etc.) are specified functions of temperature and they may be taken as constants, for an isothermal incompressible superfluid flow.

Note that equations (31-32) may be rewritten with $\omega_n = \operatorname{curl}\mathbf{v}_n$ as

$$\partial_t \mathbf{v}_s + \nabla \mu_s = \mathbf{v}_n \times \omega + \frac{\rho_s}{\rho}\mathbf{s}^0 \times \omega - \frac{\rho_n \rho_s}{\rho^2}\mathbf{F}_{ns}^0, \quad \mu_s = P'/\rho + \frac{1}{2}v_s^2, \quad (36)$$

$$\partial_t \mathbf{v}_n + \nabla \mu_n = \mathbf{v}_n \times \omega_n + \frac{\rho_s}{\rho}\mathbf{s}^0 \times \omega + \frac{\rho_s^2}{\rho^2}\mathbf{F}_{ns}^0, \quad \mu_n = P'/\rho + \frac{1}{2}v_n^2. \quad (37)$$

These equations imply an equation for the **velocity difference**,

$$\partial_t(\mathbf{v}_s - \mathbf{v}_n) + \frac{1}{2}\nabla(v_s^2 - v_n^2) = \mathbf{v}_n \times (\omega - \omega_n) - \frac{\rho_s}{\rho}\mathbf{F}_{ns}^0 \quad \text{with} \quad \omega_n = \operatorname{curl}\mathbf{v}_n. \quad (38)$$

and there is *no tendency* for mutual friction to cause any alignment in the vorticities of the superfluid and its normal component. Instead, the curl $\hat{\omega}$ part of $\mathbf{F}_{ns}^0 \neq 0$ would break any such alignment, if it were to form spontaneously. Indeed, **alignments <u>sufficient</u> for steady solutions** are

$$\mathbf{s}^0 \times (\nabla \mu_s \times \nabla \mu_n) = 0, \quad \mathbf{s}^0 \times \omega = 0 \quad \text{and} \quad \mathbf{s}^0 \times \mathbf{v}_n = 0, \qquad (39)$$
$$\text{with} \quad \mathbf{s}^0 \equiv \mathbf{v}_s + \lambda_0 \operatorname{curl} \hat{\omega} - \mathbf{v}_n ,$$

provided μ_s and μ_n are functionally unrelated. Thus, the steady equilibrium alignments imposed by mutual friction involve $\mathbf{v}_n$, $\mathbf{v}_s$ and curl $\hat{\omega}$, as well as the independent gradients of μ_s and μ_n. For example, one class of equilibria has $\mathbf{s}^0$, $\mathbf{v}_n$, ω all aligned tangent to intersections of level surfaces of μ_n and μ_s.

3   Rotating Frame Renormalized HVBK Equations

We transform to a rotating frame with relative velocities denoted with an asterisk as $\mathbf{v}_s^* = \mathbf{v}_s - \mathbf{R}(\mathbf{x})$, etc., and curl $\mathbf{R} = 2\Omega$. After a calculation involving Legendre transformations, we obtain the Hamiltonian for the relative motion, cf. the Hamiltonian in (61) of the Appendix,

$$h(\mathbf{M}^*, \rho, S, \mathbf{u}, \mathbf{A}^*, n) = \int \left\{ \frac{1}{2} \rho \, |\mathbf{v}_n^* + \mathbf{R}(\mathbf{x})|^2 + (\mathbf{M}^* - \rho \mathbf{A}^* - \rho \mathbf{v}_n^*) \cdot \mathbf{v}_n^* \quad (40) \right.$$
$$\left. + \varepsilon_0(\rho, S, \mathbf{v}_s^* - \mathbf{v}_n^*, \omega^* + 2\Omega) - \mathbf{R} \cdot \left[\rho(\mathbf{v}_n^* + \mathbf{R}) + (\rho - n)(\mathbf{A}^* - \mathbf{R}) \right] \right\} d^3 x .$$

Here $\mathbf{M}^* - \rho \mathbf{A}^* = \mathbf{J}^* = \mathbf{J} - \rho \mathbf{R}$ and $\mathbf{v}_s = \mathbf{u} - (\mathbf{A}^* - \mathbf{R})$. The equations resulting from the Lie-Poisson bracket (55) of the Appendix in these relative variables keep their forms and the condition $n = \rho$ is still preserved. We conclude with the following three remarks.

Superfluid Coriolis force couples to the vortex line velocity. The Hamiltonian evolution equation for the superfluid velocity in the rotating frame is expressed as

$$\partial_t \mathbf{v}_s^* + (\mathbf{v}_s^* \cdot \nabla)\mathbf{v}_s^* = -\nabla \Big(\mu - \frac{1}{2}|\mathbf{v}_s^* - \mathbf{v}_n^*|^2 - \frac{1}{2}|\mathbf{R}|^2 \Big) + (\mathbf{v}_\ell^* - \mathbf{v}_s^*) \times \omega^* + \mathbf{v}_\ell^* \times 2\Omega .$$

The last term is the Coriolis force and it involves the relative vortex line velocity. The curl of this equation yields

$$\partial_t(\omega^* + 2\Omega) = \operatorname{curl} \left(\mathbf{v}_\ell^* \times (\omega^* + 2\Omega) \right).$$

The form of the vortex dynamics equation is **invariant** under passing to a steadily rotating frame, and the superfluid Coriolis force contains the renormalized vortex line velocity, rather than the superfluid velocity. Therefore, this is not merely a kinematic force. The vortex line velocity appearing in the superfluid Coriolis force includes the interaction between the vortex lines and the superfluid component. It also includes the interaction with the normal component, since $\mathbf{v}_\ell$ depends on the relative momentum density and contains finite temperature effects. The superfluid Coriolis force is essential in the spin up problem in He-II, see, e.g., Reissenegger [1993].

Superfluid Taylor-Proudman theorem. For steady, or slow motions and rapid rotation we have

$$0 = \operatorname{curl}\left(\mathbf{v}_\ell^* \times 2\Omega\right).$$

If the rotation is uniform ($\nabla\Omega = 0$) and oriented vertically ($\Omega = |\Omega|\hat{\mathbf{z}}$) this becomes

$$0 = 2|\Omega|\left(\partial_z\mathbf{v}_\ell^* - \hat{\mathbf{z}}\operatorname{div}\mathbf{v}_\ell^*\right) = 2|\Omega|\left(\partial_z v_{\ell x}^*,\ \partial_z v_{\ell y}^*,\ -\partial_x v_{\ell x}^* - \partial_y v_{\ell y}^*\right)^T,$$

where $(\ )^T$ denotes transpose of a row vector into a column vector. Thus, for steady, or slow motions and rapid uniform rotation, we find that **vortex line motion is columnar**. That is, the lateral vortex line velocity is nondivergent and independent of the axial coordinate, and the axial velocity decouples from the lateral motion. Therefore, under sufficiently rapid rotation, the superfluid vortex filaments will straighten and become parallel to the axis of rotation as they approach a steady state. However, they may still undergo nondivergent motion in the lateral plane. This superfluid Taylor-Proudman theorem explains why steady superfluid vortices tend to be aligned with the rotation axis under rapid uniform rotation. The same conclusion applies, if the velocity $\mathbf{v}_\ell^*$ in the Hamiltonian formulation is replaced by the phenomenological relative velocity $\mathbf{v}_L^* = \mathbf{v}_L - \mathbf{R}$. Similar considerations are discussed in Sonin [1987] from a more microscopic viewpoint.

Relative total momentum is not conserved for rotating compressible flows. Since the Hamiltonian depends explicitly on spatial position, instead of conserving relative total momentum, we have the balance

$$\partial_t J_i^* + \partial_j T_i^{*j} = -\left.\frac{\partial\mathsf{h}}{\partial x^i}\right|_{explicit} = \frac{\rho}{2}\partial_i|\mathbf{R}|^2,$$

where h is the Hamiltonian density in equation (40). This relative momentum balance is the effect of centrifugal force. Here we have dropped terms proportional to $\rho - n$, since $\rho = n$ is still preserved in a rotating frame. Consequently, the stress tensor in the relative momentum equation also keeps its form in passing to a rotating frame, although the total relative momentum is no longer conserved if the flow is compressible.

Acknowledgments

I am grateful to H. R. Brand, A. Brandenburg, P. Constantin, R. Donnelly, V. V. Lebedev, F. Lund, J. E. Marsden, J. Niemela, A. Reisenegger, L. Skrbek, K. Sreenivasan and W. F. Vinen for stimulating discussions and encouragement. I am also grateful for hospitality at the UC Santa Barbara Institute for Theoretical Physics where this work was initiated during their Hydrodynamic Turbulence program in spring 2000. This research was supported by the U.S. Department of Energy under contracts W-7405-ENG-36 and the Applied Mathematical Sciences Program KC-07-01-01.

Appendix: Lie-Poisson Hamiltonian Formulation

Conservation of the number of quantum vortices moving through superfluid ^{4}He (and across the streamlines of the normal fluid component) is expressed by

$$\frac{d}{dt} \iint_S \omega \cdot \hat{n}\, dS = 0, \tag{46}$$

where the superfluid vorticity ω is the areal density of vortices and $\hat{n}$ is the unit vector normal to the surface S whose boundary ∂S moves with the vortex line velocity $\mathbf{v}_\ell$. When $\omega = \mathrm{curl}\,\mathbf{v}_s$ this is equivalent to a **vortex Kelvin theorem**

$$\frac{d}{dt} \oint_{\partial S(\mathbf{v}_\ell)} \mathbf{v}_s \cdot d\mathbf{x} = 0, \tag{47}$$

which in turn implies the fundamental relation

$$\partial_t \mathbf{v}_s - \mathbf{v}_\ell \times \omega = \nabla \mu. \tag{48}$$

The superfluid velocity naturally splits into $\mathbf{v}_s = \mathbf{u} - \mathbf{A}$, where $\mathbf{u} = \nabla \phi$ and (minus) the curl of $\mathbf{A}$ yields the superfluid vorticity ω. The phase ϕ is then a regular function without singularities. This splitting will reveal that the Hamiltonian dynamics of superfluid ^{4}He with vortices may be expressed as an invariant subsystem of a larger Hamiltonian system in which $\mathbf{u}$ and $\mathbf{A}$ have independent evolution equations.

We begin by defining a phase frequency in the normal velocity frame as

$$\partial_t \phi + \mathbf{v}_n \cdot \nabla \phi = \nu. \tag{49}$$

The mass density ρ and the phase ϕ are canonically conjugate in the Hamiltonian formulation. Therefore, one may set $\nu = -\delta h/\delta \rho$ for a given Hamiltonian h and the phase gradient $\mathbf{u} = \nabla \phi$ satisfies

$$\partial_t \mathbf{u} + \mathbf{v}_n \cdot \nabla \mathbf{u} + (\nabla \mathbf{v}_n)^T \cdot \mathbf{u} = -\nabla \frac{\delta h}{\delta \rho}, \tag{50}$$

where $(\ )^T$ denotes transpose, so that $(\nabla \mathbf{v}_n)^T \cdot \mathbf{u} = u_j \nabla v_n^j$. The mass density ρ satisfies the dual equation

$$\partial_t \rho + \nabla \cdot (\rho \mathbf{v}_n) = -\nabla \cdot \frac{\delta h}{\delta \mathbf{u}}. \tag{51}$$

Perhaps not surprisingly, the rotational and potential components of the super-fluid velocity $\mathbf{v}_s = \mathbf{u} - \mathbf{A}$ satisfy similar equations, but the rotational component is advected by the vortex line velocity $\mathbf{v}_\ell$, instead of the normal velocity $\mathbf{v}_n$. Absorbing all gradients into $\mathbf{u}$ yields

$$\partial_t \mathbf{A} + \mathbf{v}_\ell \times \omega = 0 \,. \tag{52}$$

Taking the difference of the equations for $\mathbf{u}$ and $\mathbf{A}$ then recovers equation (48) as

$$\partial_t \mathbf{v}_s - \mathbf{v}_\ell \times \omega = -\nabla\left(\mathbf{v}_n \cdot \mathbf{u} + \frac{\delta h}{\delta \rho}\right) \quad \text{with} \quad \mathbf{v}_s = \mathbf{u} - \mathbf{A} \,, \tag{53}$$

in which one uses regularity of the phase ϕ to set $\operatorname{curl}\mathbf{u} = 0$. It remains to determine $\mathbf{v}_\ell$ from the Hamiltonian formulation. Including the additional degree of freedom $\mathbf{A}$ associated with the vortex lines allows them to move relative to both the normal and super components of the fluid, and thereby introduces an additional reactive force without introducing any additional inertia. This Hamiltonian approach thus yields **renormalized HVBK equations**.

Proposition: *Upon splitting the superfluid velocity into* $\mathbf{v}_s = \mathbf{u} - \mathbf{A}$ *(with* $\mathbf{u} = \nabla\phi$ *so that* $\omega = -\operatorname{curl}\mathbf{A}$*) the (renormalized) HVBK equations in the Galilean frame of the normal fluid form an* **invariant subsystem of a Lie-Poisson Hamiltonian system**,

$$\frac{\partial f}{\partial t} = \{f, h\} \quad \text{with} \quad f, h \in (\mathbf{M}, \rho, S, \mathbf{u}, \mathbf{A}, n)\,,$$

and **Lie-Poisson bracket** *given by*

$$\{f, h\} =$$
$$-\int \left\{ \frac{\delta f}{\delta M_j}\left[(M_k \partial_j + \partial_k M_j)\frac{\delta h}{\delta M_k} + \rho\,\partial_j \frac{\delta h}{\delta \rho} + S\partial_j \frac{\delta h}{\delta S} + (\partial_k u_j - u_{k,j})\frac{\delta h}{\delta u_k}\right] \right.$$
$$+ \left[\frac{\delta f}{\delta \rho}\partial_k \rho + \frac{\delta f}{\delta S}\partial_k S + \frac{\delta f}{\delta u_j}(u_k \partial_j + u_{j,k})\right]\frac{\delta h}{\delta M_k} + \left[\frac{\delta f}{\delta \rho}\partial_k \frac{\delta h}{\delta u_k} + \frac{\delta f}{\delta u_j}\partial_j \frac{\delta h}{\delta \rho}\right]$$
$$\left. - \frac{\delta f}{\delta A_j}\left[\partial_j \frac{\delta h}{\delta n} + \frac{A_{j,k} - A_{k,j}}{n}\frac{\delta h}{\delta A_k}\right] - \frac{\delta f}{\delta n}\partial_k \frac{\delta h}{\delta A_k}\right\} d^3 x \,. \tag{55}$$

Remarks: Here $\mathbf{M}$ is the total momentum density, the total mass density is ρ and the entropy density is S. We shall interpret the density n later, after we develop the Hamiltonian equations of motion. It shall emerge that $n = \rho$ is an invariant condition and, hence,

$$\mathbf{M} - n\mathbf{A} = \mathbf{J} = \mathbf{p} + \rho\mathbf{v}_n \,,$$

for $n = \rho$, where $\mathbf{p} = \mathbf{J} - \rho\mathbf{v}_n = \rho_s(\mathbf{v}_s - \mathbf{v}_n)$ is the relative momentum density of the superfluid in the Galilean frame of the normal fluid. The momentum density associated with the vortex fluid will be $\mathbf{N} = -n\mathbf{A}$. The Hamiltonian will be the energy E in (7).

The Lie-Poisson bracket in the Proposition appeared first in Holm and Kupershmidt [1987] in a study of various approximate equations for the dynamics of multicomponent superfluids with charged condensates. The mathematical structure of this Lie-Poisson bracket and its association with the dual of a certain Lie algebra is discussed in Holm and Kupershmidt [1987]. Our re-interpretation of this Poisson bracket introduced and studied earlier shall now yield a extension of the HVBK equations that enables the vortex line velocity $\mathbf{v}_\ell$ and hence the vortex reactive force and mutual friction force to be expressed at finite temperature. Identifying this Poisson bracket as being dual to a Lie algebra establishes that it satisfies the Jacobi identity, $\epsilon_{ijk}\{f_i, \{f_j, f_k\}\} = 0$. The term in the Poisson bracket responsible for the reactive vortex force will turn out to be $\{A_i, A_j\} \neq 0$. The Poisson bracket $\{v_{si}, v_{sj}\}$ would vanish (as does $\{u_i, u_j\} = 0$) and thus the reactive vortex force would be absent, in any Hamiltonian formulation for which $\mathbf{A}$ and n were not independent degrees of freedom from $\mathbf{M}$, ρ, S. Volovik and Dotsenko [1979, 1980] obtain a different result and provide no Lie-algebraic justification.

A Lagrangian formulation of these equations is also available. However, it involves an equation for $\delta l/\delta \nu$ about which nothing is known physically.

Corollary #1: *The Lie-Poisson bracket is equivalent to the following separate Hamiltonian matrix forms for the dynamical equations*

$$\frac{\partial}{\partial t}\begin{bmatrix} M_i \\ S \\ \rho \\ u_i \end{bmatrix} = -\begin{bmatrix} M_j\partial_i + \partial_j M_i & S\partial_i & \rho\partial_i & \partial_j u_i - u_{j,i} \\ \partial_j S & 0 & 0 & 0 \\ \partial_j \rho & 0 & 0 & \partial_j \\ u_j\partial_i + u_{i,j} & 0 & \partial_i & 0 \end{bmatrix}\begin{bmatrix} \delta h/\delta M_j \\ \delta h/\delta S \\ \delta h/\delta \rho \\ \delta h/\delta u_j \end{bmatrix}, \qquad (57)$$

and, upon defining $\mathbf{N} = -n\mathbf{A}$,

$$\frac{\partial}{\partial t}\begin{bmatrix} N_i \\ n \end{bmatrix} = -\begin{bmatrix} N_j\partial_i + \partial_j N_i & n\partial_i \\ \partial_j n & 0 \end{bmatrix}\begin{bmatrix} \delta h/\delta N_j \\ \delta h/\delta n \end{bmatrix}. \qquad (58)$$

These are individually expressed as

$$\begin{aligned}
\partial_t S &= \{S, h\} = -\operatorname{div}(S\,\delta h/\delta\mathbf{M}), \\
\partial_t n &= \{n, h\} = -\operatorname{div}(n\,\delta h/\delta\mathbf{N}), \\
\partial_t \rho &= \{\rho, h\} = -\operatorname{div}(\rho\,\delta h/\delta\mathbf{M} + \delta h/\delta\mathbf{u}), \\
\partial_t \mathbf{u} &= \{\mathbf{u}, h\} = -\nabla\big(\delta h/\delta\rho + (\delta h/\delta\mathbf{M})\cdot\mathbf{u}\big) + (\delta h/\delta\mathbf{M})\times\operatorname{curl}\mathbf{u}, \\
\partial_t (\mathbf{N}/n) &= \{(\mathbf{N}/n), h\} \\
&= -\nabla\big(\delta h/\delta n + (\delta h/\delta\mathbf{N})\cdot(\mathbf{N}/n)\big) + (\delta h/\delta\mathbf{N})\times\operatorname{curl}(\mathbf{N}/n), \\
\partial_t (M_j + N_j) &= \{M_j + N_j, h\} = -\partial_k T_j^{\,k}.
\end{aligned}$$

Corollary #2: *Consider a translation invariant Hamiltonian density with dependence*

$$h(\mathbf{M}, \rho, S, n, \mathbf{v}_s, \omega, \mathbf{A}),$$

where $\mathbf{v}_s = \mathbf{u} - \mathbf{A}$, $\mathbf{A} = -\mathbf{N}/n$ *and* $\omega = \operatorname{curl}\mathbf{v}_s$. *The stress tensor* $T_j^{\,k}$ *is expressed in terms of this Hamiltonian as*

$$T_j^{\,k} = M_j\frac{\partial h}{\partial M_k} + v_{sj}\left(\frac{\partial h}{\partial v_{sk}} + \left(\operatorname{curl}\frac{\partial h}{\partial\omega}\right)_k\right) - v_{sl,j}\epsilon_{mlk}\frac{\partial h}{\partial\omega_m} + \delta_j^k P - A_j\frac{\partial h}{\partial A_k}\bigg|_{\mathbf{v}_s}.$$

where

$$P = M_l \frac{\partial h}{\partial M_l} + \rho \frac{\partial h}{\partial \rho} + S \frac{\partial h}{\partial S} + n \frac{\partial h}{\partial n} - h \, ,$$

as in the Euler relation for pressure.

Remark. One notes many parallels and correspondences among these equations. Note especially the expected similarities in the equations for $\mathbf{u}$ and $\mathbf{N}/n$. Recall that $\mathbf{A} = -\mathbf{N}/n$, so that the superfluid velocity is given by $\mathbf{v}_s = \mathbf{u} - \mathbf{A} = \mathbf{u} + \mathbf{N}/n$. The evolution of the superfluid velocity is consistently composed as the sum of these two separate dynamical pieces.

Proof of the Proposition: The following Hamiltonian h (and conserved energy) will yield the HVBK equations in the frame of the normal fluid upon using this Lie-Poisson bracket

$$h = \int d^3x \left[-\frac{1}{2} \rho v_n^2 + (\mathbf{M} - \rho \mathbf{A}) \cdot \mathbf{v}_n + \varepsilon_0(\rho, S, \mathbf{v}_s - \mathbf{v}_n, \omega) \right]. \tag{61}$$

The variational derivatives of the Hamiltonian h are computed in this reference frame by using the thermodynamic first law (2). Namely,

$$\delta h = \int d^3x \left[\left(\mu - \frac{1}{2} v_n^2 - \mathbf{A} \cdot \mathbf{v}_n \right) \delta \rho + T \delta S + \mathbf{v}_n \cdot \delta \mathbf{M} + (\mathbf{p} + \mathrm{curl}\,\lambda) \cdot \delta \mathbf{u} \right.$$
$$\left. - \left(\mathbf{p} + \mathrm{curl}\,\lambda + \rho \mathbf{v}_n \right) \cdot \delta \mathbf{A} + \left(\mathbf{M} - \mathbf{p} - \rho \mathbf{v}_n - \rho \mathbf{A} \right) \cdot \delta \mathbf{v}_n \right].$$

Here we have used the velocity split $\delta \mathbf{v}_s = \delta \mathbf{u} - \delta \mathbf{A}$ and assumed the boundary condition $\hat{\mathbf{n}} \cdot \omega \times \lambda = 0$ when integrating by parts. This boundary condition is satisfied identically, since $\lambda = \lambda \hat{\omega}$ in the HVBK theory. Upon substituting these variational derivatives into the Lie-Poisson bracket, Corollary #1 yields the following equations expressed in the normal fluid reference frame,

$$\begin{aligned}
\partial_t S &= \{S, h\} = -\,\mathrm{div}(S\mathbf{v}_n)\,, \\
\partial_t n &= \{n, h\} = -\,\mathrm{div}(\rho \mathbf{v}_n + \mathbf{p} + \mathrm{curl}\,\lambda)\,, \\
\partial_t \rho &= \{\rho, h\} = -\,\mathrm{div}(\rho \mathbf{v}_n + \mathbf{p} + \mathrm{curl}\,\lambda)\,, \\
&\quad (\textit{Hence, the condition } n = \rho \textit{ is preserved.}) \\
\partial_t \mathbf{u} &= \{\mathbf{u}, h\} = -\,\nabla(\mu - \frac{1}{2} v_n^2 + \mathbf{v}_n \cdot \mathbf{v}_s) + \mathbf{v}_n \times \mathrm{curl}\,\mathbf{u}\,, \\
&\quad (\textit{Hence, } \mathrm{curl}\,\mathbf{u} = 0 \textit{ is preserved.}) \\
\partial_t \mathbf{A} &= \{\mathbf{A}, h\} = n^{-1}(\rho \mathbf{v}_n + \mathbf{p} + \mathrm{curl}\,\lambda) \times \mathrm{curl}\,\mathbf{A}\,, \\
&\quad (\textit{Hence, } \mathbf{v}_\ell = \mathbf{v}_n + \rho^{-1}(\mathbf{p} + \mathrm{curl}\,\lambda) \textit{ when } n = \rho \textit{ is used.}) \\
\partial_t (M_j - nA_j) &= \{M_j - nA_j, h\} = -\,\partial_k T_j^k
\end{aligned}$$

Remarks:

(1.) Preservation of the condition $n = \rho$ by these equations allows the introduction of the momentum-carrying field $\mathbf{A}$ as an independent degree of freedom without introducing additional material inertia, provided the dynamically preserved condition $n = \rho$ holds initially. This is reminiscent of the preservation of Gauss's Law by the continuity equation for mass conservation in a fluid plasma.

(2.) The curl of the dynamical equation for the field $\mathbf{A}$ implies the **vortex line velocity**

$$\mathbf{v}_\ell = -\frac{1}{n}\frac{\delta h}{\delta \mathbf{A}} = \bar{\mathbf{v}} + \rho^{-1}\mathrm{curl}\,\lambda, \quad \text{where} \quad \bar{\mathbf{v}} = \mathbf{v}_n + \rho^{-1}\mathbf{p} = \rho^{-1}\mathbf{J}.$$

The velocity $\bar{\mathbf{v}}$ is the mass averaged velocity. The vortex slip velocity $\mathbf{s}$ corresponding to the vortex line velocity $\mathbf{v}_\ell$ is the basis for the phenomenological reactive and mutual friction forces $\mathbf{f}$ and Rayleigh dissipation function R in the HVBK system. Namely,

$$\mathbf{s} = \mathbf{v}_\ell - \mathbf{v}_n = \rho^{-1}(\mathbf{p} + \mathrm{curl}\lambda), \quad \text{with} \quad \lambda = \lambda\hat{\omega}.$$

As expected, this expression agrees with BK [1961] at zero temperature. Note that **the renormalized HVBK equations introduce no new parameters**.

(3.) The corresponding equation for $\mathbf{v}_s = \mathbf{u} - \mathbf{A}$ is then obtained as

$$\partial_t \mathbf{v}_s = -\nabla(\mu - \frac{1}{2}v_n^2 + \mathbf{v}_n \cdot \mathbf{v}_s) + \mathbf{v}_\ell \times \omega, \quad \text{where} \quad \mathbf{v}_\ell = \bar{\mathbf{v}} + \rho^{-1}\mathrm{curl}\,\lambda\hat{\omega}.$$

This may be expressed equivalently in manifestly Galilean invariant form as

$$\partial_t \mathbf{v}_s + (\mathbf{v}_s \cdot \nabla)\mathbf{v}_s = -\nabla\left(\mu - \frac{1}{2}|\mathbf{v}_s - \mathbf{v}_n|^2\right) + \mathbf{f}', \quad \text{where} \quad \mathbf{f}' = (\mathbf{v}_\ell - \mathbf{v}_s) \times \omega.$$

The term $\mathbf{f}'$ is the **Hamiltonian contribution to the reactive vortex force**. This contribution would vanish if the vortex lines moved with the superfluid velocity.

(4.) The stress tensor $T_j^k = \pi_j^k + \tau_j^k$ for total momentum conservation is given by summing

$$\pi_j^k = \left(\rho_s v_{sj} v_s^k + \rho_n v_{nj} v_n^k\right) + P\delta_j^k,$$
$$\tau_j^k = \partial_l \epsilon_{klm}(v_{sj}\lambda_m) - \lambda_j \omega^k + \omega \cdot \lambda\delta_j^k.$$

The divergence of τ_j^k defines the **vortex line tension T** as

$$\partial_k \tau_j^k = -\omega \cdot \nabla\lambda + \nabla(\omega \cdot \lambda) = -\rho_s \mathbf{T} + \nabla(\omega \cdot \lambda),$$

In the stress tensor π_j^k the pressure P is defined by the Euler relation,

$$P = -\varepsilon_0 + \mu\rho + TS,$$

so that in the normal-fluid frame the pressure satisfies

$$dP = \rho d\mu + S dT - \mathbf{p} \cdot d(\mathbf{v}_s - \mathbf{v}_n) - \lambda \cdot d\omega.$$

The stress tensor $T_j^k = \pi_j^k + \tau_j^k$ may be derived by using Corollary #2 for the Hamiltonian formulation.

Implications of the HVBK vortex dynamics.

The new Hamiltonian formulation of the renormalized HVBK equations presented in the Proposition provides a formula for the slip velocity of a vortex line element in a turbulent superfluid *at finite temperature.* Namely, for the Hamiltonian h in equation (61), one finds

$$\mathbf{s} = \mathbf{v}_\ell - \mathbf{v}_n = \rho^{-1}\left(\mathbf{p} + \operatorname{curl}\lambda\right). \tag{69}$$

This formula for $\mathbf{v}_\ell$ recovers the HVBK expression in BK [1961] at zero temperature. Otherwise, it provides an **extension to finite temperature** of the HVBK vortex force

$$\mathbf{f} = \left(\mathbf{v}_L - \mathbf{v}_s\right) \times \omega, \quad \text{with} \quad \mathbf{v}_L = \mathbf{v}_n - \rho_s\left(\alpha\,\mathbf{s} + \beta\,\hat{\omega} \times \mathbf{s}\right), \tag{70}$$

where the **renormalized vortex slip velocity** is given by

$$\mathbf{s} = \mathbf{v}_\ell - \mathbf{v}_n = \frac{\rho_s}{\rho}\left(\mathbf{v}_\ell^0 - \mathbf{v}_n\right), \quad \text{for constant } \rho_s. \tag{71}$$

The corresponding heating rate R is given by

$$R = \left(\mathbf{J} - \rho\mathbf{v}_n + \operatorname{curl}\lambda\right) \cdot \omega \times \left(\mathbf{v}_L - \mathbf{v}_n\right) = \rho\rho_s\beta\,\omega\left|\mathbf{s} \times \hat{\omega}\right|^2,$$

which is positive, as it must be.

References

1. Bekarevich, I. L. and I. M. Khalatnikov [1961] Phenomenological derivation of the equations of vortex motion in He II, *Sov. Phys. JETP* **13** 643-646.
2. Donnelly, R. J. [1999] Cryogenic fluid dynamics, *J. Phys.: Condens. Matter* **11** 7783-7834.
3. Gorter, C. J. and J. H. Mellink [1949] *Physica* **15** 285.
4. Henderson, K. L. and C. F. Barenghi [2000] The anomalous motion of superfluid helium in a rotating cavity, *J. Fluid Mech.* **406** 199-219.
5. Hills, R. N. and P. H. Roberts [1977] Superfluid mechanics for a high density of vortex lines, *Arch. Rat. Mech. Anal.* **66** 43-71.
6. Holm, D. D. and B. Kupershmidt [1987] Superfluid plasmas: multivelocity nonlinear hydrodynamics of superfluid solutions with charged condensates coupled electromagnetically, *Phys. Rev. A* **36** 3947-3956.
7. Nemirovskii, S. K. and W. Fiszdon [1995] Chaotic quantized vortices and hydrodynamic processes in superfluid helium, *Rev. Mod. Phys.* **67** 37-84.
8. Putterman, S. J. [1974] *Superfluid Hydrodynamics*, North Holland, Amsterdam.
9. Reissenegger, A. [1993] The spin up problem in Helium-II *J. Low Temp. Phys.* **92** 77-106.
10. Skrbek, L., J. J. Niemela and R. J. Donnelly [1999] Turbulent flows at cryogenic temperatures: a new frontier, *J. Phys.: Condens. Matter* **11** 7761-7783.
11. Sonin, E. B. [1987] Vortex oscillations and hydrodynamics of rotating superfluids, Rev. Mod. Phys. **59** 87-155. 1987
12. Volovik, G. E. and V. S. Dotsenko [1979] Poisson brackets and continuous dynamics of the vortex lattice in rotating He-II, *JETP Lett.* **29** 576-579.
13. Volovik, G. E. and V. S. Dotsenko [1980] Hydrodynamics of defects in condensed media in the concrete cases of vortices in rotating Helium-II and of disclinations in planar magnetic substances, *Sov. Phys. JETP* **58** 65-80.

Magnus Force, Aharonov–Bohm Effect, and Berry Phase in Superfluids

Edouard Sonin

Racah Institute of Physics, Hebrew University of Jerusalem

1 Introduction

If the vortex moves with respect to a liquid, classical [1] or quantum, there is a force on the vortex normal to the relative vortex velocity with respect to the liquid. This is the Magnus force, which plays an important role in modern condensed-matter physics. In particular, it determines the mutual fiction in superfluids [2–4] and the Hall effect in superconductors [5].

An obvious generalization of the classical Magnus force in the superfluid seemed to be a force proportional to the superfluid density ρ_s [2]:

$$\rho_s[(\boldsymbol{v}_L - \boldsymbol{v}_s) \times \boldsymbol{\kappa}] = \boldsymbol{F} , \tag{1}$$

where $\boldsymbol{v}_L$ is the vortex velocity, $\boldsymbol{v}_s$ is the superfluid velocity, and $\boldsymbol{\kappa}$ is a vector along the vortex line with $\kappa = h/m$ being the circulation of the superfluid velocity around the vortex. In absence of the external force $\boldsymbol{F}$ on a superfluid, the vortex moves with the superfluid velocity. However, quasiparticles (rotons [6] and phonons [7]) produce an additional force transverse to the vortex velocity and

$$\boldsymbol{F} = -D(\boldsymbol{v}_L - \boldsymbol{v}_n) - D'[\hat{z} \times (\boldsymbol{v}_L - \boldsymbol{v}_n)] . \tag{2}$$

Here $\boldsymbol{v}_n$ is the normal velocity. It was shown [8] that the additional transverse force $\propto D' = -\kappa\rho_n$ (the Iordanskii force) is connected with the analogue of the Aharonov–Bohm effect [9] in superfluids. A similar force on the magnetic-flux tube exists in the original Aharonov–Bohm effect for electrons [10].

Ao and Thouless [11] have pointed out that the Magnus force is connected with the Berry phase [12], which is the phase variation of the quantum-mechanical wave function of a quantum liquid generated by the adiabatic transport of the vortex round a close loop. From the Berry-phase analysis Ao and Thouless [11,13] concluded that the only force transverse to $\boldsymbol{v}_L$ is the force proportional to ρ_s, i.e., $D' = 0$ in Eq. (2). This conclusion disagreed with the previous calculations and therefore was vividly discussed [14,15].

The present paper is an attempt to bring together two points of view in order to find a source of disagreement [16]. I restrict myself with the problem of the Galilean invariant quantum Bose-liquid described by the Gross–Pitaevskii theory [17]. At large scales the theory yields equations of the hydrodynamics of an ideal inviscous liquid. In presence of an ensemble of sound waves (phonons) with the Planck distribution, which is characterized by a locally defined normal velocity, one obtains the two-fluid hydrodynamics. The momentum balance in the area around a moving vortex demonstrates the existence of the Iordanskii force. I also discuss the Berry phase. According to Refs. [11,13], the Berry phase and the Magnus forces are proportional to the total current circulation at large distances. But the total current circulation contains a normal-fluid contribution, which is proportional to the Iordanski force. Taking this contribution into account, the Berry-phase analysis agrees with the momentum-balance approach.

2   Gross–Pitaevskii Theory and Two-Fluid Hydrodynamics

In the Gross–Pitaevskii theory [17] the ground state and weakly excited states of a Bose-gas are described by the nonlinear Schrödinger equation

$$i\hbar\frac{\partial\psi}{\partial t} = -\frac{\hbar^2}{2m}\nabla^2\psi + V|\psi|^2\psi \tag{3}$$

for the condensate wave function $\psi = a\exp(i\phi)$. The nonlinear Schrödinger equation is the Euler–Lagrange equation for the Lagrangian

$$L = \frac{i\hbar}{2}\left(\psi^*\frac{\partial\psi}{\partial t} - \psi\frac{\partial\psi^*}{\partial t}\right) - \frac{\hbar^2}{2m}|\nabla\psi|^2 - \frac{V}{2}|\psi|^4 . \tag{4}$$

The Noether theorem yields the momentum conservation law $\partial j_i/\partial t + \nabla_j \Pi_{ij} = 0$ where $j = \text{Im}\{\hbar\psi^*\nabla\psi\}$ is the mass current, and the momentum-flux tensor is

$$\Pi_{ij} = \frac{\hbar^2}{2m}\left(\nabla_i\psi\nabla_j\psi^* + \nabla_i\psi^*\nabla_j\psi\right) + \delta_{ij}\left(\frac{V}{2}|\psi|^4 - \frac{\hbar^2}{4m}\nabla^2|\psi|^2\right) . \tag{5}$$

Using the Madelung transformation [3], one obtains from complex Eq. (3) two real equations for the liquid density $\rho = ma^2$ and the liquid velocity $v = (\hbar/m)\nabla\phi = (\kappa/2\pi)\nabla\phi$. Far from the vortex line these equations are hydrodynamic equations for an ideal inviscous liquid:

$$\frac{\partial\rho}{\partial t} + \nabla(\rho v) = 0 , \tag{6}$$

$$\frac{\partial v}{\partial t} + (v\cdot\nabla)v = -\nabla\mu . \tag{7}$$

Here $\mu = Va^2/m$ is the chemical potential. Equation (5) becomes the hydrodynamic momentum-flux tensor $\Pi_{ij} = P\delta_{ij} + \rho v_i(r)v_j(r)$.

A plane sound wave propagating in the liquid generates the phase variation $\phi(r,t) = \phi_0\exp(ik\cdot r - i\omega t)$. Then $\rho(r,t) = \rho_0 + \rho_{(1)}(r,t)$ and $v(r,t) = v_0 + v_{(1)}(r,t)$, where ρ_0 and v_0 are the average density and velocity in the liquid, whereas $\rho_{(1)}(r,t)$ and $v_{(1)}(r,t) = (\kappa/2\pi)\nabla\phi$ are periodical variations of the density and the velocity due to the sound wave ($\langle\rho_{(1)}\rangle = 0$, $\langle v_{(1)}\rangle = 0$). Equations (6) and (7) linearized with respect to $\rho_{(1)}$ and $v_{(1)}$ yield the sound equation for ϕ with the sound velocity $c_s = \sqrt{Va^2/m}$ and the spectrum $\omega = c_s k + k\cdot v_0$.

The total mass current expanded up to the terms of the second order with respect the wave amplitude and averaged over time is

$$\boldsymbol{j} = \rho_0 \boldsymbol{v}_0 + \langle \rho_{(1)} \boldsymbol{v}_{(1)} \rangle = \rho_0 \boldsymbol{v}_0 + \rho_0 \phi_0^2 \frac{\kappa^2 k}{8\pi^2 c_s} \boldsymbol{k} \ . \tag{8}$$

If there is an ensemble of phonons with the Planck distribution

$$n_0(E, \boldsymbol{v}_n) = \left[\exp \frac{E(\boldsymbol{p}) - \boldsymbol{p} \cdot \boldsymbol{v}_n}{T} - 1 \right]^{-1} = \left[\exp \frac{c_s p + \boldsymbol{p} \cdot (\boldsymbol{v}_0 - \boldsymbol{v}_n)}{T} - 1 \right]^{-1} , \tag{9}$$

the total mass current linearized with respect to $\boldsymbol{v}_0 - \boldsymbol{v}_n$ is

$$\boldsymbol{j} = \rho_0 \boldsymbol{v}_0 + \frac{1}{h^3} \int n_0(\boldsymbol{p}) \boldsymbol{p} \, d_3 \boldsymbol{p} = \rho_0 \boldsymbol{v}_0 + \rho_n (\boldsymbol{v}_n - \boldsymbol{v}_0) \ . \tag{10}$$

Here $\boldsymbol{p} = \hbar \boldsymbol{k}$ is the phonon momentum, $E = c_s p + \boldsymbol{p} \cdot \boldsymbol{v}_0$ is the phonon energy, and $\boldsymbol{v}_n$ is the drift velocity of phonons. This expression is equivalent to the two-fluid expression assuming that $\rho = \rho_0 = \rho_s + \rho_n$, $\boldsymbol{v}_0 = \boldsymbol{v}_s$, and the normal density is given by the usual two-fluid expression $\rho_n = -(1/3h^3) \int [\partial n_0(\varepsilon, 0)/\partial E] p^2 \, d^3\boldsymbol{p}$.

Expanding the momentum-flux tensor up to terms of the second-order with respect to the sound wave amplitude one obtains:

$$\Pi_{ij} = P_0 \delta_{ij} + \rho_0 v_{0i} v_{0j} + \left(\frac{c_s^2}{\rho_0} \frac{\langle \rho_{(1)}^2 \rangle}{2} - \rho_0 \frac{\langle v_{(1)}^2 \rangle}{2} \right) \delta_{ij}$$
$$+ \langle \rho_{(1)} (v_{(1)})_i \rangle v_{0j} + \langle \rho_{(1)} (v_{(1)})_j \rangle v_{0i} + \rho_0 \langle (v_{(1)})_i (v_{(1)})_j \rangle \ . \tag{11}$$

For the Planck distribution this yields the two-fluid momentum flux tensor

$$\Pi_{ij} = P\delta_{ij} + \rho_s v_{si} v_{sj} + \rho_n v_{ni} v_{nj} \ . \tag{12}$$

Taking into account phonon-phonon interaction, which establishes the local Planck distribution of phonons, we can derive all equations of the two-fluid hydrodynamics, as shown in Ref. [18]. Thus the two-fluid hydrodynamics can be derived from the hydrodynamics of an ideal inviscous fluid, and the latter follows from the Gross–Pitaevskii theory.

3 Interaction of Phonons with a Vortex in Hydrodynamics

In presence of a vortex the sound equation is (see Refs. [8,14] for more details)

$$\frac{\partial^2 \phi}{\partial t^2} - c_s^2 \boldsymbol{\nabla}^2 \phi = -2\boldsymbol{v}_v(\boldsymbol{r}) \cdot \boldsymbol{\nabla} \frac{\partial \phi}{\partial t} \ , \tag{13}$$

where

$$\boldsymbol{v}_v(\boldsymbol{r}) = \frac{\boldsymbol{\kappa} \times \boldsymbol{r}}{2\pi r^2} \tag{14}$$

is the circular velocity field induced by a vortex line. Here r is the position vector in the plane xy. The sound wave produces the density variation

$$\rho_{(1)} = -\frac{\rho_0}{c_s^2}\frac{\kappa}{2\pi}\left(\frac{\partial\phi}{\partial t} + v_v \cdot \nabla\phi(r)\right) \ . \tag{15}$$

One can calculate the differential cross-section in the Born approximation, but since it is quadratic in κ this does not yield a transverse force [7,8,14]. Instead we consider a quasiclassical solution of the sound equation:

$$\phi = \phi_0\exp\left(-i\omega t + ik \cdot r + \frac{i\delta S}{\hbar}\right) = \phi_0\exp(-i\omega t + ik \cdot r)\left(1 + \frac{i\kappa k}{2\pi c_s}\theta\right) \ , \tag{16}$$

where $\delta S = -(\hbar k/c_s)\int^r v_v \cdot dl = \hbar\theta\kappa k/2\pi c_s$ is the variation of the action due to interaction with the circular velocity around the vortex. The angle θ is an azimuth angle for the position vector r measured from the direction opposite to the wave vector k. This choice provides that the quasiclassical correction vanishes for the incident wave far from the vortex. One can check directly that Eq. (16) satisfies the sound equation (13) in the first order of the parameter $\kappa k/c_s$. The velocity generated by the sound wave around the vortex is

$$v_{(1)} = \frac{\kappa}{2\pi}\nabla\phi = \frac{\kappa}{2\pi}\phi_0\exp(-i\omega t + ik \cdot r)\left(ik - \frac{ik}{c_s}v_v\right) \ . \tag{17}$$

According to Eq. (16) the phase ϕ is multivalued, and one must choose a cut for an angle θ at the direction k, where $\theta = \pm\pi$. The jump of the phase on the cut line behind the vortex is a manifestation of the Aharonov–Bohm effect [9]: the sound wave after its interaction with the vortex has different phases on the left and on the right of the vortex line. This results in an interference [8,14]. The width of the interference region is $d_{int} \sim \sqrt{r/k}$.

Now we consider the momentum balance using the condition that $\int dS_j \Pi_{\perp j} = 0$ for a cylindrical surface around the vortex line. The subscript $\perp$ points a component normal to the wave vector k of the incident wave. The total momentum-flux tensor can be obtained from Eq. (11) assuming $v_0(r) = v_v(r) + v_s$:

$$\Pi_{ij} = -\rho_0(v_s - v_L) \cdot v_v\delta_{ij} + \rho_0 v_{0i}v_{0j}$$
$$+\langle\rho_{(1)}(v_{(1)})_i\rangle v_{vj} + \langle\rho_{(1)}(v_{(1)})_j\rangle v_{vi} + \rho_0\langle(v_{(1)})_i(v_{(1)})_j\rangle \ . \tag{18}$$

The first two terms in this expression yield the momentum flux without phonons, which produces the Magnus force for a liquid with the density ρ_0 and the velocity v_s. The rest terms cancel except for the contribution from the term $\rho_0\langle(v_{(1)})_i(v_{(1)})_j\rangle$ in the interference region where $v_{(1)\perp} = (\kappa/2\pi r)\partial\phi/\partial\theta$. The contribution depends on the phase jump and for a single sound wave is:

$$\int\rho_0\langle(v_{(1)})_\perp(v_{(1)})_r\rangle rd\theta = \frac{1}{8\pi^2}\rho_0\phi_0^2\frac{\kappa^2 k}{\hbar}\left[\delta S(-) - \delta S(+)\right] = \frac{1}{8\pi^2}\rho_0\phi_0^2\frac{\kappa^3 k^2}{c_s} \ , \tag{19}$$

where $\delta S(\pm) = \mp\hbar\kappa k/2c_s$ are the action variations at $\theta \to \mp\pi$.

For the Planck phonon distribution the condition $\int dS_j \Pi_{\perp j} = 0$ yields:

$$\rho_0[(\boldsymbol{v}_L - \boldsymbol{v}_s) \times \boldsymbol{\kappa}] - [\langle \rho_{(1)} \boldsymbol{v}_{(1)} \rangle \times \boldsymbol{\kappa}] = \rho_s[(\boldsymbol{v}_L - \boldsymbol{v}_s) \times \boldsymbol{\kappa}] + \rho_n[(\boldsymbol{v}_L - \boldsymbol{v}_n) \times \boldsymbol{\kappa}] = 0 \ . \quad (20)$$

The term $\propto (\boldsymbol{v}_L - \boldsymbol{v}_n)$ is the Iordanskii force, which corresponds to $D' = -\kappa \rho_n$ in Eq. (2). Uniting in Eq. (20) terms linear in $\boldsymbol{v}_L$, we see that the total transverse force (effective Magnus force) is proportional to the *total* density $\rho_0 = \rho_s + \rho_n$.

4   Momentum Balance in the Two-Fluid Hydrodynamics

Up to now we analyzed spatial scales much less than the mean-free-path l_{ph} of phonons (ballistic region). Now we shall see what is going on at scales much larger than l_{ph} where the two-fluid hydrodynamics is valid.

Interaction between phonons and the vortex in the ballistic region produces a force concentrated along the vortex line in the hydrodynamic region. The response of the normal fluid to this force is described by the Navier-Stokes equation with the dynamic viscosity η_n:

$$\frac{\partial \boldsymbol{v}_n}{\partial t} + (\boldsymbol{v}_n \cdot \boldsymbol{\nabla}) \boldsymbol{v}_n = \nu_n \Delta \boldsymbol{v}_n - \frac{\boldsymbol{\nabla} P}{\rho} - \frac{\rho_s S}{\rho_n \rho} \boldsymbol{\nabla} T \ , \quad (21)$$

where $\nu_n = \eta_n / \rho_n$ is the kinematic viscosity, S is the entropy per unit volume, and T is the temperature. At $r < r_m \sim \nu_n / |\boldsymbol{v}_n - \boldsymbol{v}_L|$ one may neglect the nonlinear inertial (convection) term $(\boldsymbol{v}_n \cdot \boldsymbol{\nabla}) \boldsymbol{v}_n$. The line force $-F_i = \oint \tau_{ij} dS_j$ on the normal fluid produces a divergent velocity field (the Stokes paradox [1]):

$$\boldsymbol{v}_n(r) = \boldsymbol{v}_n + \frac{\boldsymbol{F}}{4 \pi \eta_n} \ln \frac{r}{l_{ph}} \ , \quad (22)$$

where $\boldsymbol{v}_n$ is the normal velocity at distances $r \sim l_{ph}$, which separate the ballistic and the hydrodynamic regions. Here $\tau_{ij} = -\eta_n (\nabla_i v_{nj} + \nabla_j v_{ni})$ is the viscous stress tensor. Due to viscosity the normal velocities $\boldsymbol{v}_{n\infty}$ and $\boldsymbol{v}_n$ at large $(r \sim r_m)$ and small $(r \sim l_{ph})$ distances from the vortex line are different (the viscous drag [2]):

$$\boldsymbol{F} = \frac{4 \pi \eta_n}{\ln(r_m / l_{ph})} (\boldsymbol{v}_{n\infty} - \boldsymbol{v}_n) \ . \quad (23)$$

The momentum balance in the two-fluid region must include the viscous stress tensor: $\oint \Pi_{ij} dS_j + \oint \tau_{ij} dS_j = 0$, where Π_{ij} is given by Eq. (12). Since the normal velocity field does not contain the circular velocity $\boldsymbol{v}_v$, the flux $\oint \Pi_{ij} dS_j$ yields the superfluid Magnus force, i.e., the force $\boldsymbol{F}$ satisfies Eq. (1).

However, at very large distances $r \gg r_m$ the nonlinear convection term is more important than the viscous term. Thus the scale r_m separates the viscous and convection subregions. In the convection subregion the viscosity becomes ineffective and the momentum flux related to the linear force on the normal fluid should be connected with the normal part of the momentum flux tensor Π_{ij} given by Eq. (12). The normal momentum transmission requires a circulation of the normal velocity [19] with magnitude determined by the transverse force on the normal fluid. This is confirmed by the solution of the Navier-Stokes equation obtained by Thouless *et al.* [20]. However, separation on longitudinal and transverse components of a force should be done with respect to the normal velocity $v_{n\infty} - v_L$, but not $v_n - v_L$. Using Eq. (23), Eq. (2) for the force from the normal fluid can be rewritten (neglecting the longitudinal force $\propto D$) as

$$
F = -\frac{1}{1 + \left[\frac{D'\ln(r_m/l_{ph})}{4\pi\eta_n}\right]^2} \left\{ \frac{D'\ln(r_m/l_{ph})}{4\pi\eta_n}(v_L - v_{n\infty}) + D'[\hat{z} \times (v_L - v_{n\infty})] \right\} .
$$

(24)

The transverse component of this force determines the normal circulation at very large distances:

$$
\kappa_n = \oint dl \cdot v_n = -\frac{D'}{\rho_n \left\{ 1 + \left[\frac{D'\ln(r_m/l_{ph})}{4\pi\eta_n}\right]^2 \right\}} = \frac{\kappa}{1 + \left[\frac{\kappa\rho_n \ln(r_m/l_{ph})}{4\pi\eta_n}\right]^2} ,
$$

(25)

where we used the value $D' = -\kappa\rho_n$ for the Iordanskii force. In the limit of a strong viscous drag $\kappa\rho_n \ln(r_m/l)/4\pi\eta_n \gg 1$ the transverse force and related normal circulation are suppressed [20]. But the effect of the superfluid Magnus force and the longitudinal force $\propto D$ is also suppressed in this limit.

5   Magnus Force and the Berry Phase

Let us consider now the Berry phase in the hydrodynamic description using the Lagrangian obtained from Eq. (4) with the Madelung transformation:

$$
L = \frac{\kappa\rho}{2\pi}\frac{\partial\phi}{\partial t} - \frac{\kappa^2\rho}{8\pi}\nabla\phi^2 - \frac{V}{2}\rho^2 .
$$

(26)

The first term with the first time derivative of the phase ϕ (Wess-Zumino term) is responsible for the Berry phase $\Theta = \Delta S_B/\hbar$, which is the variation of the phase of the quantum-mechanical wave function for an adiabatic motion of the vortex around a closed loop [12]. Here

$$
\Delta S_B = \int dr\, dt\, \frac{\kappa\rho}{2\pi}\frac{\partial\phi}{\partial t} = -\int dr\, dt\, \frac{\kappa\rho}{2\pi}(v_L \cdot \nabla_L)\phi .
$$

(27)

is the classical action variation around the loop and $\nabla_L\phi$ is the gradient of the phase $\phi[r - r_L(t)]$ with respect to the vortex position vector $r_L(t)$. However, $\nabla_L\phi = -\nabla\phi$, where $\nabla\phi$ is the gradient with respect to r. Then the loop integral $\oint dl$ yields the circulation of the total current $j = (\kappa/2\pi) < \rho\nabla\phi >$ for points inside the loop, but vanishes for points outside. As a result [13],

$$
\Theta = \frac{\Delta S_B}{\hbar} = V\frac{\kappa}{2\pi\hbar}\oint(dl \cdot j) .
$$

(28)

where V is the volume inside the loop (a product of the loop area and the liquid height along a vortex). Contrary to Eq. (27), the integral in Eq. (28) is related with the variation of r, the vortex position vector r_L being fixed.

If the circulation of the normal velocity vanished, the current circulation would be $\oint(dl \cdot j) = \rho_s \kappa$, and the Berry phase and the transverse force would be proportional to ρ_s (see Geller $et\ al.$ [13]). However, according to Sec. 4, at very large distances the normal circulation κ_n does not vanish and $\oint(dl \cdot j) = \rho_s \kappa + \rho_n \kappa_n$. Using a proper value of κ_n given by Eq. (25), the Berry-phase analysis yields the same transverse force as the momentum balance. But one cannot find κ_n from the Berry-phase analysis, since the latter deals only with very large distances, whereas κ_n is determined by the transverse force, which arises at small distances (in the ballistic region) and appears in the small-distance boundary condition for the Navier-Stokes equation in the two-fluid region. This conclusion agrees with the recent analysis by Thouless $et\ al.$ [20].

Discussions with Lev Pitaevskii and David Thouless during the present workshop had a great impact on my view presented in this article. I appreciate also interesting discussions with Andrei Shelankov. The work was supported by the grant of the Israel Academy of Sciences and Humanities.

References

1. H. Lamb, $Hydrodynamics$ (Cambridge University Press, New York, 1975).
2. H.E. Hall and W.F. Vinen, Proc. Roy. Soc. **A238**, 204 (1956).
3. R.J. Donnelly, $Quantized\ vortices\ in\ helium\ II$ (Cambridge University Press, Cambridge, 1991), Sec. 2.8.3.
4. E.B. Sonin, Rev. Mod. Phys. **59**, 87 (1987).
5. P. Nozières and W.F. Vinen, Phil. Mag. **14**, 667 (1966).
6. E.M. Lifshitz and L.P. Pitaevskii, Zh. Eksp. Teor. Fiz. **33**, 535 (1957) [Sov. Phys.-JETP **6**, 418 (1958)].
7. S.V. Iordanskii, Zh. Eksp. Teor. Fiz. **49**, 225 (1965) [Sov. Phys.-JETP **22**, 160 (1966)].
8. E.B. Sonin, Zh. Eksp. Teor. Fiz. **69**, 921 (1975) [Sov. Phys.-JETP **42**, 469 (1976)].
9. Y. Aharonov and D. Bohm, Phys. Rev. **115**, 485 (1959).
10. A.L. Shelankov, Europhys. Lett., **43**, 623 (1998).
11. P. Ao and D.J. Thouless, Phys. Rev. Lett. **70**, 2158 (1993).
12. M.V. Berry, Proc. R. Soc. London A **392**, 45 (1984).
13. M.R. Geller, C. Wexler, and D.J. Thouless, Phys. Rev. B **57**, R8119 (1998).
14. E.B. Sonin, Phys. Rev. B **55**, 485 (1997).
15. H.E. Hall and J.R. Hook, Phys. Rev. Lett., **80**, 4356 (1998); E.B. Sonin, $ibid.$ **81**, 4276 (1998); C. Wexler $et\ al.$, $ibid.$ **80**, 4357 (1998); **81**, 4277 (1998).
16. A more detailed report on the present analysis is to be published in $Proceedings$ $of\ the\ the\ workshop$ "$Microscopic\ structure\ and\ dynamics\ of\ vortices\ in\ unconventional\ superconductors\ and\ superfluids$" Dresden, Germany, March 2000, cond-mat/0104221.
17. E.P. Gross, Nuovo Cimento **20**, 454 (1961); L.P. Pitaevskii, Zh. Eksp. Teor. Fiz. **40**, 646 (1961) [Sov. Phys.-JETP **13**, 451 (1961)].
18. S.J. Putterman and P.H. Roberts, Physica **117** A, 369 (1983).
19. Existence of the normal circulation at large distances in the presence of the transverse force on the vortex was pointed out by Pitaevskii (unpublished).
20. D.J. Thouless, M.R. Geller, W.F. Vinen, J-Y. Fortin, and S.W. Rhee, cond-mat/0101297.

Using the HVBK Model to Investigate the Couette Flow of Helium II

Karen L. Henderson

Faculty of Computer Studies & Mathematics, University of the West of England, Bristol, BS16 1QY, U.K.

Abstract. We review the application of the two-fluid HVBK equations to helium II in Couette geometry, that is flow between concentric, rotating cylinders. This application is particularly interesting as a large number of experiments have been carried out in this geometry and also because Couette flow is an exact solution of the HVBK equations for both the normal fluid and superfluid.

1   Introduction

When the temperature of liquid helium drops below the transition temperature of $T_\lambda = 2.172\,\mathrm{K}$ a phase transition occurs and it becomes a quantum liquid called helium II. Helium II can be described macroscopically by Landau's two-fluid model in which it is considered to be a mixture of a viscous normal fluid and an inviscid superfluid. In addition, vortex lines appear in the superfluid component when helium II rotates or when it moves along a tube faster than a small critical velocity. Feynman [1] showed that the circulation about each individual vortex line is quantised, taking the value of $\Gamma = 9.97 \times 10^{-4}\mathrm{cm}^2/\mathrm{sec}$. The most generally accepted equations for modelling the macroscopic flow of helium II are the Hall-Vinen-Bekharevich-Khalatnikov (HVBK) equations which were derived by a number of people over the years [2–6]. These equations extend Landau's two-fluid model to take into account the presence of quantized vortex lines in the flow. The derivation of the equations is based on a continuum approximation, assuming a high density of vortex lines, all aligned roughly in the same direction.

The incompressible isothermal HVBK equations of motion of the two fluids are

$$\frac{\partial \mathbf{v}_n}{\partial t} + (\mathbf{v}_n \cdot \boldsymbol{\nabla})\mathbf{v}_n = -\boldsymbol{\nabla}p^n + \nu_n \nabla^2 \mathbf{v}_n + \frac{\rho_s}{\rho}\boldsymbol{F}, \tag{1a}$$

$$\frac{\partial \mathbf{v}_s}{\partial t} + (\mathbf{v}_s \cdot \boldsymbol{\nabla})\mathbf{v}_s = -\boldsymbol{\nabla}p^s - \nu_s \boldsymbol{T} - \frac{\rho_n}{\rho}\boldsymbol{F}, \tag{1b}$$

$$\boldsymbol{\nabla} \cdot \mathbf{v}_n = 0, \qquad \boldsymbol{\nabla} \cdot \mathbf{v}_s = 0. \tag{1c}$$

where $\mathbf{v}_n$ and $\mathbf{v}_s$ are the velocity profiles of the normal fluid and superfluid respectively, ρ_n and ρ_s the normal fluid and superfluid densities, $\rho = \rho_n + \rho_s$ helium's total density, p^n and p^s effective pressures and ν_n the normal fluid kinematic viscosity. The relative amount of normal fluid and superfluid present in the flow depends on the temperature T of the liquid: if $T \to T_\lambda$ then $\rho_s/\rho \to 0$ and $\rho_n/\rho \to 1$; if $T \to 0$ then $\rho_s/\rho \to 1$ and $\rho_n/\rho \to 0$.

The mutual friction force, $\boldsymbol{F}$, describes the interaction between the normal fluid and the vortex lines and is given by

$$\boldsymbol{F} = \tfrac{1}{2}B\widehat{\boldsymbol{\omega}}^s \times (\boldsymbol{\omega}^s \times (\mathbf{v}_n - \mathbf{v}_s - \nu_s \boldsymbol{\nabla} \times \widehat{\boldsymbol{\omega}}^s)) + \tfrac{1}{2}B'\boldsymbol{\omega}^s \times (\mathbf{v}_n - \mathbf{v}_s - \nu_s \boldsymbol{\nabla} \times \widehat{\boldsymbol{\omega}}^s), \quad (2)$$

where $\boldsymbol{\omega}^s$ represents the superfluid vorticity and is a measure of the number and direction of vortex lines contained in a given small region of fluid. The term $\widehat{\boldsymbol{\omega}}^s = \boldsymbol{\omega}^s/|\boldsymbol{\omega}^s|$ represents the unit vector in the direction of superfluid vorticity. B and B' are temperature dependent coefficients which describe the interaction between the normal fluid and the vortices [7,8].

The vortex tension force, $-\nu_s \boldsymbol{T}$, describes the energy in the vortex lines and is such that

$$\boldsymbol{T} = \boldsymbol{\omega}^s \times (\boldsymbol{\nabla} \times \widehat{\boldsymbol{\omega}}^s), \quad (3)$$

where $\nu^s = (\Gamma/4\pi)\log(b_0/a_0)$ is the vortex tension parameter, a_0 is the vortex core radius and $b_0 = (|\boldsymbol{\omega}^s|/\Gamma)^{-1/2}$ is the intervortex spacing.

In this paper we review the application of the HVBK equations to Taylor-Couette flow, that is flow between two concentric rotating cylinders. Taylor-Couette flow has been used as a bench-mark for fluid mechanics since Taylor's [9] pioneering work to investigate the transition from Couette flow to Taylor vortices, which established a firm ground for using the Navier-Stokes equations and the no-slip boundary conditions. Progress in helium II has been slower than for classical fluids due in part to problems of flow visualization at such low temperatures. In considering a classical fluid, introduction of flakes or other small particles into the working fluid (usually oil or water), results in the Taylor vortices being clearly evident. In contrast there are only limited visualisation techniques available to the experimentalist at temperatures close to absolute zero. Recent attempts have been made to reveal the flow pattern of helium II by adding small particles [10]. However this was only successful at high rotation rates (40 times the critical angular velocity at which linear stability analysis predicts Couette flow becomes unstable).

Experiments on helium II between concentric cylinders were first performed by Kapitza [11] in 1941 and Donnelly & LaMar [12] have written a review of experiments involving helium II in Couette apparatus. We shall expand on two types of experiments performed, which have been used to compare theoretical predictions with. Early Taylor-Couette experiments were concerned with determining the viscosity of helium II by measuring the torque exerted by the flow on the stationary cylinder. A break in the linear dependence of the torque with the angular velocity of the rotating cylinder is taken to denote a transition from one solution to another. The second experimental technique, that of measuring the extra attenuation of a second sound wave, can be used to probe the superfluid vorticity. Second sound waves occur when there is a periodic counterflow between the normal fluid and superfluid, which corresponds to a wave of heat. Angular velocity is plotted against the attenuation factor and breaks in the curve are interpreted as transitions in the flow. By measuring the extra attenuation of second sound waves in the axial, azimuthal and radial directions it is theoretically possible to get an idea of the number and direction of the quantized vortex lines. In practice the information obtained is less complete than this.

In this paper we restrict our attention to rotation of the inner cylinder only, keeping the outer cylinder fixed. The usual simplifying assumption of infinite cylinders is adopted in Sects. 2 and 3.1 where we consider the stability of helium II Couette flow and the nonlinear flow of helium II beyond this transition respectively. In Sect. 3.2 we consider the basic flow of helium II between finite cylinders with stationary endcaps.

2   Linear Theory

In the case of flow between infinite cylinders with inner radius R_1 rotating with angular velocity Ω and stationary outer radius R_2, Couette flow, whose velocity profile is given by:

$$\boldsymbol{v}_c = (A + C/r)\,\boldsymbol{e}_r \tag{4}$$

is an exact solution of the HVBK equations for both the normal fluid and superfluid. This is provided that Ω is greater than a small critical value at which vortex lines first appear in the gap. A and C are constants depending on R_1, R_2 and Ω determined by the no-slip condition imposed on the normal fluid at the cylinder walls. The superfluid, being inviscid is not required to satisfy such boundary conditions. The only restriction is that there is no penetration through the boundary. From (4) it can be seen that the superfluid vorticity is purely axial and has magnitude $2|A|$, thus the vortex lines are aligned in the direction of rotation.

Early attempts to theoretically examine the stability of Couette flow theoretically were made by Chandrasekhar & Donnelly [13], however the issue has only recently been resolved fully by Barenghi & Jones [14] and Barenghi [15]. They performed a linear stability analysis on the HVBK equations. The Couette state was linearly perturbed and they numerically calculated the critical angular velocity, Ω_c, and corresponding critical axial wavenumber at which Couette flow becomes linearly unstable. We summarize their findings below:

- the axisymmetric mode onsets first,
- the stability of helium II Couette flow is more stable than classical Couette flow in the high temperature region but less stable in the low temperature one,
- the critical axial wavenumber decreases as the temperature decreases, becoming zero at relatively high temperatures.

For a classical fluid the critical axial wavelength at which Couette flow becomes unstable and Taylor vortices form occurs at $k_c \approx 3.1$, in other words the resulting Taylor vortices are approximately square. This is not the case for helium II. As the temperature decreases below the lambda temperature of $T_\lambda = 2.172\mathrm{K}$ the critical axial wavenumber decreases resulting in an elongation of the Taylor cells.

This effect is strongly temperature dependent and the critical axial wavenumber becomes zero at a relatively high temperature for certain parameter ranges (*e.g.* $T = 2$ K).

Compared to the classical case $T = 2.172$K, the stability of helium II is initially enhanced as the temperature decreases below the transition temperature, due to the tension in the vortex lines. However the stability is dramatically reduced as the temperature drops further. These results prompted further experiments which were performed by Swanson & Donnelly [16]. They carried out a series of second sound experiments at temperatures close to T_λ. Comparison between theory [15] and experiments [16] gave excellent agreement, particularly for temperatures close to the lambda temperature. Although the qualitative picture was correct at lower temperatures it is not realistic to expect such good quantitative agreement here, due to the breakdown of the infinite cylinder assumption. For $T < 2.07$ K linear theory predicts that the critical axial wave number is zero. In such a region, the Taylor cells would be so elongated that only a few would be present in the apparatus and end effects would undoubtedly become important.

It was only at this stage that the validity of the HVBK equations was confirmed. This led the way for further research in this area, namely numerically solving the nonlinear two-fluid HVBK equations of motion of helium II.

3   Nonlinear Solutions

3.1   Infinite Cylinder Assumption

Using the infinite cylinder assumption, Henderson, Barenghi & Jones [17] numerically solved the HVBK equations for the first time to obtain the nonlinear flow of helium II between infinite cylinders. The aim of the work was to obtain solutions for helium II corresponding to what would correspond to Taylor vortices in a classical flow and to investigate what happens to the vortex lines. Axisymmetric solutions were considered since linear theory predicts that the axisymmetric mode onsets first. To solve equations (1a-1c), boundary conditions are required. The nonlinear problem is 6th order in both the normal fluid and superfluid, however the linear problem is only 2nd order in the superfluid. Thus two further boundary conditions for the superfluid are needed in addition to the no penetration of the boundary used successfully in the linear stability analysis. The extra boundary conditions employed were

$$\omega_\phi^s = 0 \quad \text{at} \quad r = R_1 \text{ and } R_2 \tag{5a}$$
$$v_\phi^s = \Omega r \quad \text{at} \quad r = R_1 \text{ and } R_2. \tag{5b}$$

Equations (5a,5b) force the superfluid vorticity to be purely axial at the cylinder walls. This is consistent with Couette flow, in which the vortex lines are purely axial throughout the flow and results in the mutual friction being small at the boundaries which is an advantage numerically. The normal fluid satisfies the standard no slip boundary conditions as for the linear model.

The HVBK equations were solved numerically using a pseudospectral method, based on expansions in Chebychev polynomials in the radial direction and trigonometric functions in the axial direction [18]. Results were obtained for angular velocities of up to 15% above the critical angular velocity at which linear theory predicts that Couette flow becomes unstable. Apart from the axial stretching of the Taylor cells, the normal fluid displays a velocity profile similar to that of a classical fluid, However the superfluid velocity profile is markedly different to the classical case; instead of a meridional flow consisting of single pair of cells in each period, we find a more complex pattern of eddies and counter-eddies. Perhaps what is of most interest is the orientation of the vortex lines. The numerical results show that the superfluid vorticity is still predominantly axial and the deflection in the azimuthal direction is smaller than that in the radial direction. Considering the (r, z) plane the vortex lines are most densely situated near the inner cylinder at positions of maximum inflow and are deflected towards the outer cylinder at the centre of the cell, where there is outflow.

There are two ways of comparing the numerical results with experiments. Firstly by comparing the additional attenuation of a second sound wave due to the vortex lines and secondly by measuring the torque exerted on the outer cylinder. We compared the relative change in the azimuthal attenuation coefficient at an angular velocity of 5.4% above the onset of Taylor vortices and found an order of magnitude correspondence between the experimental value of Swanson & Donnelly [16]. Our predicted value was lower than the observed value which could be due to the following points:

- The observed attenuation will depend on the spatial structure of the mode used to probe the flow.
- The observed attenuation does not discriminate the sense in which the vortex lines point, however the HVBK equations are derived using an averaged approach to the vorticity of the superfluid.

Although many experiments have been carried out to measure the torque in helium II, few experiments have been carried out in a parameter range such that the stability curve has a minimum at non-zero axial wavenumber. We compare our torque measurements with those of the experiment of Donnelly [19] in which values of the torque above the transition are reported. At small angular velocity, in the Couette flow regime, the torque is proportional to the viscosity μ of the fluid. At higher angular velocities there is a break in the curve, corresponding to the onset of Taylor vortices, the torque exerted on the outer cylinder increases as the axisymmetric taylor vortices are more efficient in transferring angular momentum than the azimuthal Couette motion. Excellent agreement was found [20] between the calculation and the experimental data in the nonlinear regime for $\Omega > \Omega_c$.

These two results validated the HVBK equations for the first time in the nonlinear regime. Although comparisons with available experiments in the nonlinear regime are encouraging, it is clear that end effects become important, even at relatively high temperatures.

3.2   Unit Aspect Ratio

In order to be able to compare with further experimental data, end effects need to be included in the model. Henderson & Barenghi [21] considered helium II contained within a cylindrical annulus of inner radius R_1, outer radius R_2, height H where the gap between the cylinders has been chosen such that $H = R_2 - R_1$. Thus the Couette annulus has unit aspect ratio, in that the gap between the cylinders is equal to the height between the endcaps. The inner cylinder rotates with constant angular velocity Ω, whilst the outer cylinder and two end plates are stationary. This simple flow configuration enabled us to study how the vortex lines respond to a shear in the presence of boundaries which are both parallel and perpendicular to the natural axial direction of the vortex lines. The axisymmetric form of the HVBK equations (1a-1c) were solved using a finite difference approach taking a regular grid in both the r and z direction. The boundary conditions on the curved cylinder walls were taken to be the same as for the infinite cylinder case. However extra boundary conditions are also needed on the two endcaps, $z = 0, H$. For the normal fluid standard no slip boundary conditions were imposed. Whilst for the superfluid the following were used

$$v_z^s(r, 0) = v_z^s(r, H) = 0, \tag{6a}$$

$$\omega_r^s(r, 0) = \omega_r^s(r, H) = 0, \tag{6b}$$

$$\omega_\phi^s(r, 0) = \omega_\phi^s(r, H) = 0. \tag{6c}$$

The first condition (6a) ensures that there is no penetration of the superfluid through the boundary, whilst the last two conditions (6b,6c) correspond to perfect sliding of the vortex lines as discussed by Khalatnikov [5].

The main result of this investigation is the anomalous motion of helium II when compared to the motion of a classical fluid. The velocity profile obtained is a superposition of an azimuthal motion v_ϕ around the inner cylinder and a toroidal motion v_r and v_z in the vertical plane. The latter motion is in the form of a pair of cells similar to a Taylor vortex pair, but being caused by boundaries rather than a centrifugal instability, it is hereafter referred to an Ekman cell pair. The first interesting finding is that v_ϕ^s is almost z-independent, that is the superfluid moves around the cylinders in a column-like fashion, which is due to the tension in the vortex lines. This effect becomes more pronounced at lower temperatures when the superfluid component is higher as is illustrated in Fig. 1a,b. Each figure extends over the whole computational domain with the inner cylinder and outer cylinder on the left and right respectively. In contrast, v_ϕ^n exhibits strong z-dependence due to the no-slip boundary conditions imposed on the normal fluid at the ends and walls of the cylinders and has a similar profile to that of a classical fluid, see Fig. 1c.

Fig. 1. Azimuthal motion v_ϕ of (a) the superfluid at $T = 1.8$ K, (b) the superfluid at $T = 2.17$ K, (c) the normal fluid at $T = 2.17$ K. Lighter regions correspond to larger magnitude.

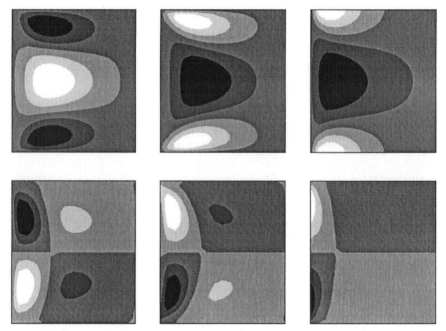

Fig. 2. Motion of helium II compared to a classical fluid. Classical: (a) radial velocity v_r; (b) axial velocity v_z. Helium II at $T = 2.11$ K: (c) v_r^n; (d) v_z^n; (e) v_r^s; (f) v_z^s. Lighter/darker regions correspond to positive/negative velocities.

The second interesting finding comes from looking at the Ekman cells in both the normal fluid and superfluid. In a classical fluid the two Ekman cells form with outflow at the centre and inflow at the ends of the cylinder, as in Fig. 2a,b. The results for helium II are quite different. We find that the superfluid Ekman cells always rotate in a counter-classical way due to the mutual friction force, that is outflow occurs at the ends of the cylinder with inflow at the centre, see Fig. 2c,d. It is also seen that the normal fluid Ekman cells rotate in a counter-classical way at lower temperatures, see Fig. 2e,f, but revert to a classical direction close to the transition temperature $T_\lambda = 2.172\mathrm{K}$ as one would expect. We also investigated the magnitude and direction of the superfluid vorticity in order to gain a picture of how the vortex lines are situated in the flow. As for the infinite cylinder case, the superfluid vorticity is primarily axial with small deflections in both the r and ϕ direction. However the vorticity is concentrated near the inner rotating cylinder, which is a measurable result.

4   Discussion

The excellent agreement between the linear stability analysis [15] and experimental data [16] was a rigorous test of the validity of the HVBK equations at least in the linear regime. The HVBK model has been validated further by the good agreement between the nonlinear calculation [18] and experimental data [19]. Obtaining nonlinear solutions for the flow of helium II between infinite and finite cylinders has allowed us to gain more insight into the flow which, because of the low temperature environment, cannot be observed directly like a classical fluid. It has also enabled us to explore the boundary conditions for the superfluid. A possible future direction of the work would be to extend the aspect ratio in order to investigate the transition to Taylor cells.

References

1. R.P. Feynman: 'Application of quantum mechanics to liquid helium.' In *Progress in Low Temperature Physics* **1**. (C.J. Gorter, North Holland 1955)
2. H.E. Hall, W.F. Vinen: Proc. Roy. Soc. London A **238**, 215 (1956)
3. H.E. Hall: Phil. Mag. Suppl. **9**, 89 (1960)
4. I.L. Bekharevich and I.M. Khalatnikov: Sov. Phys. JETP **13** , 643, (1961)
5. I.M. Khalatnikov: *An Introduction to the Theory Superfluidity*. (Benjamin 1965)
6. R.N. Hills, P.H. Roberts: Arch. Rat. Mech. Anal. **66**, 43 (1977)
7. C.F. Barenghi, R.J. Donnelly, W.F. Vinen: J. Low Temp. Phys. **52**, 189 (1983)
8. D.C. Samuels & R.J. Donnelly: Phys. Rev. Lett. **65**, 187 (1990)
9. G.I. Taylor: Phil. Trans. Roy. Soc. Lond. A **223**, 289 (1923)
10. F. Bielert, G. Stamm: Cryogenics **33**, 938 (1993)
11. P.L. Kapitza: J. Phys. USSR **4**, 181 (1941)
12. R.J. Donnelly, M.M. Lamar: J. Fluid Mech. **186**, 163 (1988)
13. S. Chandrasekhar, R.J. Donnelly: Proc. Roy. Soc. London A **241**, 9 (1957)
14. C.F. Barenghi, C.A. Jones: J. Fluid Mech. **197**, 551 (1988)
15. C.F. Barenghi: Phys. Rev. B **45**, 2290 (1992)
16. C.J. Swanson, R.J. Donnelly: Phys. Rev. Lett. **67**, 1578 (1991)
17. K.L Henderson, C.F. Barenghi, C.A. Jones: J. Fluid Mech. **283**, 329 (1995)
18. K.L. Henderson, C.F. Barenghi, J. Low Temp. Phys. **98**, 351 (1995)
19. R.J. Donnelly: Phys. Rev. Lett. **3**, 507 (1959)
20. K.L. Henderson, C.F. Barenghi: Phys. Lett. A **191**, 438 (1994)
21. K.L. Henderson and C.F. Barenghi: J. Fluid Mech. **406**, 199 (2000)

Part IV

Turbulence Theory

An Introduction
to the Theory of Superfluid Turbulence

W.F. Vinen

School of Physics and Astronomy, University of Birmingham, Birmingham B15 2TT, UK, and
Cryogenic Helium Turbulence Laboratory, Department of Physics, University of Oregon, Eugene, Oregon 97403, USA

1 Introduction

In one sense superfluid turbulence is an old subject: it was mentioned as a theoretical possibility by Feynman in 1955[1]; and it has been known experimentally since the early 1950s that flow of the superfluid component of helium II can become turbulent when there is a steady counterflow of the two fluids, such as occurs in a steady heat current[2]. The original experimental discovery was accompanied by the beginnings of a theory[3], and this theory has been developed steadily, especially by Schwarz[4,5], so that many aspects of this type of turbulence are now well understood. However, counterflow turbulence has no classical analogue, and it has attracted little interest from those who study classical fluid mechanics. Types of flow for which classical analogues do exist were observed by low temperature physicists for many years, but the presence of the two fluids were thought to make them very complicated, and they were not therefore studied in detail. More recently experiments have been reported on the analogue of a rather simple case of classical turbulence, namely that produced by steady flow through a grid[6,7]. In the classical analogue the turbulence is approximately homogeneous and isotropic, and its study has been important in the development an understanding of classical turbulence[8]. The superfluid analogue promises to be equally important.

In this paper I shall first describe some aspects of the theory of counterflow turbulence. But I shall then devote most of the paper to grid turbulence, where the theory is less well developed, although I shall make use of an important experimental result obtained with a more complicated type of flow generated by two counter-rotating discs[9]. My aim is to stimulate interest in the theory of superfluid turbulence, particularly, at this stage, in the simple case of grid turbulence, among both low temperature physicists and those with a background in classical fluid mechanics. I shall focus on open questions and unsolved problems, questions and problems that are clearly seen in grid turbulence, but which are more widely relevant. My own background is in experimental quantum fluids, and certainly not in theoretical fluid mechanics. I shall tend to speculate about what I see as the physics of superfluid turbulence, and others will tell me where my physical intuition is unreliable or, hopefully, where it can be developed along more rigorous lines.

Superfluid grid turbulence illustrates in perhaps its simplest form what is sometimes known as "vortex coupled superfluidity"; a turbulent state in which the superfluid and normal components seem to be coupled together and behave like a single classical fluid[10]. It has gradually become apparent that this type of superfluid turbulence is quite common in cases where the two fluids are not forced to move with different velocities. The ideas that we develop here in the context of grid turbulence are likely therefore to be widely applicable.

A turbulent superfluid velocity field must take the form of a tangled array of quantized vortex lines (circulation $\kappa = h/m_4$). Motion of the lines is governed by the classical Magnus effect. A force of mutual friction, f per unit length of line, acting effectively on the core of the line normal to its length, arises from the scattering of thermal excitations[11]. We shall assume that $f = \gamma (v_n - v_L)$, where v_n is the component of the velocity of the normal fluid perpendicular to the length of the vortex and v_L is the velocity of the line; for simplicity we ignore any transverse component of f. Numerical values of γ are given in reference[12].

A number of different approaches have been used to develop our understanding of superfluid turbulence. One relies on simple physical arguments backed by dimensional analysis and ideas of dynamical similarity; another relies on computer simulations. Perhaps there will in future be one that is based on rigorously established general principles; the fact that the turbulent velocity field in the superfluid component is due entirely to discrete quantized vortex lines may facilitate such a development (see the contribution of Gary Williams to this workshop). The simulations are of two types: those pioneered by Schwarz[4] and based on the idea that, except on very short length scales, quantized vortices can be regarded as vortex filaments moving according to classical fluid mechanics, with inclusion of the force of mutual friction f, which modifies the motion through the Magnus effect; and those based on the Gross-Pitaevskii or non-linear Schrodinger equation (NLSE), which includes a quantum description of the vortex core, of the thermal excitations, and of their interaction, albeit one that cannot be expected to be quantitatively applicable to liquid helium. For many purposes the classical vortex filament approach is probably sufficient, but it cannot give a fully satisfactory description of reconnections, which play an important role in superfluid turbulence.

We shall confine our discussion to turbulence in superfluid ^{4}He; the case for experiments on turbulence in superfluid ^{3}He is mentioned briefly in reference[13]. Experiments in which superfluid turbulence has probably been observed in ^{3}He are reported by Fisher at this Workshop.

2   Counterflow Turbulence

In counterflow turbulence the vortex tangle is believed to be at least approximately homogeneous, provided that the average velocities of the two fluids, V_n and V_s, are spatially uniform. The turbulence is maintained by the mutual friction. If the vortex lines move on average with the superfluid, the total average force of mutual friction per unit volume, F_{ns}, is equal to $\gamma L (V_n - V_s)$, where L

is the length of line per unit volume, and where we have ignored factors of order unity arising from the random orientation of the lines. We can illustrate one approach to superfluid turbulence by deriving the dependence of L on $(V_n - V_s)$ from a principle of dynamical similarity[3,14,15].

Let us assume that the vortex tangle is characterized by a single length, $\ell = L^{-1/2}$, characteristic of both the vortex line spacing and the vortex radius of curvature. Taking into account the Magnus effect and the force f, we can easily show that the velocity with which any element of line moves is given by

$$\mathbf{v}_L = \mathbf{V}_s + \mathbf{v}_s + \alpha\hat{\kappa} \times (\mathbf{V}_n - \mathbf{V}_s - \mathbf{v}_s),\tag{1}$$

where $\mathbf{v}_s$ is the superfluid velocity at the element due to the rest of tangle, $\hat{\kappa}$ is the unit vector parallel to the element, and $\alpha \approx \gamma/\rho_s\kappa$. There are two contributions to the magnitude of $\mathbf{v}_s$: $v_{s1} \approx \kappa/2\pi\ell$, which is due to neighbouring lines at the distance ℓ from the element; and $v_{s2} \approx (\kappa/4\pi\ell)\ln(\ell/\xi_0)$, which is due to the local curvature of the line. (We note in passing that computations based on the "localized induction approximation" take into account only the latter contribution.)

Suppose that we change the length scale in the vortex tangle by a factor g, so that $\ell \rightarrow g\ell$, and let us ignore the logarithmic factor in v_{s2}. Then $\mathbf{v}_s$ changes by the factor g^{-1}. If we change $\mathbf{V}_s$ and $\mathbf{V}_n$ by the same factor, we see from equation (1) that $\mathbf{v}_L$ is also changed by the same factor (formally, we change the length scale by g and the time scale by g^2). Therefore the whole tangle evolves in the same way as it would before scaling, except for the scaling factor g, suggesting that there is a principle of *dynamical scaling*. Application of this principle shows easily that ℓ is proportional to $(V_s - V_n)^{-1}$, so that the mutual friction per unit volume is proportional to $(V_s - V_n)^3$, as is observed to be approximately true. Taking into account the logarithmic term in v_{s2} introduces logarithmic corrections, which are indeed probably observed.

As we have mentioned, the superfluid turbulence is maintained by the mutual friction acting on individual elements of line. That the length of line can in principle increase as a result of the mutual friction is clear from the fact[3] that a vortex ring can grow if the self induced velocity of the ring is in the direction of, but less in magnitude than, $(\mathbf{V}_n - \mathbf{V}_s)$; otherwise it will shrink. Appropriately oriented parts of the tangle with low curvature can behave in a similar way, although it is not obvious that a steady turbulent state (a finite L) can be maintained. A detailed understanding of counterflow turbulence came only from the simulations of Schwarz[4,5], which showed that a steady turbulent state can be achieved through the effect of reconnections, which generate vortex configurations that favour the growth of line.

This theory of counterflow turbulence is based on the assumption that the flow of the normal fluid is laminar. The turbulent flow of the superfluid must then occur on length scales not significantly larger than ℓ; flow on larger length scales would be damped out by mutual friction. Recently it has been suggested by Melotte and Barenghi[16] that the laminar flow of the normal fluid in counterflow may not always be stable, so that both fluids may become turbulent, probably on length scales significantly larger than ℓ. The theory for such a regime presents us with a major challenge, which we mention again in Sect. 3.5.

3   Grid Turbulence in Superfluid Helium

Experiments on grid turbulence in superfluid helium at temperatures above 1K have been described by Niemela at this workshop; the original measurements were reported in reference[7]. I shall first remind you of the experimental results and of the idea that they are consistent with a quasi-classical model, as discussed by Skrbek at this workshop. Then I shall discuss why this model may work. Finally, I shall extend the discussion to lower temperatures, where there are so far no really satisfactory experimental results, but where new and potentially interesting problems arise.

3.1   Measurements of the Decay of Vortex Lines, and the Quasi-classical Model

In the experiments a grid is towed through the helium, and a measurement is made of the time dependence of the excess attenuation of second sound in a small fixed region in the helium behind the grid. The excess attenuation is caused by mutual friction associated with vortex lines in the turbulent superfluid behind the grid, and the measurements lead to data showing the decay of the line density, L, with time. The grid moves through the helium at a velocity exceeding both that required to create vortex lines (the mechanism need not concern us), and that required to create turbulence in the normal fluid. Turbulence is therefore produced in both fluids. The average velocity of each fluid must vanish (there is nothing to maintain a counterflow, and any transient counterflow would be damped rapidly by mutual friction), so the situation is quite different from that discussed in Sect. 2.

As explained by Skrbek, the observed decay in L is consistent with a *quasi-classical model*, based on the following *two* assumptions. *First*, on length scales larger than the vortex line spacing, ℓ, which turns out to be close to the length scale at which viscous dissipation is expected to occur in the turbulent normal fluid, the two fluids are coupled together in the sense that the two velocity fields are the same; the coupled fluids behave like a single classical fluid, the turbulence exhibiting an inertial range of wavenumbers in which there is negligible dissipation and in which the energy spectrum has the classical Kolmogorov form

$$E(k) = C\epsilon^{2/3}k^{-5/3}, \tag{2}$$

where C is a constant of order unity, and ϵ is the rate of energy dissipation per unit mass of helium at a high wavenumber, presumably of order or greater than ℓ^{-1}. (Roughly speaking, energy is injected from the grid into eddies with size of order the mesh size (wavenumbers of order the reciprocal of the mesh size); non-linear terms in the equation of motion cause the energy to flow to smaller eddies (higher wavenumbers) in a cascade, until it can be dissipated by viscosity.) *Secondly*, this rate of dissipation is given by

$$\epsilon = \nu' \kappa^2 L^2, \tag{3}$$

where ν' has the dimensions of kinematic viscosity. If we interpret $\kappa^2 L^2$ as an effective mean square vorticity in the superfluid, which is of doubtful validity[17,13], equation (3) is reminiscent of the rate of viscous dissipation in a turbulent classical fluid. Of course, as in a classical fluid, the Kolmogorov spectrum may represent an oversimplification, which fails to take account of, for example, intermittency (see, for example, reference[18]); but it seems reasonable to ignore this point at the present stage in the development of our understanding of superfluid turbulence.

It is important to appreciate that we can say only that the experimental data on grid turbulence are *consistent* with this quasi-classical model. Other models might also be consistent: for example, one in which both the dissipation and the energy spectrum take different forms (the energy spectrum could involve the quantum of circulation). But the quasi-classical model is perhaps the simplest model that will account for the experiments, and, as we shall argue, it can be given some theoretical backing and is consistent with other experiments.

There is clearly a need to find direct experimental evidence for the validity of the Kolmogorov spectrum (2) in superfluid grid turbulence; i.e. a direct measurement of the turbulent energy spectrum. We know of no such evidence for grid turbulence, but evidence does exist for a more complicated type of flow generated by two counter-rotating discs. This is from the important work of Maurer and Tabeling[9], who measured pressure fluctuations in this type of flow (on a rather course length scale), both above and below the superfluid phase transition. They find that over a certain range of frequencies these fluctuations do have a Kolmogorov spectrum, the spectrum being the same above and below the transition. It seems likely therefore that the value of the Kolmogorov constant, C, is the same above and below the phase transition and equal to the value expected for a classical fluid. We assume that a similar result will hold for grid turbulence. In that case the quasi-classical model of grid turbulence would account unambiguously for the experimental results.

3.2   Superfluid Turbulence on Length Scales Larger than the Vortex Line Spacing

Consider flow of the superfluid component on length scales large compared with ℓ, and suppose that this flow can be achieved by a relatively small re-arrangement of the vortex tangle, as turns out to be the case in grid turbulence[17]. Such a flow can be characterized by a velocity field that is similar to that found in a classical fluid, and we suggest that the dynamical behaviour of the superfluid is then similar to that of a classical fluid with the same density at high Reynolds number. An example, not involving turbulent flow, where the truth of this suggestion can be verified is provided by wave motion in an otherwise uniformly rotating superfluid, which contains a uniform array of lines. Waves with wavevectors directed parallel to the axis of rotation have a character that depends on the magnitude (q) of the wavevector relative to the line spacing ℓ. If $q\ell \gg 1$ the waves are Kelvin waves on the individual vortices; if $q\ell \ll 1$ the waves become indistinguishable from the classical "inertia waves" found in a classical rotating liquid (see, for example, references[19,20]. In the case of turbulent flow we must remember that the non-linear term in the Navier-Stokes equation couples motion on different length scales, so that the validity of our suggestion depends on the hypothesis of the "independence of Fourier components for distant wavevectors"[8]. But we know of no formal proof of this validity.

We emphasize that this similarity between superfluid and classical flow cannot extend to wavenumbers of order or greater than ℓ^{-1}, where the discrete nature of the vorticity cannot be ignored.

3.3   The Turbulent Energy Spectra in Superfluid Grid Turbulence

We can now start to see some theoretical justification for the quasi-classical model. We see that when the superfluid component flows with sufficient speed through a grid it could lead to turbulence in that component, characterized by the Kolmogorov spectrum for $k \ll \ell^{-1}$, as in a classical fluid. When both fluids flow through the grid two such turbulent flows could be generated. However, the presence of vortex lines in the superfluid component gives rise to mutual friction, which must tend to couple the two fluids. It can be shown[17] that, provided the normal fluid flow is not significantly affected by viscosity, and provided that $k \ll \ell^{-1}$, this friction is sufficient to ensure that the two fluids have associated with them the *same velocity fields*. (The demonstration involves a proof that the time required for mutual friction to eliminate relative motion in the two fluids on a length scale k^{-1} is considerably less than the ("turnover") time for a turbulent eddy of size k^{-1} to lose its energy by inertial transfer to other eddies.) Given, as we have seen, that viscous dissipation is expected to occur in the normal fluid only for $k \geq \ell^{-1}$, we can begin to understand the success of the quasi-classical model.

The type of coupled motion of the two fluids that we are discussing here relates of course to the vortex coupled superfluidity that we mentioned in Sect. 1. The basic idea of such coupling is presumably more widely applicable than to the simple case of grid turbulence; examples are provided by the flow investigated by Maurer and Tabeling[9], to which we have already referred, and to that round a sphere investigated recently by Smith et al[21]. The theory advanced here relies on very general arguments; support from the theory of specific types of flow, albeit very idealised, has been provided by Barenghi and his colleagues[33,23].

We emphasize that on length scales comparable with or less than ℓ, where the flow of the superfluid component is strongly constrained by the fact that vorticity can be associated only with discrete quantized vortex lines, the two velocity fields cannot be the same, even in the absence of viscous dissipation in the normal fluid.

The quasi-classical model requires not only the coupled motion of the two fluids for $k\ell \ll 1$ but also the validity of equation (3) for the total rate of dissipation of turbulent energy, both fluids contributing to this energy. It is far from obvious that equation (3) is correct.

There is also the question of the value of ν'. It turns out[7] that the existing experiments on grid turbulence yield only the ratio C^3/ν'. There has been much private discussion about the value of the Kolmogorov constant C that ought to be used. In the case of turbulence in a classical fluid C seems to be universal (within fairly large experimental error and at reasonably high Reynolds number) and equal to about 1.6[24]. The suggestion has been made that in the case of superfluid grid turbulence we are dealing with a quantum liquid, so that C might take a different value, which is perhaps temperature dependent. However, if we accept evidence from the experiments of Maurer and Tabeling[9], to which we have already referred, then C has its classical, temperature-independent, value. We conclude therefore, at least for the present, that it is sensible to use the value $C = 1.6$ to deduce the value of ν' from experiment, as is done by Niemela and Skrbek at this Workshop. We remark that this value has the same order of magnitude as η_n/ρ, where η_n is the viscosity of the normal fluid and ρ is the total density of the helium, but that its temperature dependence is quite different. Further development of our theoretical discussion requires therefore both a demonstration that equation (3) is at least reasonable and some discussion of the observed value and temperature dependence of ν'. It turns out that this discussion is best postponed until we have discussed superfluid turbulence at very low temperatures.

It may be relevant to add that the classical Kolmogorov spectrum does not depend for its validity on the Navier-Stokes equation; it depends only on the principle that the statistical properties of the turbulence are determined uniquely and universally by the length scale involved and the rate of energy dissipation at high wavenumbers. But it does depend on the absence of quantum effects, which would introduce Planck's constant, probably in the form of the quantum of circulation, κ, as a relevant parameter. If quantum effects were important, the spectrum of the turbulent energy might take the form $E(k) = \epsilon^{2/3}k^{-5/3}H(\epsilon\kappa^{-3}k^{-4})$, allowed by dimensional analysis, where H is some unknown function. Although this may not be ruled out by the experiments on superfluid grid turbulence, it does seem to be inconsistent with the experiments of Maurer and Tabeling[9].

3.4   Superfluid Turbulence at Very Low Temperatures

The experiments on grid turbulence that we have discussed so far were carried out at temperatures above 1K, where there is a significant fraction of normal fluid. We focus next on lower temperatures, including those so low that the fraction of normal fluid can be neglected. No experiments have yet been reported on superfluid turbulence produced by steady flow through a grid at low temperatures. Preliminary study of turbulence produced by an oscillating grid has been reported by Davis et al[25], and the latest results will be reported at this Workshop by McClintock; but this type of turbulence is not homogeneous and may be difficult to interpret at this stage. Two simulations relevant to low temperatures have been reported. Tsubota et al[26] have studied the decay of turbulence by the Schwarz technique. The turbulence is produced initially by simulated counterflow at a high temperature: the counterflow is then turned off; the temperature is changed to the value required; and the decay is then simulated. Nore et al[27] perform a simulation relevant to zero temperature based on a solution of the NLSE, with a Taylor-Green vortex as an initial flow. Our own approach to low-temperature grid turbulence[17], which is quite speculative, has been to try to identify the essential physics, with the help of an interesting paper by Svistunov[28], and with confirmation where possible from the simulations[28]. See also the paper by Lipniacki at this Workshop.

Consider first the case of zero temperature, when there is no normal fluid. We guess that flow through the grid leads to a Kolmogorov spectrum for $k\ell \ll 1$, as explained in Sect. 3.3. The simulations of Nore et al tend to support this guess. Energy flows towards smaller length scales (larger wavenumbers) in a classical Kolmogorov cascade until it reaches length scales of order ℓ (wavenumbers of order ℓ^{-1}), at which, as we emphasized in Sects. 3.2 and 3.3, this classical cascade can no longer exist. However, there is no mechanism for energy dissipation at this length scale: there is no mutual friction; and oscillatory motion of the lines is at such a relatively low frequency that the radiation of sound (phonon generation) is quite ineffective. Energy must therefore flow into structures smaller than ℓ, although flow at such large wavevectors cannot be described by any extension of the Kolmogorov spectrum. Such structures can form only if vortex lines come close together in some parts of the turbulent field.

This close approach of vortex lines was tentatively foreseen in reference[3], and it became very evident, even at higher temperatures, in the simulations of Schwarz[4], where it can lead, according to Schwarz, to the reconnections that are necessary to maintain counterflow turbulence (see Sect. 2). Whether or not reconnections occur, the close approach will lead to the formation of kinks on the lines (one form of small-scale structure). For reasons that we mention later, these kinks are more prominent at low temperatures, and they are seen very clearly in the simulations of Tsubota el al[26]. The evolution of the kinks must involve strongly non-linear processes, but crudely we can decompose the kinks into their Fourier components, which are harmonic Kelvin waves propagating along the lines with different wavevectors. The non-linear effects will lead to the transfer of energy to Kelvin waves of larger and larger wavenumber and frequency, until the frequency is high enough for the efficient production of phonons. Perhaps therefore the turbulence is characterized by two cascades: energy fed in at the grid first flows down a conventional Kolmogorov cascade ($k\ell < 1$) and then down a Kelvin wave cascade ($k\ell > 1$), until it is dissipated by phonon production. Recent simulations by Araki et al[29] and by Kivotides (this Workshop) give support for this view. However, the two types of cascade may be different in that energy may be fed into the Kelvin wave cascade over a wide range of wavenumbers, and not merely at the smallest relevant wavenumber.

Reconnections might lead to other processes by which turbulent energy is lost. When Schwarz demonstrated the importance of reconnections in counterflow turbulence, he did not have a microscopic theory of a reconnection; his simulations were based on a model of vortex lines as classical filaments, and he simply assumed that a reconnection would occur when two lines came sufficiently close together[4,26]. Simulations based on the NLSE can provide a microscopic theory, albeit subject to the limited extent to which the NLSE provides a good description of helium. Such simulations were performed by Koplik and Levine[30], and new simulations are reported by Adams at this Workshop. The work of Koplik and Levine suggested that reconnections can occur without energy loss; i.e. without the production of thermal excitations in the form of phonons. However, the more sensitive simulations by Adams show that phonon production probably does occur, and we guess that in real helium both phonons and rotons are produced. (Phonons are produced in the simulations of Nore et al[27], but it is not clear whether they are produced by vortex oscillations or reconnections. See also[31].) But we note that an energy loss per reconnection of order $\rho\kappa^2\xi_0$, a not unreasonable guess and one that is consistent with the Adams simulations, can be shown to have a negligible effect on the overall vortex decay rate. Reconnections can also lead to the production of small vortex rings (another form of small-scale structure), as is clear from the simulations of Tsubota et al[26]. This does not lead in itself to dissipation into thermal excitations, and in an infinite system the small rings may well be re-absorbed in subsequent reconnections. But in the case where the turbulence is confined to a finite region the rings might escape with a consequent effective loss of energy.

In this section we have so far considered only the case of strictly zero temperature. At a finite temperature there is some normal fluid, which leads to the force of mutual friction, f, introduced in Section 1. (At low temperatures the viscosity of the normal fluid is high and its density low, so turbulence in the normal fluid is unlikely to occur.) The mutual friction will lead to additional damping of the Kelvin waves and to shrinkage of any small rings. The additional damping of the Kelvin waves increases as the temperature rises, and it probably exceeds that due to phonon emission at temperatures exceeding about 0.4K. (This result is uncertain because it is based on what is really a guess at the rate of phonon emission from a Kelvin wave[17,35]. Here is another problem that needs rigorous solution.) At still higher temperatures, approaching or greater than 1K, the damping of the Kelvin waves becomes so large that they can hardly exist. Vortex motion on the length scale ℓ is then itself strongly damped by mutual friction, and energy flow into the Kelvin wave cascade is inhibited; the vortex lines become less kinked. The effect of the mutual friction in causing small rings to collapse needs further study. Indeed the role of vortex rings in general, which has been emphasized by Tsubota et al[26], requires further study.

These ideas can be used to estimate the rate of dissipation of turbulent energy, ϵ, at low temperatures[17,28]. It seems that the rate of energy loss is likely to be dominated by the rate at which energy flows from motion at length scales of order ℓ to smaller length scales. Probably therefore it is dominated by the time scale associated with vortex motion on the length scale ℓ, which is of order ℓ^2/κ. This idea leads *via* a scaling analysis to a rate of loss of vortex line given approximately by

$$\frac{dL}{dt} = -\chi_2 \frac{\kappa}{2\pi} L^2, \tag{4}$$

where χ_2 is a constant of order unity. More accurately, χ_2 is a weak (logarithmic) function of L and of other parameters such as the Kelvin wave cut-off due to phonon emission or mutual friction (this more accurate form arises from the need to take account of the relatively small length of line associated with the Kelvin waves and small rings that exist on length scales less than ℓ). The corresponding rate of energy loss is given approximately by

$$\epsilon = \nu'' \kappa^2 L^2, \tag{5}$$

where the parameter ν'' is of order κ but weakly dependent on L and temperature[17]. We see that equation (5) has the same form, approximately, as equation (3), and, as a result of a numerical accident, the magnitude of ν'' is similar to the magnitude of ν' deduced from the towed grid experiments at high temperatures. The dissipation in the turbulent helium may therefore in a sense be characterized by an effective kinematic viscosity that is not very dependent on temperature, even at low temperatures, in spite of the fact that the mechanism of energy dissipation is different at different temperatures.

A rate of loss of vortex line given by equation (4) has been observed in the simulations of Tsubota et al[26]. These simulations lack the spatial resolution required to include processes occurring on length scales much less than ℓ, and they use the assumption that small rings are removed from the turbulence by some unknown mechanism. The corresponding values of χ_2 are not therefore very reliable, although an expected increase in χ_2 with increasing temperature due to an increasing mutual friction is confirmed. One lesson that emerges from a comparison of the calculations underlying equation (4) with the simulations is that ϵ is probably rather insensitive to the details of the dissipation occurring at wavenumbers greater than ℓ^{-1}.

We emphasize that the ideas about superfluid turbulence outlined in this section are very speculative and certainly not based on rigorous analysis. There are serious gaps in the argument, relating, for example, to the fate of small vortex rings. There are no experimental results with which to compare our predictions, except in so far as the experiments of Davis et al[25] do probably confirm that energy loss occurs at low temperatures at a rate not very different from that at high temperatures. There is an urgent need for more experiments at low temperatures, but, as discussed elsewhere at this Workshop, the technical difficulties are formidable.

3.5   Dissipation at Higher Temperatures

We must now return to dissipation in grid turbulence at higher temperatures, above 1K, where our aim must be to understand the form of equation (3) and the observed magnitude of ν'. In contrast to the situation at the lower temperatures, both fluids are now turbulent, and the total energy dissipation must include contributions from both fluids. (It has recently been emphasized by Idowu *el al*[32] that viscous dissipation can occur in the normal fluid even if the normal fluid is not turbulent, because as a vortex moves through the normal fluid it drags part of that fluid with it[11]; however, the resulting viscous dissipation is normally taken into account in the drag coefficient, γ, appearing in the mutual friction force (Sect. 1). But the local drag, extending possibly over a distance of order ℓ, might affect any small-scale normal-fluid turbulence.) Viscous dissipation will occur in the turbulent normal fluid, and there will be dissipation due to mutual friction, the loss of energy in the superfluid being caused, as at somewhat lower temperatures, by the mutual friction. Estimates of the total rate of dissipation are difficult to make because the dissipation is occurring on a length scale of order ℓ, where the discrete vortex structure is important and the two velocity fields cannot be the same. Furthermore, the normal fluid is turbulent. Any simulation, for example, would require the self-consistent determination of the vortex motion and the motion of the normal fluid, taking into account the coupling due to mutual friction. The tools required for the solution of this difficult type of problem are being developed[33,34], but they have not yet been applied to realistic problems.

The situation may be more simple at temperatures below about 1.6K, but above those at which the considerations of Sect. 3.4 apply. It can be argued[35], on the basis of a consideration of the various characteristic times involved, that viscous dissipation in the normal fluid then occurs at wavenumbers that are slightly, but significantly, smaller than ℓ^{-1}, and that mutual friction has a rather small effect at these smaller wavenumbers, in spite of the fact that the motion in the two fluids has become decoupled (the arguments are similar to those used in reference[17] in connection with the coupling between the two fluids). Energy dissipation in the superfluid then takes place by mutual friction at wavenumbers close to ℓ^{-1} in a normal fluid that is at rest at these wavenumbers. The resulting dissipation from the superfluid could then be very similar to that computed in the simulations of Tsubota et al[26]. The total dissipation can then be calculated without difficulty. The result is consistent with the form of (3), and the calculated value of ν' as a function of temperature is quite close to that obtained from the experiments. But more work is required before (3) can be understood fully.

4   Summary and Conclusions

The arguments presented in this paper have been speculative, but they offer a picture of superfluid turbulence, particularly that formed behind a moving grid, that is attractive and probably consistent with the presently available experimental results. But there remain unsolved problems, and the whole speculative approach requires rigorous appraisal. The situation for temperatures below 1K is especially unsatisfactory, because as yet there exist virtually no relevant experimental results. From a practical point of view it is interesting to note that in turbulent flow superfluid helium seems to behave in many ways like a classical fluid with a kinematic viscosity similar to that of helium I, even at very low temperatures; therefore any hope that the "vanishing viscosity"of superfluid helium might allow the study of flow at phenomenally high Reynolds numbers seems misplaced.

Acknowledgements

I am very grateful to Steve Stalp, Ladislav Skrbek, Russell Donnelly, Joe Niemela, Makoto Tsubota, Carlo Barenghi, David Samuels, Peter McClintock and Edouard Sonin for allowing me to see their results (experimental or theoretical) before publication and for invaluable discussions. My work has been supported in part by NSF Grant DMR-9529609.

References

1. R. P. Feynman, in *Progress in Low Temperature Physics*, **1**, ch.2 ed. C. J. Gorter (North Holland Publishing Co., Amsterdam, 1955)
2. W. F. Vinen, Proc. Roy. Soc. **A240**, 114, 128 (1957)
3. W. F. Vinen, Proc. Roy. Soc. **A242**, 493 (1957)
4. K. W. Schwarz, Phys. Rev. **B31**, 5782 (1985)
5. K. W. Schwarz, Phys. Rev. **B38**, 2398 (1988)
6. M. R. Smith, R. J. Donnelly, N. Goldenfeld, and W. F. Vinen, Phys. Rev. Letters **71**, 2583 (1993)
7. S. R. Stalp, L. Skrbek, and J. J. Donnelly, Phys. Rev. Letters **82**, 4831 (1999)
8. G. K. Batchelor, *The Theory of Homogeneous Turbulence* (Cambridge University Press, Cambridge, 1953)
9. J. Maurer and P. Tabeling, Europhys. Letters, **43**, 29 (1998)
10. R. J. Donnelly, in *High Reynolds Number Flows Using Liquid and Gaseous Helium*(Springer-Verlag, New York, 1991)
11. H. E. Hall and W. F. Vinen, Proc. Roy. Soc. **A238**, 215 (1956)
12. C. F. Barenghi, R. J. Donnelly, and W. F. Vinen, J. Low Temp. Physics, **52**, 189 (1983)
13. W. F. Vinen, J. Low Temp. Physics, **121**, 367 (2000).
14. K. W. Schwarz, Phys. Rev. Letters, **49**, 283 (1982)
15. C. E. Swanson and R. J. Donnelly, J. Low Temp. Physics, **61**, 363 (1985)
16. D. J. Melotte and C. F. Barenghi, Phys. Rev. Letters **80**, 4181 (1998)
17. W. F. Vinen, Phys. Rev. **B61**, 1410 (2000)
18. U. Frisch, *Turbulence* (Cambridge University Press, Cambridge, 1995)
19. H. E. Hall, Proc. Roy. Soc. **A245**, 546 (1958)
20. R. J. Donnelly, *Quantized Vortices in Helium II*(Cambridge University Press, Cambridge, 1991)
21. M. R. Smith, D. K. Hilton, and S. W. Van Sciver, Phys. Fluids **11**, 751 (1999)
22. C. F. Barenghi, D.C. Samuels, G. H. Bauer, and R. J. Donnelly, Phys. Fluids **9**, 2631 (1997)
23. C. F. Barenghi and D. C. Samuels, Phys. Rev. **B60**, 1252 (1999)
24. K. R. Sreenivasan, Phys. Fluids, **7**, 2778 (1995)
25. S. I. Davis, P. C. Hendry, and P. V. E. McClintock, *Proceedings of the 22nd International Conference on Low Temperature Physics, Espoo and Helsinki* to be published in Physica B

26. M. Tsubota, T. Araki, and S. K. Nemirovskii, Phys. Rev. B**62**, 11751 (2000).
27. C. Nore, M. Abid, and M. E. Brachet, Phys. Rev. Letters, **78**, 3896 (1993), Phys. Fluids, **9**, 2644 (1997)
28. B. V. Svistunov, Phys. Rev B**52**, 3647 (1995)
29. T. Araki and M. Tsubota, J. Low Temp. Physics, **121**, 405 (2000)
30. J. Koplik and H. Levine, Phys. Rev. Letters, **71**, 1375 (1997)
31. M. Tsubota, S. Ogawa, and Y. Hattori, J. Low Temp. Physics, **121**, 435 (2000)
32. O. C. Idowu, A. Willis, C. F. Barenghi, and D, C, Samuels, Phys. Rev. B
33. C. F. Barenghi and D. C. Samuels, Phys. Rev. B**60**, 1252 (1999)
34. C. F. Barenghi, D. Kivotides, O. C. Idowu and D. C. Samuels, J. Low Temp. Physics, **121**, 377 (2000).
35. W. F. Vinen, to be published

Numerical Methods for Coupled Normal-Fluid and Superfluid Flows in Helium II

Olusola C. Idowu, Demosthenes Kivotides,
Carlo F. Barenghi, and David C. Samuels

Dept. of Mathematics, Univ. of Newcastle, Newcastle upon Tyne, NE1 7RU, UK

Abstract. Helium II at temperatures below T_λ can be described as a superposition of two interacting fluids. A normal-fluid which has non-zero viscosity, and an inviscid superfluid in which vorticity is confined to quantised vortex filaments. These two fluids are coupled together by the mutual friction force. To model the normal-fluid flow we use the forced Navier–Stokes equation while the superfluid flow is modelled as motion of quantised vortex lines. In this article we describe the numerical methods used for the coupled motion of the two fluids. We also briefly discuss the detailed flow structures observed from the two-dimensional (2-D) and the three-dimensional (3-D) simulations of this problem.

1   Introduction

Until recently, simulations of helium II flows have mainly involved calculating the motion of superfluid vortex filaments under the mutual friction forcing from an imposed normal-fluid flow [1]. A few studies have considered the reverse problem, the calculation of properties of the normal-fluid flow due to a given mutual friction forcing from the superfluid [2]. The problem of calculating the fully coupled motion of both the normal-fluid and the superfluid components of helium II is now being investigated, and we present here our first results from these simulations [3]. We present the two-dimensional calculation in some detail, since the localised nature of the mutual friction force on the normal-fluid is unusual in fluid dynamics. The results of the closely related three-dimensional calculations are also described but investigations are still preliminary in three-dimensions due to the high computational costs. We end with a short and speculative discussion of the role of these coupled flows in the generation of complex normal-fluid flows through the triggering of flow instabilities.

2   The Self-Consistent Equation of Motion

The normal-fluid is a classical fluid and it evolves with the forced Navier–Stokes equation.

$$\frac{\partial \boldsymbol{v}_n}{\partial t} + \boldsymbol{v}_n \cdot \nabla \boldsymbol{v}_n = -\frac{1}{\rho}\nabla p + \nu_n \nabla^2 \boldsymbol{v}_n + \frac{1}{\rho_n}\boldsymbol{F} \ , \tag{1}$$

where p is the pressure, $\nu_n = \mu/\rho_n$ is the normal-fluid kinematic viscosity, μ is the viscosity and $\boldsymbol{F}$ is the mutual friction force. Also, Eq. (1) must be supplemented with the incompressibility condition. The mutual friction force is related to the drag force on a superfluid vortex line which is a force per unit length of the vortex line. To obtain the mutual friction force, we divide the drag force by the normal-fluid area containing the superfluid vortex line. The length scale over which the mutual friction force acts is quite small and can be considered to be approximately a *delta function force*, non-zero only along the one-dimensional superfluid vortex lines embedded in the 3-D normal-fluid flow.

In the fully coupled calculation of the local normal-fluid flow and the superfluid vortex motion the velocity of each superfluid vortex line is given as [6]

$$v_l = h_1(V_s + v_s) + h_2 s' \times (v_n - V_s - v_s) + h_3 v_n , \tag{2}$$

where

$$h_1 = \frac{\rho_s \kappa D_0}{D_0^2 + D^2} , \tag{3a}$$

$$h_2 = \frac{\rho_s \kappa D}{D_0^2 + D^2} , \tag{3b}$$

$$h_3 = \frac{D^2 - D_0 D_t}{D_0^2 + D^2} , \tag{3c}$$

$$D_0 = \rho_s \kappa - D_t , \tag{3d}$$

are friction coefficients and s' is a unit vector in the direction of the vortex line. V_s is the averaged superfluid velocity, v_s is the self-induced velocity of the vortex line and κ is the quantum of circulation. The coefficients D and D_t are the microscopic mutual friction drag coefficients defined as

$$D = \frac{\rho_s \kappa}{q} [\alpha - \epsilon(\alpha^2 + \alpha'^2)] , \tag{4}$$

$$D_t = \frac{\rho_s \kappa}{q} [\alpha^2 + \alpha'^2 - \alpha'] , \tag{5}$$

where

$$q = 1 - 2(\alpha \epsilon + \alpha') + (\epsilon^2 + 1)(\alpha^2 + \alpha'^2) , \tag{6}$$

and $\epsilon = \rho_s \kappa / E$. E is a temperature dependent constant. The coefficients α and α' in Eq. (4) and Eq. (5) are defined from the Hall and Vinen coefficients B and B' [5–8]. $\alpha = \rho_n B/2\rho$, and $\alpha' = \rho_n B'/2\rho$. Note that the normal-fluid velocity here is not the average field normal-fluid velocity V_n but the spatially rapidly varying field v_n which is a solution of the modified Navier–Stokes equation, Eq. (1).

3   Numerical Methods for 2-D Flows

The direct numerical simulation method described in this section calculates the coupled motion of a 2-D normal-fluid flow with one or more superfluid vortex points (i.e. cross sections through the superfluid vortex lines). This is a fully-coupled, self-consistent two-fluid calculation, solving for both the normal-fluid velocity $v_n(\mathbf{x}, t)$ and the motion of the superfluid vortex filaments $v_l(\mathbf{x}, t)$. Restricting our flow to the two-dimensional case still captures the most essential physics of the problem and allows us to use a reasonably high grid resolution. We used a grid resolution of 128^2 for the results presented in this section.

In this calculation, we assume that V_s, the externally applied superfluid velocity in Eq. (2), is a uniform and constant flow. For a single vortex filament, the self-induced velocity $v_s = 0$ (in 2-D). The unit vector s is the direction vector of each superfluid vortex (either $+\hat{z}$ or $-\hat{z}$ in the two-dimensional case). Since we include the local motion of the normal-fluid in response to the superfluid vortex, we must use a mutual friction force based on this local normal-fluid velocity, not the mutual friction force based on the averaged normal-fluid velocity [6].

We used the stream function formulation to model the normal-fluid flow in 2-D. The stream function ψ_n for the normal-fluid is defined as [11,12]

$$v_n = \left(\frac{\partial \psi_n}{\partial y}, -\frac{\partial \psi_n}{\partial x}, 0 \right). \tag{7}$$

This definition ensures that the incompressibility condition is satisfied. The stream function ψ_n is determined by the Poisson equation in 2-D

$$\nabla^2 \psi_n + \omega_{n,z} = 0 \tag{8}$$

where $\omega_{n,z}$ is the z component of normal-fluid vorticity.

The equation for the normal-fluid vorticity vector $\boldsymbol{\omega}_n$ is given as

$$\rho_n \left(\frac{\partial \boldsymbol{\omega}_n}{\partial t} + (v_n \cdot \nabla)\boldsymbol{\omega}_n - (\boldsymbol{\omega}_n \cdot \nabla)v_n \right) = \mu \nabla^2 \boldsymbol{\omega}_n + \nabla \times \mathbf{F} \tag{9}$$

In 2-D flow only the $\hat{z}$ component of this equation is non-zero and the vortex stretching term $(\boldsymbol{\omega}_n \cdot \nabla)v_n$ disappears. The mutual friction force per unit area, $\mathbf{F}$, exerted on the normal-fluid by the superfluid vortex points can be written as [3,6,13]

$$\mathbf{F} = \frac{D}{A} \sum_p \hat{s} \times [\hat{s} \times (\mathbf{v}_n - \mathbf{v}_l)] + \frac{D_t}{A} \sum_p \hat{s} \times (\mathbf{v}_n - \mathbf{v}_l) \tag{10}$$

where A is the area of the normal-fluid over which the mutual friction force is distributed and p is the number of vortex lines. The force $\mathbf{F}$ is a function of position and is non-zero only in areas of size A (taken to be the computational grid spacing) which contain superfluid vortex points.

The motion of the superfluid vortex line given by Eq. (2), is an ordinary differential equation which can easily be solved numerically in 2-D by different methods. For this problem we used the Euler time-stepping method.

3.1   The Normal-Fluid Flow in 2-D

The response of the normal-fluid on the motion of the superfluid vortex line was calculated by solving equations (8) and (9). We solved the Poisson equation (Eq. 8) using a standard numerical method discussed by Press et al (1983) [12].

To solve the Navier–Stokes equation (Eq. (9)) we used the finite difference method with periodic boundary conditions. The 2-D Navier–Stokes equation in vorticity form can be written in the stream function formulation as

$$\rho_n \left[\frac{\partial \omega_n}{\partial t} + \left(\frac{\partial \psi_n}{\partial y} \frac{\partial \omega_n}{\partial x} - \frac{\partial \psi_n}{\partial x} \frac{\partial \omega_n}{\partial y} \right) \right] = \mu \left(\frac{\partial^2 \omega_n}{\partial x^2} + \frac{\partial^2 \omega_n}{\partial y^2} \right) + \left(\frac{\partial F_y}{\partial x} - \frac{\partial F_x}{\partial y} \right).$$
(11)

The above equation is for the z component since this is the only non-zero component in the 2-D formulation. We can discretised the normal-fluid vorticity equation using Eq. (11) to have,

$$\omega^{t+1}(i,j) = \omega^t(i,j) + \delta t \left[\Delta_x \psi^t(i,j) \Delta_y \omega^t(i,j) - \Delta_y \psi^t(i,j) \Delta_x \omega^t(i,j) \right]$$
$$+ \delta t \left[\nu_n (\Delta_{xx} + \Delta_{yy}) \omega^t(i,j) + \frac{1}{\rho_n} \left(\Delta_x F_y^t(i,j) - \Delta_y F_x^t(i,j) \right) \right] (12)$$

where the operators are defined as

$$\Delta_x f(i,j) = \frac{1}{2\delta x} \left[f(i+1,j) - f(i-1,j) \right] \tag{13}$$

$$\Delta_y f(i,j) = \frac{1}{2\delta y} \left[f(i,j+1) - f(i,j-1) \right] \tag{14}$$

$$\Delta_{xx} f(i,j) = \frac{1}{\delta x^2} \left[f(i+1,j) - 2f(i,j) + f(i-1,j) \right] \tag{15}$$

$$\Delta_{yy} f(i,j) = \frac{1}{\delta y^2} \left[f(i,j+1) - 2f(i,j) + f(i,j-1) \right] \tag{16}$$

The discretisation scheme here is first-order accurate in time and second-order accurate in space. The normal-fluid vorticity obtained from Eq. (12) at time step t is used to calculate the stream function using Eq. (8) at time step $t+1$.

We applied the periodic boundary condition on both the normal-fluid vorticity and the stream function defined as

$$\psi(i,j) = \psi(i+N,j) = \psi(i,j+M) \tag{17}$$

$$\omega(i,j) = \omega(i+N,j) = \omega(i,j+M) \tag{18}$$

To match the periodicity in the normal-fluid we need to define a similar boundary condition for the superfluid. To achieve this we create images vortices for each superfluid vortex line in the computational box. For the results discussed in this paper we used eight images for each vortex. The flow field produced by these image vortices have very negligible effect on the formation of the normal-fluid peak velocity.

Other boundary conditions such as the slip and the no-slip boundary conditions on the velocity can be imposed on the coupled helium II flow. In this case the method of flow computation will be different from the one described above. For instance the normal-fluid velocity and the vorticity will have to be computed differently at the boundary.

3.2   Delta Function Forcing on a Grid

The friction force exerted on the normal-fluid by the superfluid vortex line is a delta function force at the position of the superfluid vortex line. As the superfluid vortex line moves through the grid space the mutual friction force on the normal-fluid also changes. To make the transition of the mutual friction force between the grid spaces smooth, we weighted the mutual friction force over the four neighbouring normal-fluid grid points (Fig. 1). To do this we chose a simple normalised linear weighting

$$w_{i,j} = (1 - \delta x_p)(1 - \delta y_p), \tag{19}$$

where δx_p and δy_p represents the distance of the superfluid vortex line at point (x_p, y_p) from the four corner grid points (Fig. 1). The sum of these four weightings is one. With this method the mutual friction forces on the normal-fluid grid always change continuously as the superfluid vortex line moves across the grid. This method described is similar to the *Cloud-in-Cell* numerical method [11] used in modelling the motion of point vortices in classical hydrodynamics. We tested this model by changing the grid spacing and we saw no significant effect of the grid spacing on the normal-fluid velocity generated by the mutual friction forcing. Using this approach, the mutual friction force is then computed on the normal-fluid grid using Eq. (10). The value obtained is then substituted into Eq. (11) to obtain the normal-fluid vorticity.

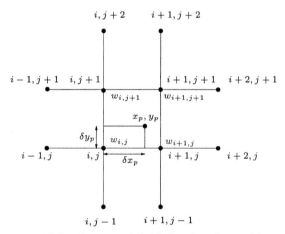

Fig. 1. Finite difference grid for the normal-fluid showing the position of the superfluid vortex line x_p, y_p and the weighting on the neighbouring grid points used for interpolation.

3.3 Extrapolation of the Normal-Fluid Flow in the Neighbourhood of the Superfluid Vortex Line

The normal-fluid flow field near a vortex line cannot be estimated from the normal-fluid grid through grid interpolation because we expect v_n to have a sharp peak at the position (x_p, y_p) of the superfluid vortex point due to the mutual friction force. To estimated the normal-fluid flow field we extrapolate the flow to the point (x_p, y_p) using the twelve neighbouring normal-fluid grid points shown in Fig. 1. A 2-D vector field $\mathbf{f}$ at point (x_p, y_p) (Fig. 1) can be extrapolated on four neighbouring grid points as

$$\mathbf{f}(x_p, y_p)\Big|_{(i,j)} = \mathbf{f}(x_i, y_j) + (\delta x_p)\mathbf{f}_x(x_i, y_j) + (\delta y_p)\mathbf{f}_y(x_i, y_j), \qquad (20)$$

$$\mathbf{f}(x_p, y_p)\Big|_{(i,j+1)} = \mathbf{f}(x_i, y_{j+1}) + (\delta x_p)\mathbf{f}_x(x_i, y_{j+1}) + (1 - \delta y_p)\mathbf{f}_y(x_i, y_{j+1}), \quad (21)$$

$$\mathbf{f}(x_p, y_p)\Big|_{(i+1,j)} = \mathbf{f}(x_{i+1}, y_j) + (1 - \delta x_p)\mathbf{f}_x(x_{i+1}, y_j) + (\delta y_p)\mathbf{f}_y(x_{i+1}, y_j), \quad (22)$$

$$\mathbf{f}(x_p, y_p)\Big|_{(i+1,j+1)} = \mathbf{f}(x_{i+1}, y_{j+1}) + (1 - \delta x_p)\mathbf{f}_x(x_{i+1}, y_{j+1}) + (1 - \delta y_p)\mathbf{f}_y(x_{i+1}, y_{j+1}),$$
$$(23)$$

where $\mathbf{f}_x = \partial \mathbf{f}/\partial x$ and $\mathbf{f}_y = \partial \mathbf{f}/\partial y$. The derivatives at the normal-fluid grid points are computed using points outside the grid area containing the superfluid vortex line (Fig 1). The derivatives are defined as

$$\begin{pmatrix} \mathbf{f}_x(x_i, y_j) \\ \mathbf{f}_y(x_i, y_j) \end{pmatrix} = \begin{pmatrix} [\mathbf{f}(i, j) - \mathbf{f}(i - 1, j)]/\delta x \\ [\mathbf{f}(i, j) - \mathbf{f}(i, j - 1)]/\delta y \end{pmatrix} \qquad (24)$$

$$\begin{pmatrix} \mathbf{f}_x(x_i, y_{j+1}) \\ \mathbf{f}_y(x_i, y_{j+1}) \end{pmatrix} = \begin{pmatrix} [\mathbf{f}(i, j+1) - \mathbf{f}(i - 1, j+1)]/\delta x \\ [\mathbf{f}(i, j+2) - \mathbf{f}(i, j+1)]/\delta y \end{pmatrix} \qquad (25)$$

$$\begin{pmatrix} \mathbf{f}_x(x_{i+1}, y_j) \\ \mathbf{f}_y(x_{i+1}, y_j) \end{pmatrix} = \begin{pmatrix} [\mathbf{f}(i+2, j) - \mathbf{f}(i+1, j)]/\delta x \\ [\mathbf{f}(i+1, j) - \mathbf{f}(i+1, j-1)]/\delta y \end{pmatrix} \qquad (26)$$

$$\begin{pmatrix} \mathbf{f}_x(x_{i+1}, y_{j+1}) \\ \mathbf{f}_y(x_{i+1}, y_{j+1}) \end{pmatrix} = \begin{pmatrix} [\mathbf{f}(i+2, j+1) - \mathbf{f}(i+1, j+1)]/\delta x \\ [\mathbf{f}(i+1, j+2) - \mathbf{f}(i+1, j+1)]/\delta y \end{pmatrix} \qquad (27)$$

The estimated normal-fluid flow at (x_p, y_p) using Eqs. (20)–(23) and the weighting function defined in Eq. (19) now becomes

$$\mathbf{f}^*(x_p, y_p) = w_{i,j} * \mathbf{f}(x_p, y_p)\Big|_{(i,j)} + w_{i+1,j} * \mathbf{f}(x_p, y_p)\Big|_{(i+1,j)} +$$
$$w_{i,j+1} * \mathbf{f}(x_p, y_p)\Big|_{(i,j+1)} + w_{i+1,j+1} * \mathbf{f}(x_p, y_p)\Big|_{(i+1,j+1)} \qquad (28)$$

The extrapolated function $\mathbf{f}^*(x_p, y_p)$ could be the normal-fluid velocity, vorticity or pressure (in a scalar form of the equation) at point (x_p, y_p).

3.4   Numerical Stability and Time Stepping

With two simultaneous calculations for the normal-fluid and the superfluid, we must choose a time step to satisfy both calculations. One limit on the time step is that we must keep it below the limit set by the normal-fluid viscosity,

$$\Delta t_{visc} \leq \frac{\rho_n \delta x^2}{\mu} . \tag{29}$$

This is a necessary stability criterion for finite difference schemes [11]. Also we want the superfluid vortex points to move smoothly through the normal-fluid computational grid, so that we sets another limit

$$\Delta t_{sf} \leq \frac{\delta x}{V_L} , \tag{30}$$

where V_L is the velocity magnitude of the superfluid vortex line. We set the time step of the calculation to be 1/10 of the minimum of these two values.

4   Results in 2-D Flows

The normal-fluid flow $\mathbf{v}_n$ was dragged by the superfluid via the mutual friction force $\mathbf{F}$. As a result of the drag, the normal-fluid forms a localised *jet-like* structure shown in Fig. 2a and c. A detailed description of this flow structure is given in [3].

At low driving superfluid velocity $\boldsymbol{V}_s$ recirculation flow can be seen around the edges of the computational box (Fig. 2a). The position of the vortex lines corresponds to the center of the normal-fluid jet. The spatial structure of the jet shown in Fig. 2a shows that the jet velocity is sharply peaked toward the center and the shape of the jet is slightly elongated in the direction of the jet velocity. This shape is maintained at all temperatures. The measured length (≈ 0.3mm) and width (≈ 0.1mm) of the jet shows very little temperature dependence. We do not currently have any predictive theory for these length scales.

The peak velocity of the normal-fluid jet however shows a very strong temperature dependence. It has a maximum at temperatures of ≈ 1.9 K and falls rapidly to zero at 1.3 K and at the lambda transition temperature. The drag on the normal-fluid by the superfluid is approximately 60% at 1.9 K

The normal-fluid energy E_n in the coupled flow is approximately proportional to V_s^2. This energy comes from the superfluid flow through mutual friction. The normal-fluid flow structure described here is very dissipative and maintains a steady flow only because it is constantly renewed by the energy gained from the superfluid through mutual friction. The simulation further reinforced this by showing that the normal-fluid jet moves along with the superfluid vortex line even when the velocity of the jet differs in magnitude and direction from that of the superfluid vortex line velocity.

Four superfluid vortex lines oriented in the same direction will rotate around each other. Starting with this initial condition, we observed the development of a net rotation in the normal-fluid (shown in Fig. 2b and d). For these

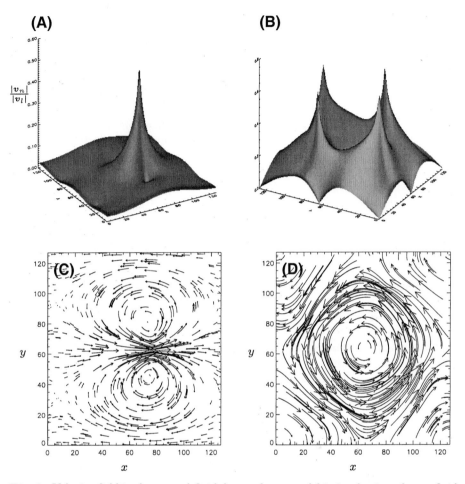

Fig. 2. Velocity field in the normal-fluid due to the mutual friction forcing of superfluid vortex lines. The calculation was made on a 128^2 grid. The size of the computational box is 1 mm. The temperature dependent parameters were taken at $T = 1.9$ K. (A) Magnitude of jet formed by a single vortex line with driving velocity of $V_S = 10^{-2}$ cm/s. (B) Velocity arrow plot of jet formed by a single vortex line. (C) Normal-fluid velocity magnitude for motion of four vortex line. (D) Normal-fluid velocity arrow plot showing the rotation of four vortex line.

multiple vortex line simulations, the vortex line motion is due to the advection by the other vortices and no external superfluid driving velocity is needed to induce the normal-fluid flow.

The localised normal-fluid flow structures formed by the mutual friction force have an unusual spatial form characterised by the length and the width of the jet. This type of forcing (usually confined to large length scales) may affect the characteristics of the normal-fluid turbulence in coupled helium II flow.

5   Numerical Methods for 3-D Flows

Our model of interacting superfluid combines the two fluid idea of Tisza-Landau with the Onsager-Feynman idea of the quantized superfluid vortices. In other words, this combines the Eulerian with the Lagrangian formulations of fluid dynamics.

Correspondingly, the numerical problem of calculating three-dimensional interacting superfluid flows can be split into three parts. The first part concerns the solution of the equations that describe the evolution of the free superfluid line vortices. The second part has to do with the solution of the system of non-linear partial differential equations that describe the free normal-fluid flow, the well known (incompressible) Navier-Stokes equations.

The third sub-problem is related to the numerical modelling of the mutual friction interaction, as well as, to the coupling of the two already mentioned calculations. This is the novel part in every method treating interacting super-fluid. In the past there have been works that calculated interacting flows, but in these works the normal-fluid was modelled as having an infinite inertia, and so one could use a kinematic prescription for it (See the article by Samuels in this volume). In this way therefore, one needs neither to employ advanced methods of numerical fluid dynamics or to model the mutual friction force acting on the classical fluid. The methods discussed in this section overcome this simplification and allowing the normal-fluid to react makes possible the investigation of novel physics which were beyond the capabilities of the previous methods.

For the calculation of the free normal-fluid we use Eulerian grid finite volume methods. For the free superfluid we use Lagrangian vortex methods. We have employed a vortex point method for the two-dimensional case and a vortex filament method for the three-dimensional one. Vortex point methods are capable of calculating three-dimensional flows too but the vortex filament method is the natural one for superfluid since in the latter vorticity has support only on very thin line vortices. In addition, this approach is already been used by Schwarz in a series of papers and it is proven to be a useful approach. Next, we describe our methods in greater detail.

5.1   The Free Normal-Fluid

The normal-fluid is described by the incompressible Navier-Stokes equation (1), with the last term at the right hand side put to zero. In this way one can use the many methods developed in the field of classical numerical fluid dynamics. We use a finite volume method which was first discussed in Harlow and Welch [14], later improved by Kim and Moin [15], and finally by Wray [16]. The details of the method can be found in these papers. A useful text in which many basic concepts are discussed is Strikwerda [17]. The method is well known in the classical turbulence community, and although it has not a very high order of accuracy, it has the advantage of being simple. This is particularly a virtue, because one wants to combine this method together with a very different one (a vortex filament method) for the superfluid. In this respect, one can first have the whole of the numerical procedure constructed, and later if needed can introduce higher order discretization schemes.

We note here in brief some important points:

(1) The convective and body-force terms are advanced in time with a Runge-Kutta scheme of an $O(\Delta t^3)$ error, and space differenced with central schemes of an $O(\Delta x^2)$ error. The stability analysis of the method requires $CFL \leq \sqrt{3}$.

(2) The diffusion terms are discretized with a Crank-Nicholson scheme of error $O(\Delta x^2)$ in space and $O(\Delta t^2)$ in time. Since this scheme is an implicit method, it requires the inversion of matrices. In order to have a method that is not too computationally expensive an operator splitting is employed which reduces the matrix inversion problem to the inversion of tridiagonal matrices (one inversion for each direction). This method of fractional steps is of error $O(\Delta t^3)$ in time (and thus is not inconsistent with the accuracy of the Runge-Kutta scheme), and by not being iterative it is fast.

(3) The pressure is advanced in time with an Euler method of error $O(\Delta t)$. The space discretization is done with central differencing and introduces errors of $O(\Delta x^2)$.

(4) The incompressibility condition is imposed by solving a Poisson equation for pressure.

We have done at present many calculations with $CFL = 1.5$. All these calculations were done with periodic boundary conditions. The calculations were robust, with no instability problems whatsoever, and with no problems in satisfying the condition for the incompressibility of the velocity field. We have used standard tridiagonal matrix inversion methods, as found in Press et al [12], and numerical fast fourier transforms for the solution of the Poisson equation for pressure (suitable for problems with periodic boundary conditions). Although, the implicit method of Crank-Nicholson allows one to integrate the diffusion operator with a time step much greater than the time step of the Runge-Kutta, we have done calculations with the much smaller viscous time step in order to resolve efficiently the viscous dominated phenomena in the normal-fluid. This is in most times a necessity in interacting superfluid as the following scaling argument suggests. If the superfluid is put to interact with a stationary normal-fluid one can (sufficiently away from the superfluid vortex core) assume that the magnitude of the mutual friction force is of the order of the quantum of circulation κ. The mutual friction forcing, in order to generate convective motion of a high Re number, must overcome the friction forces which scale with ν. Now, because of a numerical accident these constants have very close values for liquid helium. This implies that the structures seen in normal-fluid when stirred by the superfluid should be highly dissipative and thus need to be resolved all the way down to viscous time scales. One should not expect a wide range of normal-fluid scales to be excited by the mutual friction force. However, one should be careful when applying such scaling arguments close to the vortex core.

Another important point is that of space resolution. In order to be consistent with a continuum description of the normal-fluid we must resolve all possible scales for which a continuum approach should be valid (we refer to them as continuum scales). It is conceivable of course that the normal-fluid has important dynamics even at scales where the continuum theory is not valid; in such a case our model is also useless and one should resort to kinetic theory kind of calculations. One hopes that what is left out has not an important effect on what is resolved. This is a major and ever occurring problem in physics.

We do calculations with Knudsen numbers ranging from 100 to 1000 (a very big value of the Knudsen number corresponds to the continuum regime). We note here that when calculating superfluid is not enough to just resolve the Kolmogorov scale for the normal-fluid; interesting physics (in fact perhaps the only interesting physics) should appear below this scale. Finally, we might want to resolve all possible time scales at which the continuum theory holds. In such case we should choose a time step of the order 100 or 1000 mean free path times. Such a time step is expected to be smaller than the turn over time of the Kolmogorov scale size eddies.

5.2   The Superfluid

The free superfluid is modelled numerically through a vortex filament method (see the article by Samuels in this volume). Our vortex method has fewer problems than the analogous methods used in classical fluid dynamics. The main difference is in the initialisation of the vorticity. In normal-fluid one has a smooth vorticity field as an initial condition, which one has to approximate with singular concentrations of vorticity: vortex points or vortex filaments. In the case of superfluid such a problem does not exist, because in superfluid vortex filaments are not an idealisation as in classical fluids but they are what one observes macroscopically in the laboratory. From the mathematics point of view the vorticity is not defined on a superfluid vortex line. Tracking the motion of vortex lines though it is useful, since the circulation of quantum lines is not only well defined, but also always constant. This implies that the free superfluid turbulence should be simpler than the classical one. Indeed, in the classical case not only is not correct to say that the flow can be reduced to an assembly of a finite number of classical vortex filaments, but even in case such a thing could be true we would have to deal with an infinite number of classical vortex configurations corresponding to a possible infinite range of possible circulation values. These considerations, although are motivated from numerical analysis they have a value from the physics point of view too, suggesting that the relative simplicity of free quantum turbulence could help in gaining some insight into the complexities of classical turbulence.

The equation for the Lagrangian vortex line dynamics is equation (2). This equation models the physics of the balance between the Magnus, the mutual friction, and the Iordanskii forces. The last two forces are been set to zero when studying free quantum turbulence by putting zero all the terms except those corresponding to the self-induction one. Although we can do calculations with both the exact Biot-Savart law, as well as, with the Local Induction Approximation (LIA), we use the later only when the full Biot-Savart computation becomes impractical because of computational complexity.

In implementing our method we discretize each vortex line using a Lagrangian grid along each vortex line. In this way we reduce the problem from an integrod-ifferential one to a system of ordinary differential equations. For the later system we use a standard Runge-Kutta-Fehlberg method of an $O(\Delta t^4)$ accuracy. The time stepping is done with a predictor-corrector procedure. We note that the discretization along each vortex is not an arbitrary procedure. If it results in a very dense grid one sees that in order to retain stability one has to employ very small time stepping. The adequate (for stability) time step for a particular Lagrangian grid choice should come from a linear stability analysis of the full integrodifferential equation. At present we are not aware of such an analysis. So we decide first on a particular space discretization and then by trial and error we decide upon an appropriate time stepping. The Lagrangian grid size is chosen to be the same as the Eulerian one for the normal-fluid. The time stepping on the Lagrangian calculation should be appropriate for resolving the fastest possible Kelvin waves present in the system. Its calculation is as follows: the group velocity of a Kelvin wave of wavelength λ on a quantum vortex line of circulation κ and core size $a \sim 10^{-8}$ cm, is given by:

$$V_{\mathrm{wave}}(\lambda) = \frac{\kappa}{2\lambda} \ln\left(\frac{\lambda}{2\pi a}\right) \tag{31}$$

Next, we find that should the Lagrangian grid spacing be Δl, it takes time $\Delta t = \Delta l / V_{\mathrm{wave}}$ for the fastest Kelvin wave (which corresponds to $\lambda = 2\Delta l$) to travel exactly the smallest resolvable distance. Then one needs a minimum time step equal to $\Delta t/2$ in order to resolve all Kelvin waves in the system. All our calculations meet at least this requirement. A note is in order here: although the superfluid vortex ring has the tendency to develop structure at smaller and smaller scales, this structure formation has an ultraviolet cut-off. This is determined by the strongest of the following three mechanisms: acoustic damping, mutual friction [18], and (mostly active at finite temperatures) a mechanism due to Glabertson-Ostermeier instability, as discussed in Samuels and Kivotides [19]. Of course in general one might not be able to resolve all relevant scales due to space and time computational complexity. In such a case one should at least be careful to resolve scaling regimes as information about them comes available in the literature. Finally, one needs to ensure that a vortex line does not transverse more than an Eulerian grid cell at a particular time step. This condition we refer to as superfluid CFL (SFCFL) condition.

Resolution considerations are very important in calculating interacting super-fluid. Although it is easy to choose the smallest of the superfluid and normal-fluid time steps to do the calculation, the choice is not so easy in space resolution. Even if we resolve all normal-fluid continuum scales (obviously fully resolving the Kolmogorov scale) it could still be meaningful for the superfluid vortex lines (only) to develop structure at even smallest scales (should the temperature is such that there is no effective damping at continuum scales). In such a case one notes that the benefit of allowing the superfluid grid to be denser than the normal-fluid grid is not balanced by the resulting increase in the *logical depth* of the algorithms or the extra *computational complexity* . In such cases, we have chosen the same grid for both fluids.

5.3   The Interaction Modelling

The modelling of the mutual force is based on the observation that the term in the Navier-Stokes which corresponds to this force should have units of acceleration. The relation (10) gives the mutual friction force per unit length of vortex line. We use this relation as follows:

(1) We do a search first over the Eulerian domain to track which fluid volumes are threaded by a vortex line.

(2) We next (for each fluid element with the above property) calculate the integral of formula (10) over the length of line enclosed inside a particular fluid volume.

(3) Finally in order to have the correct units we divide the total force that we have obtained at step 2, with the density of the normal-fluid times the volume of the particular fluid volume. The result of this operation is appropriate for inclusion into the equation of motion (1).

Modelling the mutual friction force is not the only new aspect of any interacting superfluid calculation. One additionally considers the problem of increasing computational complexity if, for example, in a turbulence calculation one has a grid 256^3 which is quite common in well resolved turbulent flows, and in addition one must combine such a calculation with a vortex filament method. As noted before, one needs (as far as the superfluid is concerned) to resolve Kelvin wave times which are expected to be much smaller than the viscous times for the normal-fluid. It is not practical in such calculations to advance the normal-fluid with the same time step as the superfluid. We have developed a sub-cycling method which allows the superfluid to evolve inside one normal-fluid outer time step during which evolution the normal-fluid velocity and pressure fields remain frozen. We allow up to 10 sub-cycling steps but one should adjust this number according to the particular physics he is looking after. This method gave us no stability problems, although from the mathematics point of view it is possible to lead to instability.

5.4   Preliminary Results in 3-D Flows

Our program for investigating interacting superfluid flows has just started [20]. At the moment we have used our methods to investigate the motion of superfluid rings in a stationary normal-fluid. The results, are closely related to those of the 2-D flow. Fig. 3 shows a section of the normal-fluid velocity field generated by the mutual friction forcing from superfluid vortex ring. The velocity field shown is taken in the plane perpendicular to the plane of the superfluid vortex ring, so the superfluid vortex filaments pass through this plane at right angles, much as in the 2-D simulations. The normal-fluid near the superfluid vortex filament is dragged by the superfluid vortex, producing localised jets of normal-fluid.

In 3-D, this dragging of the normal-fluid by the superfluid vortex forms a pair of normal-fluid vortex rings associated with the superfluid vortex ring. The outer normal-fluid vortex ring is oriented in the same direction as the superfluid vortex ring, while the inner normal-fluid vortex ring has the opposite orientation. Without the mutual friction force from the superfluid vortex, this normal-fluid flow structure would rapidly dissipate. The Reynolds number of the normal-fluid flow, calculated using the peak normal-fluid velocity and the width of the normal-fluid jet, is of order unity, confirming the dissipative nature of the flow structure.

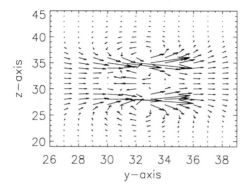

Fig. 3. The normal-fluid velocity field in the 3-D simulation. The velocity shown is measured on the plane perpendicular to the plane of the superfluid vortex ring.

6 Discussion and Conclusion

Does the mutual friction force of the superfluid vortex filaments generate any complicated, possibly turbulent flow in the normal-fluid? In the simulations reported here, it clearly does not. The normal-fluid flow structures generated are all of low Reynolds number and are thus rapidly dissipated. No significant normal-fluid flow is seen far from the source of the forcing, the superfluid vortex. In this sense, the mutual friction force does not significantly 'stir' the normal-fluid.

But even though mutual friction does not directly generate any complicated flows in the normal-fluid, there are other possibilities that must be considered. It is possible that the mutual friction force could trigger instabilities in normal-fluid shear flows. Stability calculations of a normal-fluid flow under a specified mutual friction force distribution show that such instabilities can be triggered, at sufficiently large mutual friction forcing amplitudes [2]. This behaviour has not yet been tested in the fully coupled two fluid calculations described in this paper, since we have only considered normal-fluid flows generated in an initially stationary normal-fluid. The direct effect of the mutual friction force on the normal-fluid appears to be limited to the formation of the dissipative jet. The indirect effect of this forcing on the normal-fluid flow stability now needs to be considered (See Godfrey's article in this volume).

Acknowledgements

This research is supported by University of Newcastle upon Tyne and the Leverhulme Trust.

References

1. K. W. Schwarz: Phys. Rev. B **31** 5782 (1985).
2. D. J. Melotte, C. F. Barenghi: Phys. Rev. Lett. **80** 4181 (1998).
3. O. C. Idowu, A. Willis, D. C. Samuels, C. F. Barenghi: Phys. Rev. B **62** 3409 (2000).
4. C. F. Barenghi: J. Phys., Condens. Matter **11** 7751 (1999).
5. H. E. Hall and W. F. Vinen: Proc. Roy. Soc. A **238** 215 (1956).
6. C. O. Idowu, D. Kivotides, C. F. Barenghi, D. C. Samuels: J. Low Temp. Phys. **120** 269 (2000)
7. C. F. Barenghi, R. J. Donnelly, W. F. Vinen: J. Low Temp. Phys. **52** 189 (1983).
8. W. F. Vinen: Proc. Roy. Soc. A **242** 493 (1957).
9. R. J. Donnelly: *Quantized Vortices in Helium II*, Cambridge University Press (1991).
10. R. J. Donnelly, C. F. Barenghi: J. Phys. Chem. Ref. Data, Vol.27, No.6, (1998).
11. R. Peyret, T. D. Taylor: *Computational Methods for Fluid Flow* , Springer-Verlag (1983).
12. W. H. Press, S. A. Teukolsky, W. T. Vetterling, B. P. Flannery: *Numerical Recipes in Fortran 77 second edition* , Cambridge University Press (1992).
13. R.G.K.M. Aarts, A.T.A.M. deWaele: Phys. Rev. B **50** 10069 (1994).
14. M. W. Harlow, J. E. Welch: Phys. Fluids **8** 2182 (1965).
15. J. Kim, P. Moin: J. Comp. Physics **59** 308 (1985).
16. A. A. Wray: *Very Low Storage Time-Advancement Schemes*, Internal Report, NASA Ames Research Center, Moffet Field, California (1987).
17. J. Strikwerda: *Finite Difference Schemes and Partial Differential Equations*, Wadsworth and Brooks/Cole (1989).
18. W. F. Vinen: Phys. Rev. B **61** 1410 (2000).
19. D. C. Samuels, D. Kivotides: Phys. Rev. Lett. **83** 5306 (1999).
20. D. Kivotides, C. F. Barenghi, D. C. Samuels: Science **290** 777 (2000)

From Vortex Reconnections to Quantum Turbulence

Tomasz Lipniacki

Institute of Fundamental Technological Research,
Świętokrzyska St. 21, 00-049 Warsaw, Poland

Abstract. An alternative approach to quantum turbulence is proposed in order to derive the evolution equation for vortex line-length density. Special attention is paid to reconnections of vortex lines. The summed line-length change ΔS of two vortex lines resulting from the reconnection (in the presence of counterflow V_{ns}) can be approximated in the form: $\Delta S = -at^{1/2} + bV_{ns}^2 t^{3/2}$, with $a > 0$, $b \geq 0$, at least until $\Delta S \leq 0$. For steady-state turbulence, the average line-length change $\langle \Delta S \rangle$ between reconnections has to be zero. If, for a given value of the counterflow, the line density is smaller than the equilibrium one, the reconnections occur less frequently and $\langle \Delta S \rangle$ becomes positive and the line density grows until the equilibrium is restored. When the line-density is too large, the reconnections are more frequent, the lines shorten between reconnections and the line density gets smaller. The time derivative of the total line density is proportional to the reconnection frequency multiplied by the average line-length change due to a single reconnection. The evolution equation obtained in the proposed approach resembles the alternative Vinen equation.

1 Introduction

The variety of the dynamic phenomena exhibited by superfluid ^{4}He involves the appearance and motion of quantized vortices. Due to the existence of these singularities the superfluid component is coupled dissipatively with the normal component. At low velocities He II (superfluid ^{4}He) flows in the frictionless, presumably laminar manner consistent with the ideal fluid description. When the counterflow (the relative velocity of the two helium components) $v_{ns} = v_n - v_s$ becomes sufficiently large, however, the superfluid laminar flow develops into superfluid turbulent flow in which the quantum vortices form a chaotic tangle.

The aim this work is to derive the evolution equation for vortex line-length density L. In the proposed approach special attention is paid to reconnections of vortex lines. The vortex line evolution is analyzed in the localized induction approximation, supplemented by the assumption that when two vortex lines cross, they undergo a reconnection.

Let us recall that if the curve traced out by a vortex filament is specified in the parametric form $s(\xi, t)$, then in the superfluid reference frame the instantaneous velocity of a given point of the filament is given by the equation (in scaled units in which the elementary circulation around vortex line equals to unity)

$$\dot{s} = s' \times s'' + \alpha s'' + \alpha s' \times v_{ns} , \tag{1}$$

where dot and prime denote instantaneous derivatives with respect to the scaled time τ and arc length ξ, respectively, α is the non dimensional friction coefficient and v_{ns} the counterflow.

The line-length of the vortex filament $l = \int d\xi$, which motion is given by (1), satisfies the equation

$$\frac{\partial l}{\partial \tau} = \int \left(\alpha v_{ns} \cdot (s' \times s'') - \alpha |s''|^2 \right) d\xi \ . \tag{2}$$

2 Vortex Motion Following Reconnection

We consider an idealized reconnection of two infinite straight vortex filaments. In the moment of reconnection these two filaments transform into two "new" sharply bent vortex filaments. We analyze the evolution of a vortex line forming at the initial state ($\tau = 0$) angle 2ϕ.

2.1   The Case $v_{ns} = 0$

With $v_n = v_s = 0$ Eq.(1) simplifies to

$$\dot{s} = s' \times s'' + \alpha s'' \ . \tag{3}$$

It is not difficult to show that in this case the vortex line for all times will have a similar shape (Fig.1), whose spatial scale D is growing as $\sqrt{\tau}$. This leads to the result

$$\Delta l = -A\,\tau^{1/2} \quad , \tag{4}$$

where Δl is the line change and $A = A(\alpha, \phi) > 0$.

Let us consider now the dynamical equation (3) without self the induction term $s' \times s''$. In such an equation the coefficient α can be absorbed into time τ

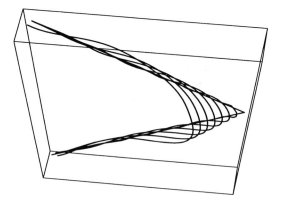

Fig. 1. Vortex line evolution after reconnection for reconnection angle $\phi = \pi/4$, and friction coefficient $\alpha = 0.1$. The line positions are shown at $\tau = i^2 \tau_o$.

and one gets

$$\frac{\partial s}{\partial(\alpha\tau)} = s'' \ . \tag{5}$$

It is clear that the line-length reduction Δl_o in the evolution governed by the above equation satisfies

$$\Delta l_o = B_o(\phi)\,(\alpha\tau)^{1/2} \ . \tag{6}$$

The simulations revealed that $\Delta l \approx \Delta l_o$. This result can be justified as follows (Lipniacki [1]) For $\alpha = 0$, the vortex evolution equation (1) reduces to the non dissipative one,

$$\frac{\partial s}{\partial\tau} = s' \times s'' \ . \tag{7}$$

During the evolution governed by Eq.(7), the following quantities are constant in time

$$l = \text{const} \quad , \quad \int |s''|^2 d\xi = \text{const.} \tag{8}$$

Now, one can evaluate the vortex line step by step using the method of fractional steps. In the first half of time step we solve Eq.(5), in the second one we solve Eq.(7) using the previous result as an initial condition. The evolution governed by non dissipative Eq.(7) does not change the line length and does not change the squared curvature - which determines the rate of line-length changes (see Eq.2) during the evolution under Eq.(5). Roughly speaking, although the first term in Eq.(3) significantly changes the evolution of the vortex (line shape), it does not significantly influence the rate of the line-length reduction. As the result we have:

$$A \approx B(\phi)\,\alpha^{1/2} \ , \quad \text{i.e.} \quad \Delta l \approx -B(\phi)\,(\alpha\tau)^{1/2} \ . \tag{9}$$

2.2   The Case $v_{ns} = const \neq 0$

In this case, for a given α and ϕ, we consider the summary line-length change ΔS of two vortex lines resulting from the reconnection:

$$\Delta S = \Delta l_1 + \Delta l_2 \ . \tag{10}$$

Let us note, that just after the reconnection, close to the reconnection point, the two vortices have the binormals $(s' \times s'')$ opposite to each other. The first term in Eq.(2) when integrated over two lines resulting from reconnection gives no contribution to total line length change. Because the second term of Eq.(2) is always negative the total length of the resulting vortices decreases in the beginning. However, during the further evolution, the characteristic curvatures (and so the absolute value of second term) get smaller, and the two vortices turn so that the average value of $v_{ns} \cdot (s' \times s'')$ becomes positive. Then the total length of vortices starts growing.

It can be shown (Lipniacki [1]) that after the idealized reconnection of two straight vortex lines the summary line-length change ΔS of two vortices resulting from the reconnection can be expanded as function of v_{ns} and τ

$$\Delta S = a_1\,\tau^{1/2} + a_2\,\tau^{3/2}v_{ns}^2 + a_3\,\tau^{5/2}v_{ns}^4 + \dots \ . \tag{11}$$

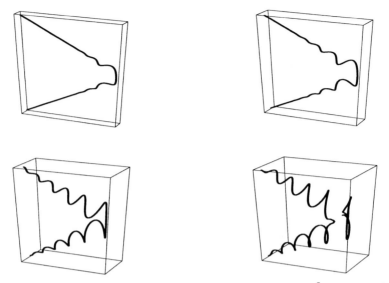

Fig. 2. The vortex evolution in configuration 2 for $\alpha = 0.3$, $\tau v_{ns}^2 = 32, 64, 96, 128$. The separated vortex ring is seen the final frame.

The coefficients a_i depend on α, ϕ and the direction of v_{ns} with respect to reconnecting lines.

Moreover, the numerical simulations revealed (Lipniacki [1]) that at least until $\Delta S \leq 0$ one can restrict to first two terms.

$$\Delta S(\tau) = -a \, \tau^{1/2} + b \, \tau^{3/2} v_{ns}^2 \,, \tag{12}$$

where $a > 0$, $b \geq 0$.

To apply the above result to the analysis of the dynamics of the whole vortex tangle one should average the function $\Delta S(\tau)$ over a representative sample of reconnection configurations. We found that when estimating the average line-length change $\Delta S_o(\tau) = \langle \Delta S(\tau) \rangle$ one can use the simplified dynamical equation

$$\dot{s} = \alpha s'' + \alpha s' \times v_{ns} \,, \tag{13}$$

(i.e. without self induction term) and get roughly the same result as for Eq.(1) (Lipniacki [1]). We should note, however, that Eq.(1) can be replaced by Eq.(13) only when calculating the line length change averaged over a representative sample of configurations. In Eq.(13) α may be absorbed into time scale; in the result we have

$$\Delta S_o = -a_o(\alpha \tau)^{1/2} + b_o(\alpha \tau)^{3/2} v_{ns}^2 \,. \tag{14}$$

To estimate a_o, b_o we assume that every line in the moment of reconnection has to be parallel or antiparallel to one of 3 "main" directions: $\hat{x}, \hat{y}, \hat{z}$ and that the direction of v_{ns} is $(1, 1, 0)$ in Cartesian coordinates. In principle, this means that there are 12 possible reconnection configurations, but it is not difficult to check that only 3 of them are essentially different. The average line-length change $\Delta S_o(\tau)$ is then estimated by averaging over these configurations. In the result we get $a_o = 2.34$, $b_o = 0.514$.

Let us notice that the coefficients a_o, b_o determine the characteristic non dimensional time $\tau^* = \tau v_{ns}^2 = a_o/b_o$ for which the average line-length change after reconnection is zero.

3   The Model

Equation (14) gives one a hint for the explanation of how the vortex tangle is sustained: for the steady-state turbulence, the average line-length change between reconnections has to be zero. If for a given value of the counterflow v_{ns}, the line density is smaller then the equilibrium value, the reconnections occur less frequently, and so, the characteristic time between reconnections is longer; hence according to Eq.(14) the line-length change between reconnections becomes positive. As the result the line density of vortex the tangle grows until the equilibrium is restored. Inversely, when the line density is too large, the reconnections are much more frequent, so the decaying term in Eq.(14) prevails and the line density gets smaller.

In the proposed approach (see Lipniacki [2] for more details) the dynamics of a vortex tangle will be considered as a sequence of reconnection followed by a "free" evolution of vortex lines resulting from each reconnection.

Let the instantenous line-length density be L. This gives one the characteristic spacing between lines l_o and the characteristic (average) line curvature $\langle |s''| \rangle$

$$l_o = L^{-1/2} , \qquad \langle |s''| \rangle = 1/l_o = L^{1/2} . \qquad (15)$$

Let us divide the vortex lines into segments with length l_o. Because the length l_o is equal to the spacing between lines we may expect that each segment moves (more or less) as a unity, but the motions of the neighboring segments are not strongly correlated. Moreover, because the length of segments is equal to the characteristic radius of curvature, when considering collisions, the segments can be roughly treated as straight ones. It is assumed that any collision of vortex segments leads to a reconnection; such assumption is justified by numerical simulations of Schwarz [5]. For the sake of simplicity we assume that all segments move with the same speed v_o which can be estimated basing on (1):

$$v_o = \langle |\dot{s}| \rangle = \sqrt{L(1+\alpha^2) + \frac{2}{3}(\alpha v_{ns})^2} . \qquad (16)$$

Having v_o and l_o (i.e. the number of segments per unit volume $n = L/l_o = L^{3/2}$) one can estimate the reconnection frequency (per unit volume) f_r,

$$f_r = \frac{v_o L^2}{2} , \qquad (17)$$

and the average time between reconnections of line segments

$$\tau_c = \frac{1}{L^{1/2}v_o} \cdot \tag{18}$$

The estimation of f_r needs some algebra since the cross section for collision of two given segments depends on their orientation with respect to their relative velocity. The time derivative of line-length density is then

$$\frac{dL}{d\tau} = f_r \, \langle \Delta S(\tau_c) \rangle \;, \tag{19}$$

where $\langle \Delta S(\tau_c) \rangle$ is the average line-length change due to a single reconnection. The resulting evolution equation is

$$\frac{dL}{d\tau} = -\frac{1}{2}a_o\alpha^{1/2} \, L^{7/4} \, v_o^{1/2} + \frac{1}{2}b_o\alpha^{3/2} \, L^{5/4} \, v_o^{-1/2} \, v_{ns}^2 \;, \tag{20}$$

with v_o given by Eq.(16).

4   Results

The obtained evolution equation (20) differs from the classical Vinen equation (Vinen [8])

$$\frac{\partial L}{\partial t} = -\beta_v L^2 \, + \, \alpha_v |v_{ns}| L^{3/2} \;, \tag{21}$$

where α_v and β_v are temperature dependent coefficients. When $\alpha \ll 1$ and $\alpha v_{ns}^2 \ll L$ the line velocity v_o can be approximated by $L^{1/2}$ and the evolution equation simplifies to

$$\frac{dL}{dt} = -\frac{1}{2}a_o\alpha^{1/2} \, L^2 + \frac{1}{2}\beta^{-1}\alpha^{3/2} \, L \, v_{ns}^2 \;. \tag{22}$$

which is know as the alternative Vinen equation. The generation term is proportional to Lv_{ns}^2 what is closer to the phenomenological theory of classical turbulence. Indeed, (Niemirowski, Fiszdon [4]) by assuming that turbulence can be characterized by a parameter, say, L, and that its time derivative dL/dt is an analytic function of L, the alternative form of Vinen equation can be interpreted as the first two terms in series expansion. Furthermore, as the generation term is the scalar function of vector argument v_{ns}, it is reasonable that the series expansion starts with this argument squared. The last comment concerns also the full form of the evolution equation obtained in the model, i.e. Eq.(20) where the counterflow velocity v_{ns} is also everywhere squared. The presence of absolute value of v_{ns} in the classical Vinen equation is a little bit strange.

   The experimental data are not precise enough to chose between Vinen, alternative Vinen equation, or the Eq.(20). It makes sense only to compare with experimental data the steady-state value of L following from (20) i.e. L_∞

$$L_\infty = v_{ns}^2 \left(\sqrt{\frac{\alpha^4}{9} + 8\left(\frac{b_o\alpha}{\pi a_o}\right)^2} - \frac{\alpha^2}{3} \right) \tag{23}$$

In a broad range of α ($\alpha \in [0.01, 1]$) the value of L_∞ predicted by (23) generally agrees with experimental data and Schwarz's [6] numerical simulations of steady-state turbulence (see [2]).

Equation (20) may be compared with data from Schwarz and Rozen[7]. The authors analyzed in numerical simulation of non-equilibrium turbulence large transients, in which the line-length density grows from small to large values, and found that the coefficients in the Vinen equation are not constant. One can calculate how those coefficients vary in time by comparing the growth term of our evolution equation (20) with the Vinen one. Then one may check [2] that Schwarz-Rozen data are considerably better fitted by Eq.(20) than by the classical Vinen equation.

Our main result is the construction of the simple model in which the microscopic analysis a of quantum tangle leads to a macroscopic evolution equation for line-length density. The numerical simulations needed to estimate the coefficients a_o, b_o in Eq.(20) are relatively simple, when compared with simulations of Schwarz. The main advantage of the presented approach is the possibility to generalize it to anisotropic flows with significant macroscopic superfluid vorticity. Such flows are expected in such phenomena as spin-up or boundary layer forming. The viscous forces in a cylinder, which starts spinning from rest, acting on the normal component, may give rise to counterflow, large enough to cause quantum turbulence which may significantly influence the dynamics of both components. In this way the angular momentum can be transferred from the cylinder via the normal component to the superfluid component. One may expect the following scenario: due to viscous forces the normal component starts spinning, and this implies counterflow which generates quantum turbulence. The mutual friction forces couple two components ($v_{ns} \rightarrow 0$) and in the end quantum vortices polarize to form a pattern of straight parallel lines and both fluids spin together. However, because of the large normal fluid velocity gradients, even in the first stages of spin-up process the arising quantum turbulence is highly anisotropic; the tangle of quantized vortices is polarized to carry considerable macroscopic superfluid vorticity. This is probably why the description of spin-up process in terms of Vinen model was found to be inconsistent (Lipniacki [3]). The line-length density calculated from Vinen equation was smaller then the line density calculated from superfluid velocity profiles. The origin of these vortices cannot be explained within the Vinen model.

References

1. T. Lipniacki: European Journal of Mechanics B/Fluids **19**, (3), 361 (2000)
2. T. Lipniacki: Arch. Mech. **53**, (1), 23 (2001)
3. T. Lipniacki: Arch. Mech. **49**, (4), 615 (1997)
4. S. K. Nemirowski, W. Fiszdon: Rev. Mod. Phys. **67** (1), 37 (1995)
5. K. W. Schwarz: Phys. Rev. B **31**, 5782 (1985)
6. K. W. Schwarz: Phys. Rev. B **38**, 2398 (1988)
7. K. W. Schwarz, J. R. Rozen: Phys. Rev. B **44**, 7563 (1991)
8. W. F. Vinen: Proc. R. Soc. London Ser. A **242**, 493 (1957)

Vortices and Stability
in Superfluid Boundary Layers

Simon P. Godfrey, David C. Samuels, and Carlo F. Barenghi

Dept. of Mathematics, Univ. of Newcastle, Newcastle upon Tyne, NE1 7RU, UK

Abstract. Boundary layer flows are of critical importance to the motion and physics of a fluid flow. In superfluid helium II, the problem of boundary layer flow highlights a significant difference between the two fluid components that make up the current model of helium II: the difference in the boundary conditions. This two fluid model also has a fundamental effect on the stability of the flow, as one must consider the stability of both of the two fluid components. We describe superfluid vortex structures which remain stable in a laminar boundary layer, and discuss their effect on the boundary layer. We also calculate the stability characteristics of the plane Poiseuille flow of helium II, and have discovered a new unstable mode caused by mutual friction, that can dominate the stability characteristics of the flow.

1 Introduction

1.1 The Two-Fluid Model

Superfluid helium II (i.e. below a temperature of 2.17°K) is best described as a superposition of two fluid components: a *normal* fluid and the *superfluid*. Both of these two components have their own velocity fields and densities (denoted by the subscripts n and s respectively). The normal fluid component is basically a classical Navier-Stokes (i.e. viscous) fluid whereas the superfluid component is an inviscid Euler fluid with some quantum effects. These two fluids interact via the mechanism of *mutual friction*, which exchanges energy and momentum between the two fluids. Mutual friction is caused by the interaction of the rotons and phonons (constituents of the normal fluid) with quantized superfluid vortex lines[1].

2 Boundary Layer Vortices

We shall first consider the laminar helium II boundary layer. The normal fluid obeys no-slip boundary conditions, whereas the superfluid is allowed to slip freely at the boundary. This difference in boundary conditions suggests there may be a region in the flow, near the boundary, where there are large differences in the two velocity fields.

The vorticity in the superfluid component is confined to quantized vortex filaments, which have velocity $\dot{\mathbf{S}}$ approximated by

$$\dot{\mathbf{S}} = \mathbf{V}_s + \alpha \mathbf{S}' \times (\mathbf{V}_n - \mathbf{V}_s) \tag{1}$$

where $\mathbf{S}$ represents a position along the vortex filament, the dot represents a time derivative, prime the derivative with respect to arclength ξ, and α is the mutual friction parameter. This is an approximate equation of motion since we are neglecting the α' term which is typically quite small. The term proportional to α gives the primary response of the vortex filament to the mutual friction. This force will be largest where the velocity difference is largest. Thus we would expect a strong mutual friction force on superfluid vortex lines near a boundary.

We model the helium II boundary layer flow by assuming that the normal fluid has a Blasius profile (with a velocity U_n far from the boundary), and the superfluid has a uniform velocity profile with a value U_s. Both velocity fields are taken to be in the $\hat{x}$ direction, along the boundary, and the parameters U_s and U_n are independent of each other.

We shall also assume that we are far enough from the leading edge of the boundary layer so that the boundary layer thickness δ is approximately constant over the region which we are considering. We seek to find non-trivial superfluid vortex filament structures that are stable, even at non-zero temperatures where energy dissipation occurs.

In an inviscid Euler flow, there exist many vortex structures which can move without changing their size or shape (vortex rings, helical vortex waves, vortex solitons) [2]. However, in superfluid helium II at non-zero temperatures (when the normal fluid component is present) energy and momentum transfer between the two fluid components causes all those examples to become unstable. Trivial vortex structures (e.g. vortex lines parallel to the direction of both the velocity fields) with the term $\alpha \mathbf{S}' \times (\mathbf{V}_n - \mathbf{V}_s)$ in (1) equal to zero at every point along the vortex structure will remain stable at non-zero temperatures. We seek to find non-trivial superfluid vortex structures that retain both their shape and size, but that have non-zero mutual friction force at almost all points along their length [3]. Such structures would constantly exchange energy and interact with the normal fluid, and as such would be important to the dynamics of the helium II boundary layer flow.

We seek to define a vortex configuration $\mathbf{S}(\xi, t)$ which will keep the same shape while moving in the streamwise direction $\hat{x}$ with a uniform and steady speed U_D relative to the mean superfluid flow U_s. Note that we are not calculating vortex structures which are pinned to the boundary. Using the Local Induction Approximation (and a lot of algebra) we derive an equation for such a structure. Further details of this calculation are given in [4]). This nondimensionalized equation is

$$
\begin{aligned}
\mathbf{S}'' = A \{ &\mathbf{S}' \times U_D + \alpha^2 [\mathbf{S}'(\mathbf{V}_n - U_s)] \\
&+ \alpha(\mathbf{V}_n - U_s) - U_D - \mathbf{S}'[\mathbf{S}' \cdot (\mathbf{V}_n - U_s - U_D)] \}
\end{aligned}
\tag{2}
$$

where A is a constant of order 1. The boundary conditions are that the vortex filament must be perpendicular to the boundary on the boundary, so that the condition of zero flow through the boundary is not violated. There are some solutions of (2) that do not return to the boundary, but we are not concerned with these. This equation was solved numerically using a third order Adams-

Bashforth-Moulton predictor-corrector method. Equation (2) can be used to find steady state superfluid vortex configurations in any geometry. The geometry of the flow is determined by the boundary conditions on the end of the superfluid vortex filament, and by the steady normal fluid velocity $V_n(x)$.

2.1   Properties of the Vortex Line Solutions

We hold two of the system parameters (α and the ratio U_s/U_n) constant, and vary the other two (U_D and U_n), and thus map out the solutions to (2) of this plane of the parameter space. A typical plot of this is shown in Fig. 1.

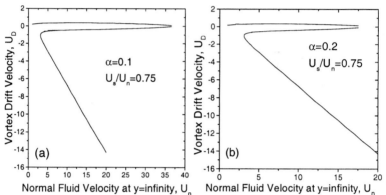

Fig. 1. Values of U_D and U_n which give give solutions to (2) and satisfy the boundary conditions. The middle branch of the plot is stable, and the upper and lower branches were found to be unstable. Parameter values are given in the figure.

For other parameter ranges (α and U_s/U_n) the plot is qualitatively the same, with three branches. Using another vortex simulation (based on the Biot-Savart law, which provides an independent test of these vortices) we have determined that the middle branch is the stable branch, with the lower branch solutions growing to become their middle branch counterparts, and the upper branch solutions shrinking rapidly to zero length.

Plots of these middle branch solutions are given in Fig. 2. These vortices move with a constant velocity through the boundary layer flow without changing their shape or form. They extend to a typical height of 2δ to 4δ (the boundary layer extends to a height of approximately 5δ). These structures are physically similar to the classical horseshoe vortex (which are associated with turbulent boundary layers, laminar flow around obstacles, interactions between boundary layers and jets and have been linked with turbulent bursting phenomena[5]), both in size and velocity (although our superfluid vortices move with greater speed), yet the underlying physics are completely different in all of these cases.

These superfluid vortex structures are not static objects, but have a dynamical equilibrium between vortex stretching and decay. The lower part of the structure is growing, whereas the upper section is shrinking. For further details

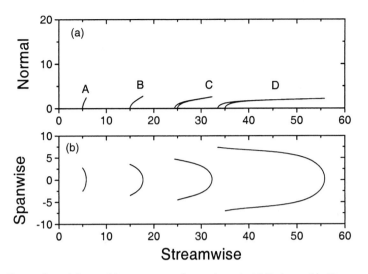

Fig. 2. Examples of the stable vortex configurations (middle branch). Parameter values are $\alpha = 0.2$, $U_s/U_n = 0.75$ and (A) $U_n=5$, (B) 10, (C) 15, (D) 17. Note the increase in length as the velocity U_n increases (b), but they never extend up out of the boundary layer (a).

of this mechanism, see Ref. [4]. This implies a constant energy transfer between the normal fluid and the superfluid. Near the boundary, on the legs of the vortex the normal fluid loses energy to the superfluid. Further away from the boundary, in the centre section of the vortex, this energy is returned from the superfluid to the normal fluid. Thus, this provides a mechanism for moving energy upwards in the flow. It should be noted that this mechanism takes energy away from an already low energy region of the flow.

2.2 Discussion

We have seen that these vortices are attractors, and thus we would expect remnant vortex filaments within a boundary layer flow to develop into these structures. The existence of such structures should be considered in boundary layer experiments, such as flow past a sphere [6], oscillating u-tube, oscillating discs and spheres [7], as these geometries all have non-dimensional velocities within the range where we would expect our structures to be present. These superfluid vortex structures also act as conduits of energy in the normal fluid, transporting kinetic energy away from the boundary. This transport of kinetic energy away from the boundary is likely to have a destabilising effect on the normal fluid.

3   Stability Analysis

The question of the stability of the normal fluid flow due to vortices present in the boundary layer is now addressed. We shall consider the 2D laminar flow between two infinite boundaries. In the absence of a mutual friction coupling between the two fluid components, the normal fluid will have a plane Poiseuille profile, and the superfluid a constant profile. However, with mutual friction, the normal fluid profile will be changed. We assume a vorticity distribution (Gaussian, centred at the points where the velocity profiles are equal), and model the mutual friction force $\mathbf{F}_{mf}$ by

$$\mathbf{F}_{mf} = \mathcal{F}(z)(\mathbf{V}_n - \mathbf{V}_s) \tag{3}$$

where $\mathcal{F}(z)$ represents the vorticity distribution. With this term added to the equations of motion for the two fluid components we calculated the forced normal fluid velocity profile. Further details can be found in [8] The next step is to investigate the linear stability of this velocity profile.

3.1   Linear Stability

We follow the method of linear stability outlined in [9]. Starting with the mutual friction forced Navier-Stokes equation for the normal fluid,

$$\frac{\partial}{\partial t}\mathbf{V}_n + (\mathbf{V}_n \cdot \nabla)\mathbf{V}_n = -\nabla P + \frac{1}{Re}\nabla^2\mathbf{V}_n + \mathcal{F}[\mathbf{V}_n - \mathbf{V}_s] \tag{4}$$

we assume that the normal fluid velocity profile $\mathbf{V}_n$ is made up of a mean flow $U(z)\hat{\boldsymbol{x}}$ (previously calculated) and a perturbation velocity $\boldsymbol{u}' = (u', 0, w')$. Substituting this, and introducing the streamfunction $\varPsi$ of the form

$$\varPsi(x, z, t) = \varPhi(z)e^{i\beta(x-ct)} \tag{5}$$

we arrive at a modified Orr-Sommerfeld equation. Where β is the wavenumber of the disturbance velocity, and $c = c_r + ic_i$ is the complex wavespeed, Note that the growth rate of the perturbation is given by αc_i. The modified Orr-Sommerfeld equation is

$$(U - c)(D^2 - \beta^2)\varPhi - U''\varPhi = (i\beta Re)^{-1}(D^2 - \beta^2)^2\varPhi$$
$$+ (i\beta)^{-1}(\mathcal{F}'D + \mathcal{F}D^2 - \beta^2\mathcal{F})\varPhi \tag{6}$$

with boundary conditions $\varPhi(-1) = \varPhi(1) = 0$ and $\varPhi'(-1) = \varPhi'(1) = 0$. Both D and prime denote derivative with respect to z. This is an eigenvalue problem which was solved numerically. The neutral stability curves are given by $c_i = 0$, although the reader should note that the growth rate is given by βc_i, not just c_i alone. Since we are only considering the case of positive β, then if c_i is greater than zero, the perturbation grows exponentially, whereas is shrinks exponentially if c_i is less than zero.

3.2   Stability Results

A typical plot of the neutral stability curves is shown in Fig. 3. The effect of the mutual friction is to slightly alter the stability profile of the main branch, but it also produces an entirely new unstable mode, which occurs for very low wavenumber β. This new instability can have a critical Reynolds number which

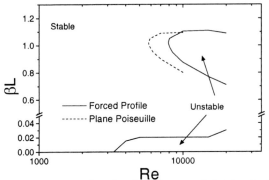

Fig. 3. Neutral stability curve for the forced normal fluid velocity profile. Forcing parameters are peak forcing f_{max}=0.005, and width of forcing $\sigma = 0.15$.

is lower than the main unstable branch. We always find that this lower branch has its most unstable eigenmode occurring at zero wavenumber. This stability analysis predicts that the channel flow of helium II goes unstable in a fundamentally different way than the classical Navier-Stokes flow. The critical Reynolds number of the lower branch may also be calculated analytically, and gives the result as

$$Re_{\text{crit}} = \frac{\pi^2}{f_{max}} \tag{7}$$

where f_{max} is the peak value of the vorticity distribution, related to the vortex line density. This analytic result agrees very well with our numerical data, as do the predicted and calculated eigenmodes. We have found that the shape of the Gaussian vorticity distribution is not critical to the stability of the flow, but the maximum value of it is important. A contour plot of the spatial structure of the streamfunction [the real part of $\Phi(z)\exp(i\beta x)$] is plotted in Fig. 4. The lower unstable mode has a much simpler spatial structure than its classical main branch counterpart, as the dependence on the downstream coordinate $\hat{x}$ is removed as $\beta \to 0$.

3.3   Discussion

From the nondimensionalization of the system, we have calculated the vortex line density required to produce our forcing magnitude f_{max}. For a flow with a length scale of 1cm, a peak velocity of 1cm/s, at $1.90°$K the required superfluid vortex line density is approximately 35cm^{-2}. This is a relatively modest line density.

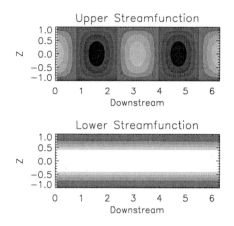

Fig. 4. Contour plot of the streamfunction Φ of the forced modes. (a) the upper mode, $\beta = 2.00$, (b) the lower mode $\beta = 0$. The Reynolds number is 20,000 with the forcing parameters as in Fig. 3.

Above a critical forcing magnitude, a new unstable mode appears, which for sufficient forcing can dominate the classical instability. This new instability has a much simpler geometry, and is not very sensitive to the width of the superfluid vorticity distribution. We would speculate that the important (in terms of the new instability) part of the mutual friction (3) is not the exact form of the vorticity distribution, but that the mutual friction is proportional to $\mathbf{V}_n - \mathbf{V}_s$. For these velocity profiles, this means that we have a strong shear effect in the mutual friction term with the force changing sign at the position where this new instability has the greatest effect.

References

1. D. C. Samuels, R. J. Donnelly: Phys. Rev. Lett. **65**, 187 (1990)
2. R. J. Donnelly: *Quantized Vortices in Helium II* (Cambridge University Press, 1991)
3. D. C. Samuels: Phys. Rev. B, **46**, (1992)
4. S. P. Godfrey, D. C. Samuels: Phys. Rev. B, **61**, 6 (2000)
5. B. J. Cantwell: Ann. Rev. Fluid Mech., **13**, (1981) C. J. Baker: J. Fluid. Mech., **95** (1979) H. Makita et al: AIAA J. **27** (1989) R. M. Kelso, A. J. Smits: Phys. Fluids, **7** (1995) G. R. Offen, S. J. Kline: J. Fluid Mech.,**70** (1975)
6. M. R. Smith, D. K. Hilton, W. W. Van Sciver: Phys. Fluids, **11** (1999)
7. R. J. Donnelly, A. C. Hollis-Hallett: Can. J. Phys., **33** (1955)
8. S. P. Godfrey, D. C. Samuels, C. F. Barenghi: Phys. Fluids, **13** 983 (2001)
9. P. G. Drazin, W. H. Reid: *Hydrodynamic Stability* (Cambridge University Press, 1984)

Grid Generated He II Turbulence
in a Finite Channel – Theoretical Interpretation

L. Skrbek and J.J. Niemela

Cryogenic Helium Turbulence Laboratory, Department of Physics,
University of Oregon, Eugene, OR 97403, USA

Abstract. Up to six orders of magnitude of He II vorticity decaying over three orders of magnitude in time in the temperature range 1.2 K $\leq T \leq$ 2 K can be described by a purely classical spectral model for homogeneous and isotropic turbulence. The He II vorticity is defined as $\omega = \kappa L$, where κ is the circulation quantum and L represents the total length of the vortex line per unit volume, an experimental observable. The model accounts for the quantum effects by introducing a temperature dependent effective kinematic viscosity. In agreement with experimental observation, the spectral decay model predicts four different regimes of the decay of vorticity in a finite channel - they switch as the energy containing and dissipative Kolmogorov length scales grow during the decay, finally both being saturated by the size of the channel.

1 Introduction

Studies of grid generated nearly homogeneous and isotropic turbulence (HIT) and its decay belong to the most important and extensively explored problems in "conventional" turbulence. Most of the experimental work is related to wind tunnels, where the turbulence is studied as it decays downstream (e.g., [1]). Turbulence without a mean flow, generated by using an oscillating grid and by towing a grid through a stationary sample of fluid has also been studied (see [2] and reference therein). Despite the long history of the subject, a general theory describing the decay of turbulence based on first principles has not yet been developed. However, experimental data containing information about the turbulent energy spectra provide a solid foundation for a phenomenological approach first outlined in [1,4], as the generally accepted forms of the three-dimensional turbulent energy spectra uniquely determine the temporal decay of turbulence. We have generalized this approach taking into account the finite size of the turbulence box, intermittency and viscosity effects and discuss the final period of decay [2,3]. In particular, we show that distinctly different decay regimes exist due to the physical restriction that eddies larger than the size of the turbulent box cannot exist[2,7].

It is interesting to compare the decay of grid generated classical and "superfluid" turbulence, which consists of a dense tangle of quantized vortex lines and normal fluid eddies. Here by "superfluid" we mean the turbulence generated by towing a grid with a velocity v_g through a stationary sample of He II in a channel, as described in the companion article in this book. We stress that our analysis is not applicable to a counterflow turbulence created by applying a heat pulse to He II.

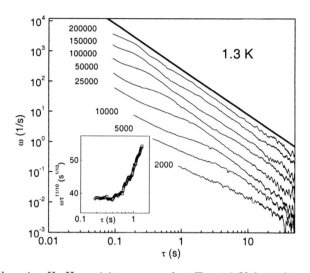

Fig. 1. The decaying He II vorticity measured at $T = 1.3$ K for indicated Re_M. Each curve represents an average of three individual decays. As the decay curves tend to collapse on the universal curve, for clarity we shifted them subsequently by factor of two downwards, the uppermost remaining unchanged. The virtual origin time t_{vol} corresponds to a virtual origin position $3.5M$ downstream in a classical wind tunnel. The inset shows the early part of the normalized vorticity decay data, displaying power law with exponent $-11/10$. After saturation, typically several orders of magnitude of decaying vorticity follow the power law with exponent $-3/2$, represented by the thick solid line.

In the framework of the phenomenological two-fluid model He II is described as consisting of the inviscid superfluid and the viscous normal fluid. In turbulent flow the presence of quantized vortices couples the two fluids together, via mutual friction[5]. Based on recent theoretical and experimental investigations[3,5,7], in the temperature range covered in the present study, 1.2 K$\leq T \leq 2$ K, turbulent He II flow resembles classical flow possessing an effective kinematic viscosity ν_{eff} of order μ/ρ, where μ is the dynamic viscosity of the normal component and ρ the total density. In particular, the usual HIT relationship between turbulent energy dissipation per unit volume and the mean square vorticity applies (limitations of applicability of this equation are discussed by Vinen[5]):

$$-dE/dt = \varepsilon = \nu_{eff}\kappa^2 L^2 = \nu_{eff}\omega^2 \tag{1}$$

Here the vorticity is defined, in analogy with a rotating bucket of He II, as $\omega = \kappa L$, where κ is the quantum of circulation and L is the length of quantized vortex line per unit volume, obtained from the experiment. At present it is not entirely clear how closely this definition of ω corresponds to the rms vorticity in classical turbulence. Discussing this problem involves careful considerations of the nature of He II turbulence arising from the quantized circulation in superfluid and details of the coupling via mutual friction and would bring us beyond the scope of this paper. Here we emphasize the similarity between classical and

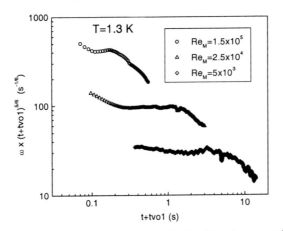

Fig. 2. Parts of the decaying He II vorticity multiplied by $(t + tvo1)^{5/6}$ measured at $T = 1.3$ K for the indicated Re_M. The plateau indicates the $-5/6$ power law, the second regime of the decay.

superfluid turbulence and show that considering them in parallel leads to a better understanding of the phenomenon of a fluid turbulence in general.

2 The Spectral Decay Model

Let us introduce a purely classical phenomenological spectral model of decaying HIT which predicts and quantitatively describes the data of the temporal decay of the quantized vortex line density obtained in the experiment where He II turbulence is created by towing a grid through a stationary sample of He II (for experimental details, see the companion article in this book). The model treats He II as a classical liquid possessing the temperature dependent effective kinematic viscosity.

The model assumes that at early times the decaying grid turbulence displays the generally accepted form of the 3D spectrum for HIT

$$E(k) = 0; k \leq k_d = 2\pi/d \tag{2}$$

$$E(k) = Ak^m; m = 2; 2\pi/d \leq k \leq k_1(t) \tag{3}$$

$$E(k) = C\varepsilon^{2/3}k^{-5/3}; k_2(t) \leq k \leq \gamma(\varepsilon/\nu_{eff}^3)^{1/4} = 2\pi\eta_{eff} \tag{4}$$

$$E(k) = 0; k \geq 2\pi\eta_{eff} \tag{5}$$

which reflects a physical restriction that eddies larger than the width of the channel, d, cannot exist and neglects intermittency. Also, the high wave number exponential tail of the spectrum is approximated by a sharp cutoff at the effective Kolmogorov length scale, η_{eff}, by introducing the dimensionless factor γ of order unity. We assume the 3D Kolmogorov constant $C = 1.62 \pm 0.17$, based on a number of classical experiments[6]. In the vicinity of the energy containing

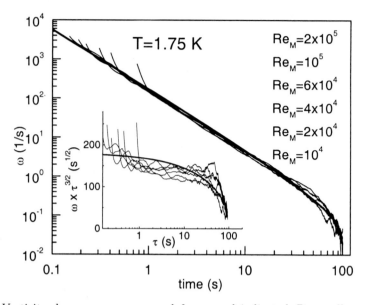

Fig. 3. Vorticity decay curves measured for several indicated Re_M collapse on the universal decay curve, represented by the thick solid line - a plot of formula (8). The inset shows the normalized data, the plateau indicates the $-3/2$ power law.

length scale $\ell_e = 2\pi/k_e(t)$, where $k_1(t) \leq k_e(t) \leq k_2(t)$, the spectral energy density displays a smooth broad maximum whose analytical form is not specified. Evaluating the total turbulent energy by integrating the 3D energy spectrum over all k leads to a differential equation for decaying turbulent energy. Applying $\varepsilon = \nu_{eff}\omega^2$ leads to a differential equation for decaying vorticity (for details of calculation including intermittency corrections, see[2]). At the early decay the spectrum stays self-similar. For $d \gg \ell_e \gg \eta_{eff}$, the energy decay is predicted to follow[2,7]

$$E(t + t_{vo1}) = E(\tau) \propto \tau^{-2\frac{m+1}{m+3}}; \ell_e \propto \tau^{\frac{2}{m+3}}; \omega \propto \tau^{\frac{3m+5}{2m+6}} \qquad (6)$$

Comparison with both the wind tunnel[1,2] and He II data[2] suggests the virtual origin position within few mesh units downstream the grid. Assuming validity of the Saffman invariant[4] (m=2) we obtain $\omega \propto \tau^{-11/10}$, i.e., the first regime of the decaying vorticity, as illustrated in Fig. 1. Experimentally, the power law is very sensitive to the exact value of virtual origin time. Due to the geometry of our apparatus, the parameter space for observing this first regime of the decay is rather limited, but clearly observable[2], see the inset in Fig.1.

As the turbulence decays further and ℓ_e grows, the lowest physically significant wave number becomes closer to the broad maximum around $2\pi/\ell_e$. The low wavenumber part of the spectrum can no longer be approximated as Ak^m with $m = 2$. Instead, it can be characterized by an effective power that decreases as the turbulence decays, such that $0 \leq m < 2$. Formula (6) then shows that the decay rate slows down. As ℓ_e approaches d, m becomes effectively zero and we

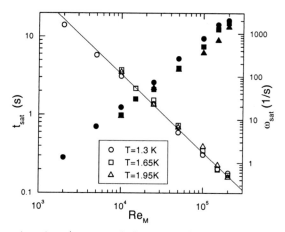

Fig. 4. The saturation time (open symbols, left axis) and corresponding vorticity values (filled symbpols, right axis) versus Re_M at different temperatures. The solid line represents $3.3 \times 10^4/Re_M$.

arrive at the second regime of the decay characterized by $\omega \propto \tau^{-5/6}$ - see Fig.2. Note that these simplified power law arguments do not hold for the decay of the turbulent energy as discussed in[2].

At the saturation time, t_{sat}, the vorticity reaches its saturation value, ω_{sat}, and the growth of ℓ_e is completed. Still neglecting the cutoff of the spectral energy at η_{eff}, the further decay - the third regime - is predicted universal:

$$\omega(\tau) = \frac{\sqrt{27}d}{2\pi}\sqrt{\frac{C^3}{\nu_{eff}}}\tau^{-3/2} \tag{7}$$

with the virtual origin time t_{vo2}. Therefore, no matter what value Re_M is (providing it is high enough to neglect viscosity corrections[2]), in the finite size box the decaying turbulence ought to reach this universal third regime of the decay. Note that the virtual origin time t_{vo2} generally differs from t_{vo1} introduced above[2], but our analysis shows that the difference is small and in our further discussion of the decay we therefore assume $t_{vo2} \cong t_{vo1} \cong 3.5M/v_g$.

We used individual decay curves to define t_{sat} and ω_{sat} as an intersection point of the power laws $\omega \propto \tau^{-5/6}$ and $\omega \propto \tau^{-3/2}$ superimposed on the decay data. The result of this fitting procedure is summarized in Fig.4. We found $t_{sat} \propto 1/Re_M$ and values of t_{sat} and ω_{sat} hardly dependent on temperature, although the normal fluid density changes over an order of magnitude. It strongly suggests that the role of quantum effects in He II turbulence in this temperature range can be at least approximately accounted for by introducing ν_{eff} and justifies applicability of a purely classical model for the decay.

So far in our discussion we neglected a role of the high wave number cutoff of the energy spectrum at η_{eff}. As the turbulent energy (or vorticity) decays and the Kolmogorov length scale grows, the relative importance of this cutoff grows and a simple power law cannot any longer describe the decay of vorticity. It is

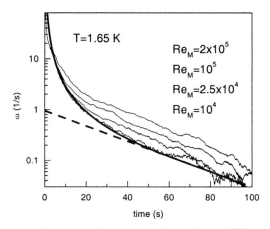

Fig. 5. The late part of the decay data measured at indicated Re_M. The thick solid line represents a plot of formula (8) with parameters $C = 1.62$, $\nu_{eff} = 2.28 \times 10^{-4}$ cm^2/s, $\gamma = 0.418$. The late decay can be characterized as exponential, of a form $\exp(t/t_0)$ with $t_0 = 29$ s, represented by a dashed line.

possible to show[2] that after saturation the universal decay of vorticity can be more accurately described by

$$\omega(\tau) = \frac{3^3 \nu_{eff}}{2^3 \gamma} \left(\frac{2\pi}{d}\right)^2 \left(\frac{t_B}{\tau}\right)^{3/2} \cos^3 \theta \qquad (8)$$

where $\cos^2(3\theta) = \tau/\tau_B$ and $t_B \cong \frac{16C\gamma^{4/3}}{9\nu_{eff}} \left(\frac{d}{2\pi}\right)^2$. Naturally, for $\gamma \to \infty$ expression (8) reduces to a simple power law (7). Formally, as the vorticity decays, η_{eff} becomes the size of the channel and the turbulent energy vanishes. Close to this stage the applicability of the spectral model is no longer justified. From Fig.4 we can estimate t_B of order 100 s and $\gamma \cong 0.4$. This matches the value of γ needed to describe the viscous corrections in classical wind tunnel data[1]. Expression (8) is used for comparison with experimental data in Figs. 4 and 5. It describes the universal decay - up to five orders of magnitude of decaying He II vorticity - measured for all Re_M at any temperature. The departure from a single power law with exponent -3/2 illustrates the increasing influence of the growing Kolmogorov length scale in the decay.

Experimentally we have found a slight increase of the experimental data from the theoretical line predicted by the model towards higher values of vorticity (see Figs. 4 and 5). There might be several reasons for this behavior. First, it might occur due to extra production of vorticity by counterflow in the channel. The turbulence created by the towed grid decays and, as a result, the temperature inside the channel grows, stimulating counterflow inside the channel due to the fountain effect. The faster the grid is pulled, the more extra vorticity is produced by counterflow. The extra heat input could also result from friction between the grid and the channel walls.

It is possible, however, that there is a deeper physical reason for this behavior originating from the quantum nature of the He II turbulence. The quantum effects were taken into account simply by introducing ν_{eff}, roughly a factor of 2-3 higher than the kinematic viscosity based on total fluid density, and ν_{eff} may depend on the Reynolds number.

As the Kolmogorov scale grows, it gradually approaches d and thus ℓ_e which is already saturated by it. Expression (8) describes the experimental data down to surprisingly low level of vorticity, of order $\omega \cong 0.1 \text{ s}^{-1}$, corresponding to a vortex line density $L = \omega/\kappa \cong 100 \text{ cm}^{-2}$ and to a mean distance between quantized vortex lines about 1 mm. Still, the essentially classical description of the decaying vorticity holds.

With no inertial scale left there is no energy transfer towards higher wavenumbers and the only possibility for further decay is the exponential viscous decay. This is the fourth and last regime of decaying vorticity in a finite channel. Note that it differs from the final period of decay observed in classical wind tunnels, as these can be essentially regarded as infinitely large[2]. The last regime is evident from Fig.5, where the late decay curves originating from various Re_M display an exponential decay of the form $\omega(t) \propto \exp(-t/t_0)$, practically indistinguishable from the spectral model prediction. Performing systematic measurements for various Re_M at $T = 1.3$ K, 1.65 K and 1.9 K, we obtained $t_0 = (27 \pm 6)$ s.

This last decay regime can be considered in analogy with the decay of the oscillatory motion in viscous fluids, characterized by exponential decay of the energy $E = E_0 \exp(-\beta t)$, where the decay coefficient $\beta = 2\nu k^2$. For $k \cong 2\pi/d$ and $\nu \cong 10^{-4} \text{ cm}^2 \text{ s}^{-1}$ it suggests a characteristic decay time close to the observed one.

Acknowledgements

We acknowledge stimulating discussions with R.J. Donnelly, G.L. Eyink, D. Holm, K. R. Sreenivasan S.R. Stalp and W.F. Vinen. This research was supported by NSF under grant DMR-9529609.

References

1. G. Comte-Bellot, S. Corrsin: J. Fluid Mech. **25**, 657 (1966); **48**, 273 (1971)
2. L. Skrbek, S.R. Stalp: Phys. Fluids **12**, 1997 (2000)
3. S.R. Stalp, L. Skrbek, R.J. Donnelly: Phys. Rev. Lett. **82**, 4831 (1999)
4. P.G. Saffman: J. Fluid Mech. **27**, 581 (1967); Phys. Fluids **10**, 1349 (1967)
5. W.F. Vinen: Phys. Rev. B **61**, 1410 (2000)
6. K.R. Sreenivasan: Phys. Fluids **7**, 2778 (1995)
7. L. Skrbek, J.J. Niemela, R.J. Donnelly: Phys. Rev. Lett. **85**, 2973, (2000)

Vortex Tangle Dynamics
Without Mutual Friction in Superfluid ^{4}He

Makoto Tsubota[1], Tsunehiko Araki[1], and Sergey K. Nemirovskii[2]

[1] Department of Physics, Osaka City University, Osaka 558-8585, Japan
[2] Institute of Thermophysics, Academy of Science, Novosibirsk 630090, Russia

1 Introduction

Recently Davis et al. observed the free decay of the vortices at mK temperatures where the normal fluid density became vanishingly small and the mutual friction did not work effectively [1]. It is unclear how the vortices decay. Motivated by this experimental work, we studied numerically the vortex dynamics without the mutual friction, thus finding some cascade process which was obscured by the normal fluid at higher temperatures. This paper reviews our recent works on this problem. The numerical procedure based on the vortex filament formulation is described in detail in Ref. [2]. Section 2 describes the dynamics of waves excited along the reconnected vortex lines under the full Biot-Savart law and the energy spectrum characteristic of the cascade process [3]. Section 3 studies the dynamics of a dense vortex tangle(VT) by the calculation under the localized induction approximation(LIA) [2]. The absence of the mutual friction makes the vortices kinked, which promotes vortex reconnections. Consequently small vortices are cut off from a large one through the reconnections. The resulting vortices also follow the self-similar process to break up to smaller ones. Although our formulation cannot describe the final destiny of the minimum vortex, the decay of the VT is found to be connected with this cascade process, which is just the cascade process at zero temperature Feynman proposed [4]. It should be noted that this cascade process in a VT includes not only the breaking up of vortices but also the vortex wave process described in Sect. 2.

2 Vortex Wave Cascade Process

Recently Vinen discussed the acoustic emission from an oscillating vortex at very low temperatures [5]. The scenario is the following. Reconnection of two vortices leave sharp kinks on them. The kinks, propagating along the vortex lines, are evolved to the vortex waves whose wavenumbers are much larger than the inverse of the average vortex spacing ℓ. Eventually the vortex waves with wavenumber larger than a critical value are strongly damped by the acoustic emission. The scenario except for the final stage can be confirmed by our formulation.

As a typical example, we calculated the collision of a straight vortex line and a moving ring by the full Biot-Savart law. Figure 1(a) shows the initial configuration of vortex lines. Toward the reconnection, the ring and the line

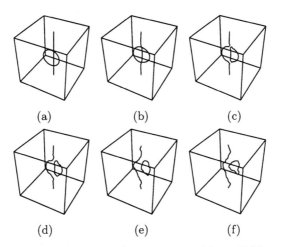

Fig. 1. Collision of a straight vortex and a ring at $t=0$(a), $t=20$(b), $t=21$(c), $t=22$(d), $t=24$(e) and $t=26$(f). At $t=20.4$, two vortices reconnect.

twist themselves so that they become locally antiparallel at the closest place (Fig.1(b)). At $t \simeq 20.4$, two vortices reconnect. After the reconnection (Fig.1(c), (d) and (e)), the resulting local cusps broaden while exciting vortex waves with various wavenumbers (Fig.1(f)).

We will introduce the energy spectrum of the velocity field v made by the vortices [6]. The kinetic energy can be defined as the integral of the square of a field:

$$E_{\text{kin}} = \frac{1}{2(2\pi)^3} \int d^3x (\sqrt{\rho}v)^2, \tag{1}$$

where ρ is the density of fluid. The energy spectrum $E_{\text{kin}}(k)$ is defined as $E_{\text{kin}} = \int_0^\infty dk E_{\text{kin}}(k)$. Using the Parseval's theorem, one gets the following energy spectrum:

$$E_{\text{kin}}(k) = \frac{1}{2} \int d\Omega_k \left| \frac{1}{(2\pi)^3} \int d^3r e^{i\mathbf{r}\cdot\mathbf{k}} \sqrt{\rho}v \right|^2, \tag{2}$$

where $d\Omega_k$ denotes the volume element $k^2 \sin\theta d\theta d\phi$ in the spherical coordinates. The energy spectrum $E_{\text{kin}}(k)$ represents the contribution from the velocity field with wave number k to the kinetic energy.

Figure 2 shows the energy spectrum in the process of Fig.1. Before the reconnection($t \leq 20.4$), the energy spectrum is almost constant in time. After the reconnection($t \geq 20.4$), the energy spectrum begins to fluctuate suddenly. The waves generated by the cusps evolve chaotically to waves with other wavenumber by the nonlinear interaction; the reconnection process results in the ergodic energy distribution of vortex waves. This is consistent with the study of the sideband instability by Samuels and Donnelly [7]. Figure 2 shows that some energy peaks move to small k region. Compared with Fig.1, this behavior may reflect the broadening of the local cusps and the ring's leaving the line. The total kinetic

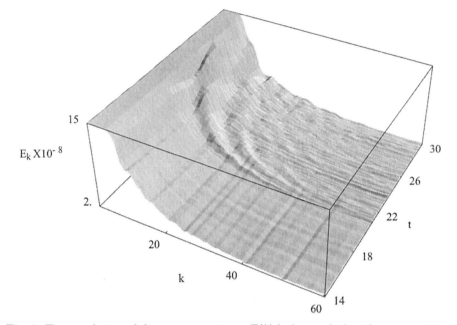

Fig. 2. Time evolution of the energy spectrum $E(k)$ before and after the reconnection.

energy is conserved despite the mode mixing. If some dissipative mechanism is introduced in the k region above a critical wave number, as in the Vinen's theory [5], the kinetic energy is transferred from small k region to large one, being dissipated at the critical value, so that the total energy decays. This is consistent with Vinen's theory.

3   Cascade Process in the Vortex Tangle

3.1   Decay of the Vortex Tangle

This section studies the free decay of the dense VT without mutual friction under the LIA. The VT in a 1cm³ cube is calculated with the space resolution $\Delta \xi = 1.83 \times 10^{-2}$cm and the time resolution $\Delta t = 1.0 \times 10^{-3}$sec. The initial VT for free decay is prepared by the development of six vortex rings subject to thermal counterflow and mutual friction [8]. After turning off the thermal counterflow, we follow the dynamics of the VT. Figure 3 shows transient VTs with and without the mutual friction. The difference is marked.

The VT subject to the mutual friction consists of relatively smooth vortex lines, while the absence of the mutual friction makes the vortices kinked, which promotes vortex reconnections [2]. The small vortices are separated from a large one through the reconnections. The resulting vortices also follow the self-similar cascade process to break up to smaller ones. Our numerical calculation cannot follow the dynamics of vortices smaller than its space resolution; such vortices are eliminated numerically and its justification will be discussed later.

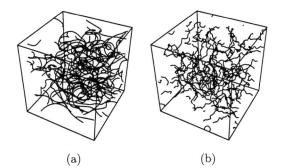

(a)   (b)

Fig. 3. Example of VTs with (a) and without (b) the mutual friction.

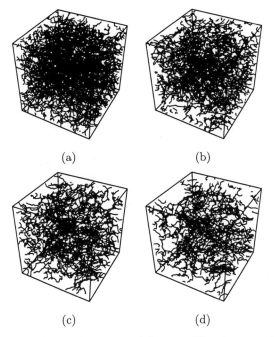

(a)   (b)

(c)   (d)

Fig. 4. Decay of a dence VT without mutual friction. The time is $t=0$(a), 30(b), 60(c) and 90(d).

The decay of the dense VT free from the mutual friction is shown in Fig. 4. The only mechanism for the decay is that cut-off procedure which eliminates the small vortices. However it should be noted that the continuous decay of the VT results in the presence of the stationary cascade process wherein large vortices break up to smaller ones through reconnections. This is because, if such cascade process is absent, even though the system is subject to that cut-off procedure, the continuous decay is never sustained. Only the cascade process that keeps supplying the small vortices can reduce the VT constantly. We confirm that the decay rate is almost independent of the space resolution when it is in the range between $4\Delta\xi$ and $\Delta\xi/4$.

It is possible to classify the kinds of reconnection in the VT dynamics. The vortex reconnection is divided topologically into three classes. The first refers to the process whereby two vortices reconnect to two vortices, which is most usual. The second is the process which divides one vortex into two vortices (the split type); the cascade process is driven by this kind of reconnection. Third is the process whereby two vortices are combined to one vortex against the cascade process (the combination type). Investigating the number of reconnection events in the VT dynamics of Fig. 4, we find that most reconnections belong to the first class but the second split type occupies about 17% of the total reconnections, being superior to that of the third combination type of about 10%. The reconnection of the split type actually promotes the cascade process, against the reverse process due to that of the combination type.

The cascade process is revealed further by investigating the size distribution of vortices. Figure 5 shows the change of the size distribution in the VT dynamics of Fig. 4. Each figure shows the number $n(x)$ of vortices as a function of their length x. The system size $a(=1\text{cm})$ and the space resolution $\Delta\xi(= 1.83\times10^{-2}\text{cm})$ are the characteristic scales in this system. The vortices longer than a are originally few, and most vortices are concentrated in the scale range $[\Delta\xi, a]$. As the cascade process progresses, every vortex generally divides into smaller ones through the split type reconnections, although some combination type reconnections may occur. As a result, the vortices between $\Delta\xi$ and a are decreased in number because they become smaller than $\Delta\xi$ and eliminated. Although the vortices larger than a become few too, some of them survive. Such vortices extend over the whole system and have small self-induced velocity, being almost straight. Of course they may happen to reconnect with other vortices, then they can be divided to smaller vortices and follow the cascade process. However their small self-induced velocity makes their reconnections uncommon, thus stabilizing them. Hence the decay of the vortex line density (VLD) is attributable mainly to the cascade process of the vortices smaller than a.

The final destiny of small vortices through the cascade process may be interpreted several ways. First, the vortices can vanish at a small scale by radiating phonons, which is discussed recently by Vinen [5]. Secondly, the vortices whose size is eventually reduced to the order of the interatomic distance no longer sustain the vortex state, probably changing into such short-wavelength excitation as roton. Since both mechanisms work only at a small scale, some process that transfers energy from a large scale to smaller scales is necessary for the decay of the VT; this is just the cascade process. Thirdly, in a real system, the small vortices may collide with the vessel walls [2]. Since only the vortices in the bulk are observed experimentally, the reconnection with the walls may reduce the observed VLD effectively.

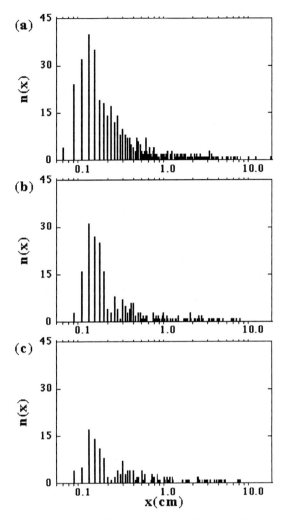

Fig. 5. Bar chart showing the number of vortices $n(x)$ as a function of the length x. The time is $t=0$(a), 50(b) and 100(c).

3.2   Comparison with the Vinen's Equation

Considering that cascade process at zero temperature proposed by Feynman [4], Vinen obtained an evolution equation for the VLD $L(t)$, what we call the Vinen's equation [9]

$$\frac{dL}{dt} = -\chi_2 \frac{\kappa}{2\pi} L^2,$$ (3)

where χ_2 is a parameter and κ the quantized circulation. Its solution is given by

$$\frac{1}{L} = \frac{1}{L_0} + \chi_2 \frac{\kappa}{2\pi} t,$$ (4)

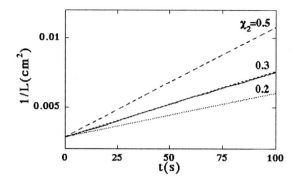

Fig. 6. Comparison of the decay of $L(t)$ and the solution of the Vinen's equation.

where L_0 is the VLD at $t = 0$.

Figure 6 shows the comparison of our numerical results and the solution of the Vinen's equation. The solid line refers to our result for the VT decay of Fig.4, while three other lines denote Eq.(4) with the parameters $\chi_2 = 0.5$, 0.3, 0.2. Our result agrees excellently with the solution of $\chi_2 = 0.3$. Since the Vinen's equation is based closely on the cascade process, this agreement supports that the cascade process occurs really in the numerical simulation. The values of χ_2 obtained at finite temperatures [5] are extrapolated towards zero temperature, then seeming to reach reasonably to $\chi_2 \simeq 0.3$; the value obtained numerically may be consistent quantitatively.

Acknowledgements

We acknowledge W.F. Vinen and P.V.E. McClintock for useful discussions.

References

1. S.L.Davis, P.C.Hendry, P.V.E.McClintock: Physica B **280**, 43(2000)
2. M.Tsubota, T.Araki, S.K.Nemirovskii: Phy. Rev. B Vol.62, Num.17(2000)
3. T.Araki, M.Tsubota: J. Low Temp. Phys.(to be published)
4. R.P.Feynman: 'Application of Quantum Mechanics to Liquid Helium'. In: *Progress in Low Temperature Physics Vol. 1.* ed. by C.J.Gorter(North-Holland, Ameterdam, 1955) pp. 17–53
5. W.F.Vinen: Phy. Rev. B **61**, 1410(2000)
6. C.Nore, M.Abid, M.E.Brachet: Phys. Fluids **9**, 2644(1997)
7. D.C.Samuels, D.J.Donnelly: Phys. Rev. Lett. **64**, 1385(1990)
8. K.W.Schwarz: Phys. Rev. B **38**, 2398(1988)
9. W.F.Vinen, Proc. R. Soc. London A **242**, 493(1957)

Applications of the Gaussian Model of the Vortex Tangle in the Superfluid Turbulent He II

Sergey K. Nemirovskii and Mikhail V. Nedoboiko

Institute of Thermophysics, 630090 Novosibirsk, RUSSIA

Abstract. In spite of an appearance of some impressive recent results in understanding of the superfluid turbulence in HeII they fail to evaluate many characteristics of vortex tangle needed for both applications and fundamental study. Early we reported the Gaussian model of the vortex tangle in superfluid turbulent HeII. That model is just trial distribution functional in space of vortex loop configurations constructed on the basis of well established properties of vortex tangle. It is designed to calculate various averages taken over stochastic vortex loop configurations. In this paper we use this model to calculate some important characteristics of the vortex tangle. In particular we evaluate the average superfluid mass current J induced by vortices and the average energy E associated with the chaotic vortex filament.

1 Introduction

The presence of vortex tangle appearing in the superfluid turbulent HeII essentially changes hydrodynamic properties of the latter (see e.g.[1] ,[2],[3]). According phenomena are studied in frame of so called Phenomenological Theory (PT) pioneered by Vinen [4] and greatly modified by Schwarz [5]. The PT describes superfluid turbulence (ST) in terms of the total length of vortex lines (per unit of volume) or the vortex line density (VLD) $\mathcal{L}(t)$ and of the structure parameters of the VT. Knowledge of these quantities allows to calculate some of hydrodynamic characteristics of superfluid turbulent HeII such as a mutual friction, sound attenuation etc. Meanwhile there exist many other physical quantities connected to distribution of the filaments and their interaction related with other physical phenomena which can not be expressed in terms of the PT. The relevant phenomena should be covered by appropriate stochastic theory of chaotic vortex filaments. Of course, the most honest way to develop such theory is to study stochastic dynamics of vortex filaments on the base of equations of motion with some source of chaos. However due to extremely involved dynamics of vortex lines this way seems to be almost hopeless. Thus, a necessity of a developing an advanced phenomenological approach appeared. We offer one variant of such approach. The main idea and the main strategy are the following. Although the phenomenological theory of the superfluid turbulence deals with macroscopical characteristics of the vortex tangle, it conveys the rich information concerning the *instantaneous* structure of the vortex tangle. Namely we know that the VT consists of the closed loops labelled by $s_j(\xi)$, uniformly distributed in space and having the total length $\mathcal{L}(t)$ per unit of volume. From acoustical experiments it follows that filaments are distributed in anisotropic manner and quantitative characteristics of this anisotropy can be expressed by some structure parameters (see [1], [3], [5]). Beside this usual anisotropy there is more subtle anisotropy connected with averaged polarization of the vortex loops. Furthermore there are some proofs that the averaged curvature of the vortex lines is proportional to the inverse interline space and coefficient of this proportionality (which is of order of unit) was obtained in numerical simulations made by Schwarz [4].

The master idea of our proposal is to construct a trial distribution function (TDF) in the space of the vortex loops of the most general form which satisfies to all of the properties of the VT introduced above. We assume that this trial distribution function will enable us to calculate any physical quantities due to the VT. In the paper we describe typical shape of vortex loop obtained from evaluating of the correlation functions. We also calculate the average hydrodynamic impulse (or Lamb impulse) $\mathbf{J}_V$ in the counterflowing superfluid turbulent HeII and the average kinetic energy E associated with the chaotic vortex loop.

2   Constructing the Trial Distribution Function

According to general prescriptions the average of any quantity $\langle \mathcal{B}(\{\mathbf{s}_j(\xi_j)\}) \rangle$ depending on vortex loop configurations is given by

$$\langle \mathcal{B}(\{\mathbf{s}_j(\xi_j)\}) \rangle = \sum_{\{\mathbf{s}_j(\xi_j)\}} \mathcal{B}(\{\mathbf{s}_j(\xi_j)\}) \mathcal{P}(\{\mathbf{s}_j(\xi_j)\}). \qquad (1)$$

Here $\mathcal{P}(\{\mathbf{s}_j(\xi_j)\})$ is a probability of the vortex tangle to have a particular configuration $\{\mathbf{s}_j(\xi_j)\}$. Index j distinguishes different loops. The meaning of summation over all vortex loop configurations $\sum_{\{\mathbf{s}_j(\xi_j)\}}$ in formula (1) will be clear from further presentation . We put the usual in the statistical physics supposition that all configuration corresponding to the same macroscopic state have equal probabilities. Thus the probability $\mathcal{P}(\{\mathbf{s}_j(\xi_j)\})$ for vortex tangle to have a particular configuration $\{\mathbf{s}_j(\xi_j)\}$ should be proportional to $1/N_{\text{allowed}}$, where N_{allowed} is the number of allowed configurations, of course infinite

$$\mathcal{P}(\{\mathbf{s}_j(\xi_j)\}) \propto \frac{1}{N_{\text{allowed}}}. \qquad (2)$$

Under term "allowed configurations" N_{allowed} we mean only the configurations that will lead to the correct values for all average quantities known from experiment and numerical simulations. Formally it can be expressed as a path integral in space of three-dimensional (closed) curves supplemented with some constrains connected to properties of the VT.

$$N_{\text{allowed}} \propto \prod_j \int \mathcal{D}\{\mathbf{s}_j(\xi)\} \times \text{constraints } \{\mathbf{s}_j(\xi)\}. \qquad (3)$$

The constraints entering this relation are expressed by delta functions expressing fixed properties of the VT. For instance constrain $\delta((\mathbf{s}'_j(\xi))^2 - 1)$ expresses that parameter ξ is the arc length. However this condition will lead to not tractable theory. We will use a trick known from the theory of polymer chains (see e.g. [6]) , namely we will relax rigorous condition and change delta function by continuous (Gaussian) distribution of the link length with the same value of integral. This trick leads to the following expression for number of way:

$$N_{\text{allowed}} \propto \prod_j \int \mathcal{D}\{\mathbf{s}_j(\xi)\} \times e^{-\lambda_1 \int_0^{\mathcal{L}} |\mathbf{s}'|^2 d\xi} . \tag{4}$$

In the same manner we are able to introduce and treat other constrains connected to the known properties of the VT structure. The detailed calculations are exposed in paper of one of the author [7] , now we write down final expression for probability of configurations

$$N_{\text{allowed}} \propto \int \mathcal{D}\{\mathbf{s}(\kappa)\} \exp\left(-\mathcal{L}\{\mathbf{s}(\kappa)\}\right). \tag{5}$$

Here $\mathbf{s}(\kappa)$ is one-dimensional Fourier transform of variable $\mathbf{s}(\xi)$ and Lagrangian $\mathcal{L}\{\mathbf{s}(\kappa)\}$ is a quadratic form of the components of the vector variable $\mathbf{s}(\kappa)$

$$\mathcal{L}\{\mathbf{s}(\kappa)\} = \sum_{\kappa \neq 0} \mathbf{s}_x^\alpha(\kappa) \Lambda^{\alpha\beta}(\kappa) \mathbf{s}_x^\beta(\kappa). \tag{6}$$

In practice to calculate various averages it is convenient to work with the characteristic (generating) functional (CF) which is defined as a following average:

$$W(\{\mathbf{P}_j(\kappa)\}) = \left\langle \exp\left(-\sum_j \sum_{\kappa \neq 0} \mathbf{P}_j^\mu(\kappa) \mathbf{s}_j'^\mu(-\kappa)\right) \right\rangle.$$

Due to that our Lagrangian 6is a quadratic form (in $\mathbf{s}(\kappa)$) and, consequently, the trial distribution function is a Gaussian one, calculation of the CF can be made by accomplishing the full square procedure to give a result

$$W(\{\mathbf{P}_j(\kappa)\}) = \exp\left(-\sum_j \sum_{\kappa \neq 0} \mathbf{P}_j^\mu(\kappa) N_j^{\mu\nu}(\kappa) \mathbf{P}_j^\nu(-\kappa)\right). \tag{7}$$

Elements of matrix $N_j^{\mu\nu}(\kappa)$ are specified from calculation of total length, anisotropy coefficient, curvature and polarization. The explicit form of them is written down in [7].

Thus we reached the put goal and have written the expression for trial CF which, we repeat, enables us to calculate any averaged of the vortex filament configuration. For instance calculating some of the correlation functions we are able to describe a typical shape of the averaged curve. It is sketched out in Fig.1

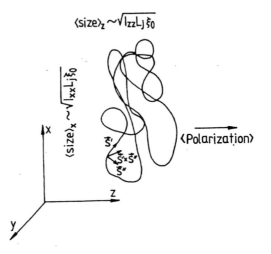

Fig. 1. A snapshot of the averaged vortex loop obtained from analysis of the statistical properties. Position of the vortex line element is described as $\mathbf{s}_j(\xi_j)$, where ξ_j is arc length, $\mathbf{s}'_j(\xi_j) = d\mathbf{s}_j(\xi_j)/d\xi_j$ is a tangent vector, unit vector along the vortex line; $\mathbf{s}''_j(\xi_j) = d^2\mathbf{s}_j(\xi_j)/d\xi_j^2$ is the local curvature vector; vector production $\mathbf{s}'_j(\xi_j) \times \mathbf{s}''_j(\xi_j)$ is binormal which is responsible for mutual orientation of the tangent vector and vector of curvature. Close ($\Delta\xi \ll R$, where R is the mean curvature) parts of the line are separated in $3D$ space by distance $\Delta\xi$. The distant part ($R \ll \Delta\xi$) are separated in $3D$ space by distance $\sqrt{2\pi R \Delta\xi}$ (with correction due to the closeness). The latter property reflects a random walk structure of the vortex loops. As a whole the loop is not isotropic having a "pancake" form. In addition it has a total polarization $\langle \int \mathbf{s}'_j(\xi_j) \times \mathbf{s}''_j(\xi_j) d\xi_j \rangle$ forcing the loop to drift along vector $\mathbf{V}_n$ and to produce nonzero superfluid mass current in z-direction

3   Hydrodynamic Impulse of the Vortex Tangle

As an one more illustration to the developed theory we discuss hydrodynamic impulse of the vortex tangle $\mathbf{J_V}$ which is defined as

$$\mathbf{J_V} = \langle \frac{\rho_s \widetilde{\kappa}}{2} \sum_j \int \mathbf{s}_j(\xi_j) \times \mathbf{s}'_j(\xi_j) \, d\xi_j \rangle \tag{8}$$

The quantity $\mathbf{J_V}$ is closely related to momentum of fluid (see [8]). The averaged $< \mathbf{s}_j(\xi_j) \times \mathbf{s}'_j(\xi_j) >$ is immediately evaluated by use of CF (7) to give the following result:

$$\mathbf{J_V^z} = - \left[\frac{\rho \widetilde{\kappa} I_l \alpha_v}{\rho_n c_2^2 \beta_v} \right] \rho_s \mathbf{V_s} \tag{9}$$

Note that the coefficient includes no fitting parameters but only characteristics known from the Phenomenological Theory (see [5]). Relation (9) shows that the vortex tangle induces the superfluid current directed against the external super-fluid current. It should be expected since there is some preferable polarization of the vortex loops. In the experiments this additional superfluid current should display itself as suppression of the superfluid density. This effect is 3D analog to the famous Kosterlitz-Thoulless effect except of that distribution of the vortex lines is not calculated but is obtained appealing to the experimental data.

Since superfluid density enters an expression for second sound velocity, it seems attractive to detect it using transverse second sound testing. To do it we have firstly to evaluate transverse change of the ρ_s and, secondly, to develop the theory to match it to nonstationary case. The general theory asserts that while applying a harmonic external second sound field suppression of superfluid density becomes the function of frequency ω of the following form:

$$\Delta\rho_s^x(\omega) = \left(\frac{\delta J_V^x}{\delta V_s^x}\right)_{transv} \frac{1}{1 + i\omega\tau_J}. \tag{10}$$

Here transverse $(\delta J_V^x/\delta V_s^x)_{transv}$ is half of the one given by rel. (9). The quantity τ_J is the time of relaxation of the superfluid current J_V. which is to be found from dynamical consideration. First, we have to derive dJ_V^x/dt with help of the equation of motion of the vortex line elements and, second, to evaluate various averaged appearing in right-hand side. function we obtain the following final result for change of the second sound velocity. Performing all of described procedures one obtains that the relative change $\Delta u_2/u_2$ of the second sound velocity is given by

$$\frac{\Delta u_2}{u_2} = -f(T)\frac{V_{ns}^4}{\omega^2}. \tag{11}$$

Here the function $f(T)$ is composed of the structure parameters of the vortex tangle

$$f(T) = \frac{4\rho\widetilde{\kappa}I_l^2\alpha^2(1 - I_{xx})^2}{\rho_n c_2^4 \beta^3}. \tag{12}$$

Decreasing of the second sound velocity in the counterflowing HeII has been really observed about two decades ago by Vidal with coauthors [9]. Let us compare our result (11) with the Vidal's experiment. Using the data on the structure parameters one obtains that e.g. for the temperatures $1.44K$ the value of function $f(T)$ is about $620s^2/cm^4$. Taking the frequency $\omega = 4.3$ rad/s, used in [9], and $V_{ns} = 2$ cm/s one obtains that $\Delta u_2/u_2 \approx 4\times 10^{-4}$, which is very close to the observed value.

4   Energy of the Vortex Tangle

In this section we calculate the averaged energy of the stochastic vortex loop distributed according trial distribution function (5) . The general expression for the energy associated with linear vortices can be written as (see e.g. [8])

$$E = \left\langle \frac{1}{2}\int \rho_s v_s^2 \, d^3\mathbf{r} \right\rangle = \left\langle \frac{\rho_s \kappa^2}{8\pi} \sum_{j,i} \int_0^{L_i}\int_0^{L_j} \frac{\mathbf{s}_i'(\xi_i)\mathbf{s}_j'(\xi_j)}{|\mathbf{s}_i(\xi_i) - \mathbf{s}_j(\xi_j)|} d\xi_i d\xi_j \right\rangle. \tag{13}$$

In 3D Fourier space the average energy E (13) can be rewritten as

$$E = \left\langle \frac{\rho_s \kappa^2}{2} \sum_{i,j} \int_{\mathbf{k}} \frac{d^3 \mathbf{k}}{(2\pi)^3 \mathbf{k}^2} \int_0^{L_i} \int_0^{L_j} \mathbf{s}'_i(\xi_i) \mathbf{s}'_j(\xi_j) d\xi_i d\xi_j \, e^{i\mathbf{k}(\mathbf{s}_i(\xi_i) - \mathbf{s}_j(\xi_j))} \right\rangle. \quad (14)$$

Comparing (14) and (7) it is possible to express the energy E in terms of the characteristic Functional

$$\langle E \rangle = \frac{\rho_s \kappa^2}{2} \sum_{i,j} \int_{\mathbf{k}} \frac{d^3 \mathbf{k}}{(2\pi)^3 \mathbf{k}^2} \int_0^{L_i} \int_0^{L_j} d\xi_i d\xi_j \, e^{i\mathbf{k}(\mathbf{s}_i(0) - \mathbf{s}_j(0))} \times \frac{\delta^2 W}{i\delta \mathbf{P}_i^\alpha(\xi_i) \, i\delta \mathbf{P}_j^\alpha(\xi_j)} \quad (15)$$

Here set of $\mathbf{P}_n(\xi'_n)$ in CF $W(\{\mathbf{P}_n(\xi'_n)\})$ is again determined with help of the θ-functions

$$\mathbf{P}_i(\xi'_i) = \mathbf{k}\theta(\xi'_i)\theta(\xi_i - \xi'_i), \quad \mathbf{P}_j(\xi'_j) = \mathbf{k}\theta(\xi'_j)\theta(\xi_j - \xi'_j),$$
$$\mathbf{P}_n(\xi_n) = 0, \quad n \neq i, j \quad (16)$$

The relation (16) implies that we have to choose in integrand in exponent of CF only points lying in interval from 0 to ξ_i on i-curve and from 0 to ξ_j on j-curve. While evaluation of self-energy of the same loop , $i = j$, one has to distinguish points ξ_i , and to put them to be e.g. ξ'_i and ξ''_i. Further results concern the case of the only loop of length L. Omitting tremendous calculations we write down the final answer in the following form:

$$E = \frac{\rho \kappa^2 L}{4\pi} \ln \frac{R}{a_0} + \frac{\rho \kappa^2 L}{4\pi} \left(1 - \frac{2}{\sqrt{\pi}} (f_2 - f_1)\right) \ln \frac{R}{a_0} \quad (17)$$
$$+ \frac{\rho \kappa^2 L}{4\pi} \left[\frac{1}{(\sqrt{\pi} - 1)^{1/2}} \frac{2 f_3}{\pi^{5/2} c_2^2 3} \cdot I_l^2 + \frac{f_2}{\pi^{3/2} (\sqrt{\pi} - 1)^{1/2}}\right],$$

where the quantities f (of order of unit) are expressed via the structure parameters of the VT as follows (below $\beta = \sqrt{I_x - I_z / I_x}$)

$$f_1(\beta) = \sqrt{2(3 - \beta^2)}(\arcsin(\beta)/\beta), \quad (18)$$

$$f_2(\beta) = \left(\sqrt{1 - \beta^2} + (2 - \beta^2) \arcsin(\beta)/\beta\right) / \sqrt{2(3 - \beta^2)}, \quad (19)$$

$$f_3(\beta) = (2(3 - \beta^2))^{3/2} \left(\sqrt{1 - \beta^2} - (\arcsin \beta)/\beta\right) / \beta \quad (20)$$

Let us comment expression (17). The first term in the right-hand side of (17) is just the energy of unit of length of a straight vortex filament (see e.g. [1]) multiplied by its length. In this form it is frequently used in theory of superfluid turbulence (see e.g. [3]) and in other applications. But there are additional terms. The third and forth terms appeared from long-range interaction, they are smaller then logarithmic ones (about ten percents). The third term is of especial interest. It appeared due to polarization of the vortex loop and its presence implies that there is some elasticity of the vortex tangle in $\mathbf{V}_{ns}$ direction. Results of the previous section showed that the VT induces some additional superfluid flow. Therefore one can expect that combination of longitudinal elasticity combining with inertia of additional will lead to appearing of elastic waves, 3D analog of the Tkachenko waves.

The second one is also logarithmically large. Logarithmic behaviour points out that this contribution came from denominator $|\mathbf{s}(\xi) - \mathbf{s}(\xi')|$. But it was the first (local) term which collected contributions from neighbor points along the line. Therefore the third term appeared from accidental self-crossing of remote (along the line) parts of the vortex filament. The fact that this term is proportional to L and is of of the first (local) contribution is due to that the line is fractal object with Haussdorf H_d dimension equal $H_d = 2$. According to general theory of fractal lines it has an infinite number of self-crossing with cardinal number $2H_d - 3$ i.e. it is equivalent to line.

5   Conclusion

We briefly exposed an essence of Gaussian model of the vortex tangle and give several examples how it can be used for evaluation of important physical characteristics such as induced momentum and energy of interaction. These characteristics has been discussed early (see e.g. [1]), however their evaluation has not been performed because of lack of a proper theory. We think that these illustrations convince that Gaussian model can serve as effective tool to study chaotic vortex filaments.

The work was partly funded by Russian Foundation for Basic Research, Grant N 99-02-16942

References

1. R.J. Donnelly:*Quantized Vortices in Helium II,* (Cambridge University Press, Cambridge 1991)
2. J.T. Tough: 'Superfluid Turbulence'. In: *Progress in Low Temperature Physics, VIII* ed.by D.F. Brewer (North Holland, Amsterdam 1982), pp. 133-219
3. S.K.Nemirovskii, W.Fiszdon: Rev. Mod. Phys., **67** , 37 (1995)
4. W.F. Vinen:Proc. R. Soc. London A **243**, 400(1958)
5. K.W.Schwarz:, Phys. Rev. **B38**, 2398 (1988)
6. M. Doi and S.F. Edwards: *The theory of polymer dynamics* (Clarendon Press, Oxford 1986)
7. S.K. Nemirovskii:Phys. Rev **B57**, 5792 (1997)
8. G.K. Batchelor, *An introduction to fluid mechanics,* (Cambridge University Press 1967)
9. F.C. Vidal:C.R. Acad Sci., **B 275**, , 609 (1972)

Stochastic Dynamics of a Vortex Loop. Thermal Equilibrium

Sergey K. Nemirovskii[1], L.P. Kondaurova[1], and M. Tsubota[2]

[1] Institute for Thermophysics, Lavrentyeva,1, 630090 Novosibirsk, Russia
[2] Department of Physics, Osaka City University, Osaka, Japan

Abstract. We study stochastic behavior of a single vortex loop appeared in imperfect Bose gas. Dynamics of Bose-condensate is supposed to obey Gross-Pitaevskii equation with additional noise satisfying fluctuation-dissipation relation. The corresponding Fokker-Planck equation for probability functional has a solution $\mathcal{P}(\{\psi(\mathbf{r})\}) = \mathcal{N}\exp(-H\{\psi(\mathbf{r})\}/T)$, where $H\{\psi(\mathbf{r})\}$ is a Ginzburg-Landau free energy. Considering a vortex filaments as a topological defects of the field $\psi(\mathbf{r})$ we derive a Langevin-type equation of motion of the line with correspondingly transformed stirring force. The respective Fokker-Planck equation for probability functional $\mathcal{P}(\{\mathbf{s}(\xi)\})$ in vortex loop configuration space is shown to have a solution of the form $\mathcal{P}(\{\mathbf{s}(\xi)\}) = \mathcal{N}\exp(-H\{\mathbf{s}\}/T)$, where $\mathcal{N}$ is a normalizing factor and $H\{\mathbf{s}\}$ is energy of vortex line configurations. In other words a thermal equilibrium of Bose-condensate results in a thermal equilibrium of vortex loops appeared in Bose-condensate. Some consequences of that fact and possible violations are discussed.

1   Introduction and Scientific Background

Quantized vortices appeared in quantum fluids have been an object of intensive study for many years (see for review and bibliography the famous book by Donnelly[1]). The greatest success in investigations of dynamics of quantized vortices has been achieved in relatively simple cases such as vortices in rotating helium (where they form a vortex array orientated along an axis of rotation) or vortex rings. However these simple cases are rather exception than a rule. Due to extremely involved dynamics initially straight lines or rings evolve to form highly entangled chaotic structure. Thus a necessity of use of statistic methods to describe chaotic vortex loop configurations arises. A most tempting way is to treat "gas" vortices as a kind of excitation and to use thermodynamic methods. One of first examples of that way was an use of the Landau criterium for critical velocity where vortex energy and momentum were applied to relation having pure thermodynamic sense. More extended examples would be the famous Kosterlitz-Thouless description of 2D vortices or its 3D variant intensively elaborated currently (for review and bibliography see e.g.[2]).

In the examples above and in many other it is assumed that chaotic vortex loop configurations are in a thermal equilibrium and their statistics obeys the Gibbs distribution . That supposition is based on fundamental physical principles and can be justified in a standard way considering vortex lops as a subsystem submerged into a thermostat and exchanging energy with the latter. The role of thermostat in the case of vortices is played by the other excitations (phonons and rotons) of an underlying physical field. In the case of Bose-Einstein condensate (BEC), which we consider in this paper, that field is an order parameter $\psi(\mathbf{r},t)$. Vortex lines are just the crossings of surfaces where both real and imaginary part of $\psi(\mathbf{r},t)$ vanish. Excitations of order parameter (phonons and rotons) interact with vortices driving the latter to statistic distribution which in accordance with general principles should be the Gibbs distribution.

It is well known however that the Gibbs distribution can be alternatively obtained in the frame of some reduced model like kinetic equations or Fokker–Planck equation. That way of course is not of such great generality as a principle of maximum entropy, but instead it allows us to clarify the mechanisms of how the Gibbs distribution is established and to discuss possible deviations and violations of equilibrium state. We choose that way to examine how a thermal equilibrium of chaotic vortex loops follows from a thermal equilibrium of BEC.

2   Langevin Equation

We perform our consideration on the basis of the Gross–Pitaevskii model[3],[4]. To describe stochastic behavior of BEC let us add to the right–hand side of the Gross–Pitaevskii equation the random stirring force $\zeta(\mathbf{x}, t)$.

$$\frac{\hbar}{m}\frac{\partial \psi}{\partial t} = -\left(\Lambda + i\right)\frac{\delta H(\psi)}{\delta \psi^*} + \zeta(\mathbf{x}, t). \tag{1}$$

Here $H\{\psi\}$-is the Ginzburg–Landau free energy functional

$$H\{\psi\} = \int d^3x \left[\frac{\hbar^2}{2m^2}|\nabla \psi|^2 - \frac{\mu}{m}|\psi|^2 + \frac{V_0}{2m}|\psi|^4\right]. \tag{2}$$

The thermal noise obeys the following fluctuation–dissipation theorem

$$\langle \zeta(\mathbf{x}_1, t_1)\zeta^*(\mathbf{x}_2, t_2)\rangle = \frac{2k_B T \Lambda}{(m/\hbar)}\,\delta(\mathbf{x}_1 - \mathbf{x}_2)\,\delta(t_1 - t_2)\ . \tag{3}$$

The stochastic problem introduced by equations (1)–(3) has a solution describing thermal equilibrium, where probability of some configuration of ψ–filed is proportional to $\exp(-H\{\psi(\mathbf{r})\}/T)$. This readily follows from the correspondent Fokker–Planck equation[5].

Let's go on to the problem of obtaining of the vortex dynamics appearing in BEC, whose own dynamics obeys relations (1)–(3). That problem, as well as more general problems of that kind has been considered many times by many methods[6] starting from pioneering papers by Pitaevskii [7] and Fetter [8]. We develop one more method, neither too rigorous nor principally new but convenient for purpose of this paper. Details of that method will be published elsewhere, here we briefly describe it to an extent to comprehend the put goal.

As any motion of topological defects the vortex line dynamics is determined by the one of the underlying field theory. On the other hand if one ignores a presence of other excitations except of the vortices one can say that all motion of

BEC is determined by the one of quantum vortices. Thus there is mutual correspondence and the order parameter $\psi(\mathbf{x}, t)$ can be considered as some functional
of a whole vortex loop configuration $\psi(\mathbf{x} \mid \{\mathbf{s}(\xi, t)\})$. Temporal dependance of
field ψ is connected with motion of lines and its rate of change (at some point
$\mathbf{x}$) is expressed by the following chain rule:

$$\frac{\partial \psi(\mathbf{x}, t)}{\partial t} = \int_{\Gamma} \frac{\delta \psi(\mathbf{x} \mid \{\mathbf{s}(\xi, t)\})}{\delta \mathbf{s}(\xi', t)} \frac{\partial \mathbf{s}(\xi', t)}{\partial t} d\xi'. \tag{4}$$

Vortices in quantum fluids are very slender tubes (except of vicinity of phase
transition) and their dynamics is analogous to the one of the strings. That implies
that we have to aim our efforts at integrating out radial degrees of freedom. It
can be reached by the following procedure. Let us further multiply equation(1)
by $\frac{1}{-\Lambda+i} \delta \psi^* / \delta \mathbf{s}(\xi_0, t)$, where ξ_0 some chosen point on the curve. Combining the
result obtained with the complex conjugate and integrating over whole space we
have

$$\frac{\hbar}{m} \int d^3 \mathbf{x} \left(\frac{\Lambda - i}{\Lambda^2 + 1} \frac{\partial \psi}{\partial t} \frac{\delta \psi^*}{\delta \mathbf{s}(\xi_0, t)} + \frac{\Lambda + i}{\Lambda^2 + 1} \frac{\partial \psi^*}{\partial t} \frac{\delta \psi}{\delta \mathbf{s}(\xi_0, t)} \right) =$$
$$-\int d^3 \mathbf{x} \left(\frac{\delta H(\psi, \psi^*)}{\delta \psi^*} \frac{\delta \psi^*}{\delta \mathbf{s}(\xi_0, t)} + \frac{\delta H(\psi, \psi^*)}{\delta \psi^*} \frac{\delta \psi}{\delta \mathbf{s}(\xi_0, t)} \right) + \tag{5}$$
$$\int d^3 \mathbf{x} \left(\frac{\Lambda - i}{\Lambda^2 + 1} \zeta(\mathbf{x}, t) \frac{\delta \psi^*}{\delta \mathbf{s}(\xi_0, t)} + \frac{\Lambda + i}{\Lambda^2 + 1} \zeta^*(\mathbf{x}, t) \frac{\delta \psi}{\delta \mathbf{s}(\xi_0, t)} \right).$$

The first integral in the right–hand side of (5) expresses a chain rule for
functional derivative $\delta H(\mathbf{s})/\delta \mathbf{s}(\xi_0, t)$ where $H(\mathbf{s})$ is the energy of moving BEC
expressed via vortex line position. Consequently considering them to be very
slender tubes (which is justified when the radius of the curvature R is much
larger of the core size r_0) and neglecting an energy associated with the core, the
quantity $H(\mathbf{s})$ is just kinetic energy of the superfluid flow created by vortices
(see e.g. [9],[10])

$$H(\mathbf{s}) = \frac{\rho_s \kappa^{-2}}{8\pi} \int_{\Gamma} \int_{\Gamma'} \frac{\mathbf{s}'(\xi)\mathbf{s}'(\xi)}{|\mathbf{s}(\xi) - \mathbf{s}(\xi')|} d\xi d\xi'. \tag{6}$$

Here $\mathbf{s}'(\xi)$ is tangent vector, double integration is performed along a whole
line, $\tilde{\kappa}$ is a quantum of circulation equal $2\pi\hbar/m$. Calculation of functional derivative

$$\delta H(\mathbf{s})/\delta \mathbf{s}(\xi_0, t)$$

is straightforward and leads to result

$$\frac{\delta H(\mathbf{s})}{\delta \mathbf{s}(\xi_0, t)} = \rho_s \tilde{\kappa} \mathbf{s}'(\xi_0) \times \mathbf{B}(\xi_0). \tag{7}$$

Quantity $\mathbf{B}(\xi_0)$ is the velocity of the line element $\dot{\mathbf{s}}(\xi_0)$ expressed by well
known Biot–Savart law. The terms in the left–hand side of equation (5) (first

line) can be evaluated in general form by observing that the major contribution into integrals appears from vicinity of the vortex filament (see e.g. [11]). Thus to evaluate integral we replace $\psi(\mathbf{x} \mid \mathbf{s}(\xi, t))$ by $\psi_v(\mathbf{x}_\perp) = \psi_v(\mathbf{s}(\xi_{cl}, t) - \mathbf{x})$ where ψ_v is well studied 2D vortex and integration over $d^3\mathbf{x}$ by $d^2\mathbf{x}_\perp d\xi_{cl}$. Functional derivative $\delta\psi^*/\delta\mathbf{s}(\xi', t)$ should be evaluated by a following rule: $\delta\psi^*/\delta\mathbf{s}(\xi', t) = \nabla_\perp \psi_v(\mathbf{x}_\perp)\delta(\xi' - \xi_{cl})$. Here ξ_{cl} is the label of point of the line closest to point $\mathbf{x}$. Using the said above and calculating integrals of squared gradients of $\psi_v(\mathbf{x}_\perp)$ we conclude that left–hand side of equation (5) transforms into

$$\frac{\hbar}{m}\frac{2\pi\rho_s}{\Lambda^2+1}\,\dot{\mathbf{s}}\,(\,\xi_0)\times\mathbf{s}'(\xi_0) + \frac{\hbar}{m}\frac{2\pi\rho_s\sigma\Lambda}{\Lambda^2+1}\,\dot{\mathbf{s}}\,(\,\xi_0). \qquad (8)$$

Let us now discuss the rest terms of equation (5) including random force $\zeta(\mathbf{x}, t)$ (the third line). Consequently considering that the all motion of BEC is connected to motion of line, we have to consider Langevin force $\zeta(\mathbf{x}, t)$ as some secondary quantity stemming from random displacements of filaments. Connection between displacements (random) of filaments $\delta\mathbf{s}$ and deviations (random) of $\delta\psi(\mathbf{x}, t)$ may be written in form similar to (4) with formal substitution $\partial\psi(\mathbf{x}, t)/\partial t \to \delta\psi(\mathbf{x}, t)$ and $\dot{\mathbf{s}} \to \delta\mathbf{s}$. Taking into account that $\delta\psi(\mathbf{x}, t) = \zeta(\mathbf{x}, t)\delta t$, and $\delta\mathbf{s} = \zeta(\xi, t)\delta t$ we conclude that quantities $\zeta(\mathbf{x}, t)$ and $\zeta(\xi, t)$ are connected to each other by a chain rule

$$\zeta(\mathbf{x}, t) = \int\limits_\Gamma \frac{\delta\psi(\mathbf{x} \mid \{\mathbf{s}(\xi, t)\})}{\delta\mathbf{s}(\xi', t)}\zeta(\xi', t)d\xi'. \qquad (9)$$

That implies that to take into consideration random displacements of line we have to change the last term in equation (5) by the one similar to (8) with substitution $\dot{\mathbf{s}}\,(xi_0) \to \zeta(\xi_0, t)$. Gathering all terms we obtain a vector equation, which can be resolved up to tangential velocity $\mathbf{s}_\parallel\,(\,\xi_0)$ along the curve. The latter does not have any physical meaning and can be removed by suitable parameterization of the label variable ξ. Solving that vector equation we arrive at

$$\dot{\mathbf{s}}\,(xi_0) = \frac{1+\Lambda^2}{1+\Lambda^2\sigma^2}\mathbf{B}(\xi_0) + \frac{(1+\Lambda^2)\Lambda\sigma}{1+\Lambda^2\sigma^2}\mathbf{s}'(\xi_0)\times\mathbf{B}(\xi_0) + (m/\hbar)\zeta(\xi_0, t). \qquad (10)$$

Equation (10) describes motion of vortex line in terms of the line itself. It is remarkable fact (not obvious in advance that noise $\zeta(\xi_0, t)$ acting on line is also additive (does not depend on line variables).

The last effort we have to do is to ascertain both the statistic properties of noise $\zeta(\xi_0, t)$ and its intensity. Shortly, it can be done by comparison of equation(8) with substitution $\dot{\mathbf{s}}\,(xi_0) \to \zeta(\xi_0, t)$ with the last term of equation (5). Clearly the former appeared as result of transformation of the latter. Equating them and taking the scalar productions of both parts of the resulting relation we arrive at

$$\langle\zeta_{\eta_1}(\xi_1, t_1)\zeta_{\eta_2}(\xi_2, t_2)\rangle = \frac{k_B T}{\rho_s\pi(\hbar/m)}\frac{(\Lambda^2+1)\Lambda\sigma}{1+\Lambda^2\sigma^2}\,\delta(\xi_1-\xi_2)\,\delta(t_1-t_2)\delta_{\eta_1,\eta_2}\,. \qquad (11)$$

Here ζ_{η_1} and ζ_{η_2} are components of random velocities in η_1, η_2 directions lying in the plain normal to the line.

Thus starting from dynamics of BEC (equation (1)) with the fluctuation–dissipation theorem (3) we derive equation (10) describing motion of vortex line in terms of line itself with the additive noise obeying the fluctuation–dissipation theorem (11). These relations complete a stochastic problem of quantized vortex dynamics under thermal noise stemming from the one stirring the underlying field. In the next section we demonstrate that this problem has an equilibrium solution given by Gibbs distribution $\exp(-H\{s\}/k_B T)$, where $H\{s\}$ is the functional of energy due to vortex loop (equation (6)) and T is the temperature of Bose–condensate.

3   Fokker–Planck Equation

To show it we, first, derive the Fokker–Planck equation corresponding to Langevin type dynamics obeyed (10) and (11). Let us introduce probability distribution functional (PDF)

$$\mathcal{P}(\{s(\xi)\}, t) = \langle \delta\left(s(\xi) - s(\xi, t)\right)\rangle. \tag{12}$$

Here δ is delta functional in space of vortex loop configurations. Averaging is fulfilled over ensemble of random force. The Fokker–Planck equation can be derived in standard way (see e.g. [12])

$$\frac{\partial \mathcal{P}}{\partial t} + \int d\xi \frac{\delta}{\delta s(\xi)} \left[\frac{1+\Lambda^2}{1+\Lambda^2\sigma^2}\mathbf{B}(\xi) + \frac{(1+\Lambda^2)\Lambda\sigma}{1+\Lambda^2\sigma^2}s'(\xi) \times \mathbf{B}(\xi)\right]\mathcal{P} + \tag{13}$$

$$\int\int d\xi d\xi' \frac{k_B T}{2\pi\rho_s(\hbar/m)^2} \frac{(\Lambda^2+1)\Lambda\sigma}{1+\Lambda^2\sigma^2} \delta(\xi-\xi')\, \delta(t_1-t_2)\delta_{\eta_1,\eta_2} \frac{\delta}{\delta s(\xi)}\frac{\delta}{\delta s(\xi')}\mathcal{P} = 0$$

Equation (13) possesses the equilibrium solution in form of the Gibbs distribution $\mathcal{P}(\{s(\xi)\}) = \mathcal{N}\exp(-H\{s\}/T)$, where $\mathcal{N}$ is a normalizing factor. Let us show that the first integral term vanishes identically for that solution. To do it we exploit relation (7) and parametrization of label variable ξ in which velocity $\dot{s}(\xi_0)$ is normal to the line. Using a tensor notation we rewrite the first term in integrand in form (we omit the coefficient and factor $\exp(-H\{s\}/k_B T)$)

$$\epsilon^{\alpha\beta\gamma}\left\{\frac{\delta s'_\beta(\xi)}{\delta s_\alpha(\xi)}\frac{\delta H(s)}{\delta s_\gamma(\xi,t)} + s'_\beta(\xi)\frac{\delta^2 H(s)}{\delta s_\gamma(\xi,t)\delta s_\alpha(\xi)} + s'_\beta(\xi)\frac{\delta H(s)}{\delta s_\gamma(\xi,t)}\frac{\delta H(s)}{\delta s_\alpha(\xi,t)}\right\}$$

The functional derivative $\delta s'_\beta(\xi)/\delta s_\alpha(\xi) \propto \delta_{\beta\alpha}$ therefore all terms vanish due to symmetry. Thus the reversible term gives no contribution to flux of probability (in the configuration space) equation (13), one says it is divergence free. Furthermore exploiting again relation 7) one convinces himself that second (dissipative) term in (13) and third (due to stirring force) term exactly compensate each other (locally) as it should be in the thermal equilibrium.

4   Possible Violation of Thermal Equilibrium

Thus we have proved that the thermal equilibrium of BEC results in the thermal equilibrium of vortex loop. We are now in position to discuss how it can be destroyed. Analyzing the proof one can see that the following steps were crucial. 1. Additive white noise $\zeta(\mathbf{x}, t)$ acting on field $\psi(\mathbf{x}, t)$ is transformed into additive white noise $f(\xi, t)$ acting on vortex line position $\mathbf{s}(\xi, t)$. 2. Intensity of noise $f(\xi, t)$ expressed by (11) is that it locally compensates dissipative flux of probability distribution functional in the Fokker–Planck equation (13).

That observation points out how the thermal equilibrium in space of vortex loops can be destroyed for real vortex tangles appeared e.g. in counterflowing HeII or formed in quenched superfluids. Being a macroscopical objects vortex loops inevitably undergo large scale perturbations generated e.g. by nonuniform flow or by action of other vortex loops, randomly placed with respect to the studied loop. One more essential source of large scale perturbations might be long wave instabilities of vortex filament motion. That type of random action drastically differs from δ–correlated in ξ space thermal noise considered above. It obviously cannot compensate dissipative flux of probability, which is proportional curvature and acts accordingly in small scales. Instead the following scenario seems to be realized [13]. Due to nonlinear character of the equation of motion the large–scale perturbations on an initially smooth filament interact creating higher harmonics. They in turn generate harmonics with larger κ, where κ is one–dimensional wave vector arisen in 1D Fourier transform (with respect to label variable ξ) of quantity $\mathbf{s}(\xi, t)$. One can say that an additional curvature created by large scale stirring force propagates in region of small scales. In real space that corresponds to entangling of vortex loop and creation of vortex line segments with large local curvature. Then dissipative processes come into play, their role now is that they remove from the system harmonics with very large wave vectors κ. So the Kolmogorov cascade–like solution with a flux of a curvature in space of κ is established and, as a result, the stochastic distribution is far from equilibrium. That scenario is quite similar to the one which is realized in classical turbulence or in the so called weak (wave) turbulence. A difference is that if in case of the wave and classical turbulence) an exchange an energy between harmonics is realized, whereas in our case there is exchange a curvature $\langle \kappa^4 \mathbf{s}(\kappa) \mathbf{s}(-\kappa) \rangle$.

The scenario described above is especially relevant in the so called low temperature superfluid turbulence case, when the normal component is very small and the usual (considered here) dissipation monotonically increasing with the curvature is absent. Some recent experiments and numerical simulations [14] show that vortex tangle in HeII decays at extremely low temperature (about 1 mK), where dissipation due to normal component is negligibly small. Obviously some other strong mechanisms of dissipation must take place. It can be e.g. emission of phonons and rotons from speedy moving parts of line or just collapse of "hairpin" segments of the filament. That mechanisms are concentrated on very small scales (of order of the vortex core size) or equivalently in region of very large wave numbers κ. Therefore regions of the pumping and the sink

S.K. Nemirovskii, L.P. Kondaurova, and M. Tsubota

of an additional curvature are greatly remote in κ space and a system must be essentially nonequilibrium.

One more reason of a violation of thermal equilibrium might be a reconnection of lines. In real vortex tangle consisting of many loops the vortex filaments undergo frequent collisions and reconnections. Just after reconnection there appear kinks on the curves disappearing later on. From mathematical point of view a kink on the curve can be described as a discontinuity of tangent vector $\mathbf{s}'(\xi, t)$ which has Fourier transform of type $\mathbf{s}'(\kappa) \propto \kappa^{-1}$. Thus the reconnection processes supply the selected curve with discontinuities having a spectrum $\mathbf{s}(\kappa) \propto \kappa^{-2}$. Taking into account a random nature of vortex line collision the reconnection processes can be modelled, in some measure as a random stirring of filament with spectrum of type $\langle \zeta(\kappa)\zeta(-\kappa) \rangle \propto \kappa^{-4}$. Remember now that the establishing of a thermal equilibrium requires that the random force correlation function is proportional $\propto \delta(\xi_1 - \xi_2)$ and, consequently the spectrum does not depend on wave number κ, $\langle \zeta(\kappa)\zeta(-\kappa) \rangle = const$. Therefore the colored noise $\langle \zeta(\kappa)\zeta(-\kappa) \rangle \propto \kappa^{-4}$ coming from reconnection processes can also lead to the Kolmogorov type nonequilibrium state. To clarify which of mechanisms forming nonequilibrium state prevails in real vortex tangle one has to investigate much more involved problem.

This work was partly supported under grant N 99-02-16942 from Russian Foundation of Basic research.

References

1. R.J. Donnelly, *Quantized Vortices in Helium II* (Cambridge University Press, Cambridge 1991).
2. G.A.Williams, J. Low Temp. Phys., **101**, 421, (1993)
3. L.P. Pitaevskii: Zh. Exp. Teor. Fiz. (U.S.S.R.) **35**, 408 (1958) [Sov. Phys. JETP **8**, 282 (1959)].
4. E.P. Gross: Nuovo Cimento **20**, 454 (1961)
5. P.C. Hohenberg and B.I. Halperin: Rev.Mod.Phys **49**, 435 (1972)
6. More or less full bibliography can be found in book by Pismen (see later). Some more recent papers would be E. Schroder and O.Tornkvist, Phys.Rev.Lett. **40**, 1908 (1997).
7. L.P. Pitaevskii: Sov. Phys. JETP **13**, 451 (1961)
8. A.L. Fetter: Phys.Rev. **151**, 100, (1966)
9. G.K. Batchelor, *An introduction to fluid mechanics* (Cambridge University Press, Cambridge 1967)
10. P.G. Saffman: *Vortex Dynamics* (Cambridge University Press, Cambridge 1992).
11. L.M. Pismen: *Vortices in Nonlinear Fields*, (Claberson Press, Oxford, 1999)
12. Jean Zinn-Justin: *Quantum Field Theory and Critical Phenomena* (Claberson Press, Oxford,1992)
13. S.K. Nemirovskii, J. Pakleza, W. Poppe:Stochasti behaviour of a vortex filament. Notes et Documents LIMSI N91-14, Orsay (1991)
14. M. Tsubota, Tsunehiko Araki, S.K. Nemirovskii: Phys Rev.**B**, 2000, (to be published).

Stochastic Dynamics of a Vortex Loop. Large-Scale Stirring Force

S.K. Nemirovskii and A.Ja. Baltsevich

Institute of Thermophysics, 630090, Novosibirsk, Lavrent'eva, 1.

Abstract. Stochastic dynamics of a vortex filament obeying local induced approximation equation plus random agitation is investigated by analytical and numerical methods. The character of a stirring force is supposed to be a white noise with spatial correlator concentrated at large distances comparable with size of the loop. Dependence of the spectral function $\langle s_\kappa^\alpha s_\kappa^\beta \rangle$ of the vortex line on both the one-dimensional wave vector κ and intensity of the external force correlator $\langle \zeta_\kappa^\alpha \zeta_\kappa^\alpha \rangle$ was studied. Here s_κ^α is the Fourier transform of the line element position $s^\alpha(\xi, t)$. It is shown that under the influence of an external random force a vortex ring becomes a small tangle whose mean size depends on external force intensity. The theoretical predictions and the numerical results are in reasonable agreement.

1 Introduction

In the previous paper [1] we discuss how the large-scale perturbations can destroy the thermal equilibrium state in the space of vortex loop configurations. In this paper we elaborate that idea and present results of the both analytical and numerical investigations on stochastic dynamics of a vortex filament in HeII undergoing an action of the large scale random displacements. Moreover we consider the case when the smooth dissipation connected with normal component is small (that correspond to the case of very small temperature) and the only strong dissipative mechanisms appear at very small scales comparable with the core radius of vortex. Thus detailed (at which scale) balance between the pumping and dissipation required for thermal equilibrium is violated and, as it has been discussed in [1], essentially nonequilibrium picture state must develop.

From a formal point of view that problem is significantly more involved, therefore we restrict ourselves to a consideration of the local induction approximation [2],[3] in the equation of motion and omit processes of reconnection. This statement of problem is, of course, far from real superfluid turbulence in He II. A value of that work is that it enables us to understand mechanisms of entanglement of vortex filament and of appearing the strongly nonequilibrium state. We would remind that idea of the vortex tangle had been launched by Feynman more than 40 years ago [2] and only about 10-15 years ago Schwarz demonstrated and confirmed that idea in his famous numerical simulations [3]. To our knowledge a similar success in analytic study is absent.

A structure of the paper is following. In the first part of this paper we develop nonequilibrium. diagram technique analogous to the one elaborated by

Wyld [4] for classical turbulence. Using further method of direct interaction approximation we derive a set of Dyson equation for the pair correlators and for the Green functions. Assuming that region of stirring force and dissipation are widely separated in space of scales we seek for a scale invariant solution in the so called inertial interval. We also present results of direct numerical simulation of the vortex tangle dynamics. Numerical results confirm the ones obtained in the analytical investigations, however some discrepancies remained.

2   Analytical Investigation

In the local induction approximation (LIA) the equation of motion of quantized vortex filament in HeII reads

$$\frac{d\,\mathbf{s}(\xi,t)}{d\,t} \;=\; \tilde{\beta}\mathbf{s}\,' \times \mathbf{s}\,'' + \delta + \zeta(\xi,t). \tag{1}$$

Here $\mathbf{s}(\xi,t)$ is a point of the filament labeled by variable ξ, $0 \le \xi \le 2\pi$, which coincides here with the arclength. The quantity δ stands for dissipation, which is small for usual scales and large for marginally small scales comparable with the core size r_0. External Langevin force $\zeta(\xi,t)$ is supposed to be Gaussian with correlator

$$\langle \zeta^\alpha(\xi_1,t_1)\zeta^\beta(\xi_2,t_2)\rangle \;=\; F^\alpha(\xi_1-\xi_2)\,\delta(t_1-t_2)\,\delta^{\alpha\beta},\quad \alpha,\beta \;=\; 1,2,3\,, \tag{2}$$

where $F^\alpha(\xi_1-\xi_2)$ is changing on the large scale of order of the line length ($\sim 2\pi$). The quantity $\tilde\beta$ is $\tilde\beta \;=\; \frac{\tilde\kappa}{4\pi}\log\frac{R}{r_0}$, with circulation $\tilde\kappa$ and cutting parameters R (external size, e.g. averaged radius of curvature) and r_0. For our numerical calculations we have chosen $\tilde\kappa \;=\; 10^{-3}cm^2/s$. This value corresponds to the case of superfluid helium.

In Fourier space equation (2) has the form

$$-i\omega s_q^\alpha \;=\; \int \Gamma^{\alpha\beta\gamma}_{\kappa\kappa_1\kappa_2} s_{q_1}^\beta s_{q_2}^\gamma \delta(q-q_1-q_2)dq_1 dq_2 + \delta_q + \zeta_q^\alpha\,. \tag{3}$$

Here s_q^α is the spacial and temporal Fourier component of $\mathbf{s}^\alpha(\xi,t)$, defined as follows:

$$\mathbf{s}_q^\alpha \;=\; \int\int \mathbf{s}^\alpha(\xi,t)\,e^{i(\omega t - \kappa\xi)}\,dt d\xi\,. \tag{4}$$

The vertex $\Gamma^{\alpha\beta\gamma}_{\kappa\kappa_1\kappa_2}$ responsible for nonlinear interaction has the form

$$\Gamma^{\alpha\beta\gamma}_{\kappa\kappa_1\kappa_2} \;=\; \frac{i\tilde\beta}{2\sqrt{2\pi}}\epsilon^{\alpha\beta\gamma}\kappa_1\kappa_2(\kappa_2-\kappa_1), \tag{5}$$

where $\epsilon^{\alpha\beta\gamma}$ it the antisymmetric unit tensor. One can show the vertex $\Gamma^{\alpha\beta\gamma}_{\kappa\kappa_1\kappa_2}$ to satisfy the so called Jacoby identities

$$\left[\kappa^n\,\Gamma^{\alpha\beta\gamma}_{\kappa\kappa_1\kappa_2} + \kappa_2^n\,\Gamma^{\gamma\alpha\beta}_{\kappa_2\kappa\kappa_1} + \kappa_1^n\,\Gamma^{\beta\gamma\alpha}_{\kappa_1\kappa_2\kappa}\right]\delta(\kappa+\kappa_1+\kappa_2) \;=\; 0,\quad n=2,4. \tag{6}$$

This relations express are tightly connected with the laws of conservation of total length L and curvature K

$$L = \int_0^{2\pi} \mathbf{s}'\mathbf{s}' \, d\xi = const, \quad K = \int_0^{2\pi} \mathbf{s}''\mathbf{s}'' \, d\xi = const. \quad (7)$$

Conservation of these quantities is readely derived from either of relations (1),(3). It is understood that conservation law is held in absence of the both dissipation and stirring force.

One of the regular approaches to describe random fields is based on the Wyld diagram technique [4], originally developed to study hydrodynamic turbulence. Following this technique we introduce for the description of random processes the following averages: the spectral density tensor (or correlator, or simply spectrum) $S_q^{\alpha\beta}$ and the Green tensor $G_q^{\alpha\beta}$ (or simply Green function) which are defined by

$$S_q^{\alpha\beta} \, \delta(q + q_1) = \langle \mathbf{s}_q^\alpha \mathbf{s}_{q_1}^\beta \rangle, \quad (8)$$

$$G_q^{\alpha\beta} \, \delta(q + q_1) = \left\langle \frac{\delta s_q^\alpha}{\delta \zeta_{q_1}^\beta} \right\rangle. \quad (9)$$

Analysis of diagrams shows that due to the antisymmetry of tensor $\epsilon^{\alpha\beta\gamma}$ contained in the expression for the vertex $\Gamma_{\kappa\kappa_1\kappa_2}^{\alpha\beta\gamma}$, both $S_q^{\alpha\beta}$ and $G_q^{\alpha\beta}$ are proportional to $\delta_{\alpha\beta}$, i.e. $S_q^{\alpha\beta} \equiv S_q^\alpha$ and $G_q^{\alpha\beta} \equiv G_q^\alpha$. Details of that technique are described in [5].

The renormalized quantities S_q^α and G_q^α (taking into account interactions) satisfy a Dyson set of diagram equations:

$$G_q^\alpha = {}^\circ G_q^\alpha + {}^\circ G_q^\alpha \Sigma_q^\alpha G_q^\alpha, \quad (10)$$

$$S_q^\alpha = G_q^\alpha \left(F_q^\alpha + \Phi_q^\alpha \right) G_q^{\alpha\star}. \quad (11)$$

Here ${}^\circ G_q^\alpha$ is the "bare" Green function which is equal to $(\omega - \delta_\kappa)^{-1}$. The mass operators Φ_q^α and Σ_q^α can be written in form of diagram series: These series frequently used in nonequilibrium processes have a standard form, explicit form of them is given in [4],[5].

3   Conservation Laws and Pair Correlators

Dyson equations have shapes indicating a cumbersome handling, therefore they can be studied for some special cases. One of them is considered in the present paper. It is connected with conservation laws for the total length and the curvature expressed by (7). Let us consider conservation of total curvature (for total length there is the same consideration). In Fourier space the conservation laws for total curvature κ can be expressed in the following form

$$\frac{\partial K_\kappa}{\partial t} + \frac{\partial P_\kappa^K}{\partial \kappa} = I_+^\kappa(\kappa) - I_-^\kappa(\kappa) \quad (12)$$

where $K_\kappa = \frac{1}{\sqrt{2\pi}} \int_0^{2\pi} s\,''s\,''e^{-i\kappa\xi}d\xi$ is the curvature density and P_κ^K is the flux of this quantity in Fourier space (or, equally, in space of scales). The right-hand side of equation (12) describes creation of additional curvature (with rate $I_+^K(\kappa)$) due to external force and annihilation of it due to dissipative mechanism (with rate $-I_-^K(\kappa)$). In the equilibrium case the flux P_κ^K is absent and source and sink terms must compensate each other locally for each κ, i.e. $I_+^K(\kappa) = I_-^K(\kappa)$. In the case under consideration when source and sink terms are widely separated in κ-space that condition is obviously violated. Therefore a flux of curvature P_κ^K in Fourier space appears. In region of wave numbers κ remote from both region of the pumping κ_+ and of the sink κ_-, $\kappa_+ \ll \kappa \ll \kappa_+$, the so called inertial interval, derivative $\partial P_\kappa^K/\partial \kappa = 0$, so P_κ^K is constant equal to P^κ. Resuming we conclude that the problem reduces to study the set of Dyson equation (10)-(11) in inertial interval under condition of constant flux of the curvature. In this case S_q^α and G_q^α are expected to be independent on the concrete type of both the source and the sink but to be dependent on value of P^K. Furthermore, the vertices $\Gamma_{\kappa\kappa_1\kappa_2}^{\alpha\beta\gamma}$ are homogeneous functions of its arguments. This property, as well as the condition $\kappa_+ \ll \kappa_-$ by virtue of which one can put $\kappa_+ = 0$ and $\kappa_- = \infty$, leads to the assumption that the problem is the scale invariant, i.e. it has no characteristic scale for κ. This suggests a power-law form of S_q^α and G_q^α

$$S_q^\alpha = \frac{1}{\kappa^{r+p}}f\left[\frac{\omega}{\kappa^r}\right] , \qquad G_q^\alpha = \frac{1}{\kappa^r}g\left[\frac{\omega}{\kappa^r}\right] . \tag{13}$$

Here both f and g are dimensionless functions of their arguments. We aim now to find the scaling indices r and p.

The first relation between indices r and p can be found from an analysis of diagram series, claiming all terms to have the same powers of argument κ. This leads to the first scaling condition

$$2r + p = 7 . \tag{14}$$

Another relation between r and p can be obtained from the Dyson equations (10), (11) which can be rewritten in the form (see e.g. [6])

$$\int d\omega\, Im\left\{S_q^\alpha \Sigma_q^\alpha - \Phi_q^\alpha G_q^{\alpha\star}\right\} = 0 . \tag{15}$$

This relation plays the role of kinetic equations for systems with a weak interaction . It has been obtained with help of the expression for the Green function $G_q^\alpha = (\omega - \Sigma_q^\alpha)^{-1}$; the external force correlator F_κ^α disappears in the inertial interval. To find a relation of interest between r and p we rewrite relation (15), disclosing expressions for mass operators Φ_q^α, Σ_q^α and restricting ourselves to first order terms in diagram series. That procedure called direct interaction approximation is frequently used in classical turbulence (see e.g. [6]). After some calculation we arrive at the following relation (see also [8]):

$$Im \int d\omega d\omega_1 d\omega_2 d\kappa_1 d\kappa_2 \delta(q + q_1 + q_2) \times \Gamma_{\kappa\kappa_1\kappa_2}^{\alpha\beta\gamma}$$
$$\times \left\{\Gamma_{\kappa\kappa_1\kappa_2}^{\alpha\beta\gamma} G_q^\alpha S_{q_1}^\beta S_{q_2}^\gamma + \Gamma_{\kappa_2\kappa\kappa_1}^{\gamma\alpha\beta} G_{q_2}^\gamma S_q^\alpha S_{q_1}^\beta + \Gamma_{\kappa_1\kappa_2\kappa}^{\beta\gamma\alpha} G_{q_1}^\beta S_{q_2}^\gamma S_q^\alpha\right\} = 0 . \tag{16}$$

To move further we perform conformal transformations in the second and third term within the braces in integrand, known as Zakharov transformations (see e.g. Zakharov [7] and Kuznetsov and L'vov [8]). For example for the second term these transformations have the form

$$\kappa = \kappa''(\kappa/\kappa''), \quad \kappa_1 = \kappa'(\kappa/\kappa''), \quad \kappa_2 = \kappa(\kappa/\kappa''), \tag{17}$$

$$\omega = \omega''(\kappa/\kappa'')^r, \quad \omega_1 = \omega'(\kappa/\kappa'')^r, \quad \omega_2 = \omega(\kappa/\kappa'')^r. \tag{18}$$

The third term is transformed in similar manner. As a result the integrand in (16) becomes

$$\Gamma^{\alpha\beta\gamma}_{\kappa\kappa_1\kappa_2} G^\alpha_q S^\beta_{q_1} S^\gamma_{q_2} \left\{ \Gamma^{\alpha\beta\gamma}_{\kappa\kappa_1\kappa_2} + \left[\frac{\kappa}{\kappa_2}\right]^x \Gamma^{\gamma\alpha\beta}_{\kappa_2\kappa\kappa_1} + \left[\frac{\kappa}{\kappa_1}\right]^x \Gamma^{\beta\gamma\alpha}_{\kappa_1\kappa_2\kappa} \right\} \tag{19}$$

where

$$x = 7 - r - 2p. \tag{20}$$

Due to Jacoby identities (6) the integrand vanishes when $x = -2$ for conservation of total length and $x = -4$ for conservation of total curvature. Substiting these values into (20) and solving equations (14), (20) we obtaine a set of couples of indices r, p corresponding to nonequilibrium. states with fluxes of the length ($r = 5/3$, $p = 11/3$) and of the curvature ($r = 1$, $p = 5$). One time correlators can be found then integrating over frequences ω

$$S^\alpha_\kappa = \int d\omega \frac{1}{\kappa^{r+p}} f\left\{\frac{\omega}{\kappa^s}\right\} \propto \begin{cases} \kappa^{-\frac{11}{3}} & \text{for length} \\ \kappa^{-5} & \text{for curvature} \end{cases} \tag{21}$$

So we have got solutions for the correlators S^α_κ which correspond to different conservation laws in (6). Since there are no sources and sinks acting in the intermediate range these solutions guarantee that the according fluxes are constant. Depending on the way of agitation of the system one can get the real spectrum as some mixture of the obtained solutions in which the fluxes of length and curvature are present simultaneously. A similar situation for wave systems has been

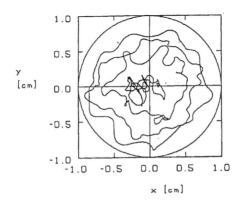

Fig. 1. See text for explanations

discussed earlier (see e.g. [6] and bibliography therein) and it is called multi-flux solution.

Having in mind to compare our result with the both numerical and experimental investigation we have to take into account a presence of δ-correlated (in ξ-space) random force considered in our previous paper [1]. In the local induction approximation the energy $H\{s\}$ of line is proportional of its length and in parametrization when ξ is arclength. can be expressed as

$$H\{s\} = \frac{\rho_s \kappa}{4\pi}^{\sim 2} \ln \frac{R}{r_0} \int_\Gamma s'(\xi)s'(\xi)d\xi \tag{22}$$

It is easy to see that the equilibrium distribution described in [1] leads in that case to correlator $S_\kappa^\alpha \propto 1/\kappa^2$. The final solution is a mix of equilibrium solution and of the ones expressed by relation (21). Because of nonlinearity it, in general, is not a simple superposition except of the cases when one of stirring action prevails and the other can be considered as small deviations. For instance if a large-scale random stirring is small in comparison with δ-correlated (in ξ-space) action we have

$$S_\kappa^\alpha = \frac{A}{\kappa^2} + \frac{B}{\kappa^{11/3}} + \frac{C}{\kappa^5}. \tag{23}$$

The second and third terms in the right-hand side of 23 are small. The constants A, B, C entering are connected with both intensity of random stirring and its structure. The further specification requires some additional analysis.

4   Some Numerical Results

In this section we present some preliminary results on a direct numerical simulations of a vortex ring evolution under action of a random stirring displacements. The large scale character of noise was guaranteed by calculating it from

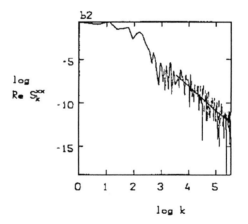

Fig. 2. See text for explanations

a Fourier series taking into account only the first few harmonics. Besides some (uncontrolled) white noise due to numerical procedure has been excited. Figure 1 shows the projection of the line in the x, y - plane (where the ring was placed initially) for several times. As predicted, an consequent arising of higher harmonics takes place leading eventually to an entanglement of the initially smooth vortex loop.

Another numerical results is shown in Fig. 2 where logarithm of quantity S_κ^{xx} averaged over several realizations is depicted as a functions of $\log \kappa$. The average slopes the graphs depend on intensity large-scale stirring force. In several realizations the slope lies between -2.5 and -3.5, which agrees with theoretical prediction 23.

This work was partly supported under grant N 99-02-16942 from Russian Foundation of Basic research.

References

1. S.K. Nemirovskii, L.P. Kondaurova, M. Tsubota: ' Stochastic Dynamics of Vortex Loop. Thermal Equilibrium'. In: *The same issue????*
2. R.J. Donnelly:*Quantized Vortices in Helium II* (Cambridge University Press, Cambridge 1991)
3. K. W. Schwarz: Phys. Rev. **B 18,** 245 (1978), K. W. Schwarz, Phys. Rev. **B 38** 2398 (1988)
4. H. W. Wyld, Ann. Phys. **14,** 134 (1961)
5. S.K. Nemirovskii, J. Pakleza, W. Poppe:Stochasti behaviour of a vortex filament. Notes et Documents LIMSI N91-14, Orsay (1991)
6. V. E. Zakharov, V. S. L'vov, G. Falkovich: *Kolmogorov Spectra of Turbulence I,* (Springer-Verlag, 1992)
7. V. E. Zakharov: Zh. Experim. Theor. Phys. **51,** 688 (1966)
8. E. A. Kuznetsov, V. S. L'vov: Phys. Lett. **64A,** 157 (1977)

Nonequilibrium Vortex Dynamics in Superfluid Phase Transitions and Superfluid Turbulence

Han-Ching Chu and Gary A. Williams

Department of Physics and Astronomy,
University of California,
Los Angeles, CA 90095 USA

Abstract. The nonequilibrium dynamics of superfluid phase transitions and super-fluid turbulence are calculated using vortex renormalization methods. In two dimensions the Kosterlitz–Thouless recursion relations are solved in conjunction with the Fokker–Planck equation for the vortex–pair dynamics, yielding complete solutions for quenched 2D superfluid transitions that are in agreement with scaling predictions and numerical simulations. For the case of 2D superfluid turbulence the steady–state solution is found to be a forward cascade of enstrophy, and it is argued that the energy spectrum varies as k^{-3}. The extension of the theory to three dimensions using vortex–loop renormalization is discussed.

1 Introduction

The topics of superfluid turbulence and superfluid phase transitions are not usually discussed together, although both are known to involve complex arrangements of tangled vorticity (vortex pairs [1] in two dimensions (2D) and vortex loops [2] in three dimensions (3D)). Here we illustrate that there is a fundamental connection between the two problems. The same vortex renormalization techniques that are used to solve the phase transition problem are applied directly to find solutions for the case of superfluid turbulence.

We show this initially in two dimensions, considering a thin superfluid helium film on a flat substrate, which is coupled to a thermal reservoir at temperature T. The vortex pairs in the film are characterized by the distribution function $\Gamma(r, t)$, which is the density of pairs of separation between r and $r+\mathrm{d}r$. It is determined from the 2D Fokker–Planck equation [3],

$$\frac{\partial \Gamma}{\partial t} = \frac{2D}{a_o^2} \frac{\partial}{\partial r'} \cdot \left(\frac{\partial \Gamma}{\partial r'} + \Gamma \frac{\partial}{\partial r'} \left(\frac{U - \boldsymbol{p} \cdot (\boldsymbol{v}_n - \boldsymbol{v}_s)}{k_B T} \right) \right) \quad , \tag{1}$$

where $r' = r/a_o$ with a_o the vortex core size, U and $\boldsymbol{p}$ are the pair interaction energy and impulse, $\boldsymbol{v}_n$ and $\boldsymbol{v}_s$ are any externally applied superfluid and normal fluid flow fields, and D is the diffusion coefficient characterizing the mutual friction drag force on the vortex cores of a pair. It can be seen that the time in this equation is scaled by $\tau_o = a_o^2/2D$, the diffusion time of the smallest pairs of separation a_o. In equilibrium where $\frac{\partial \Gamma}{\partial t} = 0$ the solution of (1) is just the usual Boltzmann distribution,

$$\Gamma = \frac{1}{a_o^4} \exp \left(\frac{-(U + 2E_c) + \boldsymbol{p} \cdot (\boldsymbol{v}_n - \boldsymbol{v}_s)}{k_B T} \right) \tag{2}$$

where E_c is the vortex core energy. The vortex interaction energy U in (1) is determined using the Kosterlitz–Thouless (KT) vortex renormalization methods [1]. The Kosterlitz recursion relations can be written in the form

$$\frac{1}{k_B T} \frac{\partial U}{\partial r} = \frac{2\pi\,K}{r} \tag{3}$$

and

$$\frac{\partial K}{\partial r} = -\,4\,\pi^3\,r^3 K^2\,\Gamma\ , \tag{4}$$

where $K = \hbar^2 \sigma_s / m^2 k_B T$ is the dimensionless areal superfluid density. In thermal equilibrium (2), (3), and (4) lead to the well-known result that the superfluid density has a universal jump to zero at the transition temperature T_{KT}.

2   Quenched Superfluid Transition

To study the quenched 2D superfluid transition [4] we first equilibrate the film with the heat bath at temperature T, generating an equilibrium distribution of vortex pairs. The top curve in Fig. 1 at $t = 0$ is the distribution at T_{KT} found by iterating (2), (3), and (4); it varies asymptotically as $(r/a_o)^{-2\pi K}$. The temperature of the heat bath is then reduced abruptly to a low temperature, $0.1\,T_{KT}$, where in equilibrium the vortex density is over 20 orders of magnitude smaller. However, the pairs cannot suddenly disappear, since the only way they can be extracted by the heat bath is if the plus–minus pairs annihilate at $r = a_o$, where the remaining core energy is converted to phonons. This can take quite a long time to occur, however, since the pairs only slowly lose kinetic energy to the diffusive frictional force of the heat bath as they move in towards annihilation.

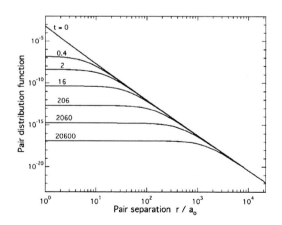

Fig. 1. Time dependence (in units of the diffusion time τ_o) of the pair distribution function Γ (in units a_o^{-4}) for an instantaneous quench from T_{KT} to $0.1\,T_{KT}$.

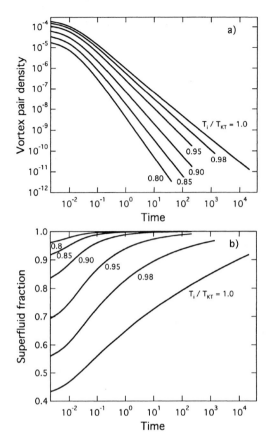

Fig. 2. a) Decay of the pair density (in units a_o^{-2}) and b) recovery of the superfluid fraction in time after a quench to $0.1\,T_{KT}$ from different starting temperatures T_i.

To solve for the time dependence of the distribution function after the quench we combine (1) and (3), which with no applied flows becomes

$$\frac{\partial \Gamma}{\partial t} = \frac{2D}{r}\frac{\partial}{\partial r}\left(r\frac{\partial \Gamma}{\partial r} + 2\pi K\,\Gamma\right) \quad . \tag{5}$$

We solve this in conjunction with (4) using the method of lines with a third–order Runge–Kutta technique on a finite domain, to a maximum pair separation $R/a_o = e^{10} = 2.2\times10^4$. The boundary conditions used are that the flux of pairs $J = -2D(r\frac{\partial \Gamma}{\partial r} + 2\pi K\,\Gamma)$ be continuous across the boundary $r = R$, and that Γ drop abruptly to zero at $r = a_o$, which is equivalent to putting a delta-function sink term at a_o on the right-hand side of (5) to generate the annihilation at that point.

Figure 1 shows the time dependence of the distribution function for an instantaneous quench from T_{KT} to $0.1\,T_{KT}$. The smallest pairs decay away first,

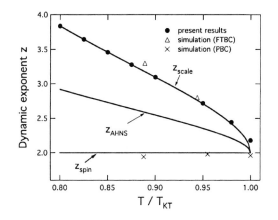

Fig. 3. Dynamic exponent z calculated from the slope of the curves in Fig. 2a, compared with various theories and simulations.

in the classic pattern of a phase–ordering transition where at long times the largest scales become more dominant. By integrating the distribution function over d^2r the pair density is obtained, shown as the top curve in Fig. 2a. At long times the density decreases nearly as t^{-1}, in agreement with scaling theories and simulations [5]. We find, however, that the exponent only approaches -1 logarithmically slowly, being -1.12 at $t = 100$ and -1.09 at $t = 2 \times 10^4$. This is a consequence of the slow recovery of the superfluid density after the quench (bottom curve of Fig. 2b), which has a logarithmic approach to its equilibrium value.

Quenches from starting temperatures below T_{KT} also display power-law variation of the vortex density, since the KT transition is a line of critical points. The plots in Fig. 2 show this for a series of initial temperatures between $0.8\,T_{KT}$ and T_{KT}. At the lower starting temperatures the superfluid density recovers more quickly, and the vortex decay then varies accurately with the form $t^{-z/2}$ where z is a temperature–dependent constant. The increasing slopes in Fig. 2a give the increasing values of z plotted in Fig. 3. We find complete agreement (except near T_{KT} due to the logarithmic slowness) with the scaling prediction of Minnhagen and co-workers [6] for the dynamic exponent

$$z_{scale} = 4 \frac{\sigma_s(T)}{\sigma_s(T_{KT})} \frac{T_{KT}}{T} - 2 \quad , \tag{6}$$

and disagreement with other predictions [3,5]. Minnhagen has recently shown [6] that simulations [5,7] of the vortex decay can be misleading, since different results for z are obtained depending on the boundary conditions used: periodic boundary conditions (PBC) yield only the spin-wave result $z = 2$, while fluctuating–twist boundary conditions (FTBC) give z_{scale}. Our results in agreement with z_{scale} confirm that this is in fact the correct exponent characterizing the vortex dynamics.

3   Superfluid Turbulence

We can apply the same KT renormalization techniques to 2D superfluid turbulence. For this the film is held at a low temperature $0.1\,T_{KT}$, and vortex pairs of large separation R are injected externally into the film at random positions and random orientations, at a rate $\dot{Q}$ pairs per unit area per second. These pairs will begin to cascade down to smaller separations under the action of the frictional force, finally reaching the annihilation scale a_o where they are absorbed by the thermal bath. Eventually a steady–state will be achieved where the distribution function becomes constant in time, with the annihilation current at a_o being equal to the injected current at R. This is a forward enstrophy cascade, which is a conserved quantity in 2D [8].

The steady-state distribution can be found by solving the Fokker–Planck equation with a source term added,

$$\frac{\partial\,\Gamma}{\partial\,t} = 0 = \frac{2D}{r}\,\frac{\partial}{\partial r}\left(r\frac{\partial\Gamma}{\partial r} + 2\pi K\,\Gamma\right) + \dot{Q}\delta(\boldsymbol{R} - \boldsymbol{r}) \quad, \tag{7}$$

in conjunction with (4), and again assuming as above that Γ drops to zero at a_o. Integrating this over $\mathrm{d}^2 r$ gives

$$r\frac{\partial\Gamma}{\partial r} + 2\pi K\,\Gamma = \frac{\alpha}{4\pi^2 a_o^4}H(R - r) \quad, \tag{8}$$

where we have defined a dimensionless injection rate $\alpha = a_o^2\dot{Q}\tau_o$, and H is the Heaviside unit step function. For low injection rates the solutions of (8) and (4) are a constant distribution function $\Gamma = \Gamma_o H(R - r)$ and a superfluid fraction unchanged from unity, where $\Gamma_o = \alpha/4\pi K_o a_o^4$ and K_o is the value of K at the scale a_o (where the areal superfluid density is just the areal density of the fluid at low T). At higher injection rates the density of the injected vortices can start to approach densities found at T_{KT}, and this has the effect of driving the superfluid density to zero at some length scale r_o. Setting $K = 0$ in (8) then gives the variation of the distribution at larger scales, $\Gamma(r) - \Gamma(r_o) = (\alpha/4\pi a_o^4)\ln(r/r_o)$. Full numerical solutions of (8) and (4) are shown in Fig. 4 for a wide range of injection rates α. At the lowest rates the injected distribution adds to the thermal vortices at the bath temperature of $0.1\,T_{KT}$, while at intermediate rates the distribution is the constant value Γ_o. As the injected distribution starts to approach values comparable to the thermal distribution at T_{KT}, the superfluid density is reduced by the pairs, finally dropping to zero, where Γ then takes on the logarithmic increase noted above.

The work done in forming a given distribution of the pairs is the negative of the free energy, since the system is coupled to the bath. The work per unit mass can be calculated from the Kosterlitz free energy [1], which in our variables is

$$W = -\frac{F}{\sigma_s^o} = \frac{2\pi k_B T}{\sigma_s^o}\int_{a_o}^{R}\Gamma(r)\,r\,\mathrm{d}r = \pi\frac{\hbar^2}{m^2}\frac{\Gamma_o}{K_o}(R^2 - a_o^2) \quad, \tag{9}$$

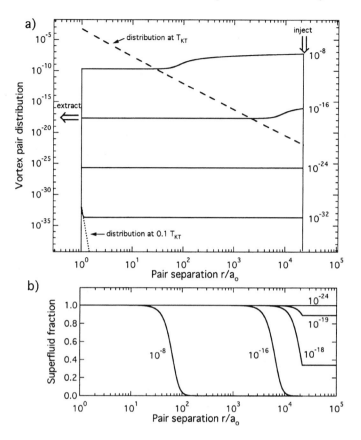

Fig. 4. a) Pair distribution resulting from injection of pairs at $R/a_o = 2.2 \times 10^4$, for a wide range of injection rates α as indicated along the right side, b) superfluid fraction versus length scale for several of the higher injection rates.

where the last equality holds at the intermediate injection rates where Γ is a constant. Defining a k-space energy spectrum $W(k)$ by $W = \int W(k)\, dk$, it can be seen that over the relevant range $1/R < k < 1/a_o$ the result of (9) is achieved for $W(k) \propto k^{-3}$. This is the same dependence found for the forward cascade in classical 2D Navier-Stokes turbulence [8]. The superfluid cascade differs from the classical case, however, since the spectrum varies linearly with the injected enstrophy flux $\eta = 2(h/m)^2 \dot{Q} a_o^{-2}$ and not as $\eta^{2/3}$. The spectrum takes the form $W(k) = C \eta \tau_o k^{-3}$ with C a dimensionless constant, where it can be seen that the k^{-3} variation is the only dimensionally consistent possibility, apart from possible $\ln(kR)$ corrections. The occurrence of τ_o (which is "universal" for the cascade since these identical vortices all have the same drag force on them) is easily understood physically: the longer it takes for a pair to cascade down the phase space the more the vortices pile up and increase the total energy. We note that for the superfluid system the enstrophy cascade is the only cascade that forms, since energy is not conserved due to the coupling to the thermal bath. The reverse energy cascade that is found in the classical turbulence [8] is not possible because the quantized circulation prevents the merger of like–sign vortices which is necessary for such a cascade. It will be of great interest if computer simulations [8] can check these predictions for the cascades in 2D superfluid turbulence.

In further work it should be possible to study the time decay of the turbulence by switching off the injection term in (7), and then solving (5) in the same manner as for the quenched transitions. In fact (5) is then readily identified as the 2D equivalent of the phenomenological Vinen equation [9] governing the dynamics of superfluid turbulence.

4   Three Dimensions

The analytic solutions described above for both 2D quenched superfluid transitions and 2D superfluid turbulence are the result of the simplicity of the KT renormalization methods. It should be quite feasible to extend these techniques to three dimensions, where the relevant excitations are vortex loops. A simple theory [2] using the KT ideas in 3D has been able to accurately model many features of the superfluid λ–transition. The renormalization allows the complex tangle of interacting noncircular loops to be reduced to a simpler problem, by averaging out the effects of shorter length scales, until only a dilute gas of loops of size equal to the coherence length are left. For the turbulence problem rings of large size will be injected, followed by a cascade to smaller loops due to the mutual friction force, until the smallest loops are removed at the scale a_o. It should be noted that in this approach the very complicated motion and reconnections of the individual loops do not need to be considered (just as they were not for the 2D pairs above); all that needs to be taken into account is a statistical "snapshot" of the distribution function at each timestep. It seems quite possible that the 3D Fokker–Plank equation [10] will again prove to be the source of the Vinen equation for the turbulent dynamics.

Useful discussions with G. Eyink are acknowledged. This work is supported by the US National Science Foundation, grant DMR 97-31523.

References

1. J.M. Kosterlitz, D. Thouless: J. Phys. C **6**, 1181 (1973) J.M. Kosterlitz: J. Phys. C **7**, 1046 (1974)
2. G. A. Williams: Phys. Rev. Lett. **83**, 2347 (1999)
3. V. Ambegaokar, B. Halperin, D. Nelson, E. Siggia: Phys. Rev. B **21**, 1806 (1980)
4. H. Chu, G. A. Williams: Physica B **284-288**, 115 (2000)
5. A. Bray, A. Briant, D. Jervis: Phys. Rev. Lett. **84**, 1503 (2000)
6. B. Kim, P. Minnhagen, P. Ollson: Phys. Rev. B **59**, 11506 (1999)
7. H. Luo, B. Zheng: Mod. Phys. Lett. **11**, 615 (1997)
8. R. Kraichnan: Phys. Fluids **10**, 1417 (1967) E. Lindborg, K. Alvelius: Phys. Fluids **12**, 945 (2000)
9. W. F. Vinen: Proc. Roy. Soc. London A **242**, 493 (1957)
10. G. A. Williams: J. Low Temp. Phys. **93**, 1079 (1993)

Part V

The NLSE and Superfluidity

The Nonlinear Schrödinger Equation as a Model of Superfluidity

Paul H. Roberts and Natalia G. Berloff

Department of Mathematics
University of California, Los Angeles, CA, 90095-1555
roberts@math.ucla.edu, nberloff@math.ucla.edu

Abstract. The results of theoretical and numerical studies of the Gross-Pitaevskii (GP) model are reviewed. This model is used to elucidate different aspects of superfluid behaviour: the motion, interactions, annihilations, nucleation and reconnections of vortex lines, vortex rings, and vortex loops; the motion of impurities; flow through apertures; superfluid turbulence and the capture of impurities by vortex lines. The review also considers some generalizations of the model.

1 Introduction

One of the most useful ways of describing superfluid helium at zero temperature begins with Schrödinger's equation for the one-particle wave function Ψ. Since liquid helium is a strongly correlated system dominated by collective effects, the form of the Hamiltonian in Schrödinger's equation cannot be derived starting from first principles. At zero temperature ^{4}He has large interatomic spacing and low density, which suggests a description in terms of a weakly interacting Bose gas for which such a derivation can be made rigorous. Continuum mechanical equations for ^{4}He are usually built on the assumption that a Bose condensate gives the exact description at zero temperature.

The imperfect Bose condensate in the Hartree approximation is governed by equations that were derived by Ginsburg and Pitaevskii [22] and by Gross [26], and generally called "GP equations". In terms of the single-particle wavefunction $\Psi(\mathbf{x}, t)$ for N bosons of mass M, the time-dependent self-consistent field equation is

$$i\hbar\Psi_t = -\frac{\hbar^2}{2M}\nabla^2\Psi + \Psi \int |\Psi(\mathbf{x}', t)|^2 V(|\mathbf{x} - \mathbf{x}'|)\, d\mathbf{x}', \tag{1}$$

where $V(|\mathbf{x} - \mathbf{x}'|)$ is the potential of the two-body interactions between bosons. The normalization condition is

$$\int |\Psi|^2\, d\mathbf{x} = N. \tag{2}$$

For a weakly interacting Bose system, (1) is simplified by replacing $V(|\mathbf{x} - \mathbf{x}'|)$ with a δ - function repulsive potential of strength V_0. This does not alter the nature of the results since the characteristic length of the weakly interacting Bose gas is larger than the range of the force [26]. Equation (1) for such a potential is

$$i\hbar\Psi_t = -\frac{\hbar^2}{2M}\nabla^2\Psi + V_0|\Psi|^2\Psi. \tag{3}$$

Equations (1)-(3) define Hamiltonian systems, the following integrals being conserved: mass

$$\mathcal{M} = M \int |\Psi|^2 \, d\mathbf{x}, \tag{4}$$

momentum density

$$\mathbf{p} = \frac{\hbar}{2i} \int [\Psi^* \nabla \Psi - \Psi \nabla \Psi^*] \, d\mathbf{x}, \tag{5}$$

and energy; in the case of (3) this is expressed by

$$\mathcal{E} = \frac{\hbar^2}{2M} \int |\nabla \Psi|^2 \, d\mathbf{x} + \frac{V_0}{2} \int |\Psi|^4 \, d\mathbf{x}. \tag{6}$$

Our starting point is a condensate everywhere at rest in what we shall call the "laboratory frame", so that $\Psi = \exp(iE_v/\hbar)$ in that frame, where E_v is the chemical potential of a boson (i.e., the increase in ground state energy when one boson is added to the system). We then consider deviations from that state by studying the evolution of ψ where

$$\psi = \Psi \exp(iE_v t/\hbar), \tag{7}$$

so that (3) becomes

$$i\hbar\psi_t = -\frac{\hbar^2}{2M} \nabla^2 \psi + V_0 |\psi|^2 \psi - E_v \psi. \tag{8}$$

It is usually convenient to model phenomena in an infinite domain in which, prior to the onset of a disturbance, $\psi = \psi_\infty$ everywhere, where, by (8)

$$\psi_\infty = (E_v/V_0)^{1/2}. \tag{9}$$

We then modify (4)-(6) to forms measuring departures from this uniform state (see, e.g. [44])

$$\int (|\psi|^2 - \psi_\infty^2) \, d\mathbf{x} = 0, \tag{10}$$

$$\mathbf{p} = \frac{\hbar}{2i} \int [(\psi^* - \psi_\infty) \nabla \psi - (\psi - \psi_\infty) \nabla \psi^*] \, d\mathbf{x}, \tag{11}$$

$$\mathcal{E} = \frac{\hbar^2}{2M} \int |\nabla \psi|^2 \, d\mathbf{x} + \frac{V_0}{2} \int (|\psi|^2 - \psi_\infty^2)^2 \, d\mathbf{x}. \tag{12}$$

The mass density and flux are

$$\rho = M\psi\psi^*, \qquad \mathbf{j} = \frac{\hbar}{2i}((\psi^* - \psi_\infty) \nabla \psi - (\psi - \psi_\infty) \nabla \psi^*). \tag{13}$$

2   The Fluid Equations

One of the principal aims of this review is to interpret the consequences of the condensate model, not from the perspective of solid state physics, but through the eyes of a fluid dynamicist. This is achieved by recasting the nonlinear Schrödinger equation by the Madelung transformation. We write

$$\psi = Re^{iS}, \tag{14}$$

so that

$$\rho = MR^2, \qquad \mathbf{j} = \rho\mathbf{u} = \rho\nabla\phi, \qquad \phi = (\hbar/M)S. \tag{15}$$

The real and imaginary parts of (8) then yield a continuity equation

$$\frac{\partial\rho}{\partial t} + \nabla\cdot(\rho\mathbf{u}) = 0, \tag{16}$$

and an integrated form of the momentum equation

$$\frac{\partial\phi}{\partial t} + \frac{1}{2}u^2 + c^2\left(\frac{\rho}{\rho_\infty} - 1\right) - c^2 a^2 \frac{\nabla^2\rho^{1/2}}{\rho^{1/2}} = 0, \tag{17}$$

where ρ_∞ is the density at infinity and c is the speed of sound:

$$\rho_\infty = M\psi_\infty^2, \qquad c^2 = E_v/M = (V_0/M^2)\rho_\infty. \tag{18}$$

The last term in (17) is sometimes called the "quantum pressure" although it is dimensionally a chemical potential. It is significant where "healing" of the wavefunction ψ is important, as for instance in a vortex core, or within a "healing layer" adjacent to a boundary, of approximate thickness

$$a = \frac{\hbar}{(2ME_v)^{1/2}}, \tag{19}$$

which is called the "healing length". Elsewhere, in the "bulk" of the fluid, the last term in (17) in insignificant, and (17) assumes the form appropriate for a classical inviscid fluid, in which the pressure, P, is proportional to ρ^2.

The unintegrated form of the momentum equation is obtained by taking the gradient of (17). This gives

$$\rho\left(\frac{\partial u_i}{\partial t} + u_j\frac{\partial u_i}{\partial x_j}\right) = -\frac{\partial P}{\partial x_i} + \frac{\partial \Sigma_{ij}}{\partial x_j}, \tag{20}$$

where P is pressure and the Σ_{ij} are the quantum stresses

$$P = \frac{V_0}{2M^2}\rho^2, \qquad \Sigma_{ij} = \left(\frac{\hbar}{2M}\right)^2 \rho\frac{\partial^2\ln\rho}{\partial x_i\partial x_j}; \tag{21}$$

see [24].

Curves on which the wavefunction ψ vanishes correspond to vortex lines in the superfluid with the correct unit of quantization, $\kappa = h/M$. For simple zeros, of the kind with which we shall be concerned, the circulation

$$C = \int_\Gamma \mathbf{u} \cdot d\mathbf{s} \tag{22}$$

round a closed contour Γ is equal to the number of vortex lines that Γ contains. Since $\mathbf{u} = \rho \nabla \phi$, the circulation C is preserved until a vortex line crosses Γ, whereupon C increases or decreases by κ. At the instant at which the vortex line meets Γ, the density ρ on Γ is zero at the point of intersection; Γ is no longer a closed curve, and Kelvin's theorem on the constancy of C is inapplicable.

This marks an extremely basic difference between the condensate equations and the Euler equations of classical fluid dynamics. Both describe conservative (Hamiltonian) systems, but the Euler fluid does not allow circulation to change. In describing topological changes in the vorticity of a classical fluid, it is necessary to invoke a mechanism that breaks the constraint of Kelvin's theorem. For example, one method of modeling the severing or coalescence of vortex filaments is to restore viscosity to the fluid, so allowing the vorticity to diffuse from one filament to the other. Another method is described in Sect. 5 below. Such devices are unnecessary when the condensate model is employed.

The GP model not only enjoys the advantage of comparative simplicity but also describes qualitatively correct superfluid behaviors at low temperatures T. In this paper we review how the GP model is successfully used to elucidate the motion, interactions, annihilations, nucleation and reconnections of vortex lines, vortex rings, and vortex loops; the motion of impurities; the flow of superfluid through apertures; superfluid turbulence; and the capture of impurities by vortex lines.

3   Shortcomings of the GP Model

Despite its success, several aspects of the local model (8) are qualitatively or quantitatively unrealistic for superfluid helium. The dispersion relation between the frequency, ω, and wave number, k, of sound waves according to (8) is

$$\omega^2 = c^2 k^2 + \left(\frac{\hbar}{2M}\right)^2 k^4. \tag{23}$$

The velocity, c, of long wavelength sound waves is therefore proportional to $\rho^{\frac{1}{2}}$ see (21). (We have here denoted the bulk density, ρ_∞, by ρ.) That this is unrealistic is seen from the experiments on Grüneisen constant $U_G = (\rho \partial c / c \partial \rho)_T$, which shows that, in the bulk (i.e., on length scales long compared with the healing length, κ/c), the fluid behaves as a barotropic fluid ($p \propto \rho^\gamma$) with $\gamma = 2.8$ (see [10] and references therein).

By writing $p = \hbar k$ and $\epsilon(p) = \hbar \omega$, we convert (23) into the dispersion curve of GP theory:

$$\epsilon(p) = p \left[c^2 + (p/2M)^2\right]^{1/2}. \tag{24}$$

At best, this describes the phonon branch of the excitation spectrum. The roton branch is lost. At first sight this does not seem important, because it is only the phonon branch that is thermally populated at low temperatures. Nevertheless, the lack of a roton branch is seen to be serious when, as in Sects. 6 and 7 below, we use the GP model to simulate superflow past boundaries. We shall describe in Sect. 8 a generalization of (8) that restores the roton branch.

To expose a related deficiency of the GP model, we invoke the "ghost" of a normal fluid by supposing that T is infinitesimal but nonzero. In local thermodynamic equilibrium, the density of quasi-particles in the fluid is

$$N_q = \frac{1}{\hbar^3} \int \left[\exp\left(\frac{\epsilon - \mathbf{w} \cdot \mathbf{p}}{kT} \right) - 1 \right]^{-1} d\mathbf{p}, \tag{25}$$

where $\mathbf{w} = \mathbf{u}_n - \mathbf{u}_s$, is the velocity of the normal fluid[1] relative to the superfluid; e.g. see [34]. It is clear from (24) that N_q is negligibly small when $w < c$, because then $\epsilon - \mathbf{w} \cdot \mathbf{p} > 0$ for all $\mathbf{p}$. When $w > c$, however, the integrand in (25) is negative and even infinite for $\mathbf{p}$ for which $\epsilon - \mathbf{w} \cdot \mathbf{p} \leq 0$. This unphysicality shows that the assumption of local thermodynamic equilibrium is untenable wherever $w > c$. Then quasi-particle production destroys superfluidity.

In real superfluid helium, the limiting value of w for the destruction of superflow is known as the 'Landau critical velocity'; we shall denote this by u_L. The argument just given correctly indicates that u_L is determined by a condition of tangency: $u_L = \epsilon(p)/p = d\epsilon(p)/dp$ and, because of the roton branch, this is much less than c for superfluid helium. Compressibility therefore plays a greater role in the condensate than in real helium.

GP theory has no preferred reference frame. This is hardly surprising in a model which applies at $T = 0$, where not even the "ghost" of normal fluid exists. Nevertheless, it is a non-uniformity for $T \to 0$ that is an undesirable attribute of GP theory. Carlson [13] proposed a way of removing this by augmenting (8) with a term motivated by Khalatnikov's (1952) theory of mutual friction between the superfluid and normal fluid at finite T, which involved three independent kinetic coefficients, of which only the third, $\zeta_3(> 0)$, concerns us here. Khalatnikov modified the integrated superfluid momentum equation to be

$$\frac{\partial \phi}{\partial t} + \frac{1}{2}u_s^2 + \mu = -\zeta_3 \nabla \cdot (\rho_s \mathbf{w}), \tag{26}$$

where μ is a chemical potential; Using the mass conservation equation

$$\frac{\partial \rho}{\partial t} + \nabla \cdot (\rho \mathbf{u}) = 0, \tag{27}$$

---

[1] We are attempting here to make a point of principle, and for this purpose we may ignore the fact that the mean free path of phonons at low temperatures may be so large that they are better represented by a Boltzmann equation than by a normal fluid continuum. We temporarily write the density of the superfluid and normal fluid by ρ_s and ρ_n (where $\rho_s + \rho_n = \rho$) and their velocities as $\mathbf{u}_s$ and $\mathbf{u}_n$ (where $\rho \mathbf{u} = \rho_s \mathbf{u}_s + \rho_n \mathbf{u}_n$ is the total momentum density).

we see that (26) can also be written as

$$\frac{\partial \phi}{\partial t} + \frac{1}{2}u_s^2 + \mu = -\zeta_3 \nabla \cdot \left(\frac{\partial \rho}{\partial t} + \nabla \cdot (\rho \mathbf{u}_n) \right). \tag{28}$$

Since $\rho \approx \rho_s$ for small T, this suggests that a useful generalization of (8) is

$$i\hbar \psi_t = -\frac{\hbar^2}{2M}\nabla^2 \psi + V_0|\psi|^2\psi - E_v\psi + \zeta M^2\psi \left[\partial_t|\psi|^2 + \nabla \cdot (|\psi|^2 \mathbf{u}_n) \right], \tag{29}$$

where $\zeta > 0$. This implies, in place of (17),

$$\frac{\partial \phi}{\partial t} + \frac{1}{2}u^2 + c^2\left(\frac{\rho}{\rho_\infty} - 1 \right) - c^2 a^2 \frac{\nabla^2 \rho^{1/2}}{\rho^{1/2}} = -\zeta\left(\frac{\partial \rho}{\partial t} + \nabla \cdot (\rho \mathbf{v}_n) \right), \tag{30}$$

and (16) is unchanged. Unlike (1) or (8) equations (29) and (30) single out a preferred reference frame, the one in which $\mathbf{v}_n = 0$. It is reasonable to suppose that $\zeta \to 0$ as $T \to 0$ since otherwise the normal fluid, which is subjected to the frictional force that is the counterpart of the ζ term in (30), would be infinitely accelerated in the limit.

Let us again suppose infinitesimal but nonzero T, and determine how the right-hand side of (30) modifies the dispersion relation (23). We add, to a uniform counterflow $\mathbf{w}$, perturbations proportional to $\exp(i\mathbf{k} \cdot \mathbf{x} - i\omega t)$, and quickly find from (16) and (30) that (23) is replaced by

$$\omega^2 = c^2 k^2(1 + \tfrac{1}{2}a^2 k^2) + i\zeta\rho_\infty k^2(\omega - \mathbf{w} \cdot \mathbf{k}). \tag{31}$$

For small ζ, the roots of (31) are approximately real; one of them is $\omega \approx \omega_r = ck(1 + \tfrac{1}{2}a^2k^2)^{1/2} > 0$, and its imaginary part is defined by the imaginary part of (31):

$$\omega_i \approx -(\zeta\rho_\infty k^3/2\omega_r) \left[w - c(1 + \tfrac{1}{2}a^2k^2)^{1/2} \right]. \tag{32}$$

If $w < c$, this is positive for all k, but if $w > c$ it is negative for all sufficiently small k, so establishing the breakdown of superfluidity once the Landau critical velocity (here c) is exceeded.

Although, as shown by Berloff (1999), the ζ term can be a useful practical tool in computing vortex structure, our interest in ζ in this review is that it represents a potentially significant source of non-thermal quasi-particles. Consider superflow past a body such as a positive ion. It is convenient, as in §6 below, to treat the body as an infinite potential barrier to the condensate, and if this is done the GP equation allows the superfluid to have *any* velocity along the boundary. In reality, however, ψ penetrates a very short distance into the body and collisions within the solid can generate quasi-particles that, on average, move with it. In the parlance of fluid mechanics, the normal fluid velocity $\mathbf{u}_n$ obeys the no-slip condition on the boundary $\mathbf{u}_n = 0$. The ζ−term leads to quasi-particle emission whenever the superflow along the boundary exceeds the Landau critical velocity; for the local GP model, the quasi-particles are seen as sound waves (phonons).

It may be worth pointing out that the condition $w > u_L$ for the destruction of superfluidity is invalid where healing occurs and the $\nabla^2 \rho^{1/2}$ term in (17) or (30) is significant. The superfluid vortex provides a prime example, in which $u_s \to \infty$ in the vortex core, although no instabilities are seen in numerical simulations (provided the vortex as a whole does not move with super-Landau speeds; see Berloff and Roberts (1999). For the remainder of this review, $\mathbf{u}_s$ will again be denoted by $\mathbf{u}$.

Finally, we should not ignore what is perhaps the most fundamental objection to GP theory. Although by including $\hbar$ it provides the verisimilitude of being a quantum description of superfluidity, it is in reality semi-classical. A proper operator description is, however, so complex that it cannot be usefully employed to study situations as complicated as those described in this review. We should not, however, overlook the progress that has been made towards a more realistic description of a vortex core (see [46]).

4 Vortices

A vortex line is defined by a zero of the wave function $\psi = 0$. The straight-line vortex creates a velocity around this line of $u = \hbar/Ms$, where s is the distance from the vortex line. The equation for the amplitude of the steady straight-line vortex is found by substituting $\psi = R(s)\exp(i\chi)$ into (8) written in cylindrical coordinates (s, χ, z):

$$\frac{d^2R}{ds^2} + \frac{1}{s}\frac{dR}{ds} - \frac{1}{s^2}R + R - R^3 = 0, \tag{33}$$

where distance is scaled by a and R by R_∞. For small distances from the axis, R is proportional to s; for $s \to \infty$, $R \sim 1 - 1/2s^2$. The energy per unit length of the vortex line is

$$\mathcal{E}_\ell = \frac{\kappa^2 \rho_\infty}{4\pi}\left(\int_0^\infty \left[\frac{dR}{ds}\right]^2 s\,ds + \int_0^\infty \frac{R^2}{s}\,ds + \frac{1}{2}\int_0^\infty (1 - R^2)^2 s\,ds\right). \tag{34}$$

The first term can be regarded as a "quantum energy", the second term is the classical vortex kinetic energy that diverges logarithmically unless a cut-off distance L is introduced; the third term in (34) represents the potential energy. The energy per unit length of the vortex is usually expressed in the form

$$\mathcal{E}_\ell = \frac{\rho\kappa^2}{4\pi}\left(\ln\frac{L}{a} + L_0\right), \tag{35}$$

where the constant L_0 is called the "vortex core parameter" and was determined numerically by Pitaevskii [40] as $L_0 = 0.3809$.

A dense array of straight-line vortices are encountered when a bucket of superfluid is rotated rapidly. This greatly affects its dynamics. This gives rise to HVBK theory; see, for example Hills and Roberts [29] and Holm (this meeting).

Straight-line vortices can transmit energy along their length by Kelvin waves. These have been comprehensively analyzed for incompressible Euler fluids; see for example Chapter 11 of Saffman [47]. They have been less well studied for compressible fluids such as the GP condensate, for which the Kelvin wave may be accompanied by acoustic emission that ultimately damps out the waves, unless they are sustained by a source. This emission is of interest in studies of super-fluid turbulence (§5). According to the analysis of Pitaevskii [40], the Kelvin waves that bend the rectilinear vortex do not emit sound if their wavelength is sufficiently long; more general types of wave have been studied by Rowlands [45].

Large vortex rings were investigated by Roberts and Grant [44] and they obtained the following expressions for the energy per unit length, momentum, and velocity, v:

$$\mathcal{E} = \tfrac{1}{2}\rho\kappa^2 R\left(\ln\frac{8R}{a} + L_0 - 2\right), \quad p = \rho\kappa\pi R^2, \tag{36}$$

$$v = (\kappa/4\pi R)\left(\frac{8R}{a} + L_0 - 1\right). \tag{37}$$

As expected, $\mathcal{E} \approx 2\pi R\mathcal{E}_\ell$, but also rings obey Hamilton's equation

$$v = \partial\mathcal{E}/\partial p. \tag{38}$$

Jones and Roberts [33] determined the entire sequence of vortex rings numerically for the GP model (8). They calculated the energy $\mathcal{E}$ and momentum p and showed how the location of the sequence in the $\mathcal{E}p$-plane relates to the superfluid helium dispersion curve. They found two branches meeting at a cusp where p and $\mathcal{E}$ assume their minimum values, p_m and $\mathcal{E}_m$. As $p \to \infty$ on each branch, $\mathcal{E} \to \infty$. On the lower branch the solutions are asymptotic to the large vortex rings (36)-(37). Since the GP model has a healing length (based on the sound velocity) different from the vortex core parameter there are two possible ways to introduce dimensional units and to plot the solitary wave sequence next to the Landau dispersion curve on the $p\mathcal{E}-$ plane. If the dimensional units based on the vortex core parameter are chosen, the cusp lies just above the Landau dispersion curve; if instead the healing length (sound speed) is selected the cusp meets the dispersion curve of the GP model, which (we recall) does not have a roton branch.

As $\mathcal{E}$ and p decrease from infinity along the lower branch, the solutions begin to lose their similarity to large vortex rings, and (36) - (37) determine $\mathcal{E}$, p, and v less and less accurately, although (38) still holds. Eventually, for a momentum p_0 slightly greater than p_m, the rings lose their vorticity (ψ loses its zero), and thereafter the solitary solutions may better be described as 'rarefaction waves'. The upper branch consists entirely of these and, as $p \to \infty$ on this branch, the solutions asymptotically approach the rational soliton solution of the Kadomtsev-Petviashvili (KP) equation and are unstable.

An interesting, and still incompletely answered question is whether the GP sequence of solitary waves is stable or not. The question is not frivolous since the classical circular rings in an incompressible fluid are known to be unstable (Widnall and Sullivan [54]). Jones et al. [32] concluded that the entire upper branch of rarefaction waves [33] is unstable, but that the lower branch solutions may be stable because their axisymmetric expansion or contraction is forbidden due to energy and momentum conservation. Grant [23] examined asymmetric perturbations of a large ring in the form of infinitesimal Kelvin waves, but he did not locate any that grew and, moreover, his analysis is inconclusive since it omitted the acoustic emission that accompanies the Kelvin waves. This was however included by Pismen and Nepomnyashchy [39], who found that it is stabilizing.

5   Superfluid Turbulence; Vortex Line Reconnection

Superfluid turbulence has been the focus of many experimental studies (e.g., [52], [18]), especially in "the high temperature regime," by which we mean $0.6°K \leq T < T_\lambda \approx 2.172°K$, where T_λ is the λ-point, which marks the transition between the normal and superfluid phases of helium. In this regime, the density, ρ_s, of superfluid is smaller than the normal fluid density, ρ_n, and in consequence turbulence in the superfluid is, to a large degree, determined by the turbulence taking place in the normal fluid. This is reflected by the success of theories of superfluid turbulence, such as that of Barenghi et al. [1], [2], in which the superfluid vorticity is largely tied to that of the normal fluid. In contrast, in the low temperature range ($T \leq 0.6°K$), where ρ_n is smaller than ρ_s, we may expect turbulence in the superfluid largely to determine turbulence in the normal fluid, rather than the reverse.

For the study of superfluid turbulence in the low temperature range it is necessary to follow the evolution of only two fields, the real and imaginary parts of ψ, rather than ρ and $\mathbf{v}$, or ρ and a non-single-valued velocity potential ϕ. There is however no high wavenumber sink of energy to terminate the Kolmogorov cascade. This obstacle may restrict the use of the simplest forms (1) or (8) of the GP model for turbulence studies, but the Carlson generalization, that uses (30 instead of (17), may be the remedy, since this recognizes that, in real helium even in the low temperature range, normal fluid is present that is coupled to the superfluid and which, through its viscosity, provides a high wavenumber sink. Another difficulty is that strong turbulence contains a complicated mixture of condensate, phonons, rotons, quantized vortices, vortex waves and shocks. Such a variety of processes may make it difficult to determine universal scaling laws for the turbulent correlation functions.

Recently several papers have appeared that discuss decaying Kolmogorov turbulence using GP theory. Nore et al. [38] studied superfluid turbulence using numerical simulations of the GP model. They decomposed the total energy (which is conserved) into incompressible kinetic, internal, and "quantum" components (that corresponds to acoustic excitations), and they computed the corresponding energy spectra. They found that the rate of transfer of kinetic energy into other energy components is comparable with the rate of energy dissipation through viscosity in classical turbulence. At the moment of maximum energy dissipation, the energy spectrum resembles the Kolmogorov inertial range.

The term "superfluid turbulence" is often used synonymously for the "evolution of a superfluid vortex tangle". The dynamics of the turbulent state depend crucially on the interactions of the vortex filaments. In a set of pioneering papers, Schwarz [48] developed a numerical technique to simulate the dynamics of the vortex tangle. This was based on the classical theory of vortex filaments in an incompressible Euler fluid. The reconnection of vortex lines was therefore forbidden by Kelvin's theorem, and he was therefore compelled, when studying changes in vortex line topology, to introduce *ad hoc* reconnection rules, e.g., that vortex filaments reconnect if, and only if, they approach within a distance of Δ of one another, where Δ is an *ad hoc* constant of the order of the core radius. As we saw in §2 above, Kelvin's theorem does not apply to GP models when a zero of ψ lies on the circuit Γ in (22). The coalescence of vortex lines merely corresponds to the merging of two zeros of ψ, and the creation of vortices corresponds to the appearance of a new zero line of ψ. These processes were first seen to happen in the GP calculations of Jones and Roberts [33] and later, in a more graphic form, by Koplik and Levine ([35] and [36]), who simulated the interaction between straight-line vortices and their reconnection. They also witnessed the annihilation of vortex rings of similar radii. Vortices can also be nucleated by moving bodies such as ions, but again this process cannot occur in an Euler fluid, but is allowed by the GP models, as is seen in the next Section. Another grave disadvantage of using the incompressible Euler fluid to model the superfluid is that it eliminates sound in transferring, and perhaps cascading, energy down the turbulence spectrum. Vinen (this meeting) has stressed the potential importance of this process. Although the simplest form (8) of GP theory gives c incorrectly (see §1), the generalized theory to be described in §8 below, may provide a viable route to progress. Svistunov [51] has stressed the role of Kelvin waves in the turbulent cascade.

In Fig. 1, the results of unpublished calculations by Berloff of the evolution of two-dimensional solutions of (8) are shown, in terms of the density ρ which, being depleted near a vortex core, is a clear marker for the line vortices. The walls of the computational box were taken to be reflective. It may be seen that the number of vortex lines is not conserved during the simulation; vortex nucleation and annihilation occurs. At the start there were 4 vortices of positive circulation and 4 of negative circulation. Over the period of integration the energy is increasingly transferred to sound waves and the number of vortices gradually diminishes.

6   Intrinsic Vortex Nucleation

In this section we consider the nucleation of quantized vortices from the standpoint of GP theory. Understanding vortex nucleation and the critical velocities associated with it is one of the most significant questions of superfluidity.

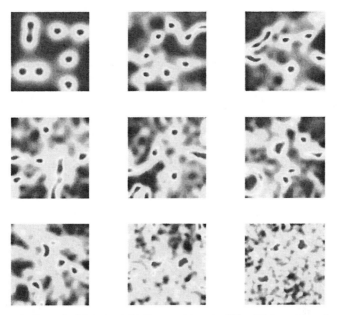

Fig. 1. The evolution of eight straight-line vortices in 2D within a box, four of each direction of circulation. As a result of the interactions, the number of vortices decreases and their energies are transmitted to sound waves

Breakdown of ideal superflow in channels and in rotating containers is usually attributed to the formation of vortices. Flow through apertures and around impurities is known to generate vortex rings and vortex loops. According to Donnelly [18] we should distinguish between 'extrinsic nucleation', which concerns the growth of pre-existing vorticity from 'intrinsic nucleation', which is creation of vorticity from "nothing". It is virtually impossible to distinguish between these two experimentally. The energy of the trapped vortex may be close to the energy of an elementary excitation such as a roton. It could have been excited thermally and remained trapped at a pinning center during the cooling down of the superfluid to low temperatures, and then be enlarged and "released" by superflow past that center.

The intrinsic nucleation is hard to understand as there is no truly microscopic theory of the superfluid. We do not know how to describe the processes on the scale of the coherence length, where the motion is governed by the full quantum many-body structure of the superfluid. We confine attention here therefore to two possible approaches to vortex nucleation. The first is based on the GP model; the second relies on semiclassical, hydrodynamic (large-scale) theory with the vortex assumed *ab initio* with tunneling introduced in an *ad hoc* way. As was emphasized by Fischer [20], and foreshadowed long ago by, for example, Schwarz and Jang [50], there are difficulties faced by the quasi-classical theory of nucleation that render it incapable of describing intrinsic nucleation. A semiclassical description is valid only if the quantum core structure of the tunneling object does not come into contact with the boundary. The principal advantage of using the GP model is that semiclassical and tunneling processes are joined seamlessly.

Vortex nucleation by an impurity such as the positive ion ^{4}He$_2^+$ moving in superfluid helium at low temperature with velocity v has been studied experimentally and theoretically (see, e.g. Donnelly, 1991), and has uncovered some interesting physics. The flow round an ion that is moving with a sufficiently small velocity, v, is well represented by one of the classical solutions of fluid mechanics, namely the flow of an inviscid incompressible fluid around a sphere. In this solution, the maximum flow velocity, $\mathbf{u}$, relative to the sphere is $3v/2$, and occurs on the equator of the sphere (defined with respect to the direction of motion of the sphere as polar axis). Above some critical velocity, v_c, the ideal superflow around the ion breaks down, leading to the creation of a vortex ring (Rayfield and Reif, 1964). Muirhead et al. [37] created a theory of vortex nucleation that allowed them to estimate three useful quantities (i) v_c, (ii) the form of the potential barrier that must be overcome for the creation of vortices both as encircling rings and vortex loops, and (iii) the nucleation rate. These calculations were carried out for a smooth rigid sphere moving through an ideal incompressible fluid using the semiclassical approach.

An important scale defined by the condensate model is the 'healing length', a, defined in (19). This determines the radius of a vortex core and the thickness of the 'healing layer' that forms at a potential barrier (such as the ion surface in GP model). The radius, b, of the ion is large compared with a, and asymptotic solutions for $\epsilon \equiv a/b \to 0$ become relevant. Such a solution has two parts, an interior or 'boundary layer' structure that matches smoothly to an exterior or 'mainstream' flow. Berloff and Roberts [6] showed that the dimensionless flow velocity $\mathbf{U} = \mathbf{u}/c$ is greatest on the equator of the sphere where it is

$$U_\theta = 3V/2 + 0.313V^3 + 0.3924V^5 + 0.648V^7 + 1.24V^9 + 2.63V^{11} + \cdots \quad (39)$$
$$+\epsilon\Big(2.12V + 1.58V^3 + 2.89V^5 + \cdots\Big),$$

where $V = v/c$. This gives $v_c \approx 0.53c$ for $\epsilon \to 0$. (We recall here that the criterion $u = c$ for criticality strictly applies only for $\epsilon = 0$; see §3.)

There is some similarity between the flow of the condensate past the ion and the motion of a viscous fluid past a sphere at large Reynolds numbers, the healing layer being the counterpart of the viscous boundary layer. There are, however, important differences. At subcritical velocities, the flow of the condensate is symmetric fore and aft of the direction of motion, and the sphere experiences no drag. In contrast, the viscous boundary layer separates from the sphere, so evading D'Alembert's paradox, destroying the fore and aft symmetry, and therefore bringing about a drag on the sphere. Moreover, when $v > v_c$, shocks form at or near the sphere, but shocks are disallowed in the condensate since they represent a violation of the Landau criterion and a breakdown of superfluidity. When $v > v_c$, the condensate evades shocks through a different mode of boundary layer separation. The sphere sheds circular vortex rings that move more slowly than the sphere and form a vortex street that trails behind it, maintained by vortices that the sphere sheds. As the velocity of the ion increases such a shedding becomes more and more irregular. Each ring is born at one particular latitude within the healing layer on the sphere. As it breaks away into the mainstream, it at first contributes a flow that depresses the mainstream velocity on the sphere below critical. As it moves further downstream however, its influence on the surface flow diminishes. The surface flow increases until it again reaches criticality, when a new ring is nucleated and the whole sequence is repeated. The vortex street trailing behind the ion creates a drag on the ion that decreases as the nearest vortex moves downstream, but which is refreshed when a new vortex is born.

Berloff [4] considered the vortex nucleation when the symmetry of the system is broken in the presence of the random noise or by introducing a plane wall at some distance from the ion. Figure 2 shows the formation and evolution of a vortex loop on the positive ion when it moves subcritically. As the velocity reaches criticality on the side of the sphere closer to the boundary, a vortex loop appears, spreads laterally, interacts with the boundary, the feet of the vortex line on the surface of the ion come closer to one another, and they detach to form a loop on the boundary.

Frisch et al. [21] and Winiecki et al. [55] have solved numerically the condensate equation for flow past a circular cylinder and Berloff and Roberts [6] for the flow around a positive ion and have confirmed the main features of the scenario just described. Below the critical velocity steady solutions of the GP model exist. Huepe and Brachet [31] numerically computed stationary steady and unsteady

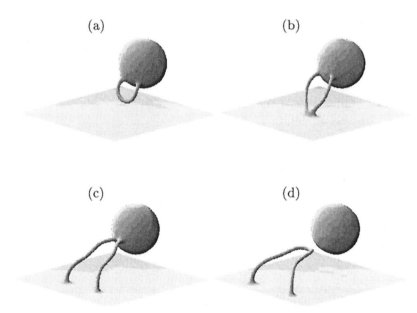

(a)   (b)

(c)   (d)

Fig. 2. The results of numerical integration of (3) for the positive ion moving with the velocity $0.51c$ for the isosurface $\varrho = 0.2\varrho_\infty$ at (a) $t = 132a/c$, (b) $t = 168a/c$, (c) $t = 216a/c$, and (d) $t = 240a/c$.(After [4])

solutions for flow around a cylinder. They confirmed the critical velocity of nucleation $v_c \approx 0.42c$ found numerically by Frisch et al. [21] and Winiecki et al. [55] and through the asymptotic expansion by Berloff and Roberts [6]. Winiecki et al. [56] carried out integrations similar to those of Huepe and Brachet [31] but for three dimensional flow around a penetrable sphere. They also found three branches of steady solutions that exist up to the nucleation point. In order of decreasing energy these are (i) irrotational flow round the sphere, (ii) a vortex loop attached to the sphere, and (iii) a vortex ring surrounding the sphere symmetrically. These three solutions are all stable. Perhaps a transition from one branch to another can be achieved by quantum tunneling in the way envisaged by Muirhead et al. [37].

In addition to the positive ion, other impurities have proved to be useful experimental probes, including neutral atoms such as ^{3}He (radius $\sim$ 4Å) and the negative ion, which is an electron in a bubble cut out of the liquid owing to the repulsive interaction between the electron and surrounding helium atoms. We now consider the negative ion.

In the Hartree approximation, the equations governing the one particle wavefunction of the condensate, ψ, and the wave function of the electron, ϕ, are a pair of coupled equations suggested by Gross [27] and by Clark [14], [15]:

$$i\hbar \frac{\partial \psi}{\partial t} = -\frac{\hbar^2}{2M} \nabla^2 \psi + (U_0 |\phi|^2 + V_0 |\psi|^2 - E)\psi, \tag{40}$$

$$i\hbar \frac{\partial \phi}{\partial t} = -\frac{\hbar^2}{2\mu} \nabla^2 \phi + (U_0 |\psi|^2 - E_e)\phi, \tag{41}$$

where M and E are the mass and single particle energy for the bosons; μ and E_e are the mass and energy of the electron. The interaction potentials between boson and electron and between bosons are here assumed to be of $\delta-$function form $U_0 \delta(\mathbf{x} - \mathbf{x}')$ and $V_0 \delta(\mathbf{x} - \mathbf{x}')$, respectively. To lowest order, perturbation theory predicts such interaction potentials to be $U_0 = 2\pi l \hbar^2 / \mu$ and $V_0 = 4\pi d \hbar^2 / M$, where l is the boson-impurity scattering length, and d is the boson diameter. The normalization conditions on the wave functions are

$$\int |\psi|^2 dV = N, \qquad \int |\phi|^2 dV = 1. \tag{42}$$

Using the system (40)-(41), Grant and Roberts [25] studied the motion of a negative ion moving with speed v using an asymptotic expansion in v/c, where c is the speed of sound, so that their leading order flow is incompressible. Treating $\epsilon \equiv (a\mu/lM)^{1/5}$ as a small parameter they calculated the effective (hydrodynamic) radius and effective mass of the electron bubble. By employing a convergent series expansion suitable for $\mathbf{u} = \mathcal{O}(c)$, Berloff and Roberts (2000) determined v_c in the limit $\epsilon \to 0$. They have shown that v_c for the negative ion is about 20% less than v_c for the positive ion, in agreement with the experimental findings of Zoll (1976); see also Table 8.2 of Donnelly (1991). This reduction may be attributed to the flattening of the electron bubble by its motion through the condensate. The "equatorial bulge" is created by the difference in pressure between the poles and equator associated with the greater condensate velocity at the latter than at the former. The existence of the bulge also enhances these differences in velocity (and pressure), as compared with a spherical impurity, with the result that, if v is gradually increased from zero, the flow u_e on the equator of the electron bubble attains the velocity of sound before u_e does for the positive ion.

Finally, we mention the nucleation of the vorticity by superfluid flow through apertures. Burkhart et al. [11] considered the properties of a vortex close to a wall as derived from the GP model and obtained a critical velocity and vortex radius at nucleation that agree with experiments.

7   Capture of Impurities by Vortex Lines

Rayfield and Reif [43] used an ion time-of-flight spectrometer to determine the dynamics of ion-quantized vortex ring complexes. They observed that above some critical velocity, v_c, ideal superflow around an ion breaks down. The moving ion produces vortex rings and the ion becomes trapped in one of these. The capture of negative ions by quantized vortex lines in a rotating bucket was first demonstrated by Careri et al. [12]. Schwarz and Donnelly [49] observed that at low temperatures (< 0.5 K) straight-line vortices can trap positive ions.

Berloff and Roberts [8] solved equations (40)-(41) numerically to observe and elucidate the process of capture; see Fig. 3. They have shown that this process can be better characterized as the reconnection of the vortex line with its pseudo-image inside the impurity. Initially the vortex line bends towards the impurity at the point where the distance between them is least. As the result of this interaction the impurity moves around the vortex axis. The process of capture continues as the vortex line terminates on the surface of impurity and its feet move to opposite poles of the impurity surface. At the same time helical waves start propagating from the impurity along the two segments of the vortex line. Such helical waves have been observed during the relaxation of the vortex angle when two vortex lines reconnect [28], and are just Kelvin waves. These waves, and associated sound waves, radiate energy to infinity.

During the time in which the vortex merges with the healing layer round the ion (now the layer of thickness $\sim a$ in which neither ϕ in (40) nor ψ in (41) can be neglected), the character of the solution alters rapidly, corresponding to the topological change that defines the capture of the ion by the vortex. Once the vortex has divided into two, with separate feet attached to the healing layer, the flow round the ion has acquired circulation that it previously could not possess.

It is clear from the discussion above that the trapped impurity is in the lower energy state than the free impurity. The difference $\triangle V$ is the 'substitution energy', also called the 'binding energy'. Donnelly and Roberts [17], using the healing model of the vortex core, estimated that

$$\triangle V = 2\pi\rho_s(\hbar/M)^2 b\left[(1 + a^2/b^2)^{1/2}\sinh^{-1}(b/a) - 1\right]. \tag{43}$$

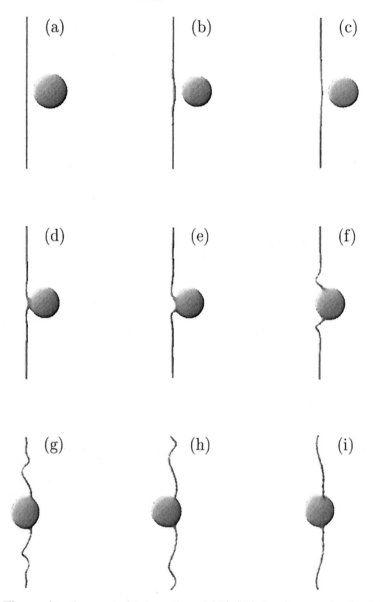

Fig. 3. The results of numerical integration of (40)-(41) for the negative ion initially placed a distance $6a$ apart from the rectilinear vortex line. The pictures show the isosurface $\varrho = 0.2\varrho_\infty$. (After [8])

The more sophisticated calculations of Berloff and Roberts [8] gave 21.7 K for a positive ion of radius $10a$, where $a = 0.47$ Å for the nondimensionalization based on the speed of sound. If instead $a = 1$ Å is used, they obtained $\triangle V/k_B \approx 46$ K, which may be compared with 33.5 K according to (43). For the electron bubble our result is $\triangle V/k_B = 55$ K, and (43) gives $\triangle V/k_B = 66$ K if $R = 16$ Å is used. The healing layers around the positive and negative ions differ dramatically and the healing layer close to the vortex feet is less depleted, which was not taken into account by (43). This explains why (43) overestimates $\triangle V$.

The capture of a penetrable sphere moving with supercritical velocity by a vortex ring that it itself created was simulated by Winiecki and Adams [57] through the full 3D numerical integration of the GP equation, supplemented with the calculation of the drag on the surface of the sphere. They have shown that after the positive ion emits the vortex ring, the drag on the ion slows it down and the ion is captured by the vortex ring; see Fig. 4.

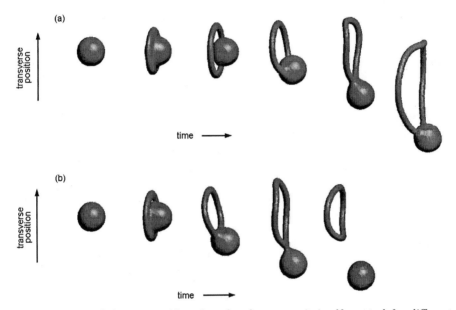

Fig. 4. Capture of the penetrable sphere by the vortex it itself created for different values of the applied force. (After [57])

Berloff and Roberts [7] performed integrations of (40)-(41) for the following configuration: a moving negative ion catches up with a vortex ring moving in the same direction. The axis of the ring does not coincide with the axis of the impurity. Such a condition is necessary to destroy the axisymmetry of the system. Figure 5 shows the process of capture of the electron by the vortex ring. Initially, the faster moving ion passes the vortex ring. The Bernoulli effect of the flow propels the ion and vortex towards one another with a force approximately proportional to s^{-3}, where s is the closest distance between them, similarly to the process of capture of the ion by a straight line vortex.

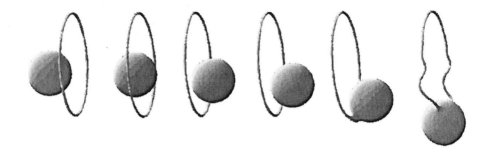

Fig. 5. Capture of the moving ion by the vortex ring: the results of numerical integration of (40)-(41) for the isosurface $\varrho = 0.2\varrho_\infty$. (After [7])

8   Nonlocal Models

In §3 we considered different shortcomings of the GP model. The natural question arises whether it is possible to remedy (3) to make it more quantitatively realistic for superfluid helium. For some time there has been a belief that, as soon as the nonlocal model (1) with a realistic two-particle potential, V, that leads to phonon-roton-like spectra is solved, the properties of superfluid helium will be well represented. The minimum requirements on such a potential would be (i) the correct position of the roton minimum and (ii) the correct speed of sound. Actually such a fit can be obtained with a variety of potentials. Pomeau and Rica [41] pioneered the use of nonlocal models for study superfluidity, but their model did not have the correct sound velocity (slope of the dispersion curve at the origin). Berloff [3] investigated the applicability of (1) with a potential that adequately represents the dispersion curve. It was shown that for liquid helium having the correct Landau dispersion curve, solutions of equation (1) develop non-physical mass concentrations. In particular, the "Eulerian part" of the momentum equation (without the quantum stress tensor) may become no longer hyperbolic in some parts of the integration volume. A virial theorem, similar to the one used to establish the catastrophic blow-up in the focusing nonlinear Schrödinger equation, can be used to establish similar catastrophes in bounded volume for (1). This indicates that the assumptions underlying the derivation of the equation break down and that higher order nonlinearities must be introduced.

A more accurate approach in modeling liquid helium is through density-functional theory done by Dalfovo et al. [16], which attempts to give an adequate microscopic description of interactions. In this approach the total energy is still written as a functional of the one-body density, but it includes short-range correlations [19]. This approach has provided a quantitatively and qualitatively reliable representation of the superfluid properties of free surfaces, helium films, and droplets (see [16] and references therein). At the same time this approach is phenomenological and results in rather complicated forms of the energy functionals with many parameters that are chosen to reproduce liquid helium properties.

Berloff and Roberts [5] attempted to modify the nonlocal model (1) in the spirit of a density - functional approach, by introducing only one additional nonlinear term in the expression for the correlation energy. This allowed them to remedy the nonphysical features of model (1), while retaining not only an adequate representation of the Landau dispersion relation, but also simplicity in the analytical and numerical studies.

The correlation energy of the Skyrme interactions in nuclei [53] is given by

$$W_c(\rho) = \frac{1}{M^2} \int \left[\frac{W_0}{2} \rho^2 + \frac{W_1}{2+\gamma} \rho^{2+\gamma} + W_2 (\nabla \rho)^2 \right],$$ (44)

where W_0, W_1, W_2 and γ are phenomenological constants. The first two terms give a local density approximation, and the gradient term corresponds to finite range interactions. The necessary nonlocality of interactions was added directly into the first term of (44) by introducing a two-body interaction potential, $V(|\mathbf{x} - \mathbf{x}'|)$, so that (44) becomes

$$W_c(\rho) = \frac{1}{M^2} \int \left[\frac{1}{2} \int \rho(\mathbf{x}) V(|\mathbf{x} - \mathbf{x}'|) \rho(\mathbf{x}') \, d\mathbf{x}' + \frac{W_1}{2+\gamma} \rho^{2+\gamma} \right] d\mathbf{x}.$$ (45)

This incorporates and generalizes the W_2 interaction term in (44), which has therefore been abandoned; $V(|\mathbf{x} - \mathbf{x}'|)$ is chosen so that the implied dispersion relation is a good fit to the Landau dispersion curve. Following private communications with C. Jones, Berloff and Roberts [5] considered a potential of the form

$$V(|\mathbf{x} - \mathbf{x}'|) = V(r) = (\alpha + \beta A^2 r^2 + \delta A^4 r^4) \exp(-A^2 r^2),$$ (46)

and also the slightly modified potential

$$V(|\mathbf{x} - \mathbf{x}'|) = V(r) = (\alpha + \beta A^2 r^2 + \delta A^4 r^4) \exp(-A^2 r^2) + \eta \exp(-B^2 r^2),$$ (47)

where $A, B, \alpha, \beta, \delta$ and η are parameters that can be chosen to give excellent agreement with the experimentally determined dispersion curve.

On adopting (45), one can see that (1) is replaced by

$$i\hbar \Psi_t = -\frac{\hbar^2}{2M} \nabla^2 \Psi + \Psi \int |\Psi(\mathbf{x}', t)|^2 V(|\mathbf{x} - \mathbf{x}'|) \, d\mathbf{x}' + W_1 \Psi |\Psi|^{2(1+\gamma)}.$$ (48)

with (7) this equation becomes

$$i\hbar \psi_t = -\frac{\hbar^2}{2M} \nabla^2 \psi + \psi \left(\int |\psi(\mathbf{x}', t)|^2 V(|\mathbf{x} - \mathbf{x}'|) \, d\mathbf{x}' + W_1 |\psi|^{2(1+\gamma)} - E_v \right).$$ (49)

This model not only produces the structure and energy per unit length of the straight-line vortex that are very close to the ones obtained from the Monte

Carlo simulations by Sadd et al. [46], but it also made it possible to bring the vortex core parameter (35) and the healing length into agreement. Figure 6 compares the experimentally determined dispersion curve with that employed by (49). The insets give the density in the core of the straight line vortex and in the healing layer at a solid boundary, both for (49) and for the GP model. The vortex rings of large radii satisfy (36)-(38). Berloff and Roberts [5] integrated (48) numerically to elucidate the behavior of the small vortex ring. The Berloff-Roberts calculations indicate that when the velocity of the vortex ring reaches the Landau critical velocity the ring becomes unstable and evanesces into sound waves. For any ring traveling with speed greater than the Landau critical velocity, the amplitude of the far-field solution will not decay exponentially at infinity, which makes the existence of such a ring impossible. One of the goals of these calculations was to clarify Onsager's concept of the roton as "the ghost of a vanished vortex ring." One can hope that the transition from the vortex ring to the sound pulse and the concomitant loss of vorticity would occur close to the roton minimum in energy-momentum space, or (more probable) close to the point where the group velocity and the phase velocity are equal (the Landau critical velocity u_L). Their calculations show that indeed there is a point on the $p\mathcal{E}-$ plane where the ring ceases to exist and where $u_L = \partial\mathcal{E}/\partial p$, but this point lies far from the roton minimum. It remains to be seen whether the idea of the roton as a ghostly vortex ring will ever be vindicated. As one has a great variety of potentials that lead to the Landau dispersion curve one can tune the parameters so that the line $\mathcal{E} = u_L p$, meets the $p\mathcal{E}-$ curve for the family of the vortex rings, to allow this sequence of vortex rings to be terminated at a lower energy and momentum level. Whether this process will lead to coalescence with the roton minimum is not yet clear.

Berloff and Roberts [9] used (49) to elucidate the differences between the processes of vortex nucleation and roton emission. They argued that vortices are nucleated when the velocity around the positive ion exceeds the velocity of sound. The moving ion generates rotons when it moves with the velocity greater than the Landau critical velocity.

9   Conclusions

Despite the fact that the interatomic spacing in helium is short compared with the coherence length, the GP equation has proved itself useful in modeling, in at least a qualitatively faithful way, many of the phenomena that have been studied experimentally. This is partially a result of the great strides taken by computer technology during the past decade that have made it possible to undertake three-dimensional simulations of processes such as vortex-vortex and vortex-ion interactions, including the nucleation of vortices by ions and the capture of ions by vortices. It has also become possible to explore more general equations of Gross-Pitaevskii type from which a more realistic dispersion curve emerges, one that possesses a roton minimum at approximately the correct momentum and energy.

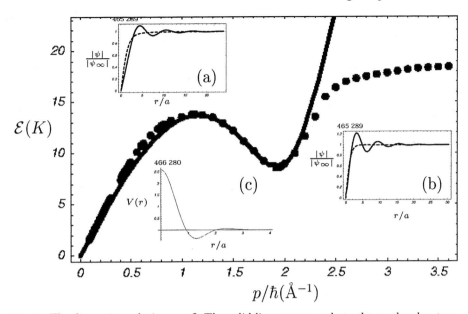

Fig. 6. The dispersion relation $p - \mathcal{E}$. The solid line corresponds to the nonlocal potential $V(|\mathbf{x} - \mathbf{x}'|)$ with $A = 0.9$, $\chi = 0.2$, and $\gamma = 1$. The dots are based on experiment. The insets show (a) the amplitude $|\psi|/\psi_{\infty}$ of the straight line vortex for the nonlocal model (49) (solid line) and the GP model (dashed line); (b) the amplitude $|\psi|/\psi_{\infty}$ of the healing layer at a solid boundary (an infinite potential barrier) placed at $r = 0$ for the nonlocal model (solid line) and the GP model (dashed line); (c) the potential $V(|\mathbf{x} - \mathbf{x}'|)$ plotted as a function of $r = |\mathbf{x} - \mathbf{x}'|$, in the nondimensional units defined in [9]. (After [9])

One topic that has not been addressed in this review concerns generalization to finite temperature. As is well known, the Landau two-fluid description of helium dynamics was devised before it was demonstrated that vorticity in the superfluid is quantized, and Landau theory encounters difficulties when quantized vortices are present . These difficulties are analogous to those encountered when the classical Euler equations are used to model the superfluid. One attempt to modify Landau theory in order to evade these difficulties was made by Hills and Roberts [30]. This theory was also, in a sense, a consequence of the one-fluid theory of Putterman and Roberts [42].

Despite the successes of Gross-Pitaevskii theory and its generalizations, much remains to be done, possibly by generalizing the theory itself, before it can realistically be used for superfluid helium. It is fortunate however that it can be legitimately applied to low density condensates, where the interatomic spacing is greater than the coherence length.

This review was written by two people whose backgrounds are in fluid mechanics. Not surprisingly therefore the perspective throughout has been fluid mechanical. Undoubtedly, we have neglected to cite work that we should have referenced. We apologize for any such omissions, which arise through ignorance and not malice.

Acknowledgments

We are grateful to "Sandy" Fetter for his criticisms of the previous version of this review, and "Lev" Pitaevskii and "Joe" Vinen for their helpful comments. This work is supported by the grant DMS-9803480 from the National Science Foundation.

References

1. C.F.Barenghi, D.C. Samuels, G.H. Bauer, R.J. Donnelly: Physics of Fluids **9**, 2631 (1997)
2. C.F.Barenghi, C.J. Swanson, R.J. Donnelly: J. Low. Temp. Phys.**100**, 385 (1995)
3. N.G. Berloff: J. Low. Temp. Phys. **116**, 359 (1999)
4. N.G. Berloff: Phys. Rev. Lett., submitted (2000)
5. N.G. Berloff, P. H. Roberts: J. Phys.: Math. Gen. **32**, 5611 (1999)
6. N.G. Berloff, P. H. Roberts: J. Phys.: Math. Gen. **33**, 4025 (2000)
7. N.G. Berloff, P. H. Roberts: J. Phys.: Math. Gen. accepted (2000)
8. N.G. Berloff, P. H. Roberts: Phys. Rev. B., accepted (2000)
9. N.G. Berloff, P. H. Roberts: Phys. Lett. A, in press (2000)
10. J. B. Brooks, R. J. Donnelly: J. Phys. Chem. Ref. Data **6**, 51 (1977)
11. S. Burkhart, M. Bernard, O. Avenel, E. Varoquaux: Phys. Rev. Lett. **72**, 380 (1994)
12. G. Careri, W. D. McCormick, F. Scaramuzzi: Phys. Lett. **1**, 61 (1962)
13. N.N. Carlson: Physica D **98**,183 (1996)
14. R.C. Clark: Phys. Lett. **16**, 42 (1965)
15. R.C. Clark: in *Superfluid Helium* ed. J. F. Allen (Amsterdam, North Holland, 1966)
16. F. Dalfovo, A. Lastri, L. Pricaupenko, S Stringari, J Treiner: Phys. Rev. B **52**, 1193 (1995)
17. R. J. Donnelly, P.H. Roberts: Proc. Roy. Soc.**A312**, 519 (1969)
18. R.J. Donnelly: *Quantized Vortices in Helium II*, (Cambridge University Press, Cambridge, 1991)
19. J. Dupont-Roc, M. Himbert, N. Pavlov, and J. Treiner: J. Low Temp. Phys. **81**, 31 (1990)
20. U.R. Fischer: Physica B **255**, 41 (1998)
21. T. Frisch, Y. Pomeau, S. Rica: Phys. Rev. Letts. **69**, 1644 (1992)
22. V.L. Ginzburg, L.P. Pitaevskii: Zh. Eksp.Teor. Fiz. **34**, 1240 (1958)
23. J. Grant: J. Phys. A: Gen. Phys. **4**, 695 (1971)
24. J. Grant: J. Phys. A: Math., Nucl. Gen. **6**, L151 (1973)
25. J. Grant, P. H. Roberts: J. Phys. A: Math., Nucl. Gen. **7**, 260 (1974)
26. E. P. Gross: J. Math. Phys. **4**, 195 (1963)
27. E. P. Gross: in *Quantum Fluids*. ed. D. B. Brewer (Amsterdam, North Holland, 1966)
28. F.R. Hama: Phys. Fluids **5**, 1156 (1962)
29. R.N. Hills, P.H.Roberts: Archiv. Rat. Mech. Anal. **66**, 43 (1977)

30. R.N. Hills, P.H.Roberts: J. Low Temp. Phys. **40**, 117 (1980)
31. C. Huepe, M. Brachet: C. R. Acad. Sci. Paris **325**, 195 (1997)
32. C.A. Jones, S. J. Putterman, P.H. Roberts: J. Phys. A: Gen. Phys. **19**, 2991 (1986)
33. C.A. Jones, P.H. Roberts: J. Phys. A: Gen. Phys. **15**, 2599 (1982)
34. I.M. Khalatnikov: *Introduction to the Theory of Superfluidity*, (W.A. Benjamin, Inc., New York Amsterdam, 1965)
35. J. Koplik, H. Levine: *Phys. Rev. Letts.* **71**, 1375 (1993)
36. J. Koplik, H. Levine: Phys. Rev. Letts. **76**, 4745 (1996)
37. C.M. Muirhead, W.F. Vinen, R.J. Donnelly: Phyl. Thans. R. Soc. Lond. A **3**, 433 (1984)
38. C. Nore, M. Abid, M. E. Brachet: Phys. Fluids **9**, 2644 (1997)
39. L.M. Pismen, A.A. Nepomnyashchy: Physica D **69**, 163 (1993)
40. L.P.Pitaevskii: Sov. Phys. JETP **13**, 451 (1961)
41. Y. Pomeau, S. Rica: Phys. Rev. Lett. **71**, 247 (1993)
42. S. J. Putterman, P. H. Roberts: Physica **117** A, 369 (1983)
43. G. W. Rayfield, F. Reif: Phys. Rev. **136**, A1194 (1964)
44. P.H. Roberts, J. Grant: J. Phys. A: Gen. Phys. **4**, 55 (1971)
45. G. Rowlands: J. Phys. A: Nucl. Gen. **6**, 322 (1973)
46. M. Sadd, G.V. Chester, L. Reatto: Phys. Rev. Lett. **79**, 2490 (1997)
47. P.G. Saffman: *Vortex Dynamics*, (Cambridge University Press, Cambridge, 1992)
48. K.W. Schwarz: Phys. Rev. B **38**, 2398 (1988)
49. K.W. Schwarz, R. J. Donnelly: Phys. Rev. Lett. **17**, 1088 (1966)
50. K.W. Schwarz, F. Jang: Phys. Rev. A **8**, 3199 (1973)
51. B. V. Svistunov: Phys. Rev. B **52**, 3647 (1995)
52. J. T. Tough: "Superfluid turbulence". In *Progress in Low Temperature Physics*, **VIII**, ed. D.F. Brewer. Amsterdam: North-Holland, 133 (1982).
53. D. Vautherin, D. M. Brink: Phys. Rev. C **5**, 626 (1972)
54. S.E. Widnall, J. Sullivan: Proc. R. Soc. Lond. A **287**, 273 (1973)
55. T. Winiecki, J. F. McCann, C. S. Adams: Phys. Rev. Letts **82**, 5186 (1999)
56. T. Winiecki, J. F. McCann, C. S. Adams: Europhys. Letts **48**, 475 (1999)
57. T. Winiecki, C. S. Adams: Europhys. Letts submitted (2000)

Vortex Nucleation and Limit Speed for a Flow Passing Nonlinearly Around a Disk in the Nonlinear Schrödinger Equation

Sergio Rica

Laboratoire de Physique Statistique,
Ecole Normale Supérieure,
24, rue Lhomond, 75231 Paris Cedex 05, France.

Abstract. I review a study over almost nine years in collaboration with T. Frisch, C. Josserand and Y. Pomeau [1–5] on the problem of critical velocities and vortex nucleation for a superflow passing around a disk in the nonlinear Schrödinger equation.

1   Introduction and Formulation of the Problem

The Gross–Pitaevskii[6] or nonlinear Schrödinger equation (NLSE for short hereafter) reads, in a dimensionless form[1]:

$$i\partial_t \psi = -\frac{1}{2}\nabla^2\psi + |\psi|^2\psi.$$

$$(1)$$

This is a partial differential equation for a complex wave function $\psi = \psi(x,y,t)$. We shall consider the boundary conditions

$$\psi(x,y,t) = 0 \quad \text{if} \quad \sqrt{x^2+y^2} < R\,, \qquad (2a)$$
$$\partial_y\psi = 0 \quad \text{at} \quad y = \pm L_y\,, \qquad (2b)$$
$$\partial_x\psi = iv_\infty\psi \quad \text{at} \quad x = \pm L_x\,. \qquad (2c)$$

The boundary term (2c) imposes a flow in the x direction and for the analysis we shall take the limits $(L_x, L_y) \to \infty$. This problem deals with two distinct independent parameters the velocity v_∞ and the disk radii R.

A direct numerical study of (1) with (2a,2b,2c), as well as simple analytical arguments that I will review in this article, show that the flow around a disk releases vortices from the perimeter of the disk creating a net drag force beyond a critical velocity, that leads to a vortical and dissipative flow on large scales. Vortices, when they appear in our simulations are produced by the system itself, without being introduced at the beginning. At low speed, after some transient the solution accomodates the boundary conditions everywhere to yield a stationary

---

[1] In their original work the nonlinear term comes as a mean field interaction term: $\psi(\mathbf{r}) \int V(|\mathbf{r} - \mathbf{r}'|)|\psi(\mathbf{r}')|^2 d\mathbf{r}'$. We have studied that model, in particular whenever interparticle interaction V allows possible a roton minimum in the dispersion relation [7].

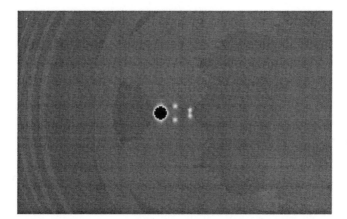

Fig. 1. Vortex shedding by a flow (from left to right) around a disk in NLSE. Two pairs of vortices are seen downstream. It appears that the vortices are nucleated when the flow becomes locally (at the edge of the disk) supersonic.

velocity field without dissipation, because of d'Alembert's paradox for perfect fluids. Beyond a critical velocity, that we shall characterize later, vortices begin to be released more or less periodically from the disk, yielding an average drag on it, this drag tends linearly to zero at threshold.

This has been a long study that has spaned several years in collaboration with T. Frisch, C. Josserand and Y. Pomeau [1–5]. The most interesting aspect is that we have determined that the criteria for vortex nucleation in the NLSE is closely related to the transonic transition [8]. Indeed, the nonlinear Schrödinger equation could be transformed into a set of "hydrodynamical" equations for a compressible fluid with a certain state equation in a long wavelength approximation. In ordinary fluid mechanics, this would lead to the formation of a shock wave inside the supersonic bubble. Nothing similar is possible in NLSE, because of the lack of built-in irreversibility, something that is necessary to balance non-linearities inside the shock wave. In NLSE, the formation of shock waves is replaced by the nucleation of vortices with a quantized circulation. In fact, these vortices are topological defect solutions of NLSE [6]. In two space dimensions they are points where the complex field $\psi(x, y, t)$ vanishes, making a $\pm 2\pi$ phase jump when one turns around the defect. The aim of this article is to present a review that summarizes the results of Refs. [1–5].

Writing $\psi = \rho^{1/2} e^{i\phi}$, we obtain two "real" hydrodynamical fields, ρ and ϕ, representing respectively the particle fluid density and the velocity potential:

$$\partial_t \rho = -\nabla \cdot (\rho \nabla \phi); \tag{3a}$$

$$\partial_t \phi = -\frac{1}{2}(\nabla \phi)^2 + \frac{1}{2\rho^{1/2}}\nabla^2 \rho^{1/2} - \rho. \tag{3b}$$

The first one is the density mass conservation equation, identifying $\nabla \phi$ by the local velocity **v**. Long wavelength and low amplitude perturbations around the homogeneous state: $\psi_0 = \sqrt{\rho_0} e^{-i\rho_0 t}$ propagate with the sound speed $c_s = \sqrt{\rho_0}$. In the second equation, the term $\frac{1}{2\rho^{1/2}}\nabla^2\rho^{1/2}$, often called quantum pressure, is negligible for large scale flows, that is for flows with a space scale much larger than the only intrinsic microscopic length, $\xi_0 = \frac{1}{\sqrt{\rho_0}}$ (ξ_0 is, also, the vortex core size).

Therefore, the problem of a flow around a disk possesses two independent dimensionless parameters, the Mach number $M = \frac{v_\infty}{c_s}$ and the ratio $\frac{\xi_0}{R}$.

After our original work of 1991/92, several papers appeared in the literature producing a great confusion in the basics of the problem. However, in a recent article, Huepe and Brachet [9] re-examined the problem of (superfluid-)vortex nucleation near an obstacle in the context of the nonlinear Schrödinger equation. This article contains good evidence for the scenario of vortex nucleation described by our previous work in the subject .

The discovery of Bose–Einstein condensation in atomic vapors [10] opens the way to test in a rather detailed fashion some predictions of the equilibrium and nonequilibrium quantum statistical mechanics. As those vapours are weakly interacting systems there is hope to compare the predictions of the Gross–Pitaevskii theory with experimental results. In particular the sound propagation was tested successfully [11]. And more recently it was observed some evidence of the existence of a superflow around an obstacle with a well defined critical speed (around one fourth the speed of sound) [12]. Ketterle's results agree quantitatively with results coming from NLSE in a rugby balloon-shape geometry in 3D, a detailed discussion on that was given by C.S. Adams in this conference and in Ref. [13].

After this introduction to the problem, in Sect. 2, I will derive the nonlinear PDE (5a,5b,5c) for the stationary flow in the limit $\frac{\xi_0}{R} \to \infty$, and discuss the limit speed, due to change of type of (5a). In Sect. 3, I will show how to solve this nonlinear PDE by series expansions in Mach number, leading to a critical Mach number in the limit $\frac{\xi_0}{R} \to \infty$:

$$M_c \approx 0.36969(7) \qquad (4)$$

Section 4 deals with the existence of vortical unstables stationary flows of (1) for $M < M_c$. Finally, in Sect. 5 I review, with the help of the Euler–Tricomi equation, the transonic transition, and how that matches with vortex shedding.

2   Critical Velocities

Far from the disk boundary, in terms of the microscopic distance, the usual hydrodynamic assumption holds: the quantum pressure term can be neglected. Then one obtains an unique nonlinear equation for the velocity potential, which is an equation for a compressible fluid[2]:

$$0 = \nabla \cdot \left(\left(\frac{1}{M^2} + \frac{1}{2}(1 - (\nabla\phi)^2) \right) \nabla\phi \right);$$ (5a)

$$\partial_r \phi = 0 \quad \text{at } r = 1$$ (5b)

$$\phi = r \cos \theta \quad \text{as } r \to \infty,$$ (5c)

Here ϕ is the velocity potential whose gradient is normalized in absolute value to one, as required by (5c). In this limit ($\frac{\xi_0}{R} \to \infty$) one may set $R = 1$, with M the Mach number at infinity. (r, θ) are polar coordinates with origin at the center of the unitary disk.

Let us point out that the boundary conditions (2b,2c) match directly with (5c), but (2a) requires a more sophisticated argument to match (5b) in the limit $\frac{\xi_0}{R} \to 0$. In fact, the inner value of the wavefunction near the disk boundary vanishes as $\psi \approx \lambda(x, y, t) \times (\sqrt{x^2 + y^2} - R)\Theta(\sqrt{x^2 + y^2} - R)e^{i\phi(x,y,t)}$, being $\lambda(x, y, t)$ a smooth non vanishing function and $\Theta(\cdot)$ the Heaviside function. Replacing this in (1) one gets at leading order that $\lambda(x, y, t)\hat{r} \cdot \nabla\phi$ must vanish, that is (5b).

At low Mach number the second order partial differential equation (5a) is elliptic everywhere outside the disk. However, as M increases equation (5a) becomes hyperbolic beyond a critical Mach number M_c. This happens whenever $\partial_v \left(\frac{v}{M^2} + \frac{v}{2}(1 - v^2) \right)$ vanishes at least at one point in the space. This criteria gives the value for the critical Mach number, as the real root of the equation:

$$\frac{1}{M^2} + \frac{1}{2} - \frac{3}{2}(\nabla\phi)^2_{max} = 0.$$ (6)

Here $(\nabla\phi)^2_{max}$ is the maximum local value of $(\nabla\phi)^2$. One can show, via an hodograph transformation (see [8]), that this maximum is reached always at the boundary of the disk. Since the explicit value $|\nabla\phi|_{max}$ depends on the geometry and the Mach number, there is no close exact solution to this problem, i.e. it is impossible to know the exact value of the critical Mach number. One may only estimate M_c perturbatively as in Sect. 3.

Recently, G. Richardson [14] solved numerically the problem (5a,5b,5c). After the numerics, he found a critical Mach number around $M_c \approx 0.373$, i.e. close to the one given by perturbation series $M_c \approx 0.36969(7)$.

Critical velocities depend strongly on the geometry because $(\nabla\phi)^2_{max}$ in (6) does depend on it. It is instructive to repeat the argument in [1] which leads to a zero-th order approximation for the critical speed for the case of a disk. The idea originates by considering as a zero-th order approximation the solution for the incompressible flow ($M \to 0$ in (5a)) satisfying the boundary conditions (5b,5c),

---

[2] For normal gas of polytropic index γ one has that (5a) reads

$$\nabla \cdot \left(\left(\frac{1}{M^2} + \frac{1}{2}(1 - (\nabla\phi)^2) \right) \nabla\phi \right) = \frac{\gamma - 2}{2} \left((\nabla\phi)^2 - 1 \right) \nabla^2\phi.$$

NLSE corresponds to $\gamma = 2$.

that is

$$\phi_0 = \left(r + \frac{1}{r}\right)\cos\theta. \tag{7}$$

At zero order one has that the maximum speed arises at $\theta = \pm\pi/2$ and $|\nabla\phi|_{max} = 2$, putting this value into (6), one gets for the critical Mach number at zero order

$$M_c^{(0)} = \sqrt{\frac{2}{11}} = 0.426401, \tag{8}$$

our first estimate in [1].

The same argument could be repeated for different geometries, it gives $M_c^{(0)} = \sqrt{\frac{8}{23}} \approx 0.589768$ for a sphere in three spatial dimensions, whenever $|\nabla\phi|_{max} = \frac{3}{2}$. This dependence becomes very dramatic for very sharp objects, as the one of the figure.

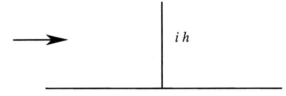

Fig. 2. Example of a flow on a sharp obstacle.

One may compute the incompressible flow using complex variables (here $z = x + iy$):

$$\phi_0 = \left(1 + \frac{z^2}{h^2}\right)^{1/2}.$$

The maximum speed diverges at the point $z = ih$; however there is a natural cut-off, ξ_0. The critical Mach number appears balancing only the first and third terms in (6): $M_c^{(0)} = \frac{2}{\sqrt{3}}\left(\frac{\xi_0}{h}\right)^{1/2}$. This dependence has been verified numerically [3].

3   Flow Around a Disk via a Janzen–Rayleigh Expansion[3]

As mentioned the problem (5a,5b,5c) deals with the compressible fluid passing around an obstacle. For the case of a disk, the effect of compressibility has been studied as a perturbation of the incompressible solution by Rayleigh [16] (the Janzen–Rayleigh expansion). He presented only the first order correction due the technical difficulties. Then, a similar expansion was extended up to order M^{16}

---

[3] Since my remark [5], it has appeared a paper by Berloff and Roberts [15] with similar expansions for a disk and a sphere in 2D and 3D.

for the case of compressible air. However the question about the convergence of the method remains (see the book of van Dyke [17]).

In this section, I realize a Janzen–Rayleigh expansion for low Mach number to compute order by order the full superfluid flow around a disk and then the critical speed is obtained.

The main difference with original method of Rayleigh is the use of the complex variable $z = x + iy$ and $\bar{z} = x - iy$, thus equation (5a) and the boundary condition (5b) become

$$\partial_{z\bar{z}}\phi = \frac{M^2}{4}\partial_z\left(\left(4|\partial_z\phi|^2 - 1\right)\partial_{\bar{z}}\phi\right) + cc., \tag{9a}$$

$$0 = z\partial_z\phi + \bar{z}\partial_{\bar{z}}\phi = 0 \quad \text{at} \quad z = e^{i\theta} \quad \& \quad \bar{z} = e^{-i\theta}. \tag{9b}$$

Equation (9a) is transformed into an integro-differential equation in the complex variables z and $\bar{z}$:

$$\phi(z,\bar{z}) = \phi_h(z) + \frac{M^2}{4}\int d\bar{z}\left(4|\partial_z\phi|^2 - 1\right)\partial_{\bar{z}}\phi + cc. \tag{10}$$

Here $\phi_h(z)$ is a boundary term (analytic on z), fixed by the boundary conditions (9b) on the disk.

Now we expand the velocity potential as

$$\phi = \phi_0 + M^2\phi_1 + M^4\phi_2 + \ldots$$

The first order ϕ_0 is the solution of the incompressible flow satisfying the boundary conditions, that is

$$\phi_0 = \frac{1}{2}\left(z + \frac{1}{z}\right) + \frac{1}{2}\left(\bar{z} + \frac{1}{\bar{z}}\right), \tag{11}$$

i.e. the same as (7) after taking $z = re^{i\theta}$. This zero order solution ensures the boundary condition at infinity (5c) for all the expansion, that is all higher order must be of lower degree than z, i.e. $\lim_{|z|\to\infty}\phi_n(z,\bar{z})/|z| \to 0$. At zero order one has that the maximum speed arises at $z = \pm i$ and $|\nabla\phi|_{max} = 2$. Replacing the later value into (6), one obtains the critical Mach number at zero order (8).

Now using $\phi = \phi_0 + M^2\phi_1$ in (10) one obtains the second order correction to the critical Mach number

$$M_c^{(1)} = \frac{\sqrt{\sqrt{233} - 11}}{2\sqrt{7}} \approx 0.390253,$$

and so on. After some iterations (which I have done with the help of *Mathematica*) one obtains the full velocity field potential ϕ as a series in M^2. In particular, for the maximum of the speed, the result is

$$
\begin{aligned}
|\nabla\phi|_{max} \approx 2 + \frac{7}{6}M^2 &+ \frac{44}{15}M^4 + \frac{1511639}{151200}M^6 \\
&+ \frac{5084105183}{127008000}M^8 + \frac{311688814107079}{1760330880000}M^{10} \\
&+ \frac{132895513753510095163}{158588208979200000}M^{12} + \ldots
\end{aligned} \tag{12}
$$

This expansion, together with equation (6), leads (after terms up to M^{50} and an improvement of the convergence, see [5] for details), the final value for the critical Mach number, $M_c \approx 0.36969(7)$.

Finally, the perturbation series (as proposed) cannot be convergent for any M. As Rayleigh suggested, it is reasonable to believe that the convergence breaks down at the transonic transition, because one cannot expand the solution for the velocity potential into an hyperbolic domain. On the other hand, the full solution possesses a multiple pole at $z = 0$. Even if this pole is inside the disk, it could be a source of problems as M increases.

A simpler task (and probably sufficient) is the study of the local convergence at the most singular point, that is the point at the maximum speed given by the series (12). The study realized in Ref. [5] suggests that the radius of convergence seems to be M_c, as expected, but the singularity seems to be non rational: $1/(M_*^2 - M^2)^{1.35}$. However, 25 terms in (12) are not sufficient and only a powerful method or faster computers could dimiss.

4   Unstable Solutions

As shown in [2,4] vortex nucleation happens via a saddle-node bifurcation[4], that is two different stationary (one stable and the other unstable) solutions collides. The stable solution was computed via the Janzen–Rayleigh series in the preceeding section. As argued in [2], unstable stationary solutions must contain a vortex because it cannot exist two potential (non-vortical) solutions for equations (5a,5b,5c). A proposed solution is one that haves two vortices in the up and down poles of the disk ($x = 0, y = \pm a$), as it is seen in Fig. 3.

One has up to zero order

$$\phi_0 = \frac{1}{2}\left(z + \frac{1}{z}\right)$$
$$+ i\frac{\Gamma}{2v_\infty R}\left[\ln(z - ia) - \ln\left(z - \frac{i}{a}\right) - \ln(z + ia) + \ln\left(z + \frac{i}{a}\right)\right] + cc. \quad (13)$$

Here Γ is the quantum of circulation ($\Gamma = 1$ in this units and $\hbar/m$ in real units), the distance a (a parameter) is fixed by imposing no motion of the vortex, *i.e.* one imposes $\frac{d}{dz}\phi(z = \pm ia) = 0$ after one removes the self-interaction. Up to zero order one has that the distance a diverges to infinity as $\frac{v_\infty R}{\Gamma}$ goes to zero, and is a monotonic decreasing function of a, that reaches asymptotically $a = 1$ as $\frac{v_\infty R}{\Gamma} \to \infty$. This dependence is slightly modified due to finite values of the Mach number and the intrinsic microscopic length ξ_0. The dimensionless parameter $\frac{\Gamma}{v_\infty R}$ can be written as the ratio of the two dimensionless parameters, in fact $\frac{\Gamma}{v_\infty R} \equiv \frac{1}{M} \times \frac{\xi_0}{R}$.

---

[4] The one dimension case has been studied by V. Hakim [18] who has shown (analytically) the existence of a stable and an unstable solution below a critical speed and a saddle-node bifurcation at threshold. The release of vortices is replaced by a periodic nucleation of one dimensional dark solitons.

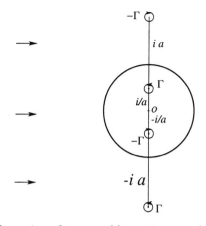

Fig. 3. Schematic configuration of an unstable stationary solution. The vortex circulations are $\pm\Gamma$, a is the distance from the upper vortex to the origin O and the images are located at $\pm i/a$.

One sees, that $\frac{\Gamma}{v_\infty R}$ goes to zero with $\frac{\xi_0}{R}$, therefore the zero-th order (13) is the same than the non-vortical solution (11). No difference is expected at all orders. Therefore one expects that as $\frac{\xi_0}{R} \to 0$ the stable and unstable branches collapse into a single one. This effect is clearly shown in figure 2 of Ref. [9], where the energy of both stable and unstable solutions are plotted as a function of the Mach number. Finally, near threshold one knows, after the cathastrophe theory, that the energy barrier is universal and behaves as $\Delta E \sim (M_c - M)^{3/2}$ (and not the one suggested by Packard in his lecture).

5   The Euler–Tricomi Equation near the Transonic Region

Near threshold for the transonic transition the series expansion in Mach number does not seem to be appropiate. We shall approximate (5a) assuming that at the dominant order the surface of the disk is flat. In this limit, the hydrodynamic boundary conditions are satisfied with an uniform tangent velocity field. Corrections have to be added to this velocity field in order to take into account the curvature of the disk boundary. The first correction is trivial, and only the second one is crucial for the analysis. This second order correction satisfies the Euler–Tricomi equation [8]:

$$- (\epsilon - y)\, \partial_{xx}\varphi + \partial_{yy}\varphi = 0 \tag{14}$$

with the boundary condition

$$\partial_y\varphi \sim -x^3; \quad \text{at} \quad y = 0. \tag{15}$$

I refer to Ref. [4] for the derivation of this equation and the boundary condition. Here, the point $(x = 0, y = 1)$ is a new origin of coordinates, the x-axis being tangential to the disk, and the y-axis perpendicular, $\epsilon \sim (M - M_c)$, and $|\epsilon| \ll 1$.

Because φ is a small correction we neglect nonlinear terms like $\partial_x\varphi\partial_{xx}\varphi$ besides $y\partial_{xx}\varphi$, and the boundary condition for φ does not require the curved boundary.

Euler–Tricomi is, in some sense, the normal form of the transonic transition, being φ its amplitude. This Euler–Tricomi equation may be interpreted as follows: $-(\epsilon - y)$ represents a generic tangential local velocity profile of an ideal compressible flow near a body, since the local main speed diminishes as y increases, that is as one moves far-away from the obstacle. The transonic transition arises at $y = \epsilon$.

The characteristics of (14) are sesqui-parabolas of the type

$$x = \pm\frac{2}{3}(\epsilon - y)^{3/2} + x_0$$

that are inside the obstacle ($y < 0$) for $\epsilon < 0$ and move out to the real space as $\epsilon > 0$.

Let us point out that the Euler–Tricomi possess multivalued solutions [8] that might give some idea of what arises when crossing the critical velocity. In fact, the roots $z(x, y)$ of the cubic polynomial (as well as any linear combination of these three roots):

$$z^3 + 3\left(y - \epsilon\right)z + 3x \tag{16}$$

are exact solutions of equation (14). Note that for $y > \epsilon$, the cubic polynomial has only one real root for all values of x, whereas, for $y < \epsilon$ there are three real roots inside a sesqui-parabola: $|x| \leq \frac{2}{3}(\epsilon - y)^{3/2}$. This multivaluedness of the real roots of the cubic equation (16) means that it is not possible, generally, to follow continuously a root of (16) along a closed path around the origin. More precisely, such solution will admit a discontinuity in the region inside the semi-cubic. Unless one can regularize the discontinuity which arises along the multivalued region, there is no hope of having a stationary solution of our problem[5]. Finally, one notes that the discontinuity should be regularized via the general time dependent nonlinear problem.

In Ref. [4], we have considered this nonlinear time dependent Euler–Tricomi equation including the quantum pressure term. The result is

$$-(\epsilon - y)\partial_{xx}\varphi + \partial_{yy}\varphi = \partial_{x^4}\varphi + \partial_{tx}\varphi + \partial_x\varphi\partial_{xx}\varphi + \ldots \tag{17}$$

which is a kind of Kadomtsev-Petviashvili equation. The Kadomtsev-Petviashvili equation possess exact solutions in 2D [19]; however, because of the inhomogeneous boundary condition one cannot use them here. We have followed a different approach using a perturbation expansion. A solvability condition leads to an amplitude equation describing a saddle-node bifurcation of the type $i\dot{Z} = -\mu + Z^2$, being Z the amplitude of φ and μ is related to ϵ. As a result, the critical velocity behaves as

$$M_c + C^{te}\sqrt{\frac{\xi_0}{R}}$$

---

[5] There is a singular exception if the discontinuity gap is 2π such a case is not generic because the phase jump depends explicitly on ϵ.

(here M_c is the $\xi_0 = 0$ limit, (4)). This behavior appears since quantum pressure stabilizes the flow above the transonic transition.

This saddle-node bifurcation gives rise to a time-dependent dynamics corresponding to vortex emission; nevertheless, for that it remains to match this outer velocity potential, solution of the Euler–Tricomi equation with an inner solution, close to the disk boundary with a vortex. To do that, one expects to relate the amplitude of φ, $i.e.$ Z, to an order parameter which parametrizes a continuous family of solutions of the full nonlinear Schödinger equation (1). Jones and Roberts [20] found the kind of solution that we are interested in, consisting in axisymmetric solitary structures, this solutions match to solution of the Kadomtsev-Petviashvili equation as $M \to M_c$.

Concluding, the author acknowledges Yves Pomeau who outlined with deep originality our collaboration over the years. He also thanks T. Frisch and C. Josserand for their participation and for several original contributions, and to M.E. Brachet and V. Hakim for constant interest and enlightened discussions.

References

1. T. Frisch, Y. Pomeau and S. Rica, Phys. Rev. Lett., **69**, 1644 (1992)
2. Y. Pomeau and S. Rica, Comptes Rendus Acad. Sc. (Paris), **t. 316** Série II, 1523 (1993)
3. S. Rica and Y. Pomeau, "Critical Velocities and Nucleation of Vortices in a Model of Superflow", in Instabilities and Nonequilibrium Structures IV, eds. E. Tirapegui & W. Zeller, Kluwer (1993)
4. C. Josserand, Y. Pomeau and S. Rica, Physica **D 134**, 111 (1999)
5. S. Rica, Physica **D 148**, 221 (2001)
6. V.L. Ginzburg and L.P. Pitaevskii, Sov. Phys. JETP **7**, 858 (1958); L.P. Pitaevskii, Sov. Phys. JETP, **13**, 451 (1961); E.P. Gross, J. Math. Phys. **4**, 195 (1963)
7. Y. Pomeau and S. Rica, Phys. Rev. Lett. **71**, 247 (1993), *Ibid.* **72**, 2426 (1994)
8. L.D. Landau and E.M. Lifshitz, Fluid Mechanics, Pergamon Press (Oxford 1987)
9. C. Huepe et M.E. Brachet, Comptes Rendus Acad. Sc. (Paris), **t. 325** Série II, 195 (1997); Physica **D 140**, 126 (2000)
10. M.H. Anderson, *et al.* Science **269**, 198 (1995); C.C. Bradley, *et al.* Phys. Rev. Lett. **75**, 1687 (1995); K.B. Davis, *et al.* Phys. Rev. Lett. **75**, 3969 (1995)
11. M.R. Andrews, *et al.* Phys. Rev. Lett. **79**, 553 (1997); Phys. Rev. Lett. **80**, 2697 (1997)
12. C. Raman, *et al.*, Phys. Rev. Lett. **83**, 2502 (1999)
13. B. Jackson, J.F. McCann and C.S. Adams, Phys. Rev. **A61**, 051603(R) (2000)
14. G. Richardson, "Vortex motion in type-II superconductor", D. Phil. Thesis Oxford University, Oxford (1995)
15. N.G. Berloff and P.H. Roberts, J. Phys. A: Math. Gen. **33**, 4025 (2000)
16. L. Rayleigh, Phil. Mag., **32**, 1 (1916)
17. M. van Dyke, "Perturbation Methods in Fluid Mechanics", the Parabolic Press (Stanford 1975), pages 4, 15 and 215
18. V. Hakim, Phys. Rev. E **55**, 2835, (1997)
19. S.V. Manakov *et al.*, Phys. Lett. **63A**, 205 (1977)
20. C.A. Jones and P.H. Roberts, J. Phys. A: Math. Gen. **15**, 2599 (1982)

Vortices in Nonlocal Condensate Models of Superfluid Helium

Natalia G. Berloff and Paul H. Roberts

Department of Mathematics
University of California, Los Angeles, CA, 90095-1555
nberloff@math.ucla.edu, roberts@math.ucla.edu,

Abstract. Nonlocal nonlinear Schrödinger equations are considered as models of superfluid helium. The models contain a nonlocal interaction potential that leads to a phonon-roton-like dispersion relation. It is shown that for any such potential the generalized Gross-Pitaevskii (GP) model has non-physical features, specifically the development of catastrophic singularities and unphysical mass concentrations. The GP equation is remedied by introducing a higher order term in the local density approximation for the correlation energy. The resulting theory is applied in two ways. The family of superfluid vortex rings is derived. The nucleation of vortex rings by a moving ion is considered.

1 Introduction

Superfluid helium at $0°K$ has a large interatomic spacing and is often described in terms of a weakly interacting Bose gas. The imperfect Bose condensate in the Hartree approximation is governed by equations that were derived by Gross and by Ginsburg and Pitaevskii. In terms of the single-particle wavefunction $\psi(\mathbf{x}, t)$ for N bosons of mass M, the time-dependent self-consistent field equation is

$$i\hbar\psi_t = -\frac{\hbar^2}{2M}\nabla^2\psi + \psi\int|\psi(\mathbf{x}', t)|^2 V(|\mathbf{x} - \mathbf{x}'|)\, d\mathbf{x}', \qquad (1)$$

where $V(|\mathbf{x} - \mathbf{x}'|)$ is the potential of the two-body interactions between bosons. The normalization condition is $\int|\psi|^2\, d\mathbf{x} = N$.

The internal energy per unit volume, $\mathcal{E}$, at point $\mathbf{x}$ and time t is given by

$$\mathcal{E}(\rho) = \frac{\hbar^2}{8M^2\rho}(\nabla\rho)^2 + \frac{1}{2M^2}\int\rho(\mathbf{x})V(|\mathbf{x} - \mathbf{x}'|)\rho(\mathbf{x}')\, d\mathbf{x}', \qquad (2)$$

and the total energy, W, is

$$W = \int\mathcal{E}(\rho)\, d\mathbf{x} = \int\frac{\hbar^2}{8M^2\rho}(\nabla\rho)^2\, d\mathbf{x} + W_c(\rho). \qquad (3)$$

The first term on the right-hand side of (3) describes the quantum kinetic energy of a Bose gas of nonuniform density; $W_c(\rho)$ is a potential or correlation energy that incorporates the effect of interactions.

For a weakly interacting Bose system (1) is simplified by replacing $V(|\mathbf{x}-\mathbf{x}'|)$ with a δ - function repulsive potential of strength $V_0 = \int V \, d\mathbf{x}'$, which leads to the local GP model. Several aspects of the local GP model are qualitatively or quantitatively unrealistic for superfluid helium. The dispersion relation between the frequency, ω, and wave number, k, of sound waves according to the local GP model is

$$\omega^2 = c^2 k^2 + \left(\frac{\hbar}{2M}\right)^2 k^4, \tag{4}$$

where $c = (V_0 \rho_\infty)^{\frac{1}{2}}/M$, $\rho_\infty = E_v/V_0$. This shows that the velocity, c, of long wavelength sound waves is proportional to $\rho^{\frac{1}{2}}$ (here we have replaced the bulk density, ρ_∞, by ρ). That this is unrealistic is seen from the experiments on Grüneisen constant $U_G = (\rho \partial c/c \partial \rho)_T \approx 2.8$ at the vapor pressure [1]. Also, the dispersion curve (4) has no roton minimum. At best, (4) describes the phonon branch of the excitation spectrum.

There is significant interest attached to the question of whether the introduction of a realistic two-particle interaction potential, V, that leads to a phonon-roton-like spectra in the GP model, gives a better description of the properties of superfluid helium than the local model [2]. The minimum requirements on such a potential would be the correct position of the roton minimum and the correct bulk normalization (see below).

2   Applicability of the Generalized Gross–Pitaevskii Model

We transform (1) by introducing the average energy level E_v, so that $\psi = \Psi \exp(-iE_v t/\hbar)$, and rescale it by

$$\mathbf{x} \to \frac{\hbar}{(2ME_v)^{1/2}}\mathbf{x}, \qquad t \to \frac{\hbar}{2E_v}t. \tag{5}$$

The dimensionless form of (1) becomes

$$-2i\frac{\partial \Psi}{\partial t} = \nabla^2 \Psi + \Psi\left(1 - \int |\Psi(\mathbf{x}',t)|^2 V(|\mathbf{x}-\mathbf{x}'|)\,d\mathbf{x}'\right), \tag{6}$$

with the bulk normalization condition $\int V(|\mathbf{x}'|)\,d\mathbf{x}' = 1$. We get the dispersion curve by linearizing about the uniform state. We write $\Psi = 1 + \Psi'$ and consider plane waves of the form $\mathrm{Re}(\Psi') = \exp i(\omega t - kx)$. Then the dispersion relation can be written as

$$\omega^2 = \frac{1}{4}k^4 + 2k\pi \int_0^\infty \sin(kr)V(r)r\,dr. \tag{7}$$

We require $\omega'(k_{\mathrm{rot}}) = 0$, $\omega(k_{\mathrm{rot}}) = \omega_{\mathrm{rot}}$, where $(k_{\mathrm{rot}}, \omega_{\mathrm{rot}})$ is the position of the roton minimum on the $k\omega-$ dispersion curve, which in dimensional units is found from experiments [3] to be $k_{\mathrm{rot}} = 1.926\mathring{A}^{-1}$, and $\omega = 8.62K^\circ k_B/\hbar$. By taking the limit of (7) for $k \to 0$ and using the normalization condition, the sound speed is found to be $1/\sqrt{2}$ as in the local model. By relating this to the known value of the sound speed at low k in He II, 238 m/s, we find that the healing length of the model (6) is fixed as $[L] = 0.47\text{Å}$, and therefore $k_{\text{rot}} = 0.907$ and $\omega_{\text{rot}} = 0.158$.

After V has been selected to give a good account of the roton minimum it is typically found that, after the normalization condition has been enforced, the convolution

$$\frac{\delta P}{\delta \rho} = \frac{1}{2M^2} \int \rho(\mathbf{x}')V(|\mathbf{x} - \mathbf{x}'|)\, d\mathbf{x}' \tag{8}$$

is not necessary positive. This convolution is the variational derivative of the mechanical pressure, P, of the hydrodynamic formulation of (1) in the semi-classical limit ($\hbar \to 0$). When the integral on the right-hand side of (8) is negative the pressure P decreases when the mass density increases, which is unphysical.

It can be shown [4] that a virial theorem, similar to the one used in establishing the catastrophic singularities of the focusing nonlinear Schrödinger equation, indicates that the solutions of the nonlocal model (1) blow up in a bounded domain.

Finally, to illustrate the development of mass concentrations we solved (6) numerically for the flow around a positive ion moving with the dimensionless velocity $U = \frac{1}{2}U_L$, where U_L is the Landau critical velocity. The ion is modeled as the infinite potential barrier so that $\Psi = 0$ on $r = b$, where b is the radius of the positive ion. The interaction potential was used in the form suggested by Jones [5]

$$V(r) = (\alpha + \beta A^2 r^2 + \gamma A^4 r^4) \exp(-A^2 r^2), \tag{9}$$

where the parameters α, β, γ, and A are chosen to give agreement with the experimentally determined dispersion curve as discussed above. Notice that the local flow velocity is below U_L everywhere. Nevertheless, persistent mass concentrations develop along the axis of symmetry; see Fig. 1.

Such an unphysical behavior indicates that assumptions made in the derivation of the equation must be unjustified. This difficulty could be overcome by

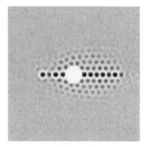

Fig. 1. The density plot in a cross-section of the solution of (6) for the flow around a sphere of radius $10a$ moving to the right with velocity $0.5v_L$. After [4].

introducing into the weakly nonlinear theory dissipation or higher order nonlinear terms. But we would like to preserve the Hamiltonian character of the GP model. Instead, we will take the density - functional approach [6], which tries to introduce an accurate microscopic picture of liquid helium.

3   Nonlocal Nonlinear Schrödinger Equation

The correlation energy of the Skyrme interactions in nuclei [7] is given by

$$W_c(\rho) = \frac{1}{M^2} \int \left[\frac{W_0}{2}\rho^2 + \frac{W_1}{2+\gamma}\rho^{2+\gamma} + W_2(\nabla\rho)^2 \right] d\mathbf{x}, \qquad (10)$$

where W_0, W_1, W_2 and γ are phenomenological constants. The first two terms give a local density approximation, and the gradient term corresponds to finite range interactions. In a somewhat similar way to [6], we add the necessary nonlocality of interactions directly into the first term of (10) by introducing a two-body interaction potential, $V(|\mathbf{x} - \mathbf{x}'|)$, so that (10) becomes

$$W_c(\rho) = \frac{1}{M^2} \int \left[\frac{1}{2} \int \rho(\mathbf{x})V(|\mathbf{x} - \mathbf{x}'|)\rho(\mathbf{x}') \, d\mathbf{x}' + \frac{W_1}{2+\gamma}\rho^{2+\gamma} \right] d\mathbf{x}. \qquad (11)$$

This incorporates and generalizes the W_2 interaction term in (10), which has therefore been abandoned.

On adopting (11), we find that the nonlinear Schrödinger equation replacing (1) is

$$i\hbar\psi_t = -\frac{\hbar^2}{2M}\nabla^2\psi + \psi \int |\psi(\mathbf{x}',t)|^2 V(|\mathbf{x} - \mathbf{x}'|) \, d\mathbf{x}' + W_1\psi|\psi|^{2(1+\gamma)}, \qquad (12)$$

and equation (6) is replaced by

$$-2i\frac{\partial\Psi}{\partial t} = \nabla^2\Psi + \Psi\left[1 - \int |\Psi(\mathbf{x}',t)|^2 V(|\mathbf{x} - \mathbf{x}'|) \, d\mathbf{x}' - \chi|\Psi|^{2(1+\gamma)}\right]. \qquad (13)$$

The bulk normalization condition becomes $\int V(|\mathbf{x}'|) \, d\mathbf{x}' = 1 - \chi$. The dispersion relation of (13) is modified in comparison with (4) by adding the term $\frac{1}{2}(1+\gamma)\chi k^2$ to its right-hand side. The bulk normalization condition gives the slope at the origin (the dimensionless speed of sound) as $\sqrt{(1 + \gamma\chi)/2}$ and the unit of length (healing length) as $[L] = 0.47\sqrt{1 + \gamma\chi}$ Å. A fit to the Landau dispersion curve can be obtained as in §2.

There are two logical choices of the parameter γ. First, we can view the term $W_1\rho^{2+\gamma}$ in (10) as the second term in the nonlinear expansion of the correlation energy in powers of ρ, and that yields $\gamma = 1$. The second possible choice is to take $\gamma = 2.8$, which gives $c \propto \rho^{2.8}$ in agreement with the experimentally determined Grüneisen constant $U_G \approx 2.8$.

Next, we shall use (13) to study the family of the vortex rings [8]. For a vortex ring of large radius R the results for the straight-line vortex can be used to give [9] the energy and momentum of such ring as

$$\mathcal{E} = \frac{1}{2}\kappa^2 \rho_\infty R \left[\ln\left(\frac{8R}{L}\right) - 2 + c \right],$$

and

$$p = \kappa \rho_\infty \pi R^2.$$

After differentiating $\mathcal{E}$ and p with respect to R and substituting into the Hamilton's equation $v = \partial \mathcal{E}/\partial p$ we get the expression for the velocity of the large vortex ring as

$$v = \frac{\kappa}{4\pi R} \left[\ln\left(\frac{8R}{L}\right) - 1 + c \right].$$

Glaberson and Donnelly [10] used the experimental results of Rayfield and Reif [11] on the relation between the energy and velocity of large vortex rings to estimate the vortex core parameter L. These estimates were based on the hollow core vortex model with $c = 0$ and produced $L \approx 0.81$Å. Jones [5] did similar calculations for the nonlocal model (6) and found $c = -0.13$, so that $L \approx 0.71$Å for the optimal choice of the parameter A. For the local GP model with $c = 0.381$ the vortex core parameter is $L \approx 1.19$Å. These values of L are much larger that the healing length found from the sound speed, which is 0.47 Å for any of the above models. Jones [5] posed the question of whether a self-consistent theory is possible, i.e., one where the vortex core parameter and the healing length are brought into harmony. The answer is "Yes." Our model (13) with $V(r) = V(r) = (\alpha + \beta A^2 r^2 + \gamma A^4 r^4) \exp(-A^2 r^2) + \delta \exp(-B^2 r^2)$ is able to bring about agreement. For $\gamma = 1$, $\chi = 3.5$, $A = 1.6$, $B = 1$, and $\delta = 1$ we numerically integrated (13) to find $c = 0.1825$, so that $L \approx 1$Å, which is the healing length of our model. This gives the energy of a vortex ring traveling at 27 cm/sec as 10 ev, which agrees with the experiments of Rayfield and Reif [11]. This choice of parameters is not very practical and, in the calculations below, we shall use instead the interaction potential (9) with $\chi = 0.2$, $\gamma = 1$, and $A = 0.9$, which also represent the roton minimum satisfactorily.

A sequence of vortex rings of small radius has been derived numerically [8]. When the velocity of the vortex ring reaches the Landau critical velocity the ring becomes unstable and evanesces into sound waves. For any ring traveling with speed greater than the Landau critical velocity, the amplitude of the far-field solution will not decay exponentially at infinity, which makes the existence of such a ring impossible. Figure 2 plots the sequence of the vortex rings on the $p\mathcal{E}$ plane together with the dispersion curve of (13). Note that, as parameters χ, γ, and A are varied, the position of a corresponding family of vortex rings on $p\mathcal{E}$-plane changes dramatically relative to dispersion curve. Actually it is even possible to bring the termination point of this family close to the roton minimum.

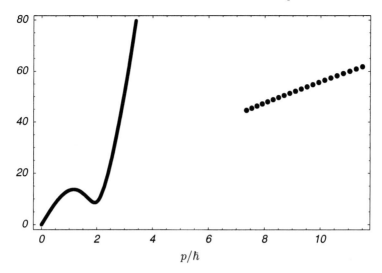

Fig. 2. The dispersion relation $p - \mathcal{E}$ and a family of the vortex rings as solutions of (13). Each dot represents the position of a vortex ring on $p\mathcal{E}$-plane.

4   Vortex Nucleation and Roton Emission

In this section we shall use the model (13) to elucidate vortex nucleation from, and roton emission by, moving ions [12]. The picture of nucleation that emerges as a result of experiments performed over the years by the McClintock group in Lancaster University (see, for instance [13]) shows that vortex nucleation and roton emission are independent processes, and that the latter is linked to v_L but the former is not. We numerically integrated (13) for the axisymmetric flow around the positive ion of radius $b = 10$ moving uniformly with velocity U. Our numerical calculations (for the details of the numerics see [4]) indicate that, provided U does not exceed the dimensionless Landau critical velocity U_L, the ion experiences no drag and the flow is steady in the frame of reference moving with the ion. Notice that the velocity on the equator of the ion may exceed U_L (for incompressible flow the velocity on the equator would be $3U/2$ but is even larger when compressibility is allowed for), but this leads neither to vortex nucleation nor roton emission.

When $U > U_L$ a modulated wave envelope is formed involving wave numbers from the neighborhood of the Landau point, where $\omega'(k_L) = \omega(k_L)/k_L$, $\omega(k) = \bar{\omega}(k) - \mathbf{U} \cdot \mathbf{k}$, and $\bar{\omega}(k)$ refers to stagnant helium. These waves radiate energy to infinity, resulting in drag on the ion. We have not so far been able to observe vortex nucleation for $U > U_L$, but we obtained insight into vortex nucleation with the help of an artificial example which tended to confirm the hypothesis [13] that roton emission and vortex nucleation are different processes.

Our artificial example is motivated by the fact that the critical velocity v_c for vortex nucleation by an electron bubble is (according to the local GP model) reduced by its shape which, when moving, is oblate [14]. The presence of ^{3}He would enhance this effect through the concomitant reduction in surface tension. We can make v_c even smaller by artificially increasing the flattening, to such an extent that v_c becomes less than v_L, so that nucleation can be studied with the model (13) without the complications of roton emission. We therefore consider an ion with an oblate spheroidal surface moving in the direction of its short (symmetry) axis with a velocity less than v_L. The ratio of lengths of axes is 5. Nucleation of vortices occurs when $v = v_c \approx 0.148 \pm 0.007c$ (when the speed of sound is reached on the equator); see Fig. 3. To compare this with the corresponding result for the Bose condensate, we performed similar calculations using the local GP model. The critical velocity of nucleation in this case was found to be $0.205 \pm 0.007c$. Such a significant drop ($\sim 30\%$) in the critical velocities between local and nonlocal model can be partially explained by the greater compressibility of the fluid, according to the nonlocal model.

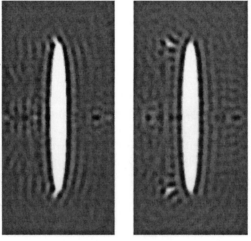

Fig. 3. The density plot in a cross-section of the solution of (13) for the flow around an oblate spheroid (see text) moving to the right with velocity $0.156c$ at $t = 100$ (left) and $t = 300$ (right). The white circles show the core of a vortex ring nucleated from the spheroid and gradually falling astern of it. After [12].

5   Conclusions

In summary, we considered a nonlocal nonlinear Schrödinger equation (13) as a model of superfluidity. The model has a finite range interaction potential that leads to a dispersion curve with a roton minimum and can accommodate a more realistic relationship between the speed of sound, the density and the pressure. The parameters of the model can be chosen to bring the healing length into agreement with the vortex core parameter. According to our model, there is no drag on a positive ion moving with $v < v_L$. As the velocity of the ion exceeds the Landau critical velocity v_L, it starts to experience drag and it creates modulated waves with wave numbers corresponding to the roton minimum. Our model failed to describe vortex nucleation in such circumstances. Nevertheless it could, through an artificial example, provide strong indications that roton emission and vortex nucleation are different processes, the former being connected to the Landau critical velocity, and the latter to the speed of sound.

This research was supported by the NSF grant DMS-9803480.

References

1. J. B. Brooks, R. J. Donnelly: J. Phys. Chem. Ref. Data, **6**, 51 (1977)
2. N.N. Bogolyubov: *J. Phys. USSR* **11** 23 (1947); E. P. Gross: J. Math. Phys., **4** 195 (1963); R. N. Hills and P.H. Roberts: Q. J. Mech. Appl. Math. **40** 279 (1987); S. Putterman and P. H. Roberts: in *Macroscopic Theories of Superfluids* (Cambridge University Press) 4 (1991); Y. Pomeau and S. Rica: Phys. Rev. Lett. **71**, 247 (1993)
3. R. J. Donnelly, J. A. Donnelly, R. N. Hills: J. Low Temp. Phys. **44**, 471 (1981)
4. N.G. Berloff: J. Low Temp. Phys. **116** 359 (1999)
5. C. A. Jones: private communications (1993)
6. F. Dalfovo, A. Lastri, L. Pricaupenko, S. Stringari, and J. Treiner: Physical Review B **52**, 1193 (1995) ; J. Dupont-Roc, M. Himbert, N. Pavlov, and J. Treiner: J. Low Temp. Phys. **81**, 31 (1990)
7. D. Vautherin, D. M. Brink: Phys. Rev. C **5**, 626 (1972)
8. N. G. Berloff, P. H. Roberts: J. Phys. A: Math. Gen. **32** 5611 (1999)
9. P.H. Roberts, J. Grant: J. Phys. A: Gen. Phys. **4**, 55 (1971); C.A. Jones, P.H. Roberts: J. Phys. A: Gen. Phys. **15**, 2599 (1982)
10. W. I. Glaberson, R. J. Donnelly: in *Progress in Low Temperature Physics IX* ed D.F. Brewer, North Holland, Amsterdam (1986)
11. G. W. Rayfield and F. Reif: Phys. Rev. A **136** 1194 (1964)
12. N. G. Berloff, P. H. Roberts: Phys. Letts. A, **274**, 69 (2000)
13. P.C. Hendry, N. S. Lawson, P.V.E. McClintock, C.D.H. Williams, and R.M. Bowley: Phys. Rev. Lett. **60** 604 (1988).
14. N. G. Berloff, P. H. Roberts: J. Phys. A: Math. Gen. (2000) submitted

Ginzburg–Landau Description of Vortex Nucleation in a Rotating Superfluid

Igor Aranson[1] and Victor Steinberg[2]

[1] Argonne National Laboratory, 9700 South Cass Avenue, Argonne, Illinois 60439
[2] Department of Physics of Complex Systems, Weizmann Institute of Science, Rehovot, 76100, Israel

Abstract. Nucleation of vortices in rotating superfluid by spin-up and rapid thermal quench is discussed in the framework of the time-dependent Ginzburg–Landau equation (TDGLE). An analysis of the instability in inhomogeneous rotationally-invariant system results in the expression for the critical rotational velocity. A stability analysis of multicharged vortices is presented. It is shown that they are very long-living objects with lifetime inversely proportional to the dissipation rate. It was found by numerical and analytical solution of the TDGLE that vortex nucleation by rapid thermal quench in the presence of superflow is dominated by a transverse instability of the moving normal-superfluid interface.

1 Introduction

There are two major ways to generate vorticity in rotating superfluid: to spin-up a bucket with superfluid helium from initially steady state without any vortices to a state with a rotation velocity $\Omega \geq \Omega_c$, where Ω_c is the critical rotation speed for a vortex nucleation, and to quench it thermally from temperature, T, above the superfluid transition temperature, T_λ, to temperature below it. These two different scenaria can be described by the same model, namely time-dependent Ginzburg–Landau equation (TDGLE) for the superfluid order parameter corrected for presence of normal component. This equation for superfluid helium was first suggested by L. Pitaevskii[1,2].

Further we review our results based on the stability analysis of the TDGLE for the rotating superfluid. The vortex nucleation in a rotating superfluid was a rather hot topic during 60th and 70th. Remnant vorticity did not permit to define correctly the critical rotational velocity for vortex nucleation at that time. We believe that next generation of experiments will overcome this problem, e.g., in the way suggested about 25 years ago[3]. Together with new experimental methods of vortex detection suggested recently[4,5], it can provide experimental test of the theory.

Formation of vortices under a rapid quench is recognized as a fundamental problem of contemporary physics [6]. Superfluid ^{3}He offers a unique "testing ground" for rapid phase transitions [7]. Recent experiments where a rotating superfluid ^{3}He was locally heated well above the critical temperature by absorption of neutrons [8] revealed vortex formation under a rapid second–order phase transition. We will discuss the dynamics of vortex nucleation under a rapid quench in the framework of the TDGLE.

2   Spin-Up and Nucleation of Vortices in Superfluid Helium

Let us consider a cell containing a superfluid helium rather close to the superfluid transition temperature T_λ. When the cell is rotated with an angular velocity Ω, the normal component is involved into a solid body rotation with $\mathbf{V_n} = \Omega \times \mathbf{r}$. The superfluid component cannot participate in the uniform rotation up to the point where the potential flow condition is satisfied, i.e. $\nabla \times \mathbf{V_s} = 0$.

A conventional approach to the spin-up problem is to describe it by a two-fluid hydrodynamic model corrected by an equation for the vortex line dynamics [10,11]. The vortex lines interact with the normal component that leads to mutual friction[11]. It is evident that this hydrodynamic description does not catch the key point of the spin-up problem, namely, the vortex nucleation which actually causes the spin-up of the superfluid component, the primary superfluid relaxation mechanism toward a steady rotation.

Another approach to the spin-up problem, which can describe both the dynamics and the nucleation of quantized vortices in a superfluid helium, is to use TDGLE together with a two-fluid hydrodynamic model[2]. Outside the vortex core which is normal, one gets the superfluid velocity circulation around a single quantum vortex $\kappa = 2\pi\hbar/m$ [11]. Corresponding set of equations, which describe the dynamics of the complex order parameter of the superfluid condensate, $\Psi = |\Psi| \exp(i\chi)$, in a rigid steady rotation of the normal component, looks in the scaled variables as follows [2](see for details [12]):

$$\partial_t \Psi = -\frac{i}{2}\left(\Delta\Psi + \Psi - |\Psi|^2\Psi\right) + \frac{\Lambda}{2}\left((\nabla - i\mathbf{V_n})^2\Psi + \Psi - |\Psi|^2\Psi\right) \qquad (1)$$

where $V_n = \Omega r$, $\mathbf{V_s} = \nabla\chi$, and Λ is a temperature-dependent parameter.

Equation (1) is reminiscent of that of the Ginzburg–Landau equation for superconductors in the London limit [13]. The role of an external magnetic field is played by the angular velocity, and of the corresponding vector-potential by the velocity of the normal component. Then by analogy one expects that at $\Omega \le \Omega_{c1}$, there exists a motionless superfluid component with no vortices. At $\Omega > \Omega_{c1}$ vortices will be nucleated and penetrate into the fluid producing a vortex lattice in the interior of a helium container. As follows from the experiments on the superfluid ^{4} He, Ω_{c1} is too low to be detected [11]. On the other hand, Ω_{c2} (which is analogous to H_{c2} in superconductors and at which superfluidity will be completely destroyed in the sample) is too high to be reached experimentally.

Equation (1) describes the spin-up of the superfluid part in a rigidly rotating flow of the normal component. This equation is asymptotically correct in the vicinity of the λ point. Moreover, one can speculate that the equation qualitatively describes some aspects of spin-up for ^4He near zero temperature with $\Lambda \to 0$. Therefore, we will consider Eq. (1) for all temperatures with $\Lambda \to 0$ for $T \to 0$ and $\Lambda \sim (T_\lambda - T)^{-1/3}$ for $T \to T_\lambda$.

The critical rotational velocity, Ω_c, for the onset of the vortex nucleation for $T \to T_\lambda$ can be found from linear stability analysis of a stationary solution. Then for this solution $\chi = 0$, and $F = |\Psi|$ is defined by the following equation:

$$\partial_r^2 F + \frac{\partial_r F}{r} + F - F^3 - \Omega^2 r^2 F = 0. \tag{2}$$

Equation (2) has to be complimented by the conditions at the $r = 0$ and the condition at the outer wall $r = R$, where R is the radius of the container. As a boundary condition at the wall we take a condition of finite suppression of the superfluid density by the wall, i.e., $\partial_r \Psi + \gamma \Psi = 0$ for $r = R$ where γ characterizes the suppression of the order parameter. For $\gamma \to 0$ we have no-flux boundary condition ($\partial_r \Psi = 0$). Solution of Eq. (2) for arbitrary Ω and R is accessible only numerically. Selected results are presented in Fig.1 of Ref[12].

The stationary solution is stable for $\Omega < \Omega_c$ and looses its stability above the critical angular velocity Ω_c. Instability of the stationary solution leads to nucleation of vortices and the corresponding spin-up of superfluid. By substitution of a solution of the form $\Psi = F(r) + W(r, \theta, t)$, where W is a small generic perturbation, one obtains a linear equation for W. Ω_c is found from the existence condition of the first nontrivial eigenmode satisfying the boundary conditions [12]. In the limit of $R \gg 1$ the solution is found by matching of the bulk solution with the solution near the wall. The analysis shows that the most unstable eigenmodes are localized in the narrow layer of the width r_b near the container wall. We obtained $r_b \sim \sqrt{R} \ll R$ for large R. The value of most unstable azimuthal number n and the critical frequency Ω_c for the container radius R is given by the expressions:

$$n = Q(\gamma) R^{3/4}, \quad \Omega_c = \frac{1}{R} \sqrt{\frac{1}{3} + \frac{\Delta(\gamma)}{R^{1/2}}}. \tag{3}$$

The dimensionless parameters $Q(\gamma), \Delta(\gamma)$ are the functions of the suppression rate γ, are obtained by the matching of outer and inner expansions, and are shown in Fig. 2 of Ref [12]. One can make estimates of Ω_c based on Eq.(3). Indeed, in dimensional variables one gets $\Omega_c = (\hbar/mR_d)[\epsilon^{2/3}/(\sqrt{3}\xi_0)] = 3.3 \times 10^3 \epsilon^{2/3}/R_d$ 1/sec, where $\epsilon = (T_\lambda - T)/T_\lambda$, $\xi_0 = 2.74 \times 10^{-8}$ cm is the correlation length far from T_λ, and the power 2/3 is introduced to assure correct scaling of the superfluid density near the λ point [14]. The conventional Feynman equation [11] has different scaling with R and no temperature dependence. On the other hand, as we discussed in Ref.[12], the temperature dependence of Ω_c found is the same as in the theory of thermal nucleation of quantized rings [11]. However, the nucleation rate in our case was calculated up to prefactor (see Ref.[12]), while in the former theory it was obtained from heuristic arguments. Moreover, we consider non-uniform distribution of ρ_s due to V_n that is impossible task for the equilibrium theory. The latter can become significant at small values of ϵ and R. Thus, at $\epsilon = 10^{-6}$ and $R_d = 0.1cm$ scaled R is of the order 10^3, and the correction to the frequency from Eq.(3) can be of the order of several percent.

We performed numerical simulations of Eq.(1), which details are presented in Ref[12]. We observed nucleation and consequent tearing off of the vortices for $\Omega \geq \Omega_c$ irrespective of Λ. However, the character of the nucleation and asymptotic states depends on Λ. For $\Lambda \to \infty$ and slightly above Ω_c we observed

nucleation of several vortices. Nucleation occurs at nonlinear stage of the instability when a set of single zeros (four zeros for $R = 65$) is torn off at the radius
R. These zeros are the seeds for the vortex cores. The vortices propagate into
the interior of the container and finally form a perfect vortex lattice, reminiscent
of that of the Abrikosov lattice [13] (see Fig. 1). Further increase of Ω results in
formation of additional vortices.

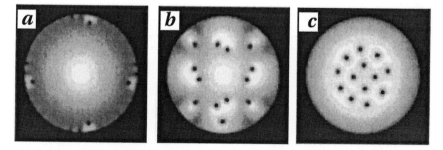

Fig. 1. Grey-coded images of $|\Psi|$ demonstrating nucleation of vortices and creation
of the vortex lattice. The dark shade corresponds to zero of $|\Psi|$; the white one to its
maximum value. Vortices are seen as black dots. The parameters are: $\Omega = 0.01, R =$
$65, \Lambda \gg 1$. The initial condition is $\Psi = 1$ plus small amplitude broad-band noise. a)
$t = 100\Lambda$, b) $t = 200\Lambda$, c) $t = 1100\Lambda$.

3   Stability of Multicharged Vortices

As we pointed out above, Eq.(1) can describe qualitatively the spin-up of superfluid helium also at $T \to 0$ with $\Lambda \to 0$. Then the TDGLE becomes the
nonlinear Schödinger equation (NLSE). Numerical simulations show the character of the vortex nucleation is drastically different for $\Lambda \to 0$. Typically whole
clusters of vortices are torn off. These clusters can be considered as a perturbed
multicharged vortex. The multicharged vortex is unstable and breaks down into
single charged vortices. However, the lifetime happens to be proportional to Λ
and diverges for $\Lambda \to 0$. The multicharged vortices with the topological charge
$\pm n$ are known to have higher energy and decay into n single-charged, or elementary vortices. Since in the NLSE with small dissipation the decay time of the
multicharged vortices can be arbitrarily large, one can expect to detect them
in experiments at very low T. Some indirect indications of the multicharged
vortices can be found in several experiments on vortex nucleation [11].

We consider the perturbative solution $\Psi = [F(r) + \eta(x, y, t)] \exp[in\theta]$ of the
NLSE with dissipation[16]

$$\partial_t \Psi = (\varepsilon + i) \left(\Delta\Psi + \Psi - |\Psi|^2 \Psi \right), \tag{4}$$

Here η is the complex function, and $\varepsilon \ll 1$ is the phenomenological parameter
which describes the bulk dissipation of superflow towards the condensate. We

assume that the only channel for the bulk dissipation at $T \to 0$ is the absorption of acoustic excitations of superflow by the normal component which presents, e.g., due to normal ^{3}He atoms. This assumption is based on consideration that the excess energy of n-charged vortex can decay at $T \to 0$ only by acoustic radiation of the bounding energy of n single charges. In the presence of the energy conservation and other integrals of motion in the NLSE this transformation of the bounding energy into acoustic field is a very slow process, i.e. multiple vortex can be long-lived.

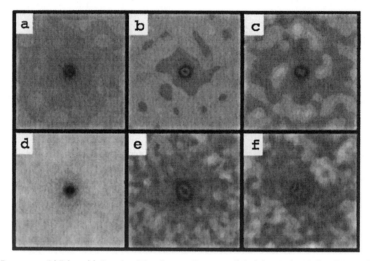

Fig. 2. Images of $|\Psi(x,y)|$ for double-charged vortex(a)-(c); and triple-charged (d)-(f) at the moments of time : (a) t=1700; (b) t=2000; (c) t=2500;(d) t=1700; (e) t=2000; and (f) t=2500. The domain size 100×100 units, number of FFT harmonics 128×128; $\varepsilon = 0.001$, boundary conditions no-flux. Single vortices are presented by black spots, the acoustic field is seen in gray shade.

Numerical analysis of the stability problem reveals that the lifetime of the multicharged vortex diverges as $\tau \sim (\varepsilon \lambda_1)^{-1}$ and is formally infinite for NLSE. However, one has slower (non-exponential) instability mode at $\epsilon = 0$. In particularly, generic perturbations grow linearly in time. In this sense the multicharged vortex is metastable and may exist for a very long time. The instability has a nonlinear nature and originates from the interaction between localized eigenmodes and the continuous spectrum of the vortex radiation. The crucial point here is that the simultaneous existence of the localized eigenmodes with $n \geq 2$ and extended excitations does not contradict to the energy conservation of the Hamiltonian NLSE, since they contribute to the energy with opposite signs. This process is similar to the growth of waves with negative energy [17]. In Fig.2 one clearly sees that a rotating double-charged and triple-charged vortices radiate away the acoustic waves. Thus, the decaying multicharged vortices are effective source of the acoustic radiation.

4   Nucleation of Vortices by Rapid Thermal Quench

Nucleation of vortices by neutron irradiation in ^{3}He-B in the presence of rotation was studied experimentally in Ref. [8]. Ignoring non-relevant complexity of the ^{3}He-B specific multicomponent order parameter, we will use the TDGLE for a scalar order parameter ψ [9]:

$$\partial_t \psi = \Delta \psi + (1 - f(\mathbf{r}, t))\psi - |\psi|^2 \psi + \zeta(\mathbf{r}, t). \qquad (5)$$

Close to T_λ the local temperature is controlled by normal-state heat diffusion and evolves as $f(\mathbf{r}, t) = E_0 \exp(-r^2/\sigma t)t^{-3/2}$, where σ is the normalized diffusion coefficient. $E_0 \gg 1$ determines the initial temperature of the hot bubble T^* due to the nuclear reaction between the neutron and ^{3}He atom and is proportional to the deposited energy $\mathcal{E}_0$. The Langevin force ζ with the correlator $\langle \zeta \zeta' \rangle = 2T_f \delta(\mathbf{r} - \mathbf{r}')\delta(t - t')$ describes thermal fluctuations with a strength T_f.

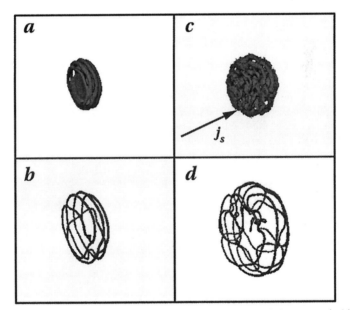

Fig. 3. 3D isosurface of $|\psi| = 0.4$ for $\sigma = 400, E_0 = 30$ and $k = 0.5$. (a-b) $T_f = 0$. Images are taken at times $t = 36, 80$. (c-d), $T_f = 0.002$, $t = 24, 80$.

Numerical simulations of Eq. (5) in 2D and 3D shows that without fluctuations ($T_f = 0$) the vortex rings nucleate upon the passage of the thermal front, Fig. 3a,b. Not all of the rings survive: the small ones collapse and only the big ones grow. Although the vortex lines are centered around the point of the quench, they exhibit a certain degree of entanglement. After a long transient period, most of the vortex rings reconnect and form the almost axisymmetric configuration. We find that the fluctuations have a strong effect at early stages: the vortices nucleate not only at the normal-superfluid interface, but also in the bulk of the supercooled region (Fig. 3c), according to the Kibble-Zurek "cosmological" mechanism [6]. However, later on, small vortex rings in the interior collapse, and only larger rings (primary vortices) survive and expand (Fig. 3d).

We conclude that the primary source of vortices in our numerical simulations is the instability of normal-superfluid interface in the presence of flow. The analysis results in the following expression for the number of survived vortices

$$N \sim \sqrt{\sigma} E_0^{1/3} \sqrt{(v_s/v_c)^2 - \beta^2 \log(T_f^{-1})/E_0^{2/3}} \qquad (6)$$

where $\beta = const$, while v_s and v_c are the imposed and critical GL superflow velocity, respectively. This estimate is in agreement with simulations, see Ref. [9]. Eq. (6) exhibits a slow logarithmic dependence of the number of vortices on the level of fluctuations. For the experimental values of the parameters our analysis results in about 10 surviving vortices per heating event. It is consistent with Ref. [8] where as many as 6-20 vortices per neutron were detected.

This research is supported by US DOE, grant W-31-109-ENG-38 I.A.), and by the Minerva Center for Nonlinear Physics of Complex Systems (V.S.).

References

1. V. L. Ginzburg and L. Pitaevskii. Sov. Phys. JETP **34**, 858 (1958).
2. L. Pitaevskii, Sov. Phys. JETP **35**, 282 (1959).
3. J. Hulin et al., Phys. Rev. **A9**, 885 (1974).
4. F. Lund and V. Steinberg, Phys. Rev. Lett., **75**, 1102 (1995).
5. H. Davidovitz and V. Steinberg, Europhys. Lett. **38**, 297 (1997).
6. G.E. Volovik, Physica B **280**, 122 (2000); T.W.B. Kibble, J. Phys. A: Math Gen **9**, 1387 (1976); W. H. Zurek: Nature **317**, 505 (1985)
7. V.B. Eltsov, M. Krusius, and G.E. Volovik: cond-mat/9809125, to be published.
8. V.M.H. Ruutu et al: Nature **382**, 334 (1996); V.M.H. Ruutu et al: Phys. Rev. Lett. **80**, 1465 (1998).
9. I.S. Aranson, N.B. Kopnin and V.M. Vinokur: Phys. Rev. Lett. **83**, 2600 (1999)
10. A. Reisenegger, J. Low Temp. Phys., **92**, 77 (1993) and references therein.
    171 (1985).
    B/Fluids **9**, 259 (1990).
11. R. J. Donnelly, *Quantized Vortices in Helium II*, Cambridge University Press, Cambridge, 1991.
12. I. Aranson and V. Steinberg, Phys. Rev. **B54**, 13072 (1996-II).
13. P. G. de Gennes, *Superconductivity of Metals and Alloys*, (Addison-Wesley, Redwood City, 1989).
14. V.L. Ginzburg and A.A. Sobaynin, Sov. Phys. Usp., **19**, 773 (1976); J. Low. Temp. Phys., **49**, 507 (1982).
15. Kramer, L., and W. Zimmermann, Physica **D 16**, 221, 1985
16. I. Aranson and V. Steinberg, Phys. Rev. **B53**, 75 (1996-I).
17. L. A. Ostrovsky et al., Usp. Fiz. Nauk **150**, 417 (1986) [Sov. Phys. Usp. **11**, 1040 (1986)].

Weak Turbulence Theory
for the Gross–Pitaevskii Equation

Sergey Nazarenko[1], Yuri Lvov[2], and Robert West[1]

[1] Mathematics Institute, University of Warwick, Coventry, CV4 7AL, UK.
[2] Rensselaer Department of Mathematical Sciences, New York, 12180-3590, USA.

1 Motivation and Background

Recent developments in the theory of Bose-Einstein condensates have resulted in a renewed interest in the Gross Pitaevskii equation (GPE),

$$i\partial_t\psi + \triangle\psi - |\psi|^2\psi - U\psi = 0, \qquad (1)$$

where the potential U is a given function of coordinate, see for example Fig. 1. In the limit $U = 0$, the GPE is called the defocusing Nonlinear Schrödinger equation (NLSE), an equation which plays a central role in the understanding of non-linear optics.

Weak Turbulence Theory (WTT) is a well developed statistical theory which has been used to great effect in understanding the behaviour of large ensembles of weakly nonlinear NLSE waves, [1–4]. A remarkable feature of the WTT is that it provides a rigorous closure and allows one to obtain Kolmogorov-type spectra (corresponding to constant spectral cascades of motion integrals) as exact analytical solutions; this being impossible in the case of Navier-Stokes turbulence. In particular, there are two spectra corresponding to the down-scale cascade of the integral of energy, $E = \int[(\nabla\psi)^2 + \frac{1}{2}\psi^4]\,d\mathbf{x}$, and the up-scale cascade of "particles", $N = \int \psi^2\,d\mathbf{x}$, respectively. The up-scale cascade of particles generates a large-scale condensate which, in turn, modifies the turbulence dynamics, changing the dominant process from being 4-wave to a 3-wave one.

The goal of this paper is to use the ideas developed for the NLSE to derive a WTT for a large set of random waves described by the GPE. An interesting picture emerges even from a naive application of the results already obtained for the NLSE case. Imagine an arbitrary initial excitation; a superposition of modes

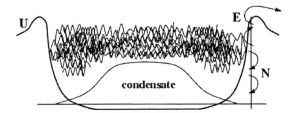

Fig. 1. Turbulent cascades of energy E and particle number N.

with energies somewhere in the middle of the potential well U, see Fig. 1. Because of the nonlinear interaction ("collisions") there will be a redistribution of energy E and particles N among the modes. Having in mind the NLSE results, one can interpret the propagation of excitations towards the low levels (condensation) as an up-scale cascade of particles. The spreading towards higher energy states and spilling over the barrier (cooling) can be interpreted as a down-scale energy cascade. Both of these processes will be described by Kolmogorov-type energy distributions over the levels (scales) which are dramatically different from any thermodynamic equilibrium distributions. Thus, the condensation and the cooling rates will also be significantly different from the ones obtained based on the assumption of a thermodynamic equilibrium and a Boltzmann distribution. Interesting WTT solutions describing the finite-time condensation were obtained in [4] and are described in another paper by Svistunov found elsewhere in this volume. Further, formation of a strong condensate will eventually modify the dynamics and change the energy distribution in a manner similar to the NLSE case.

There is of course a problem with such a naive application of the NLSE results to the GPE. Indeed, WTT for NLSE deals with spatially homogeneous wave turbulence, whereas for GPE, the turbulence is trapped by an external potential and is, therefore, intrinsicly inhomogeneous (e.g. a turbulent spot). Additional inhomogeneity of the turbulence arises because of the condensate, which in the GPE case is itself coordinate dependent.

The effects of the coordinate dependent potential and condensate can be observed already at the level of the linear dynamics. This can be most easily understood using a wavepacket (WKB) formalism that is applicable if the wavepackets' wavelengths l are much shorter than the characteristic width of the potential well L. This naturally leads it a small parameter ε such that

$$\varepsilon = \frac{l}{L} \ll 1.$$

The coordinate dependent potential and the condensate distort the wavepackets so that their wavenumbers change. This has a dramatic effect on nonlinear resonant wave interactions because now waves can only be in resonance for a finite time.

2   Weak Turbulence Theory for NLSE

Before considering WTT for GPE, let us briefly outline the results previously obtained for NLSE. The main object of WTT is the spectrum of homogeneous turbulence, $n(\boldsymbol{k})$, which in case of NLSE is defined as

$$\left\langle \hat{\psi}_k \hat{\psi}_{k'} \right\rangle = n_k \, \delta(\boldsymbol{k} - \boldsymbol{k}'), \tag{2}$$

where the angle brackets mean an ensemble average. Starting with NLSE in Fourier space and assuming that the amplitudes $\hat{\psi}_k$ are small enough (for the linear wave period to be much less than the characteristic nonlinear time) one can derive the following kinetic equation for waves (KEW) [1],

$$\partial_t n_k = \frac{1}{\pi} \int n_k n_1 n_2 n_3 \left(\frac{1}{n_k} + \frac{1}{n_1} - \frac{1}{n_2} - \frac{1}{n_3} \right) \tag{3}$$
$$\delta(\boldsymbol{k} + \boldsymbol{k}_1 - \boldsymbol{k}_2 - \boldsymbol{k}_3)\delta(\omega_k + \omega_1 - \omega_2 - \omega_3)\, d\boldsymbol{k}_1 d\boldsymbol{k}_2 d\boldsymbol{k}_3,$$

where $\omega_k = k^2$ is the linear wave frequency (we use the notation $k = |\boldsymbol{k}|$), $n_j = n(\boldsymbol{k}_j)$ and $\omega_j = \omega(\boldsymbol{k}_j)$, $j = 1, 2, 3$. This equation describes spectral redistribution due to a 4-wave resonant process, which has a characteristic time

$$\tau_{4w} = 1/n^2 k^6.$$

The right hand side of this equation is called the collision term. Like the original NLSE, KEW conserves energy E and particles N, and there exist exact analytical solutions to KEW corresponding the constant spectral fluxes of these invariants [2].

An important consequence of the up-scale particle cascade is the generation of a condensate at large scales, which can be approximated by a coordinate independent solution $\psi = \sqrt{\varrho}\exp(-i\varrho t)$, where ϱ is a constant. In this case, one can derive WTT by considering weak perturbations about the condensate as it was done in [2]. The resulting expressions are quite lengthy and for our purposes here it will suffice to note that there will be a 3-wave resonant interaction with characteristic time

$$\tau_{3w} = 2/\varrho n k^3$$

(if k^2 greater or equal ϱ which will be the only interesting case for our WKB theory). Thus, the 3-wave process will dominate the 4-wave one if the condensate is strong enough, $\varrho > n k^3$. Now, let us consider the case of inhomogeneous turbulence trapped by an external potential. The inhomogeneity affects not only nonlinear interactions but also the linear propagation and we, therefore, will start by considering the linear dynamics.

3   Linear Dynamics of the GPE

We will now develop a WKB theory for small-scale wave-packets, described by a linearised GPE, with and without a background condensate. As is traditional with any WKB-type method we assume the existence of a scale separation $\varepsilon \ll 1$, as explained in Sect. 1. In this analysis we will take $l \sim 1$ so that any spatial derivatives of a given large-scale quantity (e.g. the potential U or the condensate) are of order ε. The transition to WKB phase-space is achieved through the implementation of Gabor transforms, which are defined as,

$$\hat{g}(\boldsymbol{x}, \boldsymbol{k}, t) = \int f(\varepsilon^*|\boldsymbol{x} - \boldsymbol{x}_0|)e^{i\boldsymbol{k}\cdot(\boldsymbol{x} - \boldsymbol{x}_0)}g(\boldsymbol{x}_0, t)d\boldsymbol{x}_0, \tag{4}$$

where f is a rapidly decreasing function of $\boldsymbol{x}$, for instance a Gaussian. The parameter ε^* is small and such that $\varepsilon \ll \varepsilon^* \ll 1$. Hence, our kernel f varies at the intermediate-scale. A Gabor transform can therefore be thought of a localised Fourier transform. Physically, one can view a Gabor transform as a wavepacket distribution function over positions x and wavevectors k.

3.1   Without a Condensate

Linearising the GPE, we describe wavepackets ψ without the presence of a condensate via

$$i\partial_t\psi + \triangle\psi - U\psi = 0, \tag{5}$$

where U is a slowly varying potential. Note that (5) is just the linear Schrödinger equation commonly considered in quantum mechanics. Gabor transforming this equation and combining the result with its complex conjugate we find the following WKB transport equation,

$$D_t|\hat{\psi}|^2 = 0, \tag{6}$$

where $D_t \equiv \partial_t + \dot{x} \cdot \nabla + \dot{k} \cdot \partial_k$ represents the total time derivative along the wavepacket trajectories in phase-space. The ray equations are used to describe wavepacket trajectories in (k, x) phase-space,

$$\dot{x} = \partial_k\omega, \qquad\qquad \dot{k} = -\nabla\omega. \tag{7}$$

The frequency ω, in this case, is given by $\omega = k^2 + U$, (again we use the notation $k = |k|$). Equations (6) and (7) are nothing more than the famous *Ehrenfest* theorem from quantum mechanics. According to (7), the wavepackets will get reflected by the potential at points r_R where $U(r_R) = k_{max}^2$. We will now move on to consider linear wavepackets in the presence of a background condensate.

3.2   With a Condensate

One common misconception in the BEC theory is that the presence of a condensate acts on the higher levels by just modifing the confining potential U, see for example [5]. If this was the case the linear dynamics would still be described by the *Ehrenfest* theorem (with some effective potential), but as we will now see this is not the case. We will assume that the condensate ψ_0 is a (nonlinear) ground state solution of equation (1). We start by considering a small perturbation $\phi \ll 1$, such that $\psi = \psi_0(1 + \phi)$. Substituting this into (1) and linearising with respect to ϕ, we find

$$i\partial_t\phi + \triangle\phi + 2\frac{\nabla\psi_0}{\psi_0} \cdot \nabla\phi - \varrho(\phi + \phi^*) = 0. \tag{8}$$

where $\varrho = \varrho(x) = |\psi_0|^2$ is a slowly varying condensate density. In a similar manner to the previous section, the rest of this derivation consists of Gabor transforming (8), combining the result with its complex conjugate and finding a suitable wave-action variable such that the transport equation represents a conservation equation along the rays. This derivation is quite lengthy and can be found in [6], here we will only present the results. As one might expect from a WKB based theory, at the zeroth order of ε, we derive a dispersion relationship

$$\omega = k\sqrt{k^2 + 2\varrho}, \tag{9}$$

which is identical to the one obtained for linear waves in the presence of a homogeneous condensate $(\varepsilon = 0)$ in [2]. At the first order we obtain a transport equation given by

$$D_t n = 0, \tag{10}$$

which represents the conservation of wave-action n along wavepacket trajectories described by the ray equations (7). The wave-action is given by

$$n = \frac{1}{2} \frac{\omega\rho}{k^2} \left| \widehat{Re\phi} - \frac{ik^2}{\omega} \widehat{Im\phi} \right|^2. \tag{11}$$

Obviously, the dynamics in this case cannot be reduced to the *Ehrenfest* theorem with a renormalised potential U. It is interesting that such a wave-action agrees with that found in [2] for the homogeneous case with non-zero nonlinearity $(\varepsilon = 0, \sigma \neq 0)$. This is the opposite limit to the one considered above $(\varepsilon \neq 0, \sigma = 0)$; see Sect. 2.

4   Applicability of WKB Descriptions

In this section we will investigate the applicability of the above theory. Firstly let us consider the case of a steady weak condensate so that the effect of the nonlinear term is small in comparison to the linear ones, $|\varrho\psi_0| \lesssim |\triangle\psi_0|$. This corresponds to the eigenvalue problem $\partial_t\psi_0 = -i\Omega\psi_0$. The GPE will therefore become

$$\Omega\psi_0 + \triangle\psi_0 - \varrho\psi_0 - U\psi_0 = 0.$$

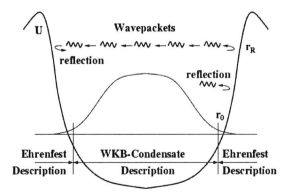

Fig. 2. Regions of applicability of WKB descriptions.

Since Ω is a constant we observe that the Laplacian term acts to balance the external potential term (like in the linear Schrödinger equation) and the nonlinear term can be at most as big as the linear ones $\Omega \sim \frac{1}{r_0^2} \sim U(r_0) \gtrsim \varrho$, where r_0 is the characteristic size of the condensate (it is defined as the condensate "reflection" point from the condition $\Omega = U(r_0)$, see below).

Now for a WKB description to be valid we require $kr_0 \gg 1$, i.e. we require the characteristic length-scale of our wavepackets to be a lot smaller than that of the large scales. Using this fact we find $k^2 \gg \frac{1}{r_0^2} \sim U(r_0) \gtrsim \varrho$. Therefore, the condensate correction to the frequency, given by (9), is small. In other words the wavepacket does not "feel" the condensate. Indeed, from $k_{\max}^2 = U(r_R)$ we have $U(r_R) \gg U(r_0)$ and this implies that $r_R \gg r_0$ (where r_R is the wavepacket reflection point, see Fig. 2). Thus, the condensate in this case occupies a tiny space at the bottom of the potential well and hence does not affect a wavepacket's motion. Therefore, a wavepacket moves as a "classical" particle described by the *Ehrenfest* equations (6) and (7). In fact, in this case it would be incorrect to try to describe the small condensate corrections via our WKB approach because these corrections are of order $\varrho \sim \varepsilon^2$ (the ε^2 terms being ignored in a WKB description).

Now consider a strong condensate such that

$$\Omega \cong U + \varrho \gg \frac{\triangle \psi_0}{|\psi_0|}, \tag{12}$$

i.e. the r dependence of the potential U is now balanced by the nonlinearity. This case is usually refered to as the *Thomas-Fermi* limit, see for example the paper by Fetter in this volume.

Furthermore, now wavepackets can "feel" the presence of a strong condensate if $\varrho \sim k^2$. We see that the WKB approach is applicable because $k^2 \sim \varrho \gg \frac{1}{r_0^2} \sim \frac{|\triangle \psi_0|}{|\psi_0|}$. According to the ray equations ω is a constant along a wavepackets trajectory, so we can find the packet's wavenumber from $k^2 = \sqrt{\varrho^2 + \omega^2} - \varrho$. One can see that k^2 remains positive for any value of ϱ which means that the presence of the condensate does not lead to any new wavepacket reflection points (at which k would turn into zero). Thus, turbulence is allowed to penetrate into the centre of the potential well. However, the group velocity increases when the condensate becomes stronger, $\partial_k \omega \sim \sqrt{\rho}$. This means that the density of wavepackets decreases towards the centre of well. Therefore, the condensate tends to push the turbulence away from the centre, towards the edges of the potential trap.

To summarize, in the presence of a strong condensate we have two regions of applicability for our WKB descriptions, see Fig. 2a. Wavepackets at a position $r < r_0$, in the central region of the potential well will evolve according to the WKB-condensate description (10). The Laplacian term only becomes important for $r > r_0$ where ϱ is exponentially small. In this case the Ehrenfest description is appropriate.

5   Weakly Nonlinear GPE Waves

Derivation of WTT for nonuniform turbulence governed by GPE consists of a combination of the WKB derivation for the linear part of dynamics and a standard WTT derivation (see e.g. [2]) where Gabor transforms are used instead of the Fourier transforms. The most straightforward example of this procedure is when the nonlinear time (described in Sect. 2) is greater than the characteristic linear time k/L (the time between consequent wavepacket reflections). This regime corresponds to an opaque gas of wavepackets. In this case, one can ignore the inhomogeneity in the derivation of the collision term found in the KEW and simply replace any Fourier transforms by Gabor transforms multiplied by $\exp(-i\boldsymbol{k} \cdot \boldsymbol{x})$ (this spatially slowly varying variable is directly analogous to the Fourier transform). Note that U will not enter into the collision term because the U-dependence of ω drops out from the 4-wave resonance condition, whereas, in the case of turbulence about the condensate, ω is completely independent of U. Therefore, KEW in this case can be readily rewritten by replacing the partial time derivative on the left hand side of the standard 4-wave or 3-wave equation, with D_t. The choice of the 4-wave or 3-wave equation depends on the relative intensity of the turbulence n and the condensate ϱ, see the condition given in Sect. 2. In particular, in the regions where the linear dynamics is described by the condensate WKB the 3-wave interaction is always dominant if $\varrho_{max} \sim k_{\max}^2 \sim 1$ (because $n \ll 1$); whereas, for the Ehrenfest type WKB, the dominant process can be either 4-wave or 3-wave. The case when the time between consequent wavepacket reflections is less than the nonlinear time is less obvious and requires further study.

References

1. A.C. Newell: Rev. Geophys. **6**, 1 (1968)
2. S. Dyachenko, A.C. Newell, A. Pushkarev, V.E. Zakharov: Physica D **57**, 96 (1992)
3. V.E. Zakharov, S.L. Musher, A.M. Rubenchik: Phys. Rep. **129**, 285 (1985)
4. B.V. Svistunov: J. Moscow Phys. Soc, **1**, 373, (1991)
5. C.W. Gardiner, M.D. Lee, R.J. Ballagh, M.J. Davis, P. Zoller: PRL **81**, 24 (1998)
6. S. Nazarenko, Y. Lvov, R.J. West: *to be published* (2000)

Dissipative Vortex Dynamics and Magnus Force

L.M. Pismen

Department of Chemical Engineering and Minerva Center for Nonlinear Physics of Complex Systems, Technion – Israel Institute of Technology, 32000 Haifa, Israel

1   Basic Equations

A single vortex in the infinite plane, emerging as a solution of a conservative PDE with global $U(1)$ and Galilean symmetry, is a very special object. Its singular properties, as well as special symmetries of the governing equation, which are broken in more realistic models, have been a source of paradoxes and erroneous conclusions during the long history of theory of vortex motion. A detailed review containing further references is found in the recent monograph by this author [1]. This communication sketches main ideas and pitfalls of singular perturbation techniques that allow to describe slow vortex motion including effects of weak dissipation.

Our starting point is the Gross–Pitaevskii or defocusing nonlinear Schrödinger (NLS) equation for complex field $u(\boldsymbol{x})$ in R^2:

$$-iu_t = \nabla^2 u + (1 - |u|^2)u. \tag{1}$$

A symmetric vortex solution with topological charge $N = \pm 1$ is expressed in polar coordinates r, ϕ as $u_0 = \rho_0(r)\mathrm{e}^{\pm i\phi}$, where the real amplitude $\rho_0(r)$ verifies

$$\rho_0'' + r^{-1}\rho_0' + (1 - r^{-2} - \rho_0^2)\rho_0 = 0. \tag{2}$$

The relevant special properties of this solution are

- logarithmic divergence of the total energy of the vortex within a circle with the radius L:

$$\mathcal{E} = \pi \int_0^L \left[|\nabla\rho_0|^2 + (\rho_0/r)^2 + \tfrac{1}{2}(1 - \rho_0^2)^2\right] \mathrm{d}r = \pi \ln(L\sqrt{\mathrm{e}}/a_0); \tag{3}$$

- invariance to planar translations and global phase shifts (the corresponding Goldstone modes are ∇u_0 and iu_0, and the conserved quantities are energy and momentum).
- Galilean invariance:

$$\boldsymbol{x} \to \boldsymbol{x} - \boldsymbol{v}_s t, \quad \nabla \to \nabla, \quad \partial_t \to \partial_t - \boldsymbol{v}_s \cdot \nabla; \quad u \to u \exp\left[i\left(\tfrac{1}{2}\boldsymbol{v}_s \cdot \boldsymbol{x} - \tfrac{1}{4}v_s^2 t\right)\right]. \tag{4}$$

As a consequence of Galilean invariance, the vortex remains symmetric when placed in a constant phase gradient $\boldsymbol{v}_s = i(u\nabla\bar{u} - \bar{u}\nabla u)/|u|^2$ and viewed in the respective Galilean frame, or, in other words, when advected with a constant velocity v_s.

The vortex motion is viewed at large distances, long periods of time and slow velocities, scaled, respectively, as ϵ, ϵ^2, and ϵ, where ϵ is the ratio of the microscopic scale of Eq. (1) – the "healing length" – to a relevant macroscopic scale L. In view of Eq. (4), the static solution remains invariant in the first order at $v_s = O(\epsilon)$. This implies a remarkably simple law of vortex motion in a slowly varying field: the vortex is embedded in the ambient flow and moves with the local velocity. In this approximation, the vortex has no inertia.

The large-scale flow that may include many interacting vortices is conveniently described in the fluid-mechanical representation using rescaled coordinates and time and the field variable presented in the Madelung form

$$u = \sqrt{\varrho}\,e^{i\vartheta/2}, \tag{5}$$

where $\varrho = |u|^2$ is density and ϑ is the flow potential, defining the flow velocity $v = \nabla\vartheta$. Then the imaginary part of Eq. (1) takes the form of the continuity equation

$$\varrho_t + \nabla \cdot (\varrho v) = 0, \tag{6}$$

and the real part is written as the Bernoulli equation

$$\vartheta_t + \tfrac{1}{2}|v|^2 + p = 0, \tag{7}$$

in which p is pressure introduced via the equation of state

$$p = 2\left[(\varrho - 1) - \epsilon^2 \varrho^{-1/2}\nabla^2\varrho^{1/2}\right]. \tag{8}$$

The far field approximation is obtained by neglecting the last term in Eq. (8), called in the superfluid context quantum pressure. Further approximation is incompressibility. Since $\varrho = 1 - O(\epsilon^2)$ outside the vortex core, the continuity equation reduces in the leading order to $\nabla \cdot v = 0$. The problem is mapped then on classical ideal fluid dynamics: the gradient of Eq. (7) yields the Euler equation

$$v_t + (v \cdot \nabla)v + \nabla p = 0, \tag{9}$$

and the far field velocity potential ϑ verifies the Laplace equation $\nabla^2\vartheta = 0$. This approximation is applicable as long as vortices are removed at distances far exceeding the core size and move with a speed far less than the speed of sound, which is of $O(1)$ on the scale of Eq. (1) or of $O(\epsilon^{-1})$ on the scale of Eqs. (6), (7). It is perturbed, however, in a singular way by small corrections to the basic model that violate Galilean invariance and energy conservation. These corrections can be properly taken into account only through matched asymptotic expansion combining different approximations at widely separated scales. Moreover, weak compressibility effects become important at long times, when acoustic dispersion drives the system to lower energy states, notwithstanding the energy conservation.

2   Magnus Force

Perturbations of different nature may cause a drift of the vortex relative to the ambient flow. A driving force of such a drift is commonly called a Magnus force. This force is in fact a universal notion, as it is required for closing the overall momentum balance whenever a vortex moves relative to the ambient flow, whatever the cause of this motion might be. This is a purely classical effect, which can be fully described in the framework of classical fluid mechanics. Consider a vortex motion with a constant velocity v_l under the action of a point force F applied at the vortex location. The flow field should be viewed in the comoving frame, where it is described by the stationary Euler equation

$$(v - v_l) \cdot \nabla v + \nabla p = F\delta^2(x). \tag{10}$$

Taking the momentum balance over a large circle yields

$$F = \oint \{np + [n \cdot (v - v_l)](v - v_l)\}ds$$

$$= \lim_{r \to \infty} r \int_0^{2\pi} \{n(v \cdot v_l) - (n \cdot v)v_l\}d\phi = 4\pi N \mathcal{J} v_l, \tag{11}$$

where $\mathcal{J}$ is the operator of rotation by $\pi/2$. The contour integral is transformed here by replacing the pressure with the help of the Bernoulli law, and observing that the velocity field induced by the vortex is purely angular and does not project on the normal vector n. The two terms, one due to pressure, and the other, due to the kinetic energy, contribute equally to the final result. The motion is across the flow in such a way that the vortex is displaced towards the region where the induced velocity is added to the extrinsic flow, and the increased average velocity results, by the Bernoulli law, in a reduced pressure.

The result, based on the momentum balance only and independent of the nature of the applied force, means in fact that a force normal to the direction of drift relative to the ambient flow must always be applied to close the overall momentum balance. It depends, however, on a presumption that the Euler equation is applicable everywhere. We know that the hydrodynamic description fails in the vortex core; thus, Eq. (11) is applicable only when the force causing the vortex drift is accumulated outside the core region. This is true, for example, when the drift is caused by scattering of long-wave phonons [2], but the action of inhomogeneities or interactions with normal fluid (incorporating also short-wave excitations) provide adverse examples, as we shall see below. Therefore the "universality" of Eq. (11), stipulating a linear mobility relation between the force and the drift velocity, is illusory. A force accumulating in the core can be recognized by a divergence in the core expansion that would typically generate a logarithmic correction to the mobility relation.

As an example, consider a model including interaction between superfluid and normal fluid components. The field equation replacing Eq. (1) can be written in the form

$$(u_t + iA v_n \cdot \nabla u) = (i + A)\left[\nabla^2 u + (1 - \tfrac{1}{4}v_n^2 - |u|^2)u\right]. \tag{12}$$

where v_n is velocity of the normal fluid and Λ is a parameter characterizing the relaxation rate. This equation retains the property of Galilean invariance (4) that has to be complemented now by the appropriate shift of v_n.

The problem of vortex motion is treated perturbatively at $|v_n| = O(\epsilon)$, i.e when the normal fluid velocity is of the same order of magnitude as the superfluid velocity v_s at typical inter-vortex separations far exceeding the size of the vortex core. We rescale the vortex velocity as $v_l = \epsilon V_l(1 + \Lambda^2)$, and the velocity of the normal fluid as $v_n = \epsilon V_n(1 + \Lambda^2)$, and carry out the Galilean transformation (4) to the coordinate frame propagating with the vortex velocity with an additional phase shift:

$$\partial_t \to \partial_t - v_l \cdot \nabla, \quad u \to u \exp\left[-\tfrac{1}{2} i \epsilon(V_l + \Lambda^2 V_n) \cdot x\right]. \tag{13}$$

Assuming that the motion is stationary in the comoving frame, Eq. (12) is rewritten to $O(\epsilon)$ as

$$\nabla^2 u + (1 - |u|^2)u + \epsilon\Lambda(V_l - V_n) \cdot \nabla u = 0. \tag{14}$$

The drift velocity V_l can be obtained from the solvability condition after the solution of this equation is expanded in ϵ. The perturbation is, however, singular, as the first-order term diverges in the far field. The solvability condition should be computed therefore in a circle of radius r_0 large compared to the core size but small on the far field scale, i.e. $1 \ll r_0 \ll \epsilon^{-1}$, and involves both area and contour integrals:

$$\text{Re}\left\{ \int_0^{r_0} r \, dr \int_0^{2\pi} \overline{\varphi}\Psi \cdot \nabla u_0 \, d\phi + r_0 \int_0^{2\pi} (\overline{\varphi}\partial_r u_1 - u_1 \partial_r \overline{\varphi})_{r=r_0} \, d\phi \right\} = 0. \tag{15}$$

where $\Psi = \Lambda(V_l - V_n) \cdot \nabla u_0$ is the inhomogeneity in the first-order expansion of Eq. (14) and $\varphi = \nabla u_0$ is the translational Goldstone mode; overline denotes the complex conjugate.

The missing value of the first-order correction u_1 on the bounding circle should be obtained by matching with the far field solution. The outer equation should have a fitting form to insure successful matching. In the far field, Eq. (12) is rewritten in extended coordinates $X = \epsilon x$, $T = \epsilon^2 t$. In the leading order, the density $\rho = 1 - O(\epsilon^2)$ is almost constant, while the phase equation reduces to the convective diffusion equation that degenerates into the Laplace equation at $\Lambda \to 0$. Assuming that the vortex speed is stationary, the phase equation in the comoving frame is

$$\Lambda(V_l - V_n) \cdot \nabla\theta + \nabla^2\theta = 0. \tag{16}$$

This equation can be solved, and the function $u_1 \approx e^{i\theta(r,\phi)}$ obtained as the limit of the solution at $r \to 0$. Computing the integrals in Eq. (15) and reversing the transformation (13) yields the mobility relation

$$v_s - v_n = \frac{1}{1 + \Lambda^2}\left[(v_l - v_n) + \mathcal{J}(v_l - v_n)N\Lambda \ln\frac{v_0(1 + \Lambda^2)}{\Lambda|v_l - v_n|}\right]. \tag{17}$$

At $\Lambda \to 0$ this reduces to $\boldsymbol{v}_l = \boldsymbol{v}_s$, i.e. to a simple advection of the vortex by the ambient superfluid. At $\Lambda \neq 0$, there is a Magnus force driving the vortex normally to the direction of superflow. This part of the mobility relation contains a logarithmic correction dependent on the core structure; in the standard model, $v_0 \approx 3.29$. The mobility relation (17) is weakly nonlinear. The nonlinearity does not appear in the logarithmic factor when divergences in the core integrals are removed by means of an artificial long-scale cut-off rather than by matching with the far region [3–5]. We can also see that inferring the absence of a transverse force due to interaction with the normal fluid [6] on the basis of Eq. (11) is erroneous.

The direction of vortex motion is generally oblique. It is directed in such a way that like vortices repel and unlike attract each other, but two interacting vortices never move directly along the connecting straight line. Due to the logarithmic correction, the angle with the connecting line decreases as the vortices accelerate at closer approach. Since the superfluid velocity in the far field obeys at $\Lambda \neq 0$ the heat equation rather than the Laplace equation, the vortex motion is generally history-dependent, and is not uniquely determined by instantaneous vortex positions. Clearly, the lifetime of a vortex pair is finite, and decreases with growing Λ.

3   Three-Dimensional Effects

The three factors determining the motion of a line vortex in 3D are the local curvature, the advection by superflow (or, in other words, the nonlocal action of the phase field), and interaction with the normal fluid. The velocity due to local curvature is computed with the help of a double matching procedure [7]. First, the equation in the core region is rewritten in the aligned coordinate frame, introducing a curvature correction, and the solvability condition is evaluated on a circle large compared with the core size as above. The flow field on this circle is obtained by solving the far field equation in two steps, matching the velocities induced by adjacent and removed parts of the line vortex. The mobility relation obtained in the non-dissipative case is

$$v_l^0 = \boldsymbol{v}_s + N \kappa \boldsymbol{b} \ln \frac{\lambda_0}{\epsilon}, \tag{18}$$

where $\boldsymbol{b}$ is the binormal to the line vortex, κ is curvature and λ_0 is a numerical parameter. One cannot, however, simply combine the curvature-induced velocity and the dissipative terms in Eq. (17). An impediment is that Eq. (17) has been obtained by matching with the solution of the far field equation (16) obtained under the assumption of stationary motion. The nonlinear logarithmic correction in Eq. (17) is the result of this matching. Generally, the argument of the logarithm obtained by matching with a non-stationary far field solution would be history-dependent, in the same way as the argument of the logarithm in Eq. (18) depends on the global shape of the line vortex. The formal analogy between the terms containing the velocity and curvature ends at this point, since, on the one hand, motion with a constant curvature is not a reasonable assumption, and, on the other hand, making the local mobility relation dependent on the history of motion is not practical. The remaining alternative is to treat the argument of the logarithm as an adjustable parameter, keeping in mind that it must be an $O(\epsilon^{-1})$ quantity. An additional *coup de force* is to allow for different logarithmic factors in the velocity and curvature terms. Then the combined mobility relation is written as

$$v_s - v_n = \frac{v_l - v_n}{1 + \Lambda^2} + N l \times \left[\frac{\Lambda(v_l - v_n)}{1 + \Lambda^2} \ln \frac{\lambda_1}{\epsilon} - \kappa n \ln \frac{\lambda_0}{\epsilon} \right]. \tag{19}$$

This relation can be further resolved with respect to the dissipative part of the vortex drift velocity, $v_l - v_l^0$, and represented in a transparent form

$$v_l = v_l^0 + \alpha \mathcal{J}_\chi (v_n - v_l^0), \tag{20}$$

where $\mathcal{J}_\chi$ is the operator of rotation through angle χ about the local tangent direction l, and

$$\alpha^2 = \frac{\Lambda^2(\nu^2 + \Lambda^2)}{1 + (2 + \nu^2)\Lambda^2 + \Lambda^4}, \quad \tan \chi = \frac{\nu(1 + 2\Lambda^2)}{\Lambda(\nu^2 - 1 - \Lambda^2)}, \tag{21}$$

where $\nu = \ln(\lambda_1/\epsilon)$. This can be interpreted as a non-isotropic friction rule, where α is the modulus of the friction coefficient, and χ is the deflection angle. The equivalent form postulated by Schwarz [8] is

$$v_l = v_l^0 + \alpha_1 l \times (v_n - v_l^0) - \alpha_2 l \times \left(l \times (v_n - v_l^0) \right). \tag{22}$$

The above derivation links this expression (albeit not in an impeccable way) to the two-fluid model, so that the phenomenological friction coefficients are expressed as

$$\alpha_1 = \alpha \cos \chi = \frac{\Lambda(\nu^2 - 1 - \Lambda^2)}{1 + (2 + \nu^2)\Lambda^2 + \Lambda^4}, \quad \alpha_2 = \alpha \sin \chi = \frac{\nu(1 + 2\Lambda^2)}{1 + (2 + \nu^2)\Lambda^2 + \Lambda^4}. \tag{23}$$

Applying this mobility relationship to a ring vortex, we can see that, due to a large logarithmic factor, the rotation angle χ is positive, and therefore the vector $v_n - v_l^0$, directed against the binormal, is rotated anticlockwise towards the direction of the normal. As the ring velocity acquires a normal component, the ring shrinks as it propagates, and its lifetime becomes finite, decreasing with growing χ or α_2.

4   Failure of Mechanistic Reduction

We can observe that logarithmic factors in mobility relations arise whenever divergent integrals appear in the perturbation scheme. Besides the above examples, logarithmic corrections arise when vortex motion is influenced by background inhomogeneities and inertia. The matching technique allows to overcome divergence problems, but a remaining troublesome point is that constants under the logarithms are not universal. One needs to know the far field configuration to compute these values, and all known analytical results are applicable, strictly speaking, only to the configurations used in these studies, which usually imply stationary motion.

The situation is still less certain when higher order acoustic effects are considered. A system of moving vortices may relax to a state of lower energy (eventually, to the quiescent state) even in a non-dissipative model Eq. (1) through emission of second sound which is either radiated out to infinity or absorbed on bounding walls. This effect is of second order in ϵ, as it is caused by local density depletion and accelerations. Acoustic radiation spreading out to infinity or adsorbed at distant walls effectively cancels the energy and momentum conservation and blurs the distinction between conservative and dissipative media. Scattering of phonons on a vortex has been originally studied [9] as a model of interaction with the normal fluid; it gives, indeed, both drag and lift components, although the respective mobility coefficients differ substantially from those produced by the two-fluid model (Sect. 2).

The common way to estimate the effects of acoustic emission and scattering is (1) to compute the flow field induced by non-dissipative vortex motion; (2) to consider the variable flow field as the source of an acoustic wave and (3) to compute the loss of energy and momentum through radiation by estimating the respective fluxes far from the source. This works only when vortices are confined to a region much smaller than the overall extent of the nonlinear medium, and allows to obtain closed equations of motion only in simplest configurations, like a single vortex advected by an oncoming wave, a pair of rotating point vortices, or an oscillating vortex ring [1]. In more complex configurations, there is no way to incorporate acoustic emission into mobility relations.

The cautionary message is that *any* factor that causes deviation of a vortex trajectory from the local superflow streamlines disrupts straightforward reduction of field equations to "mechanical" equations of vortex motion. The reduction to a "particle–field" description survives only in a weaker sense. It is still possible to replace the full field equation (1) or (12) by the phase equation, provided vortices are removed at distances far exceeding the core size, but numerical matching of the inner limit of the outer solution with the perturbed core structure is necessary to make the approximation consistent.

References

1. L.M. Pismen, *Vortices in Nonlinear Fields*, Oxford University Press, 1999.
2. E.B. Sonin, Phys. Rev. B **55**, 485 (1997).
3. H.E. Hall and W.F. Vinen, Proc. Roy. Soc. **A238**, 215 (1956).
4. A. Onuki, J. Phys. **C15** L1089 (1982).
5. E.B. Sonin, this issue.
6. C. Wexler, Phys. Rev. Lett. **79** 1321 (1997).
7. L.M. Pismen and J. Rubinstein, Physica (Amsterdam) **D47** 353 (1991).
8. K.W. Schwarz, Phys. Rev. B **31** 5782 (1985).
9. L.P. Pitaevskii, Sov. Phys. JETP **8** 888, 1959.

Transition to Dissipation
in Two- and Three-Dimensional Superflows

Cristián Huepe[1], Caroline Nore[2], and Marc-Etienne Brachet[3]

[1] James Franck Institute, University of Chicago, 5640 S. Ellis Avenue, Chicago, IL
    60637, USA
[2] Université de Paris-Sud, LIMSI, Bâtiment 508, F-91403 Orsay Cedex, France
[3] Laboratoire de Physique Statistique de l'Ecole Normale Supérieure associé au
    CNRS et aux Universités Paris 6 et 7, 24 rue Lhomond, 75005 Paris, France

Abstract. Vortex nucleation in two-dimensional (2D) and three-dimensional (3D) superflows past a cylinder is studied. The superflow is described by a Nonlinear Schrödinger
like equation. In the 2D case, a continuation method is used to characterize the bifurcation of stationary states leading to vortex formation. A saddle-node followed by a
secondary pitchfork bifurcation that leads to the branch of nucleation solutions (one
vortex in an asymmetric field) is found. The dependence of the bifurcation diagram on
the ratio of the coherence length to the disc diameter is studied. Using the 2D stationary vortex nucleation solutions to construct the initial condition, the 3D dynamics of
a vortex pinned to the surface of the cylinder is numerically studied. Quasistationary
half-ring vortices, pinned at the sides of the cylinder, are generated after a short time.
On a longer time scale, a vortex stretching may occur, inducing dissipation and drag.
The corresponding 3D critical velocity is found to be well below the 2D one.

1   Introduction

Above a certain critical velocity, superfluid flows are known to enter a dissipative
regime. The Nonlinear Schrödinger equation (NLSE) describes the dynamics of
superfluid ^{4}He, at temperatures low enough for the normal fluid to be negligible.
In the homogeneous two-dimensional NLSE flow past a disc, the existence of a
transition to dissipation due to the periodic emission of pairs of counterrotat-
ing vortices was found in [1]. Although this model system is not quantitatively
equivalent to a ^{4}He flow, it has been studied in detail for its universal properties
and mathematical interest [2–4]. A closer connection to experiments was given
by the recent experimental success in producing and manipulating dilute Bose
Einstein condensates. The NLSE has been used to accurately describe the dynamics of such systems, allowing direct quantitative comparison between theory
and experiment [5]. In particular, a recent experiment which studies the dissipation produced in a Bose Einstein condensed gas by moving a blue detuned laser
beam through it [6] has renewed the interest in dynamics of a NLSE superflow
past an obstacle.

The present paper reviews some recent work in the NLSE description of a
superflow past a cylinder in 2D and 3D [2–4,7,8]. The paper is organized as
follows. In Sect. 2 we present the hydrodynamic form of the NLSE. In Sect. 3,
the numerical tools used in this work are described. The bifurcation diagram

that governs the 2D system is explained in Sect. 4. Finally, Sect. 5 explores the possibility that a vortex stretching mechanism (as the one found in superfluid turbulence [9–11]) is responsible for low flow velocity dissipation in 3D systems.

2 Definition of the System

In this section, we briefly present the hydrodynamic form of the NLSE that models the effect of a moving cylinder of diameter D in a superfluid at rest.

Consider the following action functional

$$A = \int dt \left\{ \sqrt{2}c\xi \int d^3x \frac{i}{2} \left(\bar{\psi} \frac{\partial \psi}{\partial \tilde{t}} - \psi \frac{\partial \bar{\psi}}{\partial \tilde{t}} \right) - \mathcal{F} \right\}, \tag{1}$$

where ψ is a complex field, $\bar{\psi}$ its conjugate and $\mathcal{F}$ is the energy of the system. The coherence length ξ and the speed of sound c (for a mean fluid density $\rho_0 = 1$) are the physical parameters of the superfluid. The energy $\mathcal{F}$ reads $\mathcal{F} = \mathcal{E} - \boldsymbol{P} \cdot \boldsymbol{U}$, with

$$\mathcal{E} = c^2 \int d^3x \left([-1 + V(\boldsymbol{x})]|\psi|^2 + \frac{1}{2}|\psi|^4 + \xi^2 |\nabla \psi|^2 - \frac{1}{2} \right) \tag{2}$$

$$\boldsymbol{P} = \sqrt{2}c\xi \int d^3x \frac{i}{2} \left(\psi \nabla \bar{\psi} - \bar{\psi} \nabla \psi \right). \tag{3}$$

Here, $\mathcal{E}$ is the fluid internal energy, $\boldsymbol{P}$ the fluid impulsion and $\boldsymbol{U}$ the velocity of the moving cylinder. The potential $V(r) = (V_0/2)(\tanh[4(r - d/2)/\Delta] - 1)$ is used to represent a cylindrical obstacle of diameter D. The calculations presented below were realized with $V_0 = 10$ and $\Delta = \xi$. With these values, the fluid density inside the disc is negligible and the density boundary layer is well resolved with a mesh adapted to the coherence length. The NLSE is the Euler Lagrange equation corresponding to (1)

$$\frac{\delta \psi}{\delta t} = -\frac{i}{\sqrt{2}c\xi} \frac{\delta \mathcal{F}}{\delta \bar{\psi}} = i \frac{c}{\sqrt{2}\xi} \left([1 - V(\boldsymbol{x})]\psi - |\psi|^2 \psi + \xi^2 \nabla^2 \psi \right) + \boldsymbol{U} \cdot \nabla \psi. \tag{4}$$

It can be mapped into two hydrodynamical equations by applying Madelung's transformation [12,13] $\psi = \sqrt{\rho} \exp \left(\frac{i\phi}{\sqrt{2}c\xi} \right)$. The real and imaginary parts of the NLSE produce, for a fluid of density ρ and velocity $\boldsymbol{v} = \nabla \phi - \boldsymbol{U}$, the following equations of motion

$$\frac{\partial \rho}{\partial t} + \nabla (\rho \boldsymbol{v}) = 0 \tag{5}$$

$$\left[\frac{\partial \phi}{\partial t} - \boldsymbol{U} \cdot \nabla \phi \right] + \frac{1}{2}(\nabla \phi)^2 + c^2[\rho - \Omega(\boldsymbol{x})] - c^2\xi^2 \frac{\nabla^2 \sqrt{\rho}}{\sqrt{\rho}} = 0. \tag{6}$$

These correspond to the continuity and Bernoulli equations [14]. The last term in (6) is a dispersive supplementary *quantum pressure* term that is relevant only at length scales smaller than ξ. In the limit where $\xi/D \to 0$, the quantum pressure term vanishes and we recover the system of equations describing an Eulerian flow. Note that the NLSE admits vortical solutions of characteristic core size $\sim \xi$. These are topological defects (zeros) of ψ and thus appear as points in 2D and lines in 3D.

3   Numerical Methods

Time integration was done by using a fractional step (Operator-Splitting) method [15]. The bifurcation diagrams of stationary solutions to the NLSE describing a 2D superflow around a disc were obtained using Fourier pseudospectral methods [16] and continuation techniques [17].

When the extremum of $\mathcal{F}$ is a local *minimum*, the stationary solution ψ_S of (4) can be reached by integrating to relaxation the associated real Ginzburg-Landau equation (RGLE) [3] given by $\partial\psi/\partial t = -(1/\sqrt{2}c\xi)\delta\mathcal{F}/\delta\bar{\psi}$.

The *unstable* stationary solutions of the RGLE were found by using Newton's method [17]. In order to work with a well-conditioned system [18], we search for the roots of the following associated relation

$$f(\psi) = \Theta^{-1}\left[(1 - i\,\sigma\,\boldsymbol{U}\cdot\nabla) + \sigma\frac{c}{\sqrt{2}\xi}\left([1 - V(\boldsymbol{x})] - |\psi(t)|^2\right)\right]\psi(t) - \psi(t), \quad (7)$$

with $\Theta = \left[1 - \sigma\left(c\xi/\sqrt{2}\right)\nabla^2\right]$. Calling $\psi_{(j)}$ the value of the field ψ over the j-th collocation point, finding the roots of $f(\psi)$ is equivalent iterating the Newton step $\psi_{(j)} = \psi_{(j)} + \delta\psi_{(j)}$ up to convergence. Every Newton step requires the solution for $\delta\psi_{(k)}$ of $\sum_k \left[df_{(j)}/d\psi_{(k)}\right]\delta\psi_{(k)} = -f_{(j)}(\psi)$. This solution is obtained by an iterative bi-conjugate gradient method (BCGM) [19]. Here, σ is a free parameter that can be used to adjust the pre-conditioning of the system in order to optimise the convergence of the BCGM [18].

4   Bifurcation Diagram and Scaling in 2D

Using the numerical techniques described above, the bifurcation diagrams were computed for various values of ξ/D.

The functional $\mathcal{E}$ and energy $\mathcal{F}$ of the stationary solutions are shown in Fig. 1 for $\xi/D = 1/10$ as a function of the Mach number ($M = |U|/c$). The stable branch (a) disappears with the unstable solution (c) at a saddle-node bifurcation when $M = M^c \approx 0.4286$. The energy $\mathcal{F}$ has a cusp at the bifurcation point, which is the generic behaviour for a saddle-node. There are no stationary solutions beyond this point. When $M \equiv M^{pf} \approx 0.4282$, the unstable symmetric branch (c) bifurcates at a pitchfork to a pair of nonsymmetric branches (b). The nucleation energy barrier of these branches is given by $(\mathcal{F}_{b'} - \mathcal{F}_{a'})$ which is roughly half of the barrier for the symmetric branch $(\mathcal{F}_{c'} - \mathcal{F}_{a'})$. The arrow on the bifurcation diagram in Fig. 1 shows the Mach number value at which a vortex appears in unstable stationary solutions. When $M^n \leq M \leq M^c$, solutions are

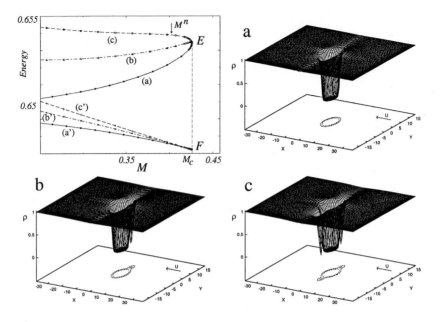

Fig. 1. Plot of the energy ($\mathcal{F}$), and functional ($\mathcal{E}$) versus Mach number ($M = |U|/c$) with $D = 10\xi$ (upper left image). The $\rho = |\psi|^2$ fields of the stable state (a), asymmetric branch (b) and symmetric branch (c) are shown for $M = 0.24$ in frames a, b and c respectively (see text)

irrotational. For $M \leq M^n$ the stable branch (a) remains irrotational (Fig. 1a) while the unstable branch (b) corresponds to a one vortex solution (Fig. 1b) and the unstable branch (c), to a two vortex solution (Fig. 1c). The distance between the vortices and the obstacle in branches (b) and (c) increases when M is decreased.

Recent 3D simulations show that equivalent solutions exist for the NLS flow past a sphere. In that case, symmetric solutions correspond to a stationary ring around the sphere while nonsymmetric ones present a vortex loop pinned to the sphere [20].

We now characterize the dependence on ξ/D of the main features of the bifurcation diagram. When ξ/D is decreased, M^c and M^{pf} become indistinguishable. In the limit where $\xi/D = 0$, the critical Mach number M^c will be that of an Eulerian flow M^c_{Euler}. Figure 2 shows the convergence of M^c to the Eulerian critical velocity. This convergence can be characterized by fitting the polynomial law $M^c = K_1(\xi/D)^{K_2} + M^c_{\text{Euler}}$ to $M^c(\xi/D)$. This fit is shown on Fig. 2 as a dotted line, yielding $K_1 = 0.322$, $K_2 = 0.615$ and $M^c_{\text{Euler}} = 0.35$. This result is compatible with the analytical approximate results presented in [21,22], that predicts a square root ($K_2 = 0.5$) polynomial dependence on ξ/D. The dashed line M^*_1 on Fig. 2 corresponds to the critical velocity $M^c = \sqrt{2/11}$ obtained in [1] by applying a local sonic criterion. It is shown in [3] that an iterative scheme

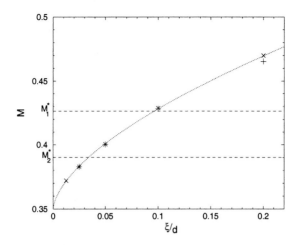

Fig. 2. Saddle-node bifurcation Mach number M^c ($\times$) and pitchfork bifurcation Mach number M^{pf} ($+$), as a function of ξ/D. The dotted curve corresponds to a fit to the polynomial law $M^c = K_1(\xi/D)^{K_2} + M^c_{\mathrm{Euler}}$. The dashed lines M^*_1 and M^*_2 correspond respectively to first and second order compressible corrections to M^c (see text)

can be used to improve the $M^c = \sqrt{2/11}$ result. The next order [23] to the critical Mach number is $M^*_2 = \sqrt{\sqrt{233} - 11}/(2\sqrt{7}) \approx 0.3903$.

5   Subcriticality and Vortex-Stretching in 3D

We used the 2D laminar stationary solution $\psi_{0V}(x, y)$ (corresponding to branch (a) on Fig. 1) and the one-vortex unstable stationary solution $\psi_{1V}(x, y)$ (branch (b)) to construct the 3D initial condition

$$\psi_{3D}(x, y, z) = f_1(z)\psi_{1V}(x, y) + [1 - f_1(z)]\psi_{0V}(x, y) \qquad (8)$$

(Fig. 3a). The function $f_1(z) = (\tanh[(z - z_1)/\Delta_z] - \tanh[(z - z_2)/\Delta_z])/2$, takes the value 1 for $z_1 \leq z \leq z_2$ and 0 elsewhere (Δ_z is an adaptation length). The resulting dynamical evolution can be schematically described in terms of short-time and long-time dynamics.

During the short-time dynamics, the initial pinned vortex line rapidly contracts, evolving through a decreasing number of half-ring-like loops, down to a single quasi-stationary half-ring (see Figs. 3b-3d). The evolution takes place near the plane perpendicular to the flow, provided that the initial vortex is long enough to contract to a quasi-stationary half-ring as shown on Fig. 3d. Otherwise, the vortex line collapses against the cylinder while moving upstream. On a longer time scale, the quasistationary half-ring can either start moving upstream or downstream. When driven downstream, the vortex loop is continuously stretched while the pinning points move towards the back of the cylinder. When the half ring moves upstream, it eventually collapses against the cylinder, generating a laminar superflow.

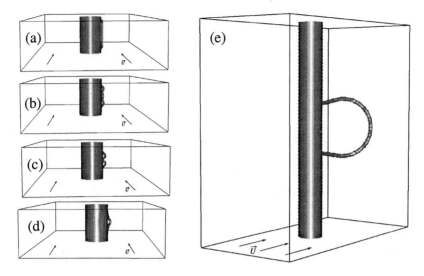

Fig. 3. Left: Short-time dynamics of a vortex pinned to the cylinder with $\xi/D = 1/40$, $|U|/c = 0.26$ and $\Delta_z = 2\sqrt{2}\xi$ in the $[L_x \times L_y \times L_z]$ periodicity box ($L_x/D = 2.4\sqrt{2}\pi$, $L_y/D = 1.2\sqrt{2}\pi$ and $L_z/D = 0.4\sqrt{2}\pi$). The surface $|\psi_{3D}| = 0.5$ is shown at times (A) $t = 0$, (B) $t = 5\xi/c$, (C) $t = 10\xi/c$ and (D) $t = 15\xi/c$. Right (E): Vortex stretching at $t = 150\xi/c$ with $|U|/c = 0.35$ and $\xi/D = 1/20$

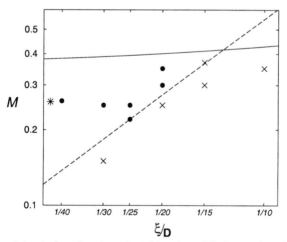

Fig. 4. Parametric study of vortex stretching in a Mach number M versus ξ/D log-log diagram. Circles: stretching, crosses: no stretching. The asterisk represents the experiment reported in [6]. The solid line is the 2D saddle-node bifurcation Mach number M_{2D}^c. The dashed line corresponds to $R_s = 5.5$ (see text)

In order to distinguish between the two situations we have carried out the 3D runs that we display schematically on Fig. 4. The runs with vortex stretching are labeled by circles and those without by ×. Note that all runs were performed at

Mach numbers below the 2D saddle-node bifurcation Mach number $M_{2D}^c(\xi/D)$. The critical Mach number for the onset of dissipation in the experiment where a blue detuned laser beam is moved through a Bose Einstein condensed gas [6] is marked by an asterisk.

Note that the numerical study presented above does not focus on the critical velocity (M_{3D}^c) at which 3D stationary solutions disappear. This problem has been recently addressed in [24,20]. Our approach is rather to propose a mechanism and criteria for dissipation through vortex stretching at $M < M_{3D}^c$.

A frontier between the dissipative and nondissipative cases can be drawn approximately as the dashed line in Fig. 4, which corresponds to the expression $R_S = 5.5$ with $R_S \equiv |U|D/c\xi = MD/\xi$. This superfluid "Reynolds" number is defined in the same way as the standard (viscous) Reynolds number Re $\equiv$ $|U|D/\nu$ (with ν the kinematic viscosity). In the superfluid turbulent regime ($R_S \gg 1$), R_S was shown to be equivalent to the standard (viscous) Reynolds number Re [9–11].

We acknowledge useful scientific discussions with V. Hakim, L. Tuckerman and E. Varoquaux. This work was supported in part by the ECOS-CONICYT program no. C96E01 and by the MRSEC Program of the National Science Foundation under Award Number DMR 9808595. Computations were performed at the Institut du Développement et des Ressources en Informatique Scientifique.

References

1. T. Frisch, Y. Pomeau, and S. Rica: Phys.Rev.Lett., **69**, 1644 (1992)
2. C. Huepe and M.-E. Brachet: C. R. Acad. Sci. Paris, **325**(II), 195–202 (1997)
3. C. Huepe and M. E. Brachet: Physica D, **140**, 126–140 (2000)
4. C. Huepe: Bifurcations et instabilités dans les condensats de Bose Einstein et les écoulements superfluides. PhD thesis, Ecole Normale Supérieure, Paris (1999)
5. Franco Dalfovo, Stefano Giorgini, Lev P. Pitaevskii, and Sandro Stringari: Reviews of Modern Physics, **71**(3), (1999)
6. C. Raman, M. Köhl, R. Onofrio, D.S. Durfee a nd C.E. Kuklewicz, Z. Hadzibabic, and W. Ketterle: Phys. Rev. Lett., **83**(13), 2502 (1999)
7. C. Nore, C. Huepe, and M. E. Brachet: Phys. Rev. Lett., **84**(10), 2191 (2000)
8. C. Huepe, S. Métens, G. Dewel, P. Borckmans, and M.-E. Brachet: Phys. Rev. Lett., **82**(2), 1616 (1999)
9. C. Nore, M. Abid, and M. Brachet: Phys. Rev. Lett., **78**, 3896 (1997)
10. C. Nore, M. Abid, and M. Brachet: Phys. Fluids, **9**(9), 2644 (1997)
11. M. Abid, M. E. Brachet, J. Maurer, C. Nore, and P. Tabeling: Eur. J. Mech. B/Fluids, **17**(4), 665 (1998)
12. R. J. Donnelly: *Quantized Vortices in Helium II.* (Cambridge Univ. Press, Cambridge 1991)
13. E. A. Spiegel: Physica D, **1**, 236 (1980)
14. L. Landau and E. Lifschitz: *Mécanique des fluides, vol. 6.* (Editions Mir, 1989)
15. R. Klein and A. J. Majda: Physica D, **53**, 267 (1991)
16. D. Gottlieb and S. A. Orszag: *Numerical Analysis of Spectral Methods.* (SIAM, Philadelphia 1977)
17. R. Seydel: *From Equilibrium to Chaos: Practical Bifurcation and Stability Analysis.* (Elsevier, New York 1988)

18. C. Mamun and L. Tuckerman: Phys. Fluids, **7**(1), 80 (1995)
19. W. Press, S. Teukolsky, W. Vetterling, and B.Flannery *Numerical Recipes.* (Cambridge Univ. Press, Cambridge, 1994)
20. T. Winiecki,J.F. McCann, C.S. Adams: Europhys. Lett., **48**(5), 475–481 (1999)
21. C. Josserand: Dynamique des Superfluides: Nucléation de vortex et Transition de phase du permier ordre. PhD thesis, Ecole Normale Supérieure, Paris (1997)
22. C. Josserand, Y. Pomeau, and S. Rica: Physica D, **134**(1), 111–125 (1999)
23. V. Hakim. private communication. 1999.
24. N.G. Berloff, and P.H. Roberts: J. Phys. A, **33**, 4025–4038, 2000.

Part VI

Bose–Einstein Condensation

Motion of Objects
Through Dilute Bose–Einstein Condensates

C.S. Adams[1], B. Jackson[1], M. Leadbeater[1], J.F. McCann[2], and T. Winiecki[1]

[1] Dept. of Physics, University of Durham, Rochester Building, South Road, Durham, DH1 3LE, England. UK
[2] Dept. of Applied Mathematics and Theoretical Physics, Queen's University, Belfast, BT7 1NN, Northern Ireland. UK

Abstract. This paper discusses the motion of objects through quantum fluids described by the Gross Pitaevskii (GP) equation. The object moves without dissipation at velocities below a threshold which corresponds to the critical velocity for vortex nucleation. Above the critical velocity, vortex shedding is the dominant mechanism of energy transfer between the object and the fluid. We compare the predictions of the GP model with experiments on an oscillating laser beam in an alkali vapour Bose Einstein condensate and ions in superfluid helium-4.

1   Introduction

A wide variety of quantum fluids are now accessible experimentally, ranging from the dilute Bose-Einstein condensates produced in atomic vapours [1] to liquid helium and superconductors [2]. The dilute Bose gas is a special case in that the interactions can be accurately represented by a mean-field leading to a relatively simple theoretical description based on the Gross Pitaevskii (GP) equation, a form of the non-linear Schrödinger equation [3]. Consequently one enticing product of the recent discovery of Bose Einstein condensation in dilute alkali vapours is the potential for refining our understanding of quantum fluids.

The dilute Bose Einstein condensate is identical to a classical Euler fluid except for the quantisation of circulation and the influence of the kinetic energy term appearing in the GP equation. The latter produces fluid healing and shear stresses which enable vortex formation without viscosity [4]. The conservation of circulation (Kelvin's theorem) requires that vortex lines are created in pairs or as rings which emerge from a point [5], or at boundaries [6]. Recent experiments on dilute atomic vapours have confirmed many of the predictions of the GP model. For example, the expected vortex solutions have been observed in rotating condensates [7,8], and experiments with a moving object [9,10] are consistent with GP predictions of the critical velocity for vortex formation [11].

In this paper, we consider the motion of objects through homogeneous and inhomogeneous condensates. In Sect. 2 we briefly review the fluid properties described by the GP equation. In Sect. 3 we consider the process of vortex formation in terms of the stationary uniform flow solutions of the GP equation. We discuss how vortices can exist in laminar flow solutions below the critical velocity, and that there is a 'smooth' transition between flows without and with vortices.

In Sect. 4 we use the uniform flow solutions to determine the critical velocity, and in Section 5 we return to the time-dependent solutions to study the link between vortex shedding and drag. In Sect. 6, we consider the motion through an inhomogeneous condensate, and show that if the object passes through lower density regions, then the critical velocity is significantly lowered, as observed experimentally [9,10]. Finally in Sect. 7 we compare the results of experiments on ions in superfluid helium and the predictions of the GP model.

2   Fluid Equations

The evolution of a dilute Bose-Einstein condensate in the limit of low temperature is given by the solution of the time-dependent Gross-Pitaevskii (GP) equation,

$$i\hbar\partial_t\psi(\mathbf{r},t) = \left[-\frac{\hbar^2}{2m}\nabla^2 + V(\mathbf{r},t) + g|\psi(\mathbf{r},t)|^2\right]\psi(\mathbf{r},t) , \tag{1}$$

where ψ is normalised to the number of condensate particles, the coefficient of the non-linear term, $g = 4\pi\hbar^2 a/m$, a is the s-wave scattering length, $V(\mathbf{r},t)$ represents external potentials arising from the trap and any moving obstacle. The link between equation (1) and the equivalent equations of fluid mechanics is made by defining a mass density, $\rho = m\psi^*\psi$, and a momentum current density, $J_k = (\hbar/2i)(\psi^*\partial_k\psi - \psi\partial_k\psi^*)$, then the fluid velocity is given by, $v_k = J_k/\rho$, or equivalently $v_k = (\hbar/m)\partial_k\varphi$, where φ is the phase of the field ψ. By definition, the velocity field corresponds to a potential flow, however, it can support circulation at singularities (vortices).

The continuity equation (conservation of mass or probability) follows from the definition of ρ and equation (1),

$$\partial_t\rho + \partial_k J_k = 0 . \tag{2}$$

The conservation of momentum equation may be found by considering the rate of change of the momentum current density,

$$\partial_t J_k + \partial_j T_{jk} + \rho\partial_k(V/m) = 0 , \tag{3}$$

where the momentum flux density tensor takes the form [12,13],

$$T_{jk} = \frac{\hbar^2}{4m}(\partial_j\psi^*\partial_k\psi - \psi^*\partial_j\partial_k\psi + \text{c.c.}) + \frac{g}{2}\delta_{jk}|\psi|^4 . \tag{4}$$

This can be rewritten as,

$$T_{jk} = \rho v_j v_k - \sigma_{jk} , \tag{5}$$

where the stress tensor σ_{jk} is given by,

$$\sigma_{jk} = -\tfrac{1}{2}\delta_{jk}g(\rho/m)^2 + (\hbar/2m)^2\rho\partial_j\partial_k\ln\rho . \tag{6}$$

The form of equations (2), (3), and (5) is identical to those for classical fluid flow [15], except for the stress, equation (6). The normal stress, $-\sigma_{jk}, j = k$, or pressure is given by,

$$p = \tfrac{1}{2}g(\rho/m)^2 - (\hbar/2m)^2 \rho \nabla^2 \ln \rho .\tag{7}$$

The first term, the interaction pressure, supports the propagation of sound. In the limit of long wavelength excitations (i.e, the linear part of the Bogoliubov spectrum), the second term, known as the quantum pressure, is neglibible, and the speed of sound is given by [14,15],

$$c = \sqrt{\partial p/\partial \rho} = (g\rho/m^2)^{\frac{1}{2}} .\tag{8}$$

For a dilute Bose-Einstein condensate shear stresses, $\sigma_{jk}, j \neq k$, arise from density gradients, due the second term in (6). This property gives rise to the possibility of vortex formation and drag without viscosity.

Combining equations (2) and (3), along with the identity, $\rho^{-1}\partial_j[\rho\partial_j\partial_k \ln \rho] = 2\partial_k[\rho^{-\frac{1}{2}}\partial_j\partial_j\rho^{\frac{1}{2}}]$ the momentum equation may be rewritten as,

$$\partial_t v_k + v_j \partial_j v_k + \partial_k[g\rho/m^2 - (\hbar^2/2m)\rho^{-\frac{1}{2}}\partial_j\partial_j\rho^{\frac{1}{2}} + V/m] = 0 ,\tag{9}$$

which corresponds to the GP form of the Euler equation. By integrating, or more directly from the real part of equation (1), one obtains the conservation of energy or Bernoulli equation,

$$\hbar\partial_t\phi + \tfrac{1}{2}mv^2 + g\rho/m - (\hbar^2/2m)\rho^{-\frac{1}{2}}\nabla^2\rho^{\frac{1}{2}} + V = 0 .\tag{10}$$

The conservation of angular momentum (Kelvin's theorem), follows from Euler's equation (9) and states that the circulation around a closed 'fluid' contour does not change in time. This means that within the fluid, vorticity must be created in the form of rings or pairs of lines which emerge from a point [5]. The exception is at boundaries, where the wavefunction is clamped to zero and no closed fluid loop can be drawn, e.g., at the surface of an impenetrable object [15] or from the edge of a trapped condensate [6].

3   Time-Independent Solutions in the Object Frame

For homogeneous fluids, where the external potential is due to the obstacle only, it is convenient to rescale length and velocity in terms of the healing length, $\xi = \hbar/\sqrt{mn_0 g}$, and the sound speed, $c = \sqrt{n_0 g/m}$, respectively, where n_0 is the number density far from the object. In this case, for an object moving with velocity U, equation (1) becomes

$$i\partial_t\tilde{\psi}(r,t) = \left[-\tfrac{1}{2}\nabla^2 + V(r - Ut) + |\tilde{\psi}(r,t)|^2\right]\tilde{\psi}(r,t) ,\tag{11}$$

where $\tilde{\psi} = \psi/\sqrt{n_0}$. Unless otherwise stated we use these units throughout. If $\psi'(r',t) = \tilde{\psi}(r,t)$ is the field in the fluid frame written in terms of the object frame coordinates, $r' = r - Ut$, then the GP equation for $\psi'(r',t)$ is

$$i\partial_t\psi' = -\tfrac{1}{2}\nabla^2\psi' + V(r')\psi' + |\psi'|^2\psi' + iU \cdot \nabla'\psi' .\tag{12}$$

Solutions of (12) of the form, $\tilde{\psi}(r', t) = \phi(r')e^{i\mu t}$, where μ is the chemical potential, are found only for $U \leq U_c$, where U_c is the critical object velocity for vortex formation [16–18]. Consequently, such time-independent solutions provides a useful technique to determine the critical velocity as illustrated in Sect. 4.

The time-independent solutions of (12) also characterise the process of vortex formation. In Fig. 1(left) we plot the energy dependence of the solutions corresponding to a penetrable sphere with radius $R = 3.3$. We define the energy relative to a laminar flow state with $V = 0$ and having the same number of particles, i.e.,

$$E = \int dr \left[\frac{1}{2} |\nabla \phi|^2 + V |\phi|^2 + \frac{1}{2} \left(|\phi|^2 - 1 \right)^2 \right] . \qquad (13)$$

Away from the object, any deviation of the local particle density $|\phi|^2$ from 1 constitutes an excitation. The three branches correspond to (in order of increasing energy) laminar flow with no vortices, a vortex ring pinned to the sphere, and a vortex ring encircling the sphere. These states are illustrated by surfaces of constant density shown in Fig. 2. A more detailed discussion is given in Ref. [17].

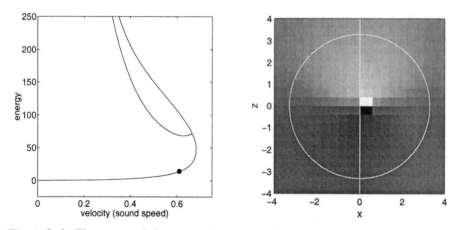

Fig. 1. Left: The energy of the time-independent solutions of the GP equation for flow past a penetrable sphere with radius $R = 3.3$ ($V(r) = 1.0$, $r \leq R$, $V(r) = 0.0$, $r > R$) as a function of the flow velocity. No time-independent solutions exist for flow above a critical velocity, $U_c = 0.68$. The three branches correspond to (in order of increasing energy) laminar flow, a pinned vortex ring, and an encircling vortex ring. The vortex ring appears in the laminar flow solution at $U = 0.62$ (as indicated), well below the critical velocity. Right: Cross sections showing the phase pattern of the laminar flow solutions at $U = 0.615$ (left side) and $U = 0.625$ (right side). The object is indicated by the white circle. At $U = 0.625$ the phase changes from $-\pi$ (black) to π (white) near the centre of the object indicating the appearance of the vortex ring. As the energy increases the vortex ring grows, eventually crossing the object boundary at $U = 0.61$ on the encircling ring branch.

Fig. 2. Surfaces of constant density ($|\phi|^2 = 0.25$) showing the three steady-state solutions associated with motion of a spherical object: from left to right, laminar flow; pinned ring; encircling ring.

Figure 1(right) shows a cross-section of the phase through the centre of the object at velocities of $U = 0.615$ and 0.625. The grid used for the numerical calculation is clearly visible. The rapid variation of the phase at $U = 0.625$ indicates the emergence of a vortex ring from a point at the centre of the object, in agreement with Kelvin's theorem. Also significant is that the vortex first appears in the laminar flow branch of the solution well below the critical velocity, $U_c = 0.68$. This point highlights the gradually nature of the transition from laminar flow to flows containing vortices.

If an object is subject to a constant force, the time-dependent evolution follows the stationary solutions as shown in Fig. 3. The external force accelerates the object along the laminar flow solution (1) up to a peak velocity, $U_c = 0.68$, where an encircling vortex ring emerges and begins to slow the object down (the back-action of the fluid on the object motion is included in the calculation [19]). Subsequently the object jumps into the vortex core exciting oscillations in the vortex ring. This process is illustrated in Roberts and Berloff, this volume, Fig. 4. Even for impenetrable objects [17], there is no direct path between laminar flow and a pinned vortex ring. Consequently, the symmetric encircling vortex ring state always forms before the ring becomes pinned.

If the force is large, the object can escape from the vortex ring, as in Fig. 3(c). In this case, the cycle repeats resulting in periodic vortex shedding. For object with infinite mass, the vortices detach immediately after nucleation and the time-independent pinned and encircling vortex ring solutions are not reached. We will return to the case of an infinitely massive object and vortex shedding in Sect. 5.

4   The Critical Velocity

The critical velocity for the breakdown of superfluidity is often expressed in terms of the Landau condition [14], $v_L = (\epsilon/p)_{min}$, where ϵ and p are the energy and momentum of elementary excitations in the fluid. In the dilute Bose gas, the long wavelength elementary excitations are sound waves and the Landau criterion predicts that $v_L = c$. However, for flow past an object, the maximum velocity difference within the fluid approaches c, when the object velocity, U, is only a fraction of the sound speed. In a GP fluid, an additional complication is that the sound speed depends on the density, equation (8), and density gradients

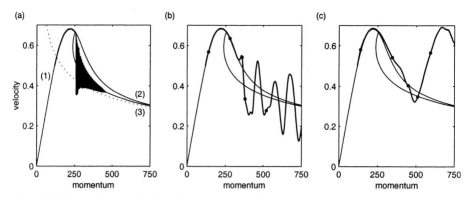

Fig. 3. The evolution of the object velocity, U, with initial value $U = 0.5$, due to a constant external force, $\boldsymbol{F}$. The time-dependent solution (bold line) is plotted on top of the corresponding time-independent solutions of (12) as a function of the total momentum $\boldsymbol{P} = \boldsymbol{F}t$ for (a) $\boldsymbol{F} = 0.05$, (b) $\boldsymbol{F} = 2$ and (c) $\boldsymbol{F} = 4$. In (a), the initial laminar flow state (1) evolves into an encircling vortex ring (2), followed by an abrupt decrease in velocity when the object jumps into the vortex core. This transition to the pinned ring state (3) excites vibrations of the ring leading to oscillations in the object velocity. As the vortex ring radius increases the object velocity tends towards the velocity of a free vortex ring, as indicated by the dotted line. In (c), the force is sufficient to detach the object from the ring and the cycle repeats. The dots in (b) and (c) correspond to the times of the isosurface plots shown in Roberts and Berloff, Fig. 4.

due to the quantum pressure term. By solving equation (12), one finds that the critical object velocity, U_c, is reached when the flow velocity, v, exceeds the local sound speed, $c = |\psi|$, in a region of the fluid where the quantum pressure is zero. This is illustrated in Fig. 4 which shows the flow velocity, the sound speed, and the quantum pressure term near the surface of a large ($R = 50$) impenetrable sphere.

In the limit of large radius, $R \to \infty$ (where one can neglect the boundary layer and hence the quantum pressure term), one can predict the critical velocity analytically using an asymptotic expansion for the speed at the equator, $v = \frac{3}{2}U + 0.313U^3 + 0.392U^5 + \ldots$ [20], and correcting for the dependence of the sound speed on density. In steady state, neglecting the quantum pressure, the Bernoulli equation (10) states that the sum of the interaction and kinetic energy terms near the object and in the far-field must be equal, i.e.,

$$1 + \frac{1}{2}U^2 = n + \frac{1}{2}v^2 \ .$$

The solution becomes unstable when the flow velocity becomes equal to the sound speed, $v = \sqrt{n}$, [21,22]. Combining these results (including terms up to U^{11}) gives $U_c = 0.53004$ for $R \to \infty$, in good agreement with our numerical value $U_c = 0.53285$ for $R = 50$. The critical velocity decreases with increasing R tending asymptotically towards the $R = \infty$ value [17,18].

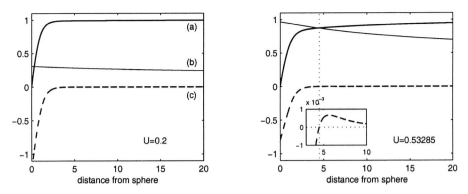

Fig. 4. Laminar flow solution for the motion of a sphere with radius $R = 50$ and velocity $U = 0.2$ (left) and $U = 0.53285$ (right). The three curves show, (a) the 'bulk' sound speed, $|\psi| = \sqrt{n}$, (b) the velocity, v, and (c) the quantum pressure term, $\nabla^2 |\psi|/2|\psi|$, as a function of position. Note that, at the critical velocity $U_c = 0.53285$ (right), $v = \sqrt{n}$, where the quantum pressure is zero.

For trapped condensates the speed of sound and hence the critical velocity depend upon position. This can lead to a significant lowering of the critical velocity if the object passes through lower density regions of the condensate as discussed in Section 6.

5   Vortex Shedding and Drag

The main significance of the critical velocity is that it marks the threshold for the breakdown of superfluidity and the on-set of dissipation. For macroscopic objects, where vortex pinning can be neglected, motion faster than the critical velocity, $U > U_c$, results in periodic vortex shedding. The energy and momentum transfer to the fluid leads to a drag force on the object. The drag force $\boldsymbol{F}$ may be calculated either from the energy transfer

$$\frac{\mathrm{d}E}{\mathrm{d}t} = \boldsymbol{F}.\boldsymbol{v} \; , \tag{14}$$

or the rate of momentum transfer: By integrating equation (3), one finds that the k-th component of the force,

$$F_k = \partial_t \int_\Omega \mathrm{d}\Omega J_k = -\int_S \mathrm{d}S \, n_j T_{jk} - \int_\Omega \mathrm{d}\Omega \, \rho \partial_k (V/m) \; , \tag{15}$$

where S is the surface of the object or control surface within the fluid [13], Ω is the volume enclosed by S, n_j is the j-component of the normal vector to S, and $\mathrm{d}S$ is a surface element. The second term on the right-hand side can be likened to the buoyancy of the fluid. In the case of homogeneous flow past an impenetrable object, Sect. 4 and 5, $\psi = 0$ on the object surface and the potential is uniform elsewhere, therefore, only the first term contributes. Conversely, for a penetrable object in a trapped condensate, Sect. 6, Ω may be chosen to encompass the entire fluid, and the first term is zero.

A plot of the drag force as a function of velocity for a cylindrical object is shown in Fig. 5. The drag is calculated by applying equation (15) to the solutions of equation (1). The vortices in the wake of a moving object interact creating fluctuations in the flow pattern and the drag, therefore the data shown in Fig. 5 correspond to an average over many vortex emission cycles. The force per unit length is measured in units of $\hbar\sqrt{n_0^3 g/m}$. The threshold velocity for the on-set of drag is 0.4 as expected for a cylindrical object [22].

The contribution of vortex shedding to the total drag force can be estimated by considering the momentum transfer due to vortex emission, i.e.,

$$\boldsymbol{F}_{\mathrm{v}} = f_{\mathrm{v}}\boldsymbol{p}_{\mathrm{v}} \ , \tag{16}$$

where f_{v} is the vortex shedding frequency and $\boldsymbol{p}_{\mathrm{v}}$ is the momentum of a vortex pair as it is created in the equatorial plane. The comparison between equations (15) and (16) shown in Fig. 5 suggests that for $U < c$ vortex shedding is the dominant dissipation mechanism, whereas for $U > c$, an increasingly significant contribution arises from sound waves. For $U > c$, the reflected sound waves create a standing wave pattern in front of the object as shown in the time-averaged density images of Fig. 6. In the far-field the orientation of the standing waves corresponds to the Mach angle as in classical acoustics.

Figure 6 also displays some interesting features in the wake. The dark streaks emerging from the shadow region behind the object correspond to sound waves emitted by interacting vortices. As a pair of vortex lines loses energy they move closer together and eventually annihilate. Far downstream, a significant fraction of the vortex energy is converted into sound. This conversion of vortex energy into sound energy due to vortex motion and reconnections [23] is an important contributary process in the the decay of superfluid turbulence (see Chapters by Vinen and McClintock).

6   The Critical Velocity in Inhomogeneous Condensates

Experiments on moving laser beams in trapped atomic condensates suggest a critical velocity of $U_{\mathrm{c}} \sim 0.1$ [9,10], much lower than the homogeneous value for an cylindrical object $U_{\mathrm{c}} \sim 0.4$ [22]. Vortex stretching [24] and enhanced phonon excitation at low velocity due to the spatial inhomogeneity [25] have been proposed as possible mechanisms for this reduction. Numerical simulations of trapped atomic condensates in three dimensions indicate that $U_{\mathrm{c}} \sim 0.1 - 0.2$ [11] in rough agreement with experiment. Fig. 7(left) shows the energy transfer to the condensate as a function of the object velocity. In the experiments this energy is measured as a heating effect [9,10]. At finite temperature, the conversion of energy into heat may be accerlated by the interaction with the thermal cloud [26], however, even in the limit $T \to 0$ one would expect a decay of the vortex structures as discussed in connection with Fig. 6. Apart from the lower critical velocity the behaviour is similar to the idealised homogenous case discussed in

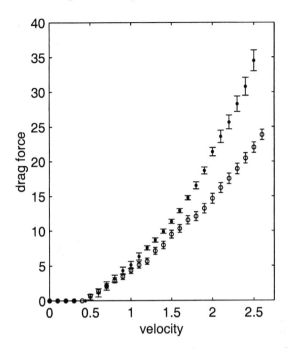

Fig. 5. The time-averaged drag force (black circles) as a function of velocity for an impenetrable cylinder with radius $R = 3$. The force per unit length is measured in units of $\hbar\sqrt{n_0^3 g/m}$. The open circles indicates the contribution due to vortex shedding. The error bars indicate the residual fluctuations in the time-averaged drag.

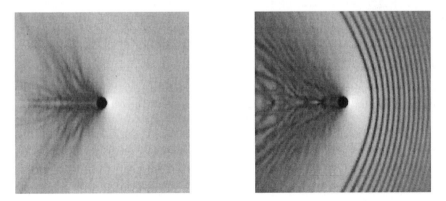

Fig. 6. Time-averaged fluid density in the frame of a cylindrical object (radius $R = 3$) moving from left to right with speeds of 0.9 (left) and 1.2 (right). Dark and light indicate low and high density respectively. For supersonic motion a standing wave pattern appears in front of the object.

Sect. 5. We suggest that the lower critical velocity is mainly due to formation of vortices where the object intersects lower density regions of the condensate. Fig. 7(right) shows how at low object speeds ($U = 0.24c$) vortices enter the condensates as half rings at the surface. These half rings penetrate towards the centre along the low density regions close to the object axis, eventually connecting to form a pair of lines. In this case, the reconnection is being driven by an external force and the vortex line length increases. This is the inverse of the usual reconnection process where the vortex line length decreases [23]. The subsequent motion of the vortex lines is determined by density gradients and the interaction with other vortices [27].

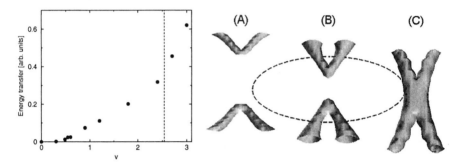

Fig. 7. Three dimensional GP simulations of a moving laser beam in a trapped Bose-Einstein condensate. In this example, distance and time are measured in terms of the harmonic oscillator length, $\sqrt{\hbar/2m\omega}$, and the inverse trap frequency, $1/\omega$. Left: The rate of energy transfer to the condensate as a function of the laser velocity. The critical velocity is $U_c \sim 0.1c$, where $c = 2.54$ is the sound speed at the centre of the condensate (indicated by the dashed line). Right: Isosurface plots ($v^2 = 10^{-3}$), showing the evolution of the vortices at times, (A) 6.8, (B) 7.6, and (C) 8.4, for an object velocity, $U = 0.24c$. Also shown in (B) is a dashed circle in the $x-y$ plane, indicating the spatial extent of the condensate given by the Thomas-Fermi radius, $R_{TF} = 3.59$. The laser beam is moving into the paper and downwards along the normal to the dashed circle. Vortices are seen to nucleate at the surface as half-rings, before penetrating into the bulk to reconnect as a vortex pair. Other parameters match those given in Fig. 5 of [11].

The notion that vortices are responsible for dissipation is further supported by the recent experimental observations of the density dependence of the critical velocity and good agreement with predictions of the heating rate [10]. More evidence could be gained by comparing with theoretical predictions of the variation of U_c with laser beam intensity [11] and radius [21].

7   Comparison to Ions in Helium

In superfluid He-4 interactions lead to a depletion of the condensate and a different dispersion curve to the dilute Bose gas [2,28]. A modified GP model which includes the non-local nature of the interactions has been used to reproduce the correct dispersion curve [29–31]. The GP equation can also be applied to describe a finite temperature system [32,33], or similarly the depletion of the condensate due to interactions [34]. However, some insight can be gained by applying the standard GP model to the motion of objects in helium. For example, existing theories of vortex nucleation cannot treat the intermediate states where the vortex is less than a healing length from the object [35]. An advantage of the GP equation is that fluid healing and the quantisation of circulation are included explicitly, making it possible to identify a complete path of vortex nucleation. Using the GP model, both the time-independent solutions (Sect. 3) and time-dependent simulations [19] suggest that an object can be accelerated up to a maximum velocity, where a vortex ring gradually emerges encircling the object, and subsequently becomes pinned to it. An important question is whether the spherical symmetry favours the encircling ring state. In Fig. 8 we show the time-independent states for a sphere with a bump. The bump lowers the critical velocity from 0.68 to 0.65, and now there is a continuous transition between laminar flow and an vortex ring encircling the bump. However, the small vortex loops of the type discussed by Muirhead *et al.* [35] are not observed. Smaller rings have a lower energy alternative, to exist as a 'virtual' ring inside the object, as in Fig. 1(right). The vortex appears only when it has a radius larger than that of the object.

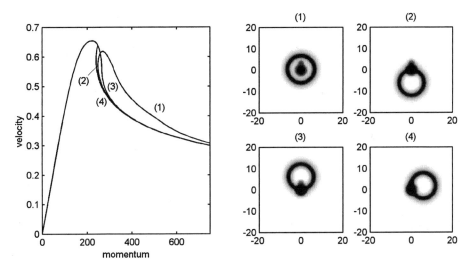

Fig. 8. Uniform flow solutions for a sphere ($R = 3.3$) with a bump ($R = 1.5$). Left: The velocity-momentum dependence of the laminar and four vortex solutions. Right: Cross sections showing the fluid density (dark corresponds to low density around the object) for the four allowed vortex configurations: (1) Vortex ring encircling the object and bump; (2) pinned to the object; (3) encircling the bump; and (4) pinned to the bump. Note that laminar flow evolves continuously into a vortex ring encircling the bump, solution (3). Subsequently, the system evolves to the lowest energy state (4).

To convert the critical velocity to a value pertaining to ions in HeII we use the measured healing length, $\xi/\sqrt{2} = 0.128$ nm [36], and the GP sound speed $c = \hbar/m\xi = 88$ ms^{-1}. For an object with radius $R = 3.3$, equivalent to 0.6 nm which is close to the size of the 'snowball' that surrounds a positive ion [28], one obtains $v_c \sim 60$ ms^{-1}, in rough agreement with experiments [37]. The surface roughness effect discussed above tends to reduce the critical velocity. In addition, there is experimental support for the GP picture of a continuous transition between a bare ion and an ion plus vortex [37].

For ions in HeII, there is also strong experimental evidence that vortex nucleation involves an energy barrier [38]. In the GP model the energy barrier corresponds to the energy required to transfer from the laminar flow to the pinned ring solution in Fig. 1(left). This energy decreases with increasing flow velocity but remains finite at the critical velocity because the pinned ring branch joins the laminar-encircling ring curve at an energy above the critical point. The size of the energy barrier depends on the object radius, mass, and penetrability. For an impenetrable sphere with radius $R = 3.3$ (see Fig. 5 in Ref. [17]), we find an energy barrier of approximately $8\hbar n_0 c\xi^2$. For HeII, this would correspond to ~ 3.8 K, similar to the value derived by Muirhead et al. using arguments based on classical fluid mechanics [35]. For HeII in the limit of low temperature, the transition across the energy barrier can still be driven by local fluctuations of the condensate or non-condensate density.

8   Conclusion

The main aim of this paper was to illustrate the use of the GP equation as a fluid dynamical model of dilute Bose-Einstein condensates, and to consider to what extent experimental and theoretical studies of the dilute quantum fluids can provide new insight into more complex superfluids. We have shown that the process of vortex formation is characterised by a 'smooth' transition between flows without and with vortices, and that vortex rings initially emerge encircling the object. For inhomogeneous condensates, the critical velocity is lowered by the formation of vortices where an object intersects lower density regions. Finally, we suggest that an extension of the GP model may provide a good description of the mechanism of vortex formation by ions in liquid helium.

Acknowledgements

We thank N. G. Berloff, S. Rica, and W. F. Vinen for stimulating discussions.

References

1. See e.g. *Bose-Einstein condensation in atomic gases*, Proc. Int. School of Physics Enrico Fermi, eds. M. Inguscio, S. Stringari and C. Wieman (IOS Press, Amsterdam, 1999).

2. D. R. Tilley and J. Tilley, *Superfluidity and superconductivity*, 3rd Ed., (IoP, Bristol, 1990).
3. V. L. Ginzberg and L. P. Pitaevskii, Zh. Eksp. Teor. Fiz. **34**, 1240 (1958) [Sov. Phys. JETP **7**, 858 (1958)]; E. P. Gross, Nuovo Cimento **20**, 454 (1961) and J. Math. Phys. **4**, 195 (1963).
4. T. Winiecki, B. Jackson, J. F. McCann, and C. S. Adams, J. Phys. B. **33**, 4069 (2000).
5. B. Jackson, J. F. McCann, and C. S. Adams, Phys. Rev. Lett. **80**, 3903 (1998).
6. B. M. Caradoc-Davies, R. J. Ballagh, and K. Burnett, Phys. Rev. Lett. **83**, 895 (1999).
7. M. R. Matthews, B. P. Anderson, P. C. Haljan, D. S. Hall, C. E. Wieman, and E. A. Cornell, Phys. Rev. Lett. **83**, 2498 (1999).
8. K. W. Madison, F. Chevy, W. Wohlleben, and J. Dalibard, Phys. Rev. Lett. **84**, 806 (2000).
9. C. Raman, M. Köhl, R. Onofrio, D. S. Durfee, C. E. Kuklewicz, Z. Hadzibabic, and W. Ketterle, Phys. Rev. Lett. **83**, 2502 (1999).
10. R. Onofrio, C. Raman, J. M. Vogels, J. Abo-Shaeer, A. P. Chikkatur, and W. Ketterle, preprint, Phys. Rev. Lett. **85**, 2228 (2000).
11. B. Jackson, J. F. McCann, and C. S. Adams, Phys. Rev. A **61**, 051603 (R) (2000).
12. T. Frisch, Y. Pomeau, and S. Rica, Phys. Rev. Lett. **69**, 1644 (1992).
13. T. Winiecki, J. F. McCann, and C. S. Adams, Phys. Rev. Lett. **82**, 5186 (1999).
14. P. Nozières and D. Pines, *Theory of Quantum Liquids Vol II* (Addison-Wesley, Redwood City, 1990).
15. *Fluid Mechanics (2nd ed.)*, L. D. Landau and E. M. Lifshitz (Pergamon, Oxford, 1987).
16. C. Huepe and M.-É. Brachet, C. R. Acad. Sci. Paris, **325**, Série 2 b, 195 (1997).
17. T. Winiecki, J. F. McCann, and C. S. Adams, Europhys. Lett. **48**, 475 (1999).
18. C. Huepe, M.-É. Brachet, Physica D, **140**, 126 (2000).
19. T. Winiecki and C. S. Adams, Europhys. Lett. **52**, 257 (2000).
20. N. G. Berloff and P. H. Roberts, J. Phys. A **33**, 4025 (2000).
21. C. Josserand, Y. Pomeau, and S. Rica, Physica D, **134** 111 (1999);
22. S. Rica, Physcia D **148**, 221 (2001).
23. M. Leadbeater, T. Winiecki, D. C. Samuels, C. F. Barenghi, and C. S. Adams, Phys. Rev. Lett. **86** no. 7 to appear (2001)
24. C. Nore, C. Huepe, and M. E. Brachet, Phys. Rev. Lett. **84**, 2191 (2000).
25. P. O. Fedichev and G. V. Shlyapnikov, cond-mat/0004430.
26. P. O. Fedichev and G. V. Shlyapnikov, Phys. Rev. A **60** R1779 (1999).
27. B. Jackson, J. F. McCann, and C. S. Adams, Phys. Rev. A **61**, 013604 (2000).
28. R. J. Donnelly *Quantized vortices in Helium II*, (CUP, Cambridge, 1991).
29. N. G. Berloff, J. Low Temp. Phys. **116**, 359 (1999).
30. N. G. Berloff and P. H. Roberts, J. Phys. A **32**, 5611 (1999).
31. N. G. Berloff and P. H. Roberts, Phys. Lett. A **274**, 569 (2000).
32. R. J. Marshall, G. H. C. New, K. Burnett, and S. Choi, Phys. Rev. A **59**, 2085 (1999).
33. M. J. Davis, S. A. Morgan and K. Burnett, cond-mat/0011431
34. M. Leadbeater and C. S. Adams, unpublished
35. C. M. Muirhead, W. F. Vinen, and R. J. Donnelly, Phil. Trans. R. Soc. Lond. A **311**, 433 (1984).
36. G. W. Rayfield and F. Reif, Phys. Rev. **136**, 1194 (1964).
37. G. W. Rayfield, Phys. Rev. Lett. **19**, 1371 (1967).
38. P. C. Hendry, N. S. Lawson, P. V. E. McClintock, and C. D. H. Williams, Phys. Rev. Lett. **60**, 604 (1988).

Stability of a Vortex in a Rotating Trapped Bose–Einstein Condensate[*]

Alexander L. Fetter and Anatoly A. Svidzinsky[**]

Geballe Laboratory for Advanced Materials, Stanford University, Stanford, CA 94305-4045, USA

1 Time-Dependent Gross–Pitaevskii Equation

The remarkable achievement of Bose-Einstein condensation in dilute trapped alkali-metal atomic gases [1–3] has stimulated the (now successful) search for quantized vortices that are usually associated with external rotation [4–8]. An essential feature of these condensates is the order parameter, characterized by a complex macroscopic wave function $\Psi(r,t)$. Theoretical descriptions of vortices in trapped low-temperature condensates have relied on the time-dependent Gross-Pitaevskii (GP) equation [9–11], which omits dissipation. For a trap rotating at angular velocity Ω about $\hat{z}$, it is a nonlinear Schrödinger equation

$$i\hbar\frac{\partial\Psi}{\partial t} = \left(T + V_{\text{tr}} - \Omega\,L_z + g|\Psi|^2\right)\Psi, \tag{1}$$

where $T = -\hbar^2\nabla^2/2M$ is the kinetic-energy operator, $V_{\text{tr}} = \frac{1}{2}M\sum_j\omega_j^2x_j^2$ is the harmonic trap potential, $L_z = xp_y - yp_x$ is the z component of angular momentum, and $g \equiv 4\pi\hbar^2a/M$ characterizes the strength of the short-range interparticle potential (here, a is the positive s-wave scattering length for repulsive two-body interactions; typically a is a few nm). Current experiments involve "dilute" systems, so that the zero-temperature condensate contains nearly all N particles, with $\int dV\,|\Psi|^2 \approx N$. For a steady solution, $\Psi(r,t) = \Psi(r)e^{-i\mu t/\hbar}$, where μ is the chemical potential.

1.1 Equivalent Hydrodynamics of Compressible Isentropic Fluid

If the condensate wave function is written as $\Psi = e^{iS}|\Psi|$, the real and imaginary parts of Eq. (1) precisely reproduce the time-dependent irrotational hydrodynamics of a compressible isentropic fluid, written in terms of the number density $n = |\Psi|^2$ and the velocity potential $\Phi = \hbar S/M$, where $v = \nabla\Phi$ [9,10,12]. In principle, the dynamics of a curved vortex line in a rotating trap follows directly from general hydrodynamics, but the trap potential and resulting nonuniform density complicate the problem considerably. Thus it is preferable to start with simple situations.

[*] This work was supported in part by NSF Grant No. 99-71518. The manuscript benefited from our participation in workshops organized by H. Stoof at the Lorentz Center, Leiden, The Netherlands and by S. Stringari at ECT*, Trento, Italy; we are grateful for their hospitality.
[**] Supported in part by Stanford University.

1.2 Thomas–Fermi Limit for Large Condensates

For a noninteracting condensate, the macroscopic wave function is the ground state of the harmonic oscillator, with spatial extent $d_j = (\hbar/M\omega_j)^{1/2}$ and $j = x, y, z$; typically d_j is a few μm. The repulsive interactions act to expand the condensate, and the relevant dimensionless interaction parameter is Na/d_0, where $d_0 = (d_x d_y d_z)^{1/3}$ is a suitable geometric mean. Recent experiments focus on the regime $Na/d_0 \gg 1$, when the quantum-mechanical energy associated with the density gradients is negligible. In this Thomas-Fermi (TF) limit [13], the nonrotating GP equation has the solution $g|\Psi_{TF}|^2 + V_{\mathrm{tr}} \approx \mu$, with a parabolic density profile

$$n_{TF}(\boldsymbol{r}) \approx n_{TF}(0) \left(1 - \frac{x^2}{R_x^2} - \frac{y^2}{R_y^2} - \frac{z^2}{R_z^2} \right). \tag{2}$$

Here $R_j = (2\mu/M\omega_j^2)^{1/2} \gg d_j$ fixes the condensate's dimensions and $n_{TF}(0) = \mu/g$ is the central density. The (large) dimensionless ratios $R_0^5/d_0^5 = 15Na/d_0$ and $\mu/\hbar\omega_0 = \frac{1}{2}R_0^2/d_0^2$ characterize the effect of the repulsive interactions, where ω_0 and R_0 are appropriate geometric means.

2 Energy of a Vortex in a Large Rotating Trap

The time-dependent GP equation (1) follows from a variational energy functional

$$E(\Omega) = \int dV \, \Psi^* \left(T + V_{\mathrm{tr}} + \tfrac{1}{2}g|\Psi|^2 - \Omega\, L_z \right) \Psi. \tag{3}$$

If Ψ represents a straight singly quantized vortex in a rotating disk-shape trap ($R_\perp \gg R_z$), the TF density yields the increase in energy $\Delta E(\boldsymbol{r}_0, \Omega)$ associated with the vortex at the transverse position $\boldsymbol{r}_0$ [14]. For a nonrotating trap with $\Omega = 0$, this energy $\Delta E(\boldsymbol{r}_0, 0)$ decreases monotonically with lateral displacement. Energy conservation requires that the allowed motion is a precession at fixed trap potential (the trajectory is elliptical for an anisotropic trap). In the presence of weak dissipation, the vortex slowly spirals outward, lowering its energy.

For arbitrary Ω and small lateral displacements, $\Delta E(\boldsymbol{r}_0, \Omega)$ has the form

$$\Delta E(\boldsymbol{r}_0, \Omega) \approx \frac{4\pi}{3} \frac{R_z n(0)\hbar^2}{M} \left\{ \left[\ln\left(\frac{R_\perp}{\xi} \right) - \frac{2M\Omega R_\perp^2}{5\hbar} \right] \right.$$
$$\left. - \frac{1}{2}\left(\frac{x_0^2}{R_x^2} + \frac{y_0^2}{R_y^2} \right) \left[3\ln\left(\frac{R_\perp}{\xi} \right) - \frac{2M\Omega R_\perp^2}{\hbar} \right] \right\}, \tag{4}$$

where $\xi = \hbar/(2M\mu)^{1/2}$ is the vortex-core radius and $2/R_\perp^2 = 1/R_x^2 + 1/R_y^2$ defines the mean transverse radius of the condensate. In the TF limit, ξ is small, with $\xi R_0 = d_0^2$, ensuring a clear separation of length scales $\xi \ll d_0 \ll R_0$.

- For small positive Ω, $\Delta E(\boldsymbol{r}_0, \Omega)$ decreases with increasing lateral displacements so that a vortex will spiral outward in the presence of weak dissipation.

- The curvature at $r_0 = 0$ vanishes at $\Omega_m = \frac{3}{2}(\hbar/MR_\perp^2)\ln(R_\perp/\xi)$, signaling the onset of metastability. Since $\Delta E(0, \Omega_m) > 0$, this state is not truly stable.
- When $\Omega > \Omega_m$, the trap center becomes a local *minimum*. For weak dissipation, a vortex that is slightly displaced from the center will spiral *inward*.
- When Ω reaches $\Omega_c = \frac{5}{2}(\hbar/MR_\perp^2)\ln(R_\perp/\xi) = \frac{5}{3}\Omega_m$, a central vortex becomes stable because $\Delta E(0, \Omega_c) = 0$. Hence Ω_c is the thermodynamic critical angular velocity for vortex creation.
- For $\Omega > \Omega_c$, the form of $\Delta E(r_0, \Omega)$ shows that a vortex at the outer edge remains metastable, but the barrier for entry into the condensate decreases. Spontaneous nucleation of a vortex may eventually occur through a surface instability [15–17]. The observed critical angular velocity for vortex creation [5,7] is $\approx 70\%$ higher than the predicted TF value Ω_c, in qualitative agreement with such a surface mechanism (but see Ref. [18] and Sec. 4.2 for an alternative explanation).

3   Small-Amplitude Excitation of a Vortex in a Rotating Trap

A macroscopic Bose condensate acts like an external particle source, in the sense that the same physical excited state (here labeled by j) can be achieved either by adding or by subtracting one particle. The actual eigenstates involve "quasiparticle" operators $\alpha_j^\dagger$ and α_j that are linear combinations of the two quantum-mechanical states. For a given normal mode, the resulting pair of coupled complex amplitudes $u_j(\mathbf{r})$ and $v_j(\mathbf{r})$ obey the Bogoliubov equations [10,19] that determine the corresponding eigenfrequency ω_j. Imposing Bose-Einstein commutation relations $[\alpha_j, \alpha_k^\dagger] = \delta_{jk}$ requires that these amplitudes obey the particular normalization $\int dV (|u_j|^2 - |v_j|^2) = 1$, and the resulting quasiparticle Hamiltonian reduces to a set of uncoupled harmonic oscillators $\sum_j \hbar\omega_j \alpha_j^\dagger \alpha_j$, summed over all modes with positive normalization. In the simplest case of a uniform condensate, the eigenfrequencies are all positive, which ensures that the system is stable because the energy then has a lower bound. For a uniform condensate moving with velocity $\mathbf{v}$, however, some of the eigenfrequencies can become negative as soon as $|\mathbf{v}|$ exceeds the speed of sound, which corresponds to the instability associated with the Landau critical velocity [20]. Physically, the system can spontaneously generate quasiparticles because the Hamiltonian is no longer bounded from below.

These general ideas have direct relevance to the stability of a singly quantized vortex in a rotating trapped condensate. For simplicity, it is convenient to consider an axisymmetric condensate in equilibrium with a trap rotating at an angular velocity Ω. In this case, states can be characterized by their azimuthal angular quantum numbers m_j, and the transformation to rotating coordinates $H \to H - \Omega L_z$ ensures that the eigenfrequencies $\omega_j(\Omega)$ in the rotating frame are simply $\omega_j(\Omega) = \omega_j - m_j\Omega$, where ω_j is the corresponding eigenfrequency in the nonrotating frame.

3.1   Stability of a Vortex

The first numerical study of the Bogoliubov equations for a singly quantized vortex in a nonrotating trap found a single negative-frequency mode (the "anomalous mode") [21,22], implying that the vortex in the condensate is unstable. Physically, this instability arises because the condensate with a vortex has a higher energy than the vortex-free condensate.

This situation is especially clear for a noninteracting Bose gas in an axisymmetric harmonic trap, when the first excited (vortex) state has an excitation energy $\hbar\omega_\perp$ and an angular momentum $\hbar$. The transition back to the true ground state involves the (negative) frequency $-\omega_\perp$ and the (negative) change in angular quantum number -1. More generally, the numerical analysis for small and medium interaction strength $Na/d_\perp$ [21] found that the anomalous frequency ω_a remained negative (with $m_a = -1$ unchanged). In a frame rotating with angular velocity Ω, the anomalous frequency becomes $\omega_a(\Omega) = \omega_a + \Omega$. Since $\omega_a < 0$, the eigenfrequency $\omega_a(\Omega)$ in the rotating frame rises toward 0 with increasing Ω and vanishes at a critical rotation speed $\Omega^* = -\omega_a = |\omega_a|$. The Bogoliubov description of small oscillations implies that a condensate with a singly quantized vortex is *unstable* for $\Omega < \Omega^*$ but becomes *stable* for $\Omega > \Omega^*$.

In contrast to the numerical study [21] for small and medium values of the dimensionless coupling parameter Na/d_0, a direct perturbation analysis is feasible for a large condensate $(Na/d_0 \gg 1)$ containing an axisymmetric singly quantized vortex (the TF limit). Detailed study of the Bogoliubov equations [14] yields the explicit expression for the anomalous frequency in the rotating frame $\omega_a(\Omega) = -\frac{3}{2}\left(\hbar/MR_\perp^2\right)\ln\left(R_\perp/\xi\right) + \Omega$, which has the expected form. The resulting critical angular velocity $\Omega^* = \frac{3}{2}\left(\hbar/MR_\perp^2\right)\ln\left(R_\perp/\xi\right)$ for the onset of local stability agrees with Ω_m inferred from Eq. (4) for the onset of metastability.

3.2   Splitting of Normal-Mode Frequencies Caused by a Vortex

The ground-state condensate can sustain dynamical oscillations driven by the mean-field repulsive interaction (analogous to plasma oscillations in a charged medium). These normal modes become particularly simple in the TF limit of a large condensate [23], and experiments have confirmed the predictions for the lowest few modes in considerable detail [11]. For an axisymmetric condensate, the normal modes can be classified by their azimuthal quantum number m, and modes with $\pm m$ are degenerate.

When the condensate contains a vortex, the asymmetric circulating flow affects the preceding normal modes. In particular, the originally degenerate modes are split by the Doppler shift of the local frequency (analogous to the splitting of magnetic sublevels in the Zeeman effect). In the TF limit, this small fractional splitting of the degenerate modes is proportional to $|m|d_0^2/R_0^2$ [24,25]; it has served to detect the presence of a vortex and to infer the circulation and angular momentum [7,26].

4   Vortex Dynamics

At zero temperature, the time-dependent GP equation (1) determines the dynamics of the condensate in a rotating trap. A vortex line will move in response to the nonuniform trap potential, the external rotation, and its own local curvature. This problem is especially tractable in the TF limit, because the small vortex core radius ξ permits a clear separation of length scales; the method of matched asymptotic expansions yields an explicit expression for the local velocity of a vortex line [27,28].

4.1   Dynamics of Straight Vortex

It is simplest to consider a straight vortex [14], which applies to a disk-shape condensate with $R_\perp \gg R_z$. Assume that the vortex is located near the trap center at a transverse position $r_0(t)$. In this region, the trap potential does not change significantly on a scale of order ξ. The solution proceeds in two steps.

(a) First, consider the region near the vortex core that is assumed to move with a transverse velocity $V \perp \hat{z}$. Transform to a co-moving frame centered on the vortex core, where the trap potential exerts a force proportional to $\nabla_\perp V_{tr}$ evaluated at $r_0(t)$. The resulting solution includes both the detailed core structure and the "asymptotic" region $|r_\perp - r_{\perp 0}| \gg \xi$

(b) Second, consider the region far from the vortex, where the core can be treated as a distant singularity. The short-distance behavior of this solution includes the region $\xi \ll |r_\perp - r_{\perp 0}|$. The two solutions must agree in the common region, which determines the translational velocity V of the vortex line.

The details become intricate, but the final answer is elegant and physical:

$$V = \frac{3\hbar}{4M\mu} \left[\ln\left(\frac{R_\perp}{\xi}\right) - \frac{2M\Omega R_\perp^2}{3\hbar} \right] \hat{z} \times \nabla_\perp V_{tr} , \tag{5}$$

where $R_\perp$ for an asymmetric trap is defined below Eq. (4). This expression has several important aspects

- The motion follows an equipotential line along the direction $\hat{z} \times \nabla_\perp V_{tr}$, conserving energy, as appropriate for the GP equation at zero temperature. For an asymmetric harmonic trap, the trajectory is elliptical.
- For a nonrotating trap, the motion is counterclockwise, in the positive sense.
- With increasing external rotation, the translational velocity V decreases and vanishes at the special value $\Omega_m = \frac{3}{2}(\hbar/MR_\perp^2)\ln(R_\perp/\xi)$ discussed below Eq. (4).
- For $\Omega > \Omega_m$, the motion as seen in the rotating frame is clockwise.
- A detailed analysis of the normalization of the Bogoliubov amplitudes shows that the positive-norm state has the frequency

$$\omega = \frac{2\omega_x\omega_y}{\omega_x^2 + \omega_y^2} (\Omega - \Omega_m) . \tag{6}$$

This normal-mode frequency is negative (and hence locally unstable) for $\Omega < \Omega_m$, but it becomes locally stable for $\Omega > \Omega_m$, in agreement with the discussion below Eq. (4).

4.2   Inclusion of Curvature

Equation (5) for the local velocity of a straight vortex oriented along $\hat{z}$ can be generalized to include the possibility of a different orientation of the vector $\hat{t}$ locally tangent to the vortex core. In addition, local curvature k of the vortex line defines a plane that includes both $\hat{t}$ and the local normal $\hat{n}$, inducing an additional translational velocity along the binormal vector $\hat{b} \equiv \hat{t} \times \hat{n}$. A detailed analysis [29] yields the translational velocity of the element located at r_0

$$\boldsymbol{V}(\boldsymbol{r}_0) = -\frac{\hbar}{2M}\left(\frac{\hat{t} \times \boldsymbol{\nabla} V_{\text{tr}}(\boldsymbol{r}_0)}{g|\Psi_{TF}|^2} + k\hat{b}\right)\ln\left(\xi\sqrt{\frac{1}{R_\perp^2} + \frac{k^2}{8}}\right) + \frac{2\boldsymbol{\nabla} V_{\text{tr}}(\boldsymbol{r}_0) \times \boldsymbol{\Omega}}{\nabla_\perp^2 V_{\text{tr}}(\boldsymbol{r}_0)},$$
(7)

where $\nabla_\perp^2$ is the Laplacian in the plane perpendicular to $\boldsymbol{\Omega}$. In the first term, $|\Psi_{TF}|^2$ vanishes near the condensate boundary; hence $\hat{t} \times \boldsymbol{\nabla} V_{\text{tr}}(\boldsymbol{r}_0)$ must also vanish there, implying that the vortex is locally perpendicular to the surface.

Equation (7) allows a study of the dynamics of small-amplitude displacements of the vortex from the z axis, when $x(z,t)$ and $y(z,t)$ obey coupled equations. In the limit $\omega_z = 0$, there is no confinement in the z direction, and the density is independent of z. The resulting two-dimensional dynamics exhibits helical solutions that are linear combinations of two plane standing waves.

More generally, for $\omega_z \neq 0$, the density near the z axis has the TF parabolic form, and the solutions become more complicated. It is convenient to define the asymmetry parameters $\alpha = R_x^2/R_z^2$ and $\beta = R_y^2/R_z^2$, where $\alpha > 1$, $\beta > 1$ indicate a disk shape and $\alpha < 1$, $\beta < 1$ indicate a cigar shape.

- For a nonrotating trap with the special asymmetry values $\alpha = 2/[n(n+1)]$ (here, n a positive integer), the effects of the nonuniform trap potential and the curvature just balance, and the condensate has stationary solutions with the vortex at rest in the xz plane. A disk-shape trap has no such states, and the first one occurs for the spherical trap with $\alpha = 1$. The next such state occurs for $\alpha = \frac{1}{3}$, when the condensate is significantly elongated. Similar considerations for β apply to stationary states in the yz plane.
- For other values of α and β, solutions necessarily involve motion of the vortex line relative to the stationary condensate.
- Analytical solutions can be found for an extremely flat disk with $\alpha \gg 1$ and $\beta \gg 1$, reproducing the frequency Ω_m found in Eq. (4).
- For small deformations of a vortex line in an axisymmetric trap with $\alpha = \beta$, a disk-shape or spherical condensate ($\alpha \geq 1$) has only a single (unstable) precessing normal mode with a negative frequency $\omega_a < 0$. In this case, an external rotation $\Omega \geq \Omega_m = |\omega_a|$ stabilizes the vortex. For these geometries, Ω_m is less than the thermodynamic critical value Ω_c. In a spherical condensate, the one anomalous mode $|\omega_a|$ agrees with the observed vortex precession frequency seen in the JILA experiments [8,18].

- In contrast, an axisymmetric cigar-shape condensate has additional negative-frequency precessing modes, and Ω_m can exceed Ω_c for sufficiently elongated condensates; such behavior seems relevant for the ENS experiments [5,7,18], where $\Omega_m \approx 1.7\Omega_c$ is close to the observed rotation speed for creating the first vortex.

- For small axisymmetric deviations from a spherical trap, a straight vortex line can execute large-amplitude periodic trajectories. In this case, the vortex line becomes invisible when it tips away from the line of sight, and it then periodically returns to full visibility. Such revivals agree with preliminary observations at JILA [26].

References

1. M. H. Anderson *et al.*, Science **269**, 198 (1995).
2. C. C. Bradley *et al.*, Phys. Rev. Lett. **75**, 1687 (1995).
3. K. B. Davis *et al.*, Phys. Rev. Lett. **75**, 3969 (1995).
4. M. R. Matthews *et al.*, Phys. Rev. Lett. **83**, 2498 (1999).
5. K. W. Madison *et al.*, Phys. Rev. Lett. **84**, 806 (2000).
6. K. W. Madison *et al.*, e-print cond-mat/0004037.
7. F. Chevy, K. W. Madison, and J. Dalibard, e-print cond-mat/0005221.
8. B. P. Anderson *et al.*, e-print cond-mat/0005368.
9. E. P. Gross, Nuovo Cimento **20**, 454 (1961).
10. L. P. Pitaevskii, Zh. Eksp. Teor. Fiz. **40**, 646 (1961) [Sov. Phys. JETP **13**, 451 (1961)].
11. F. Dalfovo *et al.*, Rev. Mod. Phys. **71**, 463 (1999).
12. A. L. Fetter, Phys. Rev. A **53**, 4245 (1996).
13. G. Baym and C. J. Pethick, Phys. Rev. Lett. **76**, 6 (1996).
14. A. A. Svidzinsky and A. L. Fetter, Phys. Rev. Lett. **84**, 5919 (2000); e-print cond-mat/9811348.
15. F. Dalfovo *et al.*, Phys. Rev. A **56**, 3840 (1997).
16. D. L. Feder, C. W. Clark, and B. I. Schneider, Phys. Rev. A **61**, 011601(R) (1999).
17. T. Isoshima and K. Machida, Phys. Rev. A **60**, 3313 (1999).
18. D. L. Feder *et al.*, e-print cond-mat/0009086.
19. A. L. Fetter, Ann. Phys. (N.Y.) **70**, 67 (1972).
20. L. D. Landau, J. Phys. (USSR) **5**, 71 (1941).
21. R. J. Dodd *et al.*, Phys. Rev. A **56**, 587 (1997).
22. D. S. Rokhsar, Phys. Rev. Lett. **79**, 2164 (1997).
23. S. Stringari, Phys. Rev. Lett. **77**, 2360 (1996).
24. A. A. Svidzinsky and A. L. Fetter, Phys. Rev. A **58**, 3168 (1998).
25. F. Zambelli and S. Stringari, Phys. Rev. Lett. **81**, 1754 (1998).
26. E. Cornell, private communication.
27. L. M. Pismen and J. Rubinstein, Physica D **47**, 353 (1991).
28. B. Y. Rubinstein and L. M. Pismen, Physics D **78**, 1 (1994).
29. A. A. Svidzinsky and A. L. Fetter, e-print cond-mat/0007139.

Kinetics of Strongly Non-equilibrium Bose–Einstein Condensation

Boris Svistunov

Russian Research Center "Kurchatov Institute", 123182 Moscow, Russia

Abstract. We consider the ordering kinetics in a strongly non-equilibrium state of a (weakly) interacting Bose gas, characterized, on one hand, by large occupation numbers, and, on the other hand, by the absence of long-range order. Up to higher-order corrections in inverse occupation numbers, the evolution is described by non-linear Schrödinger equation with a turbulent initial state. The ordering process is rather rich and involves a number of qualitatively different regimes that take place in different regions of energy space. Specially addressed is the case of evolution in an external potential.

1 Introduction

Kinetics of Bose–Einstein condensation (BEC) in a weakly interacting Bose gas is one of the most fundamental problems of non-equilibrium statistical mechanics. The exciting progress in the experiment with BEC in ultracold gases initiated by the pioneer works [1] opens up an opportunity of laboratory study of non-trivial regimes of BEC kinetics.

From the very beginning it should be realized that the statement of the problem of BEC kinetics involves a number of aspects that are of crucial importance to the very character of the evolution process. The nature of the process strongly depends on how the BEC is being achieved: Say, by slow cooling, or by self-evolution of an essentially non-equilibrium initial state. More generally, it is important to take into account whether the considerable deviation from equilibrium occurs only at some sufficiently large length scales and only in the fluctuation region (so that the kinetics is of universal character and does not reflect specifics of weakly interacting system), or the non-equilibrium situation arises far enough from the critical region and the resulting ordering kinetics is characteristic only to weakly interacting gas. Obviously, the picture of evolution can be "trivialized" if some portion of condensate is already present in the initial state. Finally, and especially importantly for the realistic case of a trapped gas, finite size of the system, or just only that of the condensate can partially or completely change the relaxation scenario, if this size turns out to be less than some correlation length relevant to the ordering process in the infinite system.

In this paper we concentrate on a statement of BEC kinetics problem that we believe to be the most characteristic of the case of weakly interacting gas. Namely, we consider the self-evolution of weakly interacting gas with a strongly non-equilibrium initial state. To maximally simplify the consideration without qualitatively changing the nature of the process, we assume that in the initial state all occupation numbers are either much larger than unity, or equal to zero, and that there are no correlations between different single-particle modes. The advantage of choosing such an initial condition is that from the very beginning one can employ classical-field description in terms of non-linear Schrödinger equation (NLSE), which in the theory of Bose gases is known as Gross-Pitaevskii equation [2], with a certain turbulent initial condition. It should be stressed that, in contrast to a wide-spread prejudice, the very description in terms of NLSE does not imply the presence of condensate, or any sort of dynamical phase transition (see, e.g., discussion in [3]). The question of the presence of condensate, or, more generally, the question of (long-range) order is the question of the (long-range) structure of corresponding classical field.

Hence, the dynamical model for our problem reads ($\hbar = 1$)

$$i\frac{\partial \psi}{\partial t} = -\frac{\Delta}{2m}\psi + V(\boldsymbol{r})\psi + U|\psi|^2\psi, \tag{1}$$

where $|\psi|^2$ is interpreted as particle density (not the condensate density!), m is the particle mass, $V(\boldsymbol{r})$ is the external potential; $U = 4\pi a/m$ is the vertex of the effective pair interaction, a is the scattering length. To introduce the initial condition to (1) one has to consider the expansion of $\psi(\boldsymbol{r}, t)$ in terms of eigen modes $\varphi_\varepsilon(\boldsymbol{r})$ $[(-\Delta/2m + V)\varphi_\varepsilon = \varepsilon\varphi_\varepsilon]$ of linear part of NLSE: $\psi(\boldsymbol{r}, t) = \sum_\varepsilon a_\varepsilon(t)\varphi_\varepsilon(\boldsymbol{r})$. Then, at the initial moment $t = 0$, the phases of the complex amplitudes a_ε can be considered as random, while $|a_\varepsilon|^2$ is identified with the occupation number n_ε of the mode ε (see, e.g., [3] for more details). To the best of our knowledge, the first formulation of BEC kinetics problem in terms of NLSE was given in [4].

A full-scale numeric simulation of NLSE with the turbulent initial condition could, in principle, cross almost all the t's in the strongly non-equilibrium BEC kinetics problem. Such a simulation has not been done yet [5]. Nevertheless, we will see that from general considerations it is possible to propose the evolution scenario and to obtain all relevant estimates.

In very general terms, the direction in which the field ψ will evolve is clear from the following considerations. First, it is natural to expect that the system must relax to a certain equilibrium state. Secondly, this equilibrium state should correspond to zero temperature, since the classical field described by the equation (1) forms a heatbath at absolute zero with respect to itself. Hence, if there is a stable groundstate (that is if $U > 0$) for a given particle density, then the system should approach it in this or that way, the excess energy being carried away (to higher and higher harmonics) by ever decreasing portion of high-frequency fraction of the field. At the final stage of evolution all the particles are condensed except for an infinitesimally small high-frequency portion.

Though the general tendency of evolution is clear, the particular relaxation scenario is not at all self-evident. A detailed analysis [6–8] leads to a rather sophisticated scenario that involves a number of qualitatively different stages. The evolution starts with an explosion-like wave in energy space, propagating from higher energies towards the lower ones, that leads to a formation of a specific power-law distribution of particles. Immediately after its formation, this distribution starts to relax. Simultaneously, in the low-energy region the so-called coherent regime sets in that leads to the formation of quasi-condensate correlation properties. Basically, the quasi-condensate state corresponds to what is known in the theory of superfluidity as the state of superfluid turbulence. It can be viewed as a condensate containing a tangle of vortex lines (plus a specific sharply non-equilibrium distribution of long-wave phonons). The formation of the quasicondensate occurs very rapidly (characteristic time is much smaller than the time of the wave formation). In contrast to it, the final stage of long-range ordering, associated with relaxing superfluid turbulence and long-wave phonons, takes a macroscopically large time.

In the present paper we render the homogeneous BEC scenario [6–8] (Sects. 2 and 3) and project it onto the case of a trapped gas. We find out that in an external potential the evolution picture can be even more rich.

2   Kinetic Regime

During some initial period of evolution the correlations between different amplitudes a_ε are vanishingly small. Such a regime (known in the theory of non-linear classical-field dynamics as weak turbulence [9]) admits a description in terms of kinetic equation. This stage is thus referred to as kinetic stage.

Kinetic equation corresponding to the weak-turbulence regime of non-linear Schrödinger equation belongs to a generic class of scale-invariant models with four-wave particle- and energy-conserving interaction, that allows an analysis of evolution kinetics in general terms (see, e.g., [6]). For the BEC kinetics the analysis suggests that there are two alternative ways of the initial evolution: (i) shrinking of the particle distribution as a whole towards $\varepsilon = 0$ (during infinite time), or (ii) a specific wave in the energetic space leading to a singularization of distribution at the point $\varepsilon = 0$ at some finite time moment $t = t_*$. The answer to the question of which scenario takes place for a given model depends only on the scaling properties of the collision integral and the density of states; and the case of NLSE corresponds to the scenario (ii).

The evolution at the beginning of the kinetic stage results in the formation of self-similar wave in the energy space propagating in an explosion-like fashion from the high-energy region (where the particles are initially distributed) towards lower energy scales. Corresponding self-similar solution of the kinetic equation has the form [6]

$$n_\varepsilon(t) = A\varepsilon_0^{-\alpha}(t)f(\varepsilon/\varepsilon_0(t)), \quad t \le t_*, \tag{2a}$$

$$\varepsilon_0(t) = B\,|\,t_* - t\,|^{1/2(\alpha-1)}. \tag{2b}$$

Here A and B are dimensional constants depending on the initial condition and related to each other by the formula $B = \text{const}(m^3 U^2 A^2)^{1/2(\alpha-1)}$. The dimensionless function f (numeric data for f see in [6]) is defined up to an obvious scaling freedom. The explosion character of the evolution guarantees that the wave reaches the point $\varepsilon = 0$ at some finite time moment $t = t_*$ [$t = 0$

corresponds to the beginning of evolution], the value of t_* being on the order of the typical time of (stimulated) collisions in the gas at $t = 0$. Physically, this explosion-like evolution is supported by the stimulation of the collision rate at the head of the wave, $\varepsilon \sim \varepsilon_0$, by ever growing occupation numbers.

Generally speaking, the index α in (2a)-(2b) cannot be established from the scaling properties of the collision term of the kinetic equation, being related thus to the particular form of the latter. It is possible, however, to specify lower and upper limits for α following from the consistency of (2a) and (2b) with the requirement that these formulae describe an explosion-like singularization of distribution (rather than infinite-time shrinking). To this end we note that from the scale invariance it follows that $f(x)$ behaves like some power of x at $x \gg 1$. At $t = t_*$ the occupation numbers have to be finite at $\varepsilon > 0$, hence

$$f(x) \to x^{-\alpha} \quad \text{at} \quad x \to \infty \,. \tag{3}$$

The requirement that the particle distribution does not shrink as a whole implies that the number-of-particles integral for the distribution (2a), (3) is divergent at $\varepsilon \to \infty$. This immediately yields $\alpha < 3/2$. The condition $\alpha > 1$ is necessary for ε_0 (2b) to approach zero at $t = t_*$. So we have $1 < \alpha < 3/2$). The most accurate up-to-date numeric analysis of α was performed in [10] with the result $\alpha \approx 1.24$.

At $t > t_*$ kinetic description is still valid for not so small energies, but to obtain an adequate sewing with the solution (2a)-(2b) one has to explicitly introduce the (quasi)condensate, employing the conservation of the total number of particles (see the discussion in [6]). The structure of the self-similar solution

$$n_\varepsilon(t) = A\varepsilon_0^{-\alpha}(t)\tilde{f}(\varepsilon/\varepsilon_0(t)) \,, \quad \varepsilon > 0 \,, \quad t \geq t_* \tag{4}$$

$[\tilde{f}(x) \to f(x)$ at $x \to \infty]$ corresponds to a back wave in the energy space, destroying the singular distribution created by the wave (2a)-(2b). The particles being released during this destruction go directly to quasicondensate. For the quasicondensate density n_0 we thus have

$$n_0(t) = \frac{A}{4\pi^2}(2m)^{3/2}\varepsilon_0^{3/2-\alpha}(t)\int_0^\infty dx\sqrt{x}[x^{-\alpha} - \tilde{f}(x)] \propto (t - t_*)^{(3-2\alpha)/4(\alpha-1)} \,. \tag{5}$$

As follows from general considerations and is supported by direct numeric analysis [6], $\tilde{f}(x) \propto 1/x$ at $x \ll 1$, which means that the back wave creates a quasi-equilibrium distribution at $\varepsilon \ll \varepsilon_0(t)$ [with infinite at $t = t_*$ and ever decreasing afterwards temperature $\propto \varepsilon_0^{-\alpha}(t)$].

To estimate the parameter A for a given initial conditions one extrapolates the solution (2a)-(2b) to a region of energies $\sim \varepsilon_{\text{init}}$, where the particles were initially concentrated with typical occupation numbers $n_{\varepsilon_{\text{init}}}$. This immediately yields $A \sim n_{\varepsilon_{\text{init}}} \varepsilon_{\text{init}}^\alpha$.

3 Coherent Regime

Strictly speaking, evolution in a kinetic regime does not lead to the ordering. It is seen from the fact that the description in terms of kinetic equation is associated with the random phase approximation (RPA) and thus valid only when the phases of a_ε's are practically uncorrelated. In such a state even local order (quasicondensate) is absent. Quasicondensation implies a strong change of the correlation properties as compared to the RPA state [7]. It occurs in the regime of *strong* turbulence (so-called coherent regime), when typical time of evolution is comparable to the time of oscillation of the phases of relevant a_ε's. The essence of the process of the quasicondensate formation is the transformation of the strong turbulence into the state known as superfluid turbulence.

The degrees of freedom associated with the quasicondensate are the same as in a genuine condensate. These are phonons and topological defects (vortex lines). That is quasicondensate can be viewed as a condensate with (i) a tangle of vortex lines and (ii) strongly non-equilibrium distribution of long-wave phonons implying strong fluctuations of the phase of the quasicondensate part ψ_0 of the field ψ at large distances.

Given the solution (2a)-(2b) of the kinetic equation, one readily estimates where and when the coherent regime sets in, and what is the typical density of the quasicondensate upon its formation. The characteristic time of evolution at the energy scale ε is the collision time $\tau_{\mathrm{coll}}^{-1}(\varepsilon) \sim m^3(U\varepsilon n_\varepsilon)^2$. For $n_\varepsilon \sim A\varepsilon^{-\alpha}$ the RPA criterion $\tau_{\mathrm{coll}}(\varepsilon)\varepsilon \gg 1$ becomes invalid at $\varepsilon \sim \varepsilon_{\mathrm{coh}} = (m^3 U^2 A^2)^{1/(2\alpha-1)}$. The distribution $n_\varepsilon \sim A\varepsilon^{-\alpha}$ at the scale $\varepsilon_{\mathrm{coh}}$ is formed at the time t_{coh} obeing an obvious relation $\varepsilon_0(t = t_* - t_{\mathrm{coh}}) \sim \varepsilon_{\mathrm{coh}}$. Hence, t_{coh} estimates the time moment when the coherent regime sets in. The time interval $|\, t_* - t_{\mathrm{coh}} \,| \varepsilon_{\mathrm{coh}}^{-1} \ll t_*$ is a typical time of the process of quasicondensate formation (in the strong turbulent regime all characteristic times and distances scale with $\varepsilon_{\mathrm{coh}}^{-1}$ and $(m\varepsilon_{\mathrm{coh}})^{-1/2}$, correspondingly). The initial quasicondensate density n_0^{init} is the density corresponding to the harmonics $\varepsilon \sim \varepsilon_{\mathrm{coh}}$. It obeys an obvious relation $n_0^{\mathrm{init}}U \sim \varepsilon_{\mathrm{coh}}$, in accordance with the fact that in strong turbulent regime kinetic and potential energies are of the same order. The initial spacing between the vortex lines in the quasicondensate scales as $(m\varepsilon_{\mathrm{coh}})^{-1/2}$.

The coherent regime and the back wave are practically independent processes, with a reservation that the coherent part of the field is being pumped with particles from the high-energy region. This pumping, however, does not affect the character of the coherent evolution.

Relaxation of the quasicondensate towards genuine condensate goes in two directions: (i) relaxation of the vortex tangle [that is relaxation of the superfluid turbulence] and (ii) relaxation of the long-wave phonons. Both processes require a macroscopically large time (an analysis of this stage of evolution see in [8]).

4   External Potential

In the experiments with trapped ultracold gases normally there takes place the Knudsen regime, when free path length of a particle with the energy ε, $l_{\mathrm{free}}(\varepsilon)$, is much larger than the typical radius of the particle's trajectory, R_ε. For definiteness we consider a parabolic trap with all the three frequencies of the same order ω_0. So that $R_\varepsilon \sim v(\varepsilon)/\omega_0$ [$v(\varepsilon)$ is the typical velocity corresponding to the energy ε], and the condition $l_{\mathrm{free}}(\varepsilon) \gg R_\varepsilon$ is equivalent to $\tau_{\mathrm{coll}}(\varepsilon)\omega_0 \gg 1$.

Knudsen regime is a very convenient starting point for analyzing kinetics in a potential. Almost in all qualitative aspects it corresponds to an isotropic homogeneous case, since the distribution of particles depends only on the two variables, ε and t (ergodic approximation). The main quantitative difference comes from the difference in the density of states, the scaling of the collision time remaining the same.

Initial picture of evolution is described by the self-similar wave (2a)-(2b) with $\alpha \approx 1.6$ [11,12]. The question then is: What happens when $t \to t_*$? The effect of the potential can be associated with two rather different reasons. The first reason is the essential discreteness of the low-lying energy levels, which drastically changes the kinetics when $\tau_{\mathrm{coll}}^{-1}(\varepsilon) < \Delta(\varepsilon)$, where $\Delta(\varepsilon)$ is the typical interlevel spacing. The second reason is the violation of the Knudsen regime at the head of the wave at some stage of evolution [because of the decreasing $\tau_{\mathrm{coll}}(\varepsilon_0(t))$ with t].

Remarkably, the above-mentioned two circumstances arise always separately, and (apart a certain cross-over region) there is nothing in between. More specifically, as it immediately follows from the non-equality $\Delta(\varepsilon) \leq \omega_0$ and the fact that the relevant collision time $\tau_{\mathrm{coll}}(\varepsilon_0(t))$ permanently decreases, if there occurs a break-down of Knudsen regime (in corresponding region of coordinate space with a typical size R_{Kn} around the center of the potential), the discreteness of levels will never become relevant.

The case, when the discreteness of levels starts to act within the Knudsen regime, is rather transparent physically and is studied to a large extent both experimentally [13] and theoretically [11,12]. The evolution scenario in this case is as follows. When the wave reaches the scale where the level discreteness becomes relevant, its further propagation is suppressed (essentially discrete harmonics practically do not interact with each other) and the back wave is formed. At $\varepsilon < \varepsilon_0(t)$, the back wave generates quasi-equilibrium distribution with time-dependent permanently decreasing temperature $T(t)$ and permanently increasing number of particles. Condensation thus occurs in a quasi-equilibrium way, without the coherent stage (interaction between low-lying harmonics is negligible), and the whole process can be described within the kinetic approach [11,12].

We are mostly interested in the case, when at some $t = t_{\mathrm{Kn}}$ the Knudsen regime breaks down for energies $\varepsilon_{\mathrm{Kn}} \sim \varepsilon_0(t = t_{\mathrm{Kn}})$. From (2a)-(2b) we estimate $\varepsilon_{\mathrm{Kn}} \sim [m^3 U^2 A^2/\omega_0]^{1/2(\alpha-1)}$. The size of corresponding spatial region is defined by $m\omega_0^2 R_{\mathrm{Kn}}^2 \sim \varepsilon_{\mathrm{Kn}}$. We argue that within the region $r < R_{\mathrm{Kn}}$ at $t > t_{\mathrm{Kn}}$ the external potential becomes *irrelevant* at least until the quasicondensate is formed, so that the most important evolution stage basically does not differ from the homogeneous case. Indeed, it is quite natural that further evolution within the region $r < R_{\mathrm{Kn}}$ will result in the formation of *anti*-Knudsen regime $l_{\mathrm{free}}(\varepsilon) \ll R_{\mathrm{Kn}}$ for $\varepsilon \ll \varepsilon_0(t_{\mathrm{Kn}})$, because of increasing collision rate with increasing the occupations numbers. In the anti-Knudsen regime the evolution during the time period on the order of collision time is insensitive to the external potential (the criteria for the Knudsen regime and for the sensitivity to the potential within the collision time coincide). But this time is enough to form the wave (2a)-(2b) in the energy space (with the exponent α corresponding to the homogeneous case) and then to form quasicondensate. During the wave evolution in the energy space, the free-path length of the particles with $\varepsilon \sim \varepsilon_0(t)$ is getting progressively smaller, which renders the proposed scenario self-consistent. A minor deviation from the pure homogeneous picture is that now the moment t_* depends on the distance from the center of the potential, so that the coherent regime first should start at $r = 0$ (the point of maximal initial density) and then gradually occupy all the anti-Knudsen region up to $r \sim R_{\mathrm{Kn}}$. By this moment the quasicondensate is formed at $r \leq R_{\mathrm{Kn}}$. In terms of the total number of particles, N, and the typical single-particle energy of the initial distribution in the potential, $\varepsilon_{\mathrm{in}}$, the estimate for R_{Kn} is: $R_{\mathrm{Kn}} \sim [U^2 N^2 m^{5-2\alpha} \omega_0^{9-4\alpha} \varepsilon_{\mathrm{in}}^{2(\alpha-3)}]^{1/4(\alpha-1)}$.

On the basis of the above discussion one can introduce the parameter

$$p = N^2 \omega_0 m^3 U^2 (\omega_0/\varepsilon_{\mathrm{in}})^{6-2\alpha} , \qquad (6)$$

that determines which of the two regimes takes place under given initial conditions: the discrete-harmonic regime ($p \ll 1$), or the superfluid-turbulence one ($p \gg 1$).

For more details on the ordering kinetics in a trapped gas see [14].

References

1. M.H. Anderson *et al.*: Science **269**, 198 (1995); C.C. Bradley *et al.*: Phys. Rev. Lett. **75**, 1687 (1995); K.B. Davis *et al.*: Phys. Rev. Lett. **75**, 3969 (1995)
2. V.L. Ginzburg and L.P. Pitaevskii: Zh. Eksp. Teor. Fiz. **34**, 1240 (1958) [Sov. Phys. JETP **7**, 858 (1958); E.P. Gross: J. Math. Phys. **4**, 195 (1963)
3. Yu. Kagan and B.V. Svistunov: Phys. Rev. Lett. **79**, 3331 (1997)
4. E. Levich and V. Yakhot: J. Phys. A: Math. Gen. **11**, 2237 (1978)
5. Note, however, an interesting preliminary study performed in [K. Damle, S.N. Majumdar, and S. Sachdev: Phys. Rev. A **54**, 5037 (1996)]; as well as the all-quantum simulation [P.D. Drummond and J.F. Corney, Phys. Rev. A **60**, R2661 (1999)], where the evidence of spontaneous vortex formation in a trapped gas was reported.
6. B.V. Svistunov: J. Moscow Phys. Soc. **1**, 373 (1991)
7. Yu. Kagan, B.V. Svistunov, and G.V. Shlyapnikov: Zh. Eksp. Teor. Fiz. **101**, 528 (1992) [Sov. Phys. JETP **75**, 387 (1992)]
8. Yu. Kagan and B.V. Svistunov: Zh. Eksp. Theor. Fiz. **105**, 353 (1994) [Sov. Phys. JETP **78**, 187 (1994)]
9. See the contribution by S. Nazarenko to this volume for details.
10. D.V. Semikoz and I.I. Tkachev: Phys. Rev. D **55**, 489 (1997)
11. C.W. Gardiner *et al.*: Phys. Rev. Lett. **81**, 5266 (1998); M.J. Davis, C.W. Gardiner, and R.J. Ballagh: Phys. Rev. A **62**, 63608 (2000)
12. M.J. Bijlsma, E. Zaremba, and H.T.C. Stoof: Phys. Rev. A **62**, 63609 (2000)
13. H.-J. Miesner *et al.*: Science **279**, 1005 (1998)
14. B.V. Svistunov: cond-mat/0009295 (to be published in Phys. Lett. A)

Quantum Nucleation of Phase Slips in Bose–Einstein Condensates

H.P. Büchler[1], V.B. Geshkenbein[1,2], and G. Blatter[1]

[1] Theoretische Physik, ETH-Hönggerberg, CH-8093 Zürich, Switzerland
[2] Landau Institute for Theoretical Physics, 117940 Moscow, Russia

Abstract. We present a theoretical study of quantum fluctuations in a Bose-Einstein condensate confined within a thin cylindrical trap and perturbed by a moving impurity. We derive an effective action which maps the problem to that of a massive particle with damping in a periodic potential. Quantum fluctuations lead to a finite nucleation rate of phase slips and we make use of known results in our determination of the transport characteristic. Real Bose-Einstein condensate are finite systems and exhibit interesting effects depending on topology: in superfluid rings we obtain a critical velocity below which the nucleation rate is quenched. In a cigar shaped condensate the low-energy action is equivalent to that of a capacitively shunted Josephson junction. The state with a well defined phase difference across the impurity then is unstable towards a decoupled state with a fixed number of particles on either side of the impurity.

1   Introduction

A spectacular phenomenon of quantum fluids is the friction free transport, often studied in terms of the Josephson effect across a constriction or orifice separating two reservoirs. Recently, new types of experiments with a moving laser beam perturbing a condensate of sodium atoms enable the observation of superfluid flow in Bose-Einstein condensates [1] and the presence of a critical velocity beyond which the nucleation of phase slips leads to dissipation. Reducing the transverse dimension of the condensate increases the effect of quantum fluctuations and leads to drastic changes in the transport characteristics. Here, we study the effect of quantum fluctuations on the superfluid flow through an impurity in a quasi one-dimensional Bose-Einstein condensate.

Transport properties in charged superfluids are conveniently studied by current biasing the circuit via external electrical devices. In particular, using such experimental setups, the quantum fluctuations of the superconducting wave function in a small Josephson junction device [2] and in thin superconducting and superfluid wires have been examined [3–5]. Recently, a new type of experiments has been set up determining the superfluid flow of a weakly interacting Bose-Einstein condensate around an impurity [1]. In this setup the flow is induced by the motion of an impurity, e.g., a laser beam repelling the atoms from its focus. In our analysis we chose this setup of a moving impurity in a Bose-Einstein condensate for studying the quantum nucleation of phase slips.

The effect of phase slips is most conveniently studied in a superfluid ring with length L where the impurity is moving with uniform velocity v around the

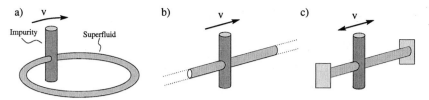

Fig. 1. Superfluid in a small thin cylindrical trap with a moving impurity: a) ring structure with periodic boundary condition for the condensate wave function, b) infinite wire, c) finite system of length L with closed ends.

ring (see Fig. 1a). The creation of a phase slip leads to an additional winding of the superfluid phase describing an acceleration of the superfluid, which in turn is related to a suppression of the order parameter, i.e., particles leave the condensate. The finite length L together with the periodic boundary conditions provide us with a quantization of the velocity of the superfluid in units $2v_L = 2\pi\hbar/Lm$, i.e., a phase slip leads to an increase of the mean velocity by $2v_L$. As a consequence, the relative velocity v_{rel} between the superfluid and the uniformly moving impurity decreases by a rate which can be estimated as

$$\partial_t v_{\mathrm{rel}} = -2\,\Gamma(v_{\mathrm{rel}})\,v_L, \tag{1}$$

where Γ denotes the quantum nucleation rate for phase slips. When the relative velocity between the superfluid and the impurity is smaller than the critical velocity necessary for the nucleation of phase slips we reach an equilibrium state. Switching on the laser beam during a time Δt the amount of energy transferred to the superfluid is given by

$$\Delta E = \int_{\Delta t} dt \Gamma(v) E_{\mathrm{PS}}(v), \tag{2}$$

with $E_{\mathrm{PS}}(v)$ the energy transferred by a single phase slip event. Two methods for measuring this energy transfer have been applied in experiments on bulk superfluids: determination of the heating of the condensate by measuring the thermal fraction [1,6] and by direct observation of the flow field in the superfluid via *in situ* non-destructive imaging of the condensate [6].

Increasing the length of the superfluid ring, finite size effects become irrelevant and the behavior of the ring approaches that of an infinitely long wire. This situation is illustrated in Fig. 1b. In contrast to the ring setup, the impurity never returns to its initial position. As a consequence, only parts of the superfluid are accelerated by a phase slip and waves traveling with a velocity close to the sound velocity carry away the energy transferred to the superfluid.

Closing the wire with impenetrable boundaries at a distance L (see Fig. 1c), these sound waves are reflected and provide a homogeneous heating of the sample due to the nucleation of phase slips. In contrast to the previous setups the impurity oscillates instead of moving with uniform velocity, and the mean relative velocity between the superfluid and the impurity does not decrease. The amount of energy transferred to the superfluid is given by Eq. (2).

2   Effective Action

We start with the Gross-Pitaevskii Lagrangian describing the dynamics of the
confined condensate wave function of the weakly interacting bosons [7]. With m
the mass of the bosons, U the strength of the repulsion, and ρ_0 the condensate
density, the Lagrangian in one dimension reads

$$\mathcal{L}_{GP} = i\hbar \,\overline{\psi}\partial_t\psi - \frac{\hbar^2}{2m}(\partial_x\overline{\psi})(\partial_x\psi) - \frac{U}{2}\left[\overline{\psi}\psi - \rho_0\right]^2 . \tag{3}$$

The repulsive interaction produces the healing length $\xi = \hbar/\sqrt{m\rho_0 U}$ and a
finite compressibility κ related to the sound velocity $c_s = \sqrt{\kappa/m\rho_0} = \sqrt{\rho_0 U/m}$.
Comparing the length scale ξ with the scale given by the condensate density $1/\rho_0$
provides us with a dimensionless parameter $K = \pi\xi\rho_0$, the number of particles
per healing volume measuring the relevance of fluctuations. The condition of
weakly interacting bosons where a mean field description is valid translates into
$K \gg 1$. The energy scale of the interaction is given by the chemical potential
$\mu = \rho_0 U$. The impurity is described by a suppression of the chemical potential

$$\mathcal{L}_{\text{int}} = -V(x - vt)\overline{\psi}\psi \tag{4}$$

with v the velocity of the impurity. We will consider a short range interaction
between the impurity and the superfluid, i.e., $V(x) = g\xi\delta(x)$, with impurity
strength $g \gg \mu$. The finite size of the system is taken into account via boundary
conditions for the condensate wave function. In the following we will consider
three different setups (see Fig. 1): a superfluid ring with periodic boundary
condition $\psi(x, t) = \psi(x+L, t)$, the infinite system, and the superfluid tube where
the flow vanishes at the ends of the tube, i.e., $\partial_x\phi(L/2, t) = \partial_x\phi(-L/2, t) = 0$,
where ϕ denotes the phase of the wave function $\psi = \sqrt{\rho_0}(1 + h)\exp(i\phi)$.

The integration over all degrees of freedom except the phase difference φ
across the impurity is carried out in the imaginary time description of the superfluid and provides us with the effective action [8]

$$\frac{S}{\hbar} = \frac{T}{\hbar}\sum_s \alpha(\omega_s)\frac{|\varphi(\omega_s)|^2}{2} + \int_0^\beta d\tau \left\{\frac{E_J}{\hbar}\left[1 - \cos\varphi(\tau)\right] - \rho_0 v\varphi(\tau)\right\} \tag{5}$$

where β denotes the inverse temperature T, while $\omega_s = 2\pi T s/\hbar$ are the Mat-
subara frequencies with $s \in \mathbf{Z}$. The static Josephson like potential with the
Josephson energy $E_J = K\mu^2/g\pi$ derives from the interaction of the superfluid
with the impurity, while the driving term is induced by the motion of the impurity. The first term describes the dynamics of φ and follows from integrating
out the low energy excitations in the leads. The kernel $\alpha(\omega)$ depends on the
boundary conditions of the condensate wave function.

We start with the simplest setup: the infinitely long tube shown in Fig. 1b.
Then the kernel $\alpha(\omega)$ reads

$$\alpha(\omega) = \alpha_d(\omega) + \alpha_m(\omega) = \frac{K}{2\pi}|\omega| + \frac{\hbar}{E_C}\omega^2. \tag{6}$$

The first term describes a Caldeira-Leggett type [9] dissipation via radiation of sound waves, while the second term introduces a kinetic term with 'charging' energy $E_C = 4\pi^2 E_J/K^2$. The characteristic time scale τ_c of the action (5) is

$$\tau_c = \frac{K\hbar}{\pi} \frac{1}{E_J} = \frac{4\pi\hbar}{K} \frac{1}{E_C} \tag{7}$$

and derives from a comparison of the different terms in the action $K/4\pi \sim \hbar/E_C\tau_c \sim E_J\tau_c/\hbar$. This scale is much slower than the time scale $\tau_0 = \hbar/\mu$ of the unperturbed superfluid, $\tau_c = (g/\mu)\tau_0$, justifying the assumption that only low energy excitations in the leads contribute to the effective action (5).

The action (5) with the kernel (6) is also known to describe a resistively and capacitively shunted Josephson junction (RCSJ-model) [10] between two charged superconductors. The quality factor defined by $Q = 2\pi\sqrt{E_J/(K^2 E_C)}$ separates two regimes: the junction is overdamped with the damping dominating over the inertia if $Q < 1$, while for $Q > 1$ we enter the underdamped regime with a hysteretic current-voltage characteristic [10]. In the uncharged situation considered here, the relation $E_C = 4\pi^2 E_J/K^2$ between the 'charging' energy and the Josephson energy leads to $Q = 1$, and the action is at the border between the overdamped and the underdamped regime. Then effects leading to hysteretic behavior become only relevant for velocities close to the critical velocity $v \sim v_c = E_J/\hbar\rho_0$.

The effect of quantum fluctuations in the RCSJ model has been extensively studied [9,11,12]. The system shows a quantum phase transition at $K = 1$ [11]. For $K < 1$ quantum fluctuations dominate and the ground state is described by a delocalized phase and the system exhibits a linear response $\Delta\mu = \hbar\langle\partial_t\varphi\rangle = \hbar\rho_0 v/K$ characteristic for a normal fluid, i.e., the phase coherence of the condensate across the impurity is suppressed and a fixed particle number $\mathcal{N}$ is established [12]. On the other hand, for $K > 1$ describing a weakly interacting Bose-Einstein condensate the damping is strong and reduces the effect of quantum fluctuations. Then the ground state is characterized by a fixed (localized) phase φ allowing for a superflow $v = v_c \sin\varphi$ across the impurity with a vanishing linear response $\Delta\mu = \hbar\langle\partial_t\varphi\rangle = 0$ at zero temperature. However, the quantum nucleation of phase slips provides us with algebraic corrections to the response function $\langle\partial_t\varphi\rangle = 2\pi\Gamma$ with the rate [13] (see Fig. 2)

$$\Gamma = \frac{y^2}{\tau_c}\left(\frac{2\pi\tau_c}{\beta}\right)^{2K-1} \sinh\left(\pi\rho_0\beta v\right) \frac{|\Gamma\left(K + i\rho_0 v\beta\right)|^2}{\Gamma(2K)}. \tag{8}$$

At zero temperature, the quantum nucleation is algebraic in v, $\Gamma \sim v^{2K-1}$, while a finite temperature and small driving forces lead to a linear quantum nucleation rate $\Gamma \sim vT^{2K-2}$ with an algebraic temperature dependence. Finally, the energy E_{PS} transferred to the superfluid by a single phase slip takes the form

$$E_{\mathrm{PS}} = \left[iv\hbar\int dx\overline{\psi}\partial_x\psi\right]_{\tau=0} = 2K\mu\frac{v}{c_s}. \tag{9}$$

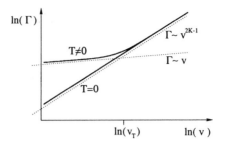

Fig. 2. Quantum nucleation rate of phase slips: algebraic behavior $\Gamma \propto v^{2K-1}$ at zero temperature, and a linear behavior $\Gamma \propto vT^{2K-2}$ for small velocities and finite temperature with a crossover at $v_T = \pi T/mc_s$.

3   Finite Size Effects

Ring: The periodic boundary conditions in a superfluid ring of length L (see Fig. 1a) provide us with a quantization of the velocity of the superfluid by $2v_L = 2\pi\hbar/Lm$. The relative velocity between the impurity and the superfluid takes the form

$$v_{\rm rel} = v_c \sin\varphi = v - \frac{v_L}{\pi}\varphi, \qquad (10)$$

where the first term is the flow induced by the motion of the impurity, while the second term describes the flow of the superfluid in the leads with $\varphi \in [2\pi k, 2\pi(k+1))$; $k \in \mathbf{Z}$ is the winding number of the phase. The relative velocity $v_{\rm rel}$ drives the system and determines the quantum nucleation rate. For large rings with $L > \xi_c = c_s\tau_c$ and $k = 0$ we can drop the second term and the relative velocity is determined by the velocity of the impurity alone. The kernel

$$\alpha_d(\omega) = K\omega\frac{\cosh\omega L/c_s + 1}{\sinh\omega L/c_s} = \begin{cases} 2Kc_s/L, & \omega < c_s/L, \\ K|\omega|, & \omega > c_s/L. \end{cases} \qquad (11)$$

is modified at low frequencies: first, the finite size produces a quantization of the available sound modes, leading to a suppression of the spectral density at low frequencies. Second, the closed structure of the setup induces a self-interaction resulting in an additional static potential $\propto (KL/c_s)\varphi^2$ which quenches the low frequency quantum fluctuations in the phase φ. For any value of K the ground state then is characterized by a fixed phase φ allowing for a superfluid flow across the impurity. For $K > 1$, the velocity v_L is a critical velocity: for $v < v_L$ the phase coherence is not reduced by quantum fluctuations and the superfluid flows free of dissipation through the impurity; at the critical velocity v_L, there is a sharp onset of phase slips which goes over to the nucleation rate for the infinite system (8) as the velocity increases.

Tube: Next, we study a superfluid in a tube of finite length L (see Fig. 1c). The finite size again produces a quantization of the modes in the leads which suppresses the spectral density and modifies the damping kernel at low frequencies,

$$\alpha_d(\omega) = K\omega \frac{\cosh \omega L/c_s - 1}{\sinh \omega L/c_s} = \begin{cases} (KL/2c_s)\,\omega^2, & \omega < c_s/L, \\ K|\omega|, & \omega > c_s/L. \end{cases} \tag{12}$$

The low frequency action then reduces to that of a capacitively shunted Josephson junction,

$$\frac{S}{\hbar} = \int d\tau \left\{ \frac{\hbar}{E_L} \frac{(\partial_\tau \varphi)^2}{2} + \frac{E_J}{\hbar}\,[1 - \cos \varphi] + \mathcal{L}_> \right\} \tag{13}$$

with the 'charging' energy $E_L = \hbar c_s/KL$. The additional term $\mathcal{L}_>$ accounts for high frequency contributions $\omega > c_s/L$ and includes a kinetic (α_m, see (6)) and a dissipative term. For long wires $L > \xi_c$ the mass term $\alpha_m(\omega) = (\hbar/E_C)\,\omega^2$ is small, while it does not appear altogether in the case of short wires $L < \xi_c$.

We start with systems where the characteristic frequency is smaller than c_s/L allowing us to drop $\mathcal{L}_>$ in determining the ground state properties of (13). This condition is satisfied for short tubes $L < \xi_c$ with the characteristic frequency $\sqrt{E_J E_L}/\hbar = \sqrt{L/\pi\xi_c}\,(c_s/L) < c_s/L$. Then, the low frequency eigenstates of the system (13) are characterized by Bloch states in the 'quasinumber' $\mathcal{N}$, see Ref. [12] (the 'quasinumber' $\mathcal{N}$ plays the role of the 'quasimomentum' k in the periodic crystals). For weakly interacting bosons with $E_J/E_L = K^2 L/\pi\xi_c \gg 1$ a tight binding analysis is applicable and the lowest energy band takes the form

$$E(\mathcal{N}) = W \cos (2\pi \mathcal{N}), \tag{14}$$

where the band width W can be calculated in the instanton approach using a single kink solution [14]

$$W \sim \frac{\hbar}{\tau_c} \frac{L}{\xi_c} \exp \left(-K\sqrt{L/\xi_c} \right), \tag{15}$$

with $\xi_c = c_s \tau_c$. Comparing to the classical Josephson effect, the fixed phase φ driving the superfluid velocity $v = v_c \sin \varphi$ in the Josephson junction is now replaced by the fixed 'quasinumber' $\mathcal{N}$ driving the chemical potential difference $\Delta\mu = 2\pi W \sin (2\pi\mathcal{N})$. Similarly, $\partial_t \varphi = \mu/\hbar \to \partial_t \mathcal{N} = \rho_0 v$. Then for a slowly moving impurity the system exhibits 'Bloch oscillations' in the chemical potential $\Delta\mu(t) = 2\pi W \sin(2\pi\rho_0 vt)$ in analogy to the ac Josephson effect. These oscillations are due to the accumulation of particles in front of the impurity, the latter allowing only discrete particles to tunnel. Each 'Umklapp' process then describes a particle tunneling through the impurity. Two mechanisms lead to the disappearance of the 'Bloch oscillations' at larger driving forces: first, transitions to higher bands, known as 'Zener tunneling', become relevant for frequencies $\rho_0 v$ larger than the band gap $\sqrt{E_J E_L}$ ($v > (v_L/K)\sqrt{L/\xi_c}$) and describe the accumulation of particles in front of the impurity, i.e., absence of 'Umklapp' processes quenches the tunneling. The flow *across* the impurity then does not match the flow of particles *towards* the impurity, producing a steady increase in the chemical potential difference $\Delta\mu$. Second, for $\rho_0 v > c_s/L$, or equivalent $v > v_L/K$, the damping terms in $\mathcal{L}_>$ localize the phase and a crossover to the classical Josephson effect appears: the dissipation projects the phase φ and renders it sharp. In an actual experiment, we start with the sample in the superfluid state characterized by a homogeneous phase. Switching on the laser decouples the Bose-Einstein condensate into two subsystems. For a vanishing velocity of the impurity, the initially well defined phase difference φ evolves according to the action (13) and spreads over the various minima within the characteristic time scale $\hbar/W$. The observed phenomena are then determined by the Bloch states and a weak driving force $v < (v_L/K)\sqrt{L/\xi_c}$ provides us with 'Bloch oscillations' in the chemical potential $\Delta\mu$. Increasing the driving force, $v > v_L/K$ the systems enters a regime where 'Zener' tunneling and dissipation compete with the physics of 'Bloch oscillations'. Note that this regime is terminated by the critical velocity v_c of the link, as $v_c < v_L$ in these short tubes.

For long wires with $L > \xi_c$ the terms $\mathcal{L}_>$ in (13) are relevant and the characteristic frequency is $\sqrt{E_J E_C}/\hbar \sim c_s/\xi_c > c_s/L$. The ground state properties of the system and low drive response then involves a more difficult interplay between dissipation and Bloch physics which requires more study. The situation simplifies at larger drives $v > v_L$: all processes are fast involving frequencies $\omega > c_s/L$ and the damping term in (12) dominates over the inertia. Then we recover the physics of the infinitely long system: the phase φ becomes localized allowing for a superfluid flow through the impurity and the quantum nucleation rate calculated via the instanton approach reduces to the result (8).

In conclusion, quantum fluctuations in the condensate wave function of Bose-Einstein condensates brings about fascinating new phenomena in superfluids and their experimental observation is a challenging task for the future providing us with a better understanding of macroscopic quantum effects.

References

1. C. Raman et al., Phys. Rev. Lett. **83**, 2502 (1999).
2. J. M. Martinis, M. H. Devoret, and J. Clarke, Phys. Rev. B **35**, 4682 (1987).
3. N. Giordano, Physica B **203**, 460 (1994).
4. A. Bezryadin, C. N. Lau, and M. Tinkham, Nature **404**, 971 (2000).
5. A. D. Zaikin, D. S. Golubev, A. van Otterlo, and G. T. Zimányi, Phys. Rev. Lett. **78**, 1552 (1997); V. A. Kashurnikov, A. I. Podlivaev, N. V. Prokof'ev, and B. V. Svistunov, Phys. Rev. B **53**, 13091 (1996); Y. Kagan, V. N. Prokof'ev, and B. V. Svistunov, Phys. Rev. A **61**, 045601 (2000).
6. R. Onofrio et al., Phys. Rev. Lett. **85**, 2228 (2000).
7. F. Dalfovo, S. Giorgini, L. P. Pitaevksii, and S. Stringari, Rev. Mod. Phys. **71**, 463 (1999).
8. H. P. Büchler, V. D. Geshkenbein, and G. Blatter, *to be published*.
9. A. O. Caldeira and A. J. Leggett, Ann. Phys. (N.Y.) **149**, 374 (1983).
10. M. Tinkham, *Introduction to Superconductivity*, McGraw-Hill, 1996.
11. A. Schmid, Phys. Rev. Lett. **51**, 1506 (1983); S. A. Bulgadaev, Sov. Phys. JETP Lett. **39**, 317 (1984).

12. G. Schön and A. D. Zaikin, Phys. Rep. **198**, 237 (1990).

13. U. Weiss, H. Grabert, P. Hänggi, and P. Riseborough, Phys. Rev. B **35**, 9535 (1987).

14. S. Coleman, Phys. Rev. D **16**, 2929 (1977); C. G. Callan and S. Coleman, Phys. Rev. D **16**, 1762 (1977); S. Coleman, *Aspects of symmetry*, Cambridge University Press, 1988.

Part VII

Vortex Reconnections and Classical Aspects

Vortex Reconnection in Normal and Superfluids

Joel Koplik

Benjamin Levich Institute and Department of Physics
City College of the City University of New York
New York, NY 10031 USA

1 Introduction

An example of vortex reconnection is shown in a time sequence in Fig. 1, wherein two distinct vortex filaments in a fluid move together, merge, and then divide into two or more filaments moving away, with part of one initial filament connected to part of the other. The physics underlying this example [1] will be presented later, but the key feature is the evident change in the topology of the vortices. In this lecture, we will discuss vortex reconnection in both normal and superfluids, emphasizing the relevance of the process to their respective turbulent flows, the similarities between the two cases, and the computational issues. The lecture is aimed at a fairly general audience: no detailed knowledge of fluid mechanics is assumed beyond a nodding acquaintance with the Navier–Stokes equation, and nothing about superfluidity beyond the idea of a two-fluid system with an quantum-mechanically condensed component, and a willingness to accept the Gross-Pitaevskii model for the latter. My emphasis will be on the superfluid case, and the reconnection process in normal fluids is discussed in more detail in the cited literature. My original work reviewed here was done in collaboration with Herbert Levine. Related and more recent work along these lines is described in the lectures by Adams and Roberts in this volume.

2 Some Vortex Generalities

The simplest example of a vortex flow field [2] is the infinite line vortex,

$$u(r) = \frac{\kappa}{2\pi r}\hat{\phi} \tag{1}$$

corresponding to steady rotation about the z-axis (in cylindrical coordinates r, ϕ, z). The vorticity is defined generally as as $\omega = \nabla \times u$, which in this case has the value $(\kappa/r)\delta(r)\hat{z}$, so that the orientation of ω is the rotation axis, and the magnitude is related to the strength of the rotation. The constant prefactor has been written in terms of the *circulation* Γ through any closed curve enclosing the z-axis,

$$\Gamma \equiv \oint_C dr \cdot u = \int_S dS \cdot \omega \tag{2}$$

where S is a surface whose boundary is C; for the line vortex (1) one finds $\Gamma = \kappa$. This velocity field is an exact solution of the Euler equation, the Navier–Stokes

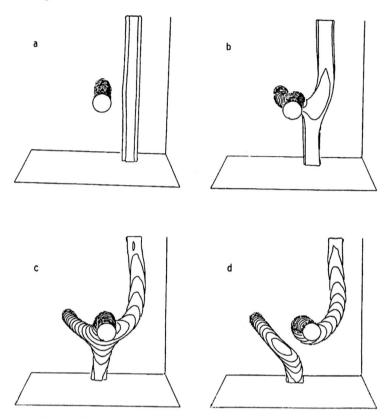

Fig. 1. Reconnection of two superfluid vortex filaments [1].

equation for normal fluids

$$\rho\left[\frac{\partial \boldsymbol{u}}{\partial t} + \boldsymbol{u} \cdot \nabla \boldsymbol{u}\right] = -\nabla p + \mu \nabla^2 \boldsymbol{u} \tag{3}$$

with the viscosity μ set to zero, with pressure $p = p_0 - \rho \kappa^2 / 8\pi^2 r^2$, and constant density ρ. The singularity along the axis can be avoided by smoothing the vorticity over a finite core region [3], and various choices are available. For example, the "Rankine vortex" has $u_\phi = \kappa r / 2\pi a^2$ inside the core region $r < a$ and the previous form for $r > a$ outside, so that inside the core the vorticity is uniform, $\boldsymbol{\omega} = \hat{\boldsymbol{z}} \kappa / \pi a^2$, but still zero outside. In the presence of viscosity, these vortices are no longer exact solutions, and in fact a vortex will decay due to friction in the absence of external forcing, but for finite intervals one can get often by with a suitable time-dependent core.

In a superfluid the situation is rather analogous. In this talk we adopt the Gross-Pitaveskii (GP) model [4] for the wave function of the condensate

$$i\hbar\frac{\partial \psi}{\partial t} = -\frac{\hbar^2}{2m}\nabla^2\psi + V_0\left[|\psi|^2 - n_0\right]\psi \tag{4}$$

As in any quantum mechanical system, we can define a density $\rho = |\psi|^2$ and a velocity $\boldsymbol{u} = \boldsymbol{j}/\rho$, where the usual probability current is $\boldsymbol{j} = \mathrm{Re}\,[(\hbar/im)\psi^*\nabla\psi]$. After some algebra [5], the real and imaginary parts of the GP equation become the equation of continuity of a compressible fluid, and the Euler equation with an unusual and singular equation of state,

$$p = \frac{\hbar^2}{2m}\frac{\nabla^2\sqrt{\rho}}{\sqrt{\rho}} + \frac{V_0}{m}(n_0 - \rho^2) \qquad (5)$$

The GP model (4) has exact, steady, stable solutions of the form $\psi_1 = f(r)e^{i\theta}$, again in cylindrical coordinates, where $f(r)$ is a smooth function with the asymptotic behaviors $f \to \sqrt{n_0}$ as $r \to \infty$ and $f \sim r$ as $r \to 0$. The crossover between the two limiting forms of f occurs at a "core radius" $a_0 = \hbar^2/2mn_0V_0$, which for parameters appropriate for $^4\mathrm{He}$ is $O(1\text{Å})$. The corresponding velocity field is

$$\boldsymbol{u} = \frac{h}{m}\frac{\hat{\phi}}{2\pi r} \qquad (6)$$

for $r \gg a_0$, which is nothing but a line vortex with circulation h/m. Solutions whose circulation is any integer multiple of this value exist, but are unstable: the circulation is thus quantized. At smaller r the velocity is messier, but the vanishing of $f(r)$ at the origin leads to a current which is finite everywhere. Furthermore, the density $\rho = f^2$ is a constant outside of the core, and tends smoothly to zero as $r \to 0$, while the vorticity is non-zero only within the core region where ρ varies. Thus, outside of a microscopic core region, a superfluid has vortex excitations whose dynamics is given by the incompressible Euler equation, and thus isolated superfluid vortices have the same hydrodynamical behavior as those of an inviscid normal fluid.

An important feature of inviscid fluids which bears on vortex stability and possible reconnection is the conservation of circulation, or Kelvin's theorem. The theorem states that the circulation around a contour which moves with the fluid is constant in time, and its relevance here is that single vortex filaments cannot just disappear, whereas two adjacent filaments of equal and opposite vorticity could annihilate locally. More generally, it appears superficially in Fig. 1 that some contours may enclose vortices initially but not later, so it is useful to present the derivation of Kelvin's theorem, simply to see where it could fail. The argument is to consider the total or Lagrangian time derivative of the circulation:

$$\frac{d}{dt}\oint_{C(t)} \boldsymbol{u}(t)\cdot d\boldsymbol{r} = \oint_{C(t)} \frac{d\boldsymbol{u}(t)}{dt}\cdot d\boldsymbol{r} + \oint_{C(t)} \boldsymbol{u}(t)\cdot\frac{d}{dt}(d\boldsymbol{r})$$

Note that because the curve $C(t)$ moves with the fluid, it is not differentiated. In the first integral, $d\boldsymbol{u}/dt$ can be replaced with the right-hand side of the Navier-Stokes equation, $\rho^{-1}[-\nabla p + \mu\nabla^2\boldsymbol{u}]$, and in the second integral we write $\boldsymbol{u}\cdot(d/dt)d\boldsymbol{r} = \boldsymbol{u}\cdot d(d\boldsymbol{r}/dt) = (1/2)d(\boldsymbol{u}^2)$. The two total derivatives integrate to zero, leaving

$$\frac{d\Gamma(t)}{dt} = \frac{\mu}{\rho}\oint_{C(t)} \nabla^2\boldsymbol{u}$$

As noted above, circulation is only conserved in the absence of viscosity. A further limitation on the theorem is that the derivation tacitly assumes that the functions involved are sufficiently regular to be differentiated inside the integral and manipulated as above. At a superfluid vortex core, the density vanishes, so that in view of (5), the pressure is singular and there is cause for concern.

The motion of a vortex filament is a combination of the background fluid motion plus any self-induced velocity it produces. In general, the velocity field due to any vorticity distribution may be recovered by a Biot-Savart integral. In simple cases, the nature of the motion may be deduced by symmetry and application of the right-hand rule, as in electrodynamics. For example, an isolated infinite straight vortex filament does not move, since there is no preferred direction. Two such *parallel* filaments rotate about their center of mass, while *anti*-parallel filaments translate in tandem, as can be easily seen by appropriate gestures with one's right hand. More generally, for vanishing filament thickness and the inviscid limit the Biot-Savart law can be reduced [3] to the "local induction approximation"

$$\frac{\partial \boldsymbol{R}}{\partial t} \approx \frac{\kappa \log L/a}{4\pi} \frac{\partial \boldsymbol{R}}{\partial s} \times \frac{\partial^2 \boldsymbol{R}}{\partial s^2} \tag{9}$$

Here, points $\boldsymbol{R}$ on the the filament are parametrized by arclength s and time t, and (a, L) are short and long-distance cutoffs, respectively. The self-induced velocity is then along the binormal to the filament axis, so for example an isolated *non*-straight filament does move, and a circular vortex ring moves along its centerline, perpendicular to the plane of the ring.

3   The Importance of Reconnection

Vorticity and vortex filaments are an important ingredient in high velocity flows of both normal and superfluids. Although the turbulent states have important similarities that motivate the study of reconnection, which we will discuss here, the corresponding mechanisms by which they arise are of course rather different.

Consider first the behavior of a bucket of a normal viscous fluid as it is spun up from rest about its axis. The fluid will begin to rotate along with its container, and at low velocities develop a laminar state with constant angular velocity and a parabolic free surface at the top. At large enough angular velocity, this uniform state is unstable and time-fluctuating structures such as spiral vortices appear. With further increases in rotation rate, these localized fluctuations occur more frequently, and grow in size and lifetime and completely overwhelm the laminar flow. At high rotation rates, one sees velocity fields dominated by fluctuations, whose length scales extend up to the size of the container, and whose time scales range from the microscopic on up. One crucial feature of fully-developed turbulent flows in normal fluids is the presence of regions of concentrated vorticity, as in the simulation example shown in Fig. 2a [6], where values of $|\boldsymbol{\omega}|$ greater than a selected threshold are shown. Experimental studies [7], using bubbles for visualization, likewise indicate the presence of regions of both line-like and sheet-like excitations, which fluctuate in time and space. The precise connection between vortical structures and turbulence in normal fluid flows is a subject of current work [8], but there is at least a school of thought that they underlie the "intermittency" phenomena observed in the turbulent energy spectrum and structure functions. In terms of the subject at hand, although the evolution of these high vorticity structures in a fully-developed turbulent flow does not correspond precisely to the mutual interaction of isolated filaments, consideration of such processes is an obvious ingredient. Aside from its relevance to turbulence, vortex reconnection can occur in many other contexts in normal fluid dynamics where vortex filaments or other structures are produced, for example in the wakes of airplane wings or ships, and would have an impact on issues such as drag reduction. A further application is to the question of whether singular solutions of the Navier-Stokes equations develop from smooth initial data [9]. Typical configurations which have been used for numerical investigations have concentrated vorticity, and reconnection may regularize them.

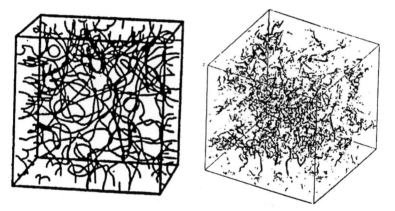

Fig. 2. (Left) Regions of high vorticity in a simulated turbulent flow [6]; (Right) Simulation of a superfluid vortex tangle [10].

If a bucket of superfluid is rotated about its axis, one sees a rather different sequence of states as the rotation rate is increased. At low angular velocity nothing happens to the fluid at all, until at a critical value a vortex filament is nucleated, parallel to the rotation axis. The configuration persists until a second critical velocity, whereupon a second filament appears, and so on. The filaments are parallel lines, distributed in the orthogonal plane at regular lattice points, and aside from stability considerations higher rotation rate just produces more filaments. Of course this configuration is unstable, and any fluctuation of one filament creates a perturbing velocity field which acts on the others. Unlike the normal fluid case, however, the individual superfluid vortices are "topologically" stable due to quantization of circulation, and they simply begin to move in response to the local velocity field. The result is a disordered and fluctuating collection of vortex filaments, resembling a pot of boiling spaghetti, the starting point for the calculations of Schwarz [10], and subsequently others, of the statistical properties of the turbulent superfluid state. There are no experimental results visualizing this vortex tangle, but a simulation snapshot is given in Fig. 2b, which evidently resembles its normal fluid counterpart Fig. 2a. Such calculations are based on the local induction approximation, which is quite reasonable for the widely separated vortices in a superfluid, but the calculation shows that as these filaments move about they are are prone to intersect each other. At this point a more microscopic calculation is needed to determine the outcome of a vortex collision.

4   Reconnection in Normal Fluids

Given an evolution equation for the appropriate fields, either Navier-Stokes or Gross-Pitaevskii, it is ostensibly a straightforward matter to integrate numerically to find the time-evolution of an initial two-vortex configuration. In the former case, although there are many well-developed numerical techniques, vortex interactions involve both long and short length scales, and the calculation is expensive to carry out in detail. Most work I'm aware of dates from ca. 1990, and is reviewed by Kida and Takaoka [11] with a different slant than in the present talk. In Fig. 3a,b, we show two representative examples, filament reconnection as studied by Melander and Hussain [12], and vortex ring merger following Kida, Takaoka and Hussain [13]. In both cases a spectral (Fourier series) numerical method was used, the figures are isosurfaces of a selected vorticity magnitude, the fluid is incompressible, and the initial condition is symmetric in that the two starting vortices have the same strength. The general result is that adjacent, equal and oppositely oriented vortex structures tend to merge and then move apart, perhaps with some transient secondary regions of high vorticity left behind. For the vortex ring case at least, there are visualization experiments [14] which show the same merger into one ring and subsequent breakup into two new rings moving in an orthogonal plane.

    In terms of understanding the mechanism, the first stage is the fact that oppositely oriented segments of concentrated vorticity move towards each other. This behavior follows intuitively from the local induction approximation (9), as explained by Siggia [15]. In the typical configuration for interaction, vortex filaments which approach are each bent into a hairpin shape with opposite orientations. When the cores begin to overlap, there is a cancelation of opposite signed vorticity, leading to the bridging seen in the figures. The words which usually accompany figures of the intermediate state are that the closest segments of adjacent opposite vortex lines annihilate in the merger region while the remainder of these lines simply connect up outside, and the more distant parts of the opposed vortex lines persist to form the thin bridge connecting the main filaments as they pull apart. The initial motion of the outer parts of the filaments then carries the reconnected configuration away.

    It would of course be desirable to have an analytic model of the reconnection process, but those available are not entirely consistent with numerical tests

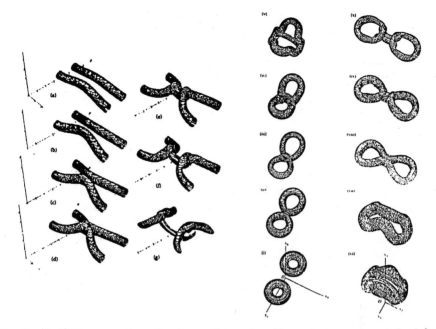

Fig. 3. (Left) Reconnection of anti-parallel vortex filaments in a normal fluid [11]; (Right) Vortex ring merger and breakup in a normal fluid [13].

[16,17]. These authors present very detailed numerical analyses of reconnection simulations, which differ somewhat in detail, but with no clear and simple insight emerging. Among the additional features which can be identified in the simulations are vortex dipole formation, flattening of the core regions, localized vorticity amplification, and jet formation. Much of the difficulty is already evident in the complicated three-dimensional spatial variation seen in the figures.

What about Kelvin's theorem? The core of a vortex will entail some viscous effects (in the line vortex case, $u \sim 1/r$ would make the viscosity term dominant at small r, and in the Rankine case velocity gradients are discontinuous at edge of the core), so the theorem is only valid if applied to contours outside the core, and therefore outside the reconnection region. If two vortex filaments were to approach with parallel vorticity, a contour encircling both would have a net circulation that is the sum of the individual ones, and it would not be possible for them to reconnect and annihilate this vorticity. Fortunately the relevant case is anti-parallel, when the locally induction flow outside the core leads to natural attraction, and there a zero total circulation initially is quite compatible with reconnection.

If the initial condition is asymmetric, with vortices of different circulation interacting, it is not possible to cleanly reconnect in this way, because Kelvin's theorem applied to a contour encircling the reconnection region implies that the final circulation is non-zero. Numerical simulations show that an assortment of final states with exotic final states can occur [18], such as a weak vortex wrapping around a stronger one. Other simulations have studied the evolution of intertwined vortex rings, which produce their own bestiary of curious shapes [19]. Likewise, other effects such as compressibility of the fluid can complicate the dynamics and alter the final topology [20].

5   Reconnection in Superfluids

The superfluid case turns out to be much simpler and cleaner. One one hand, there is a very simple and elegant numerical technique available for the nonlinear Schroedinger equation, and on the other, the fact that vorticity is quantized means that only symmetric cases are relevant, and furthermore it is not possible to have persistent residual bridges. Two cautionary remarks are needed. First, the GP model is precisely that, and is not a completely faithful description of Helium-II. However, it does exhibit vortex configurations which have reasonably correct behavior as far as is known, and we may regard the details of its wave function as representing a particular approximation for regularizing what would otherwise be a singular core region. Most of the motion is just that of an Euler fluid, and the behavior we find resembles that of normal fluids, so it is plausible to suppose that we just miss some details of the excitations in the final state. A second limitation is that we consider pure condensate and no explicit coupling to normal fluid at all, so strictly speaking we have a zero-temperature system. Again, to the extent that reconnection phenomena involve only the general aspects of the short distance properties of the superfluid, we may hope that the results are general.

The numerical method used here is a split-time-step algorithm [21] first introduced for the GP equation with the opposite sign of the nonlinearity (an equation relevant to laser focusing). We [1] first rescale length and time to get the dimensionless equation

$$\frac{\partial \psi}{\partial t} = i\nabla^2 \psi - i\psi(|\psi|^2 - 1)$$

As a variant of the simple Euler method $\psi(t+dt) = \psi(t) + dt \cdot \partial \psi / \partial t$, break the right-hand side into its two terms and treat them successively. In Fourier space, assuming periodic boundary conditions, the Laplacian term just integrates into a phase

$$\psi(\mathbf{k}, t) \rightarrow \tilde{\psi}(\mathbf{k}, t+dt) = e^{-ik^2 dt} \psi(\mathbf{k}, t) \tag{11}$$

while in real space, since *without* the Laplacian term one has $\partial/\partial t \, |\psi|^2 = 0$, the nonlinear term also integrates to a phase,

$$\tilde{\psi}(\mathbf{x}, t+dt) \rightarrow \psi(\mathbf{x}, t+dt) = e^{-i(|\tilde{\psi}|^2 - 1)dt} \, \tilde{\psi}(\mathbf{x}, t+dt)$$

of course, the wave function must be Fourier transformed twice at every time step, but with periodic boundary conditions this is efficiently done by FFT algorithms. Thus, the schematic numerical code is

```
do i=1, nsteps
    FFT ψ
    ψ → e^{-ik²dt} ψ
    inverse FFT ψ
    ψ → e^{-i(|ψ|²-1)dt} ψ
enddo
```

and the actual code is not much more complicated than this. Aside from simplicity, because at each sub-time-step the wave function is just multiplied by a phase, the norm

$$\int d^3x \, |\psi(\boldsymbol{x}, t)|^2 \sim \int d^3k \, |\psi(\boldsymbol{k}, t)|^2$$

is conserved, so total probability is constant in time. Similarly, the energy

$$E = \int d^3x \left[\frac{1}{2}|\nabla\psi|^2 + \frac{1}{4}|\psi|^4 - \frac{1}{2}|\psi|^2 + \frac{1}{4} \right]$$

is conserved; since the energy is proportional to the total length of vortex line, there is no possibility of vortex stretching here.

The remaining ingredient is the initial wave function corresponding to the particular vortex configuration of interest. Beginning with a background value $\psi = 1$, a trivial solution of (10), a single line vortex along the z-axis is obtained by multiplying it by the wave function $\psi_1 = f(r)e^{i\theta}$ discussed above. For a two-vortex configuration, the latter is multiplied by a second factor ψ_1, where now

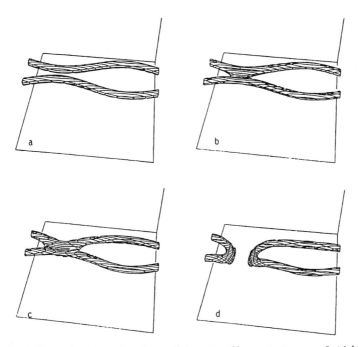

Fig. 4. Reconnection of anti-parallel vortex filaments in superfluid [1].

r and θ are defined with respect to the new axis of the second filament [22]. This provides a reasonable approximation, provided the filaments are initially separated by a distance much larger than the core size. Any error is taken care of by the integration of the GP equation itself, provided the vortices have time to relax to the correct form before they begin to interact with each other. For a vortex ring [23], a similar procedure may be used, provided the radius of the ring is much larger than the core size. At any point $\mathbf{x}$ we consider the plane passing through the point and containing the axis of the ring. The ring then intersects this plane at two points $\mathbf{x}_{\pm}$, once in the positive and once in the negative sense, and we take $\psi(\mathbf{x}, t = 0) = \psi_1(\mathbf{x} - \mathbf{x}_+)\psi_1^*(\mathbf{x} - \mathbf{x}_-)$. The test of this procedure is to observe the translation of a single ring along its center axis, and verify that the observed velocity agrees with analytic calculations.

It is now straightforward to study reconnection. In Fig. 1, the initial configuration has two filaments with axes at $90°$, and an initial separation of 4 in units of the core size. The filaments bend towards each other, merge, and then split with an obvious topological rearrangement. Figure 4 shows the case most relevant to reconnection in a vortex tangle, where the initial vortices have opposite orientation, and have a mild hairpin-like bend towards each other. In a periodic box, only certain relative orientations are consistent with a two-vortex state; we find that initial relative angles of $180°$, $135°$ and $90°$ reconnect while $0°$ and $45°$ do not. Since it is only in the former cases that the vortices tend to approach each other, unless there is a strong external flow in addition to the mutual induction velocity, we conclude that intersecting vortices almost always reconnect.

Note the similarity to the normal fluid reconnection simulation in Fig. 3a; the main difference is the absence of the bridge connecting the withdrawing final vortices. In a normal fluid it is possible to leave behind high-vorticity regions after an interaction in which viscosity enters and invalidates Kelvin's theorem, but in a superfluid this would require a new segment of quantized filament, and this is not possible. As for the reconnection event itself, the superfluid density vanishes at the vortex core, so Kelvin's theorem applies only to contours which avoid the merger region and there is no contradiction. In contrast to the Navier-Stokes simulations, nothing very interesting happens in the cores during merger processes: the density contours just smoothly merge into each other, without rapid variation in space or time. Although this calculation conserves energy globally, the reconnection process can redistribute it, and in fact there is an emission of radiation from the reconnection region [24]

Vortex ring interactions resemble the normal fluid case, but the simplicity of the numerical methods allowed us to examine a number of cases with various initial orientations for long times [23]. If we begin with two rings, moving along their axes so as to intersect at $90°$, the result is similar to that in Fig. 3b, again without the bridge joining the two final rings. The key geometrical feature observed in the normal fluid vortex ring merger is present – the fact that there are two final rings which move off in a plane perpendicular to the plane of incidence. In the GP calculations, the rings continue to move periodically through the simulation box, occasionally repeating the merger and dissociation process, whereas the Navier-Stokes calculations are dissipative and awkward to follow for long times, so the ultimate fate of the bridge is not known. If the initial condition is varied, however, we can obtain other final states. In Fig. 5, the initial intersection angle is 120°, and the final state has four rings; the difference evidently is related to the detailed shape of the intermediate, distorted single-ring state, and its tendency to close on itself. Other less-symmetric initial states where the initial rings intersect off-axis can give three final rings, or presumably any number.

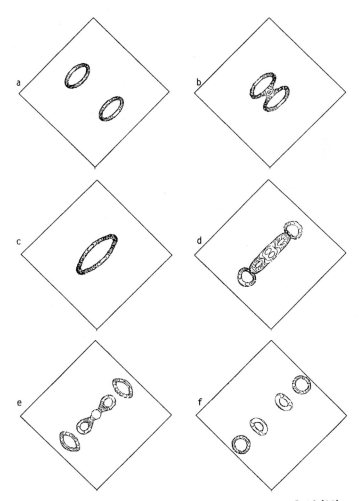

Fig. 5. Vortex ring merger and breakup in a superfluid [23].

6 Conclusions

We have discussed vortex reconnection in a comparative way for normal and superfluids. The dynamics at the scale of the vortex cores is rather different, and apparently much smoother for superfluids, whereas the large scale behavior of vortex filaments and rings is very much the same. Since both systems are essentially Euler fluids outside a vortex core, the latter behavior is unsurprising, and provides further evidence for the similarities between normal and superfluid turbulence.

The superfluid reconnection calculations discussed here incorporated the minimal physics needed to address the phenomena intelligently – the simplest reasonable model of the condensed state, zero temperature, and neglect of the normal fluid and mutual friction. It is our belief that in a model more realistic than GP one would see perhaps more structure in the core region, corresponding to a richer excitation spectrum, but no significant change in general behavior or vortex topology. The role of the GP model in this sense is to provide a physically reasonable smoothing of the core singularity present in an idealized line vortex, and that more refinements would just affect the details. The similarities between the superfluid and Navier-Stokes results discussed above already justify this assertion.

Adding the normal fluid and finite temperature is a very interesting avenue for further work at the atomic scale considered here, since issues such as thermal nucleation and vortex coupling to normal fluid are likely to be qualitatively significant. Similarly, the generation of vorticity due to interaction of the superfluid with solid walls would be better understood if further microscopic information were available. It would be a pleasure to hear these topics reviewed at the next meeting on superfluid vortices.

References

1. J. Koplik and H. Levine: Phys. Rev. Lett. **71**, 1375 (1993)
2. G. K. Batchelor: *An Introduction to Fluid Dynamics* (Cambridge University Press, Cambridge 1967)
3. P. G. Saffman: *Vortex Dynamics* (Cambridge University Press, Cambridge 1992)
4. E. P. Gross: J. Math. Phys. **4**, 195 (1963); L. P. Pitaevskii: Sov. Phys. JETP **13**, 451 (1961)
5. R. J. Donnelly: *Quantized Vortices in Helium II* (Cambridge University Press, Cambridge 1991)
6. A. Vincent and M. Meneguzzi: J. Fluid Mech. **225**, 1 (1991)
7. S. Douady, Y. Couder and M. E. Brachet: Phys. Rev. Lett. **67**, 983 (1991)
8. U. Frisch: *Turbulence* (Cambridge University Press, Cambridge 1995)
9. H. K. Moffatt: J. Fluid Mech. **409**, 51 (2000), and earlier references therein.
10. K. W. Schwarz: Phys. Rev. B **38**, 2398 (1988), and earlier references therein. For later work, see the lecture by Tsubota in this volume, and its references.
11. S. Kida and M. Takaoka: Annu. Rev. Fluid Mech. **26**, 169 (1994)
12. M. V. Melander and F. Hussain: Phys Fluids A **1**, 633 (1989)
13. S. Kida, M. Takaoka and F. Hussain, J. Fluid Mech. **230**, 583 (1991)

14. T. Fohl and J. S. Turner: Phys Fluids **18**, 433 (1975); Y. Oshima and N. Izutsu: *ibid.* **31**, 2401 (1988)
15. E. D. Siggia: Phys Fluids **28**, 794 (1985)
16. O. N. Boratov, R. B. Pelz and N. J. Zabusky: Phys Fluids A **4**, 581 (1992)
17. M. J. Shelley, D. I. Meiron and S. A. Orszag: J. Fluid Mech. **246**, 613 (1993)
18. N. J. Zabusky and M. V. Melander: Physica D **37**, 555 (1989)
19. H. Aref and I. Zawadski: Nature **354**, 50 (1991)
20. D. Virk, F. Hussein and R. M. Kerr: J. Fluid Mech. **304**, 47 (1995)
21. M. Taha and M. Ablowitz: J. Comput. Phys. **55**, 203 (1984)
22. A. L. Fetter: Phys. Rev. **138**, A429 (1965)
23. J. Koplik and H. Levine: Phys. Rev. Lett. **76**, 4745 (1996)
24. M. Leadbeater, T. Winiecki, D. C. Samules, C. F. Barenghi and C. S. Adams, cond-mat/0009060; see the paper by Adams in this volume.

Helicity in Hydro and MHD Reconnection

Axel Brandenburg[1,2] and Robert M. Kerr[3,4]

[1] Department of Mathematics, University of Newcastle upon Tyne, NE1 7RU, UK
[2] NORDITA, Blegdamsvej 17, DK-2100 Copenhagen Ø, Denmark
[3] NCAR, Boulder, CO 80307-3000, USA
[4] Department of Atmospheric Sciences, University of Arizona, Tucson, AZ 85721-0081, USA

Abstract. Helicity, a measure of the linkage of flux lines, has subtle and largely unknown effects upon dynamics. Both magnetic and hydrodynamic helicity are conserved for ideal systems and could suppress nonlinear dynamics. What actually happens is not clear because in a fully three-dimensional system there are additional channels whereby intense, small-scale dynamics can occur. This contribution shows one magnetic and one hydrodynamic case where for each the presence of helicity does not suppress small-scale intense dynamics of the type that might lead to reconnection.

1 Introduction

The term reconnection is used in both the MHD and fluids communities to describe topological changes in magnetic or vorticity fields due to resistivity or viscosity and could not occur in ideal cases where these dissipative terms are zero. In the strictly ideal limit the connectivity of the field lines would not change, but this is a singular limit and even the smallest amount of resistivity or viscosity allows the connectivity to change, albeit on small length scales. In the presence of finite dissipative terms, large amounts of energy can be converted into heat (or other forms of energy if one goes beyond the hydrodynamic approximations). MHD reconnection plays a role in understanding why the solar corona (i.e. the tenuous layers above the solar surface) are heated to $\sim 10^6\,\mathrm{K}$, even though at the surface of the sun the temperature is only $\sim 6000\,\mathrm{K}$. The other aspect is that the dissipative terms allow the field line connectivity to change. There are strong indications from observations of the solar corona in X-rays that field loops originally tied to the solar surface all of a sudden break loose and transport large amounts of flux into outer space. This raises the issue of how fast can reconnection occur. This is perhaps the single most challenging aspect of the problem.

Early work on MHD reconnection was concerned with steady state configurations, allowing a constant flux of material to pass through an X-point type configuration in two-dimensional field line configurations. However, because the reconnection site becomes very thin as the magnetic resistivity decreases, the amount of flux processed through the reconnection site decreases like the square root of the resistivity and by this mechanism finite reconnection in a dynamical timescale is not feasible for typical astrophysical values of resistivity. Other more complicated initial conditions can lead to slow shocks that increase the reconnection rate, as discussed in a recent textbook [1]. Is not clear, however, whether the various boundary conditions studied so far represent anything physical in the corona and whether the results could explain the nanosecond timescales over which hard X-ray output associated with reconnection is seen to rise.

2   Dissipation of Energy and Helicity

Magnetic reconnection has two distinct aspects. One is the speed at which magnetic energy can be converted into heat and the other is the speed at which the magnetic topology can change. The two need not be the same. The perhaps worst possible type of topology to change is one that invokes mutual linkage of flux tubes, which can be described by the magnetic helicity H defined as

$$H = \int \mathbf{A} \cdot \mathbf{B} \, dV, \tag{1}$$

where $\mathbf{B}$ is the magnetic field and $\mathbf{A}$ is the vector potential such that $\mathbf{B} = \mathbf{\nabla} \times \mathbf{A}$. Obviously, $\mathbf{A}$ is not uniquely defined, because adding an arbitrary gradient field to $\mathbf{A}$ would not change $\mathbf{B}$. However, the value of H is unaffected by this if the integral is taken over a domain where the normal component of the field vanishes on the boundaries. In that case

$$\int (\mathbf{A} + \mathrm{grad}\, \varphi) \cdot \mathbf{B} \, dV = \int \mathbf{A} \cdot \mathbf{B} \, dV + \int \varphi \mathbf{\nabla} \cdot \mathbf{B} \, dV = H, \tag{2}$$

because the magnetic field is always solenoidal, $\mathbf{\nabla} \cdot \mathbf{B} = 0$. Another conserved quantity is the cross helicity, $H_c = \int \mathbf{u} \cdot \mathbf{B} \, dV$, which describes the linkage between flux tubes and vortex tubes. In the absence of magnetic fields the hydrodynamic helicity, $H_h = \int \mathbf{u} \cdot \boldsymbol{\omega} \, dV$, is conserved by the inviscid Euler equations, and it describes the linkage of vortex tubes with themselves.

The standard example that highlights the connection between magnetic helicity and topology is an interlocked pair of flux rings (see, e.g., the first panel of Fig. 1), for which the magnetic helicity is given by twice the product of the two magnetic fluxes of each of the two flux rings. However, helicity is also associated with two orthogonal flux tubes as shown in Fig. 5.

The dramatic difference between the dissipation of magnetic energy and magnetic helicity can best be seen by contrasting the equations of the conservation of magnetic energy and magnetic helicity,

$$\tfrac{1}{2}\frac{d}{dt}\langle \mathbf{B} \cdot \mathbf{B} \rangle = -\langle \mathbf{u} \cdot (\mathbf{J} \times \mathbf{B}) \rangle - \eta \langle \mathbf{J} \cdot \mathbf{J} \rangle, \tag{3}$$

$$\tfrac{1}{2}\frac{d}{dt}\langle \mathbf{A} \cdot \mathbf{B} \rangle = -\langle \mathbf{u} \cdot (\mathbf{B} \times \mathbf{B}) \rangle - \eta \langle \mathbf{J} \cdot \mathbf{B} \rangle, \tag{4}$$

where angular brackets denote volume averages, and surface terms are assumed to vanish.

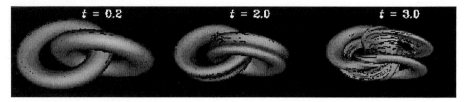

Fig. 1. Resistive evolution of an initially interlocked pair of flux rings. Isosurfaces of the magnetic field are shown at different times.

The important point to note here is that the magnetic energy can maintain a steady state where Joule dissipation, $\eta\langle J^2 \rangle$, can be finite and large if work is done against the Lorentz force, i.e. if $-\langle u \cdot (J \times B) \rangle > 0$. At the same time, however, there is no such term in the magnetic helicity equation, so in that case a steady state is only possible if the current helicity, $\langle J \cdot B \rangle$, vanishes.

In the absence of any forcing, f, the hydrodynamic helicity is conserved in a similar manner, but there are two important differences. First, because $\int u \cdot \omega \, dV$ contains one more derivative than $\int u^2 \, dV$, it dissipates faster than the energy if there is dissipation. Second, if there is forcing, the hydrodynamic helicity is no longer conserved. This difference to the hydromagnetic case can best be seen by contrasting eqs. (3) and (4) with the corresponding equations in hydrodynamics,

$$\tfrac{1}{2} \frac{d}{dt} \langle u \cdot u \rangle = \langle u \cdot f \rangle - \nu \langle \omega \cdot \omega \rangle, \tag{5}$$

$$\tfrac{1}{2} \frac{d}{dt} \langle u \cdot \omega \rangle = \langle \omega \cdot f \rangle - \nu \langle \theta \cdot \omega \rangle, \tag{6}$$

where $\theta = \nabla \times \omega$ is the curl of the vorticity, and surface terms are again assumed to vanish. Thus, unlike the magnetic counterpart, kinetic helicity conservation is only possible in the special case where the forcing is perpendicular to the vorticity and the flow is inviscid. With dissipation and the absence of forcing both kinetic energy and kinetic helicity are decaying, but kinetic helicity contains an extra derivative more than the kinetic energy and so decays faster than energy and does not pose a hard constraint.

3   Interlocked Flux Rings

In Fig. 1 we show an example of an initial flux tube configuration where, in our case, each tube has the flux $\Phi = \int B \cdot dS = 0.7 B_0 d^2$, where B_0 is the maximum field strength in the core of each tube and d is its radius. The magnetic helicity is measured to be $H = \int A \cdot B \, dV = 0.98 B_0^2 d^4$, in perfect agreement with the formula $H = 2\Phi^2$.

The subsequent evolution of this flux tube configuration is governed by the curvature force acting separately in each flux tube trying to make them contract. Eventually the two tubes come into contact and produce an intense current sheet where they touch. Figure 2 shows the peak current, $\|J\|_\infty$, plotted in two ways,

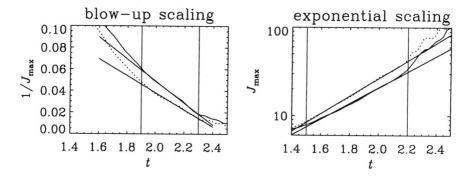

Fig. 2. Semi-logarithmic plot of $\|J\|_\infty$ for a compressible 240^3 calculation in a domain of size 4 (dotted line: filtered, and solid line: unfiltered initial conditions) together with fits to exponential growth and blow-up behavior, respectively. The blow-up scaling fits better at later times.

one showing a period of singular growth and the other showing a period of exponential growth. This initial period is represented in Fig. 1 by the first two frames showing the approach and initial deformation of the linked flux tubes. At the last time visualized, the surfaces in the outer region appear to merge into a single continuous flux tube, while the inner region appears to be annihilated in a complicated reconnected structure with writhe. Figure 3 takes another look at this time using flux lines instead of surfaces. The flux lines in the outer region that appeared to be continuous can now be seen to change direction abruptly where they plunge into the inner region. And the inner region is now seen to be continuously connected to the outer flux lines and instead of being a single flux tubes with writhe, it now appears to be the original flux lines just twisted around each other with almost no reconnection.

In the ideal case, the magnetic helicity is conserved for all time. Even in the resistive case the magnetic helicity is very nearly constant. Furthermore, the peak current increases to large values, which appear to be limited only by the numerical resolution. This is related to the newly posed millennium question of whether regularity of the Navier-Stokes equations can be shown [2]. A singularity probably does not develop for the full viscous and resistive equations due to the development of reconnection. However, singularities do seem possible for Euler and ideal MHD. Numerical calculations have been used to provide insight into the interaction of anti-parallel vortex tubes using the incompressible Euler equations [3]. The key to providing useful results was the direct comparison with hard analytic bounds for the maximum growth rate of the vorticity [4]. This initial condition was very contrived with special symmetries and no helicity, unlike real flows, and its generality remains uncertain.

A similar analytic bound has been shown to exist for the ideal MHD equations [5]. Therefore two intertwined questions have arisen. First, is there an initial condition for ideal MHD which might show similar singular growth? Second, what role might the various types of helicity play in suppressing or enhancing

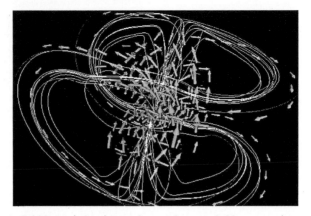

Fig. 3. Magnetic field lines (white) together with some field vectors (in grey) indicating the field orientation for a resistive flux ring calculation at $t = 4$, i.e. shortly after the time of the suspected singularity in the ideal calculation.

the reconnection rate? It has been found [6] that in the ideal case the two inter-locked magnetic flux tubes go through a phase where behavior consistent with a singularity seems plausible. This was surprising because as noted above this is a nearly maximally helical initial condition and was expected to suppress non-linearity. For hydrodynamics, it has also been claimed [7] that the helical initial condition of two orthogonal vortex tubes showed signs of a singularity, although the analytic test [4] was never applied. These two cases raise two possibilities. In hydrodynamics there might exist a mechanism whereby helicity is shed per-mitting stronger nonlinear growth, while in MHD the nonlinearity really comes from $\mathbf{J} \times \mathbf{B}$ and $\mathbf{J} \cdot \mathbf{B}$ is not conserved, so there might in fact be no constraint upon locally strong nonlinearity.

Let us consider an argument for why helicity might be required to allow, at least for a period, nearly singular growth for ideal MHD. This is based upon old arguments for why there could not be a singularity of Euler. It has been argued that a singularity would not occur for Euler because what drives the growth in the vorticity is the axial strain stretching the vorticity, and this strain must grow at the same rate as the vorticity to sustain this growth. This could only be achieved by an enormous growth in the curvature of vortex lines [8], which in turn would require a delicate balance in the growth of the pressure Hessian. This was originally thought not to be feasible, but newer analysis of the anti-parallel Euler calculations [9,10] has shown that all of the analytic requirements needed to achieve this delicate state are in fact obeyed. This is possible because the vorticity and the strain are in fact just different manifestations of the same vector field and can be strongly aligned.

While it has not been shown analytically, one would expect that for ideal MHD to show similar singular growth, a similar delicate balance would have to exist between the vorticity, the current, and the magnetic and velocity strain fields. Current and magnetic strains are just different manifestations of the same vector magnetic field, just as vorticity and strain are manifestations of the same vector velocity field. However, one is still left with the current being completely distinct from vorticity. Only if there is some property of the vector fields that strongly couples these two fields could they act in concert to give a singularity. Perhaps because helicity is conserved, for strongly helical structures there exists such a constraint. Therefore only for strongly helical magnetic structures could singular, or nearly singular, nonlinear growth occur.

New analysis has shown that the location of the peak in the current is at the juncture between the outer flux lines and the inner flux lines where the maximum in the curvature is located. This would be consistent with new mathematical analysis by J. D. Gibbon (unpublished) that strong magnetic field line curvature should be associated with any singular growth. The vorticity is also the strongest in this region, suggesting the type of symbiotic growth of current and vorticity that we believe is needed if there is to be a singularity of ideal MHD.

Figure 3 is resistive, but is very similar to visualizations of new ideal calculations that were run at higher resolution, up to the equivalent of 1296^3 mesh points if a uniform mesh had been used. However, these new calculations, while they do extend the period of seemingly singular growth, now appear to show that the singular growth eventually is suppressed in the incompressible case. The evidence relies on consistent behavior between the two highest resolution calculations.

What might be the cause of this suppression? While it will take time to fully understand these massive data sets, the initial indications are that it is occurring as the peak in the current moves outside the inner region with its strongly aligned vorticity, magnetic, and current fields. Our suspicion is that the importance of the inner region is that, through twist, the location of most of the initial helicity associated with the linked flux tubes is in the inner region. If this can be shown, then it might tell us that the secret to maintaining nearly singular growth is to maintain as high a level of local helicity for as long as possible. This would be consistent with arguments [11] that fast reconnection is associated with the entanglement of flux lines due to footpoint motion, which is known to produce the required heating rates [12]. Helicity has also been shown to play a role in coronal simulations of an arcade and a twisted flux loop [13], with nearly singular growth in current similar to what we have observed.

There are other possible mechanisms that could suppress singular growth. In a simulation [14] of nearly the same initial condition as the one used in our original paper [6], there is only exponential growth that is associated with the appearance of current sheets. More recent detailed analysis of our calculation shows that the exponential growth is actually associated with the appearance of two nearly overlapping orthogonal current sheets and the pressure barrier between them that suppresses stretching terms and growth. This is an important result because in some sense the more physical initial situation might be two flux tubes that do not overlap at all. Our simulations that show stronger growth in the current all have some overlapping between the initial flux tubes, something

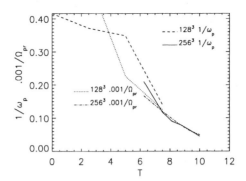

Fig. 4. $1/\|\omega\|_\infty$ and $1/\langle \omega_i e_{ij} \omega_j \rangle$ in Euler for orthogonal vortices.

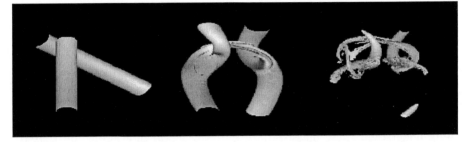

Fig. 5. Isosurfaces of vorticity as a fraction of the peak vorticity. The three frames are $t = 0, 6$ and 10. Arms are pulled out of the original vortices, become anti-parallel, then vorticity within the arms develops singular behavior.

that should not happen in an astrophysical situation where the flux tubes are initially separated by large distances.

4   Orthogonal Vortex Tubes

We now turn to the case of *straight* tubes that are orthogonal to each other. We note that also in this case there is finite helicity. The magnetic case has been studied previously [15], but here we focus on the hydrodynamic case with vortex tubes. Figure 4 shows the inverse of the peak vorticity and the inverse of the enstrophy production rate for the orthogonal vortex tubes whose inviscid evolution is shown in Fig. 5. Plotting these inverses was previously shown to be the most effective way to highlight the $1/(t_c - t)$ singular behavior. Figure 4 shows that the initial growth is weak, unlike the anti-parallel case. Then the growth of peak vorticity, ω_p, and enstrophy production, Ω_{pr}, becomes stronger with their inverses going roughly linearly to zero at the same singular time.

What is the configuration around the peak vorticity once singular growth starts? And what role does helicity play? Analysis of the three-dimensional fields shows that the peak vorticity is located in the arms that are being pulled off of the two original orthogonal vortices. The last time shows that these isosurfaces are parallel, and analysis shows that the vorticity within these surfaces is anti-parallel. That is, to develop singular growth exactly the same alignment of vorticity that was previously described as a contrived situation is actually what the dynamics generate by themselves. This is consistent with vortex filament work [16]. In terms of helicity, locally around the anti-parallel vortices there is no kinetic helicity density. Therefore in order for orthogonal vortices to develop singular growth, the flow must realign itself to be non-helical, shedding any helicity to achieve this.

In conclusion, these calculations have demonstrated that the role of helicity can be rather complex. In the hydrodynamic case it appears that the absence of helicity is required for there to be singular growth and in the MHD case helicity seems to be required. The role of helicity upon reconnection should now be investigated for these and similar configurations [15].

References

1. E. R. Priest and T. G. Forbes: Magnetic Reconnection. CUP (2000).
2. http://www.claymath.org/prize_problems/index.htm
3. R. M. Kerr: Phys. Fluids 5, 1725 (1993).
4. J. T. Beale, T. Kato, A. Majda: Comm. Math. Phys. 94, 61 (1984).
5. R. E. Caflisch, I. Klapper, G. Steele: Comm. Math. Phys. 184, 443 (1997).
6. R.M. Kerr, A. Brandenburg: Phys. Rev Lett. 83, 1155 (1999).
7. O.N Boratav, R.B. Pelz, N.J. Zabusky: Phys. Fluids A 4, 581 (1992).
8. J.D. Gibbon, M. Heritage: Phys. Fluids 9, 901 (1997).
9. R.M. Kerr: 'The role of singularities in turbulence'. In 19th ICTAM Kyoto '96, ed. by T. Tatsumi, E. Watanabe, T. Kambe (Elsevier Science 1997) pp. 57–70.
10. R.M. Kerr: 'The outer regions in singular Euler'. In Fundamental problematic issues in turbulence, ed. by A. Tsinober, A. Gyr (Birkhauser 1998) pp. 57–70.
11. E.N. Parker: 'Nanoflares and the solar X-ray corona'. Astrophys. J. 330, 474 (1988).
12. K. Galsgaard, Å. Nordlund: 'Heating and activity of the solar corona: I. boundary shearing of an initially homogeneous magnetic-field'. J. Geophys. Res. 101, 13445 (1996).
13. T. Amari, J.F. Luciani: 'Helicity redistribution during relaxation of astrophysical plasmas'. Phys. Rev. Lett. 84, 1196 (2000).
14. R. Grauer, C. Marliani: 'Current-sheet formation in 3D ideal incompressible magnetohydrodynamics'. Phys. Rev. Lett. 84, 4850-4853 (2000).
15. R.B. Dahlburg, D. Norton: 'Parallel computation of magnetic flux-tube reconnection'. In Small-scale structures in three-dimensional hydro and magnetohydrodynamic turbulence, ed. by M. Meneguzzi, A. Pouquet, P.L. Sulem (Lecture notes in physics 462 Springer-Verlag 1995) pp. 331–339.
16. A. Pumir, E. D. Siggia: 'Vortex dynamics and the existence of solutions of the Navier-Stokes equations'. Phys. Fluids. 30, 1606 (1987).

Tropicity and Complexity Measures for Vortex Tangles

Renzo L. Ricca

[1] Mathematics Department, University College London
   Gower Street, London WC1E 6BT, UK
[2] Isaac Newton Institute for Mathematical Sciences
   20 Clarkson Road, Cambridge CB3 0EH, UK
   ricca@math.ucl.ac.uk

Abstract. In this paper we introduce and discuss new concepts useful to analyse and characterize patterns of vortex lines in fluid flows. We define measures of tropicity to identify 'tubeness', 'sheetness' and 'bulkiness' of vortex lines and to measure the spreading of field lines about preferred directions. Algebraic, geometric and topological measures based on crossing number information are discussed and are put in relation to the kinetic helicity and the energy of the fluid system.

1 Vortex Structures and Tangles in Classical and Quantized Vortex Flows

Coherent structures represent an essential feature of classical turbulent flows [2], [5]. Experimental and numerical results have indeed shown that vorticity has a tendency to coalesce into highly localized regions. As modern visualizations of classical and superfluid flows show [16], [14], [15], [17], strong anisotropies emerge as vortical flows re-organize themselves to form tubes, sheets or quantized complex tangles of vortex lines. As vortex structures evolve, different types of non-linear effects and instabilities take place, until continuous break-up and re-structuring are overcome by total dissipation.

The tremendous progress in visualization techniques and real-time diagnostics of complex flow patterns [18] makes it now possible accurate recognition of vortex pattern formation and interaction; detailed mechanisms of braiding, linking and re-structuring of vortex lines can be analysed by real-time simulations to a high degree of accuracy. In this context geometric and topological information is available from data-sets of numerical simulations; properly analysed, it provides valuable help to understand fundamental properties of dynamics and energetics of turbulent flows [13], [11]. New concepts and tools based on geometry and topology are therefore being developed [12] to quantify physical information associated with structural complexity of such flows. In the following sections we shall introduce measures based on geometric and topological concepts that will provide useful tools to quantify structural complexity of vortical flows and help to develop new measures to estimate physical properties.

2 Measures of Tropicity for Vortex Tangles: Tubeness, Sheetness and Bulkiness

A first step in the application of measures of structural complexity is to develop tools for pattern recognition of vortex structures. These must be based on estimates of anisotropy and spatial extension of vortex lines in the fluid. Let us consider a fluid region $\mathcal{D}$ in $\mathbb{R}^3$ and a generic tangle $\mathcal{T}$ of n vortex lines $\mathcal{L}$, i.e. $\mathcal{T} = \bigcup_n \mathcal{L}_n$ in $\mathcal{D}$. A vortex line is given by a line of vorticity, whose support is identified with a smooth, simple space curve, not necessarily closed in $\mathcal{D}$ ($\mathcal{D}$ can be a sub-region of the entire fluid domain), with vorticity everywhere tangent to the curve. It is useful to introduce the concept of 'tropicity' (a measure of the space configuration of the vortex region) to characterize the degree of tubeness, sheetness and bulkiness of $\mathcal{T}$ in $\mathcal{D}$. Consider a single vortex line $\mathcal{L}$ as frozen in space and time. A measure of the spatial extent of $\mathcal{L}$ in $\mathcal{D}$ is given by the maximal distance D_1 between two points P_i and P_j on $\mathcal{L}$, i.e.

$$D_1 = \max_{i,j} \mathrm{d}(P_i, P_j) \equiv \overline{P_0 P_1} = \sqrt{X^2 + Y^2 + Z^2} \,, \tag{1}$$

where $P_0 = (x_0, y_0, z_0) \in \mathcal{L}$, $P_1 = (x_1, y_1, z_1) \in \mathcal{L}$ and $X = x_1 - x_0$, $Y = y_1 - y_0$, $Z = z_1 - z_0$. The unit vector $\hat{\mathbf{T}}_1 = (P_1 - P_0)/D_1$ (*first directional tropicity vector*) given by the director cosines X/D_1, Y/D_1, Z/D_1, is the principal directional axis of $\mathcal{L}$. Let us consider then the transversal spatial extension of $\mathcal{L}$. Take the maximal distance of a third point $P_i \in \mathcal{L}$ (not aligned with P_0 and P_1) from the line $l(P_0, P_1)$: a measure of maximal width spanned by $\mathcal{L}$ in $\mathcal{D}$ is given by

$$D_2 = \max_i \mathrm{d}(P_i, l(P_0, P_1)) \equiv \overline{P_2 O} \,, \tag{2}$$

where $P_2 \in \mathcal{L}$ and $O \in l(P_0, P_1)$, where O, being not necessarily a point on $\mathcal{L}$, is the footpoint of the orthogonal projection of P_2 onto $\overline{P_0 P_1}$. The *second directional tropicity vector* is the unit vector $\hat{\mathbf{T}}_2 = (P_2 - O)/D_2$, which prescribes direction and orientation of the second principal directional axis of $\mathcal{L}$ in $\mathcal{D}$. We can now define the plane $\Pi = \Pi(P_0, P_1, P_2)$ and, by taking the maximal distance of a fourth point $P_i \in \mathcal{L}$ from Π, we have a measure of the maximal three-dimensional extension of $\mathcal{L}$ in $\mathcal{D}$, i.e.

$$D_3 = \max_i \mathrm{d}(P_i, \pi(P_0, P_1, P_2)) \equiv \overline{P_3 Q} \,, \tag{3}$$

($P_3 \in \mathcal{L}$) taken along the direction of the *third directional tropicity vector* $\hat{\mathbf{T}}_3 = \hat{\mathbf{T}}_1 \times \hat{\mathbf{T}}_2$.

Let us consider the whole tangle of vortex lines $\mathcal{T}$. The spatial extension of $\mathcal{T}$ in $\mathcal{D}$ is measured by the three maximal distances D_i, this time taken with respect to the principal axes $\boldsymbol{\lambda}_i$ (along $\hat{\mathbf{T}}_i$, $i = 1, 2, 3$) determined by sampling P_i over the whole tangle $\mathcal{T}$. The tropicity volume is given by $V(\mathcal{D}) = D_1 D_2 D_3$, and measures of tropicity can be defined by taking the relative ratio of these quantities. We have

- if $D_3 = O(D_2)$ and $D_2 \ll D_1$: *tubeness* $\stackrel{\text{def}}{=} \Lambda_1 \equiv \dfrac{D_1}{D_2}$;

- if $D_3 \ll D_2$ and $D_2 = O(D_1)$: *sheetness* $\stackrel{\text{def}}{=} \Lambda_2 \equiv \dfrac{D_2}{D_3}$;

- if $D_3 = O(D_2)$ and $D_2 = O(D_1)$: *bulkiness* $\stackrel{\text{def}}{=} \Lambda_3 \equiv \dfrac{D_1 D_2 D_3}{D_3^3}$.

These quantities provide a first crude information of the space configuration of the pattern. For classical turbulent flows these quantities are particularly useful to detect regions of high tubeness and sheetness, whereas for quantized flows bulkiness may provide more useful information related to vortex line density. Moreover, as we shall see below, information on directional tropicity finds useful applications for geometric and topological estimates of structural complexity.

3   Measures of Geometric Complexity: Directional Alignment and Writhing

In classical fluids strong anisotropy is characterized by the presence of elongated, tubular regions. If $\boldsymbol{\omega} = \nabla \times \boldsymbol{u}$ denotes vorticity ($\boldsymbol{u} = \boldsymbol{u}(\boldsymbol{x})$ being the velocity field function of the position vector $\boldsymbol{x}$), and $\boldsymbol{\sigma}$ vortex stretching, defined by

$$\sigma_i = S_{i,j}\omega_j = \frac{1}{2}\left(\frac{\partial u_i}{\partial x_j} + \frac{\partial u_j}{\partial x_i} \right)\omega_j , \tag{4}$$

then a measure of how vorticity and vortex stretching lines are spread about the principal directional axis $\hat{\mathbf{T}}_1$ is given by the angles η_ω and η_σ, defined by

$$\tan \eta_\omega = \frac{|\bar{\boldsymbol{\omega}} \times \hat{\mathbf{T}}_1|}{\bar{\boldsymbol{\omega}} \cdot \hat{\mathbf{T}}_1} , \qquad \tan \eta_\sigma = \frac{|\bar{\boldsymbol{\sigma}} \times \hat{\mathbf{T}}_1|}{\bar{\boldsymbol{\sigma}} \cdot \hat{\mathbf{T}}_1} , \tag{5}$$

where $\bar{\boldsymbol{\omega}} = \langle \boldsymbol{\omega} \rangle_{\mathcal{D}}$ and $\bar{\boldsymbol{\sigma}} = \langle \boldsymbol{\sigma} \rangle_{\mathcal{D}}$ denote space averages of the field lines over the tropicity domain $\mathcal{D}$. Similarly, we can define the spread of vorticity and stretching (through the angles χ_ω and χ_σ) about a sheet-like distribution by taking

$$\cot \chi_\omega = \frac{\bar{\boldsymbol{\omega}} \cdot \hat{\mathbf{T}}_3}{|\bar{\boldsymbol{\omega}} \times \hat{\mathbf{T}}_3|} , \qquad \cot \chi_\sigma = \frac{\bar{\boldsymbol{\sigma}} \cdot \hat{\mathbf{T}}_3}{|\bar{\boldsymbol{\sigma}} \times \hat{\mathbf{T}}_3|} . \tag{6}$$

For quantized vortex tangles directional writhing provides a measure of average coiling of vortex lines in space. Let $\Pi_i = \Pi(\hat{\mathbf{T}}_i)$ denote a plane of projection with normal $\hat{\mathbf{T}}_i$ and consider the projection of the tangle $\mathcal{T}$ onto the plane Π_i. A 'good' diagram $\mathcal{T}_i = \mathcal{T}(\hat{\mathbf{T}}_i)$ (which of course depends on the line of projection $\hat{\mathbf{T}}_i$) is given by a curve graph, whose self-intersections are given by countably many points, where the segments intersect transversally. Good projections can be found by an appropriate choice of the projection plane. By keeping track of the orientation of the curve (induced by the vorticity vector), we obtain an oriented diagram (see Fig. 1) and by assigning the value $\epsilon_r = \pm 1$ to each projected

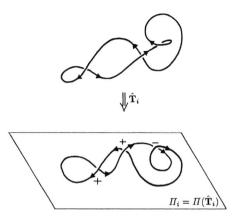

Fig. 1. Example of an oriented space curve projected onto the plane Π_i.

crossing r (according to standard convention on signs [6]), we can quantify the directional writhing in terms of algebraic sum of positive and negative crossings on Π_i (see [10])

$$Wr(\mathcal{T}_i) = \frac{1}{4\pi} \int_{\mathcal{T}_i} \frac{d\boldsymbol{X} \times d\boldsymbol{X}^* \cdot (\boldsymbol{X} - \boldsymbol{X}^*)}{|\boldsymbol{X} - \boldsymbol{X}^*|^3} = \sum_{r \in \mathcal{T}_i} \epsilon_r \, . \qquad (7)$$

The writhing number $Wr(\mathcal{T})$ is given by averaging the directional writhe over the whole solid angle. It may be computationally convenient to approximate this quantity by taking the algebraic mean over the three principal orthogonal planes as reference projection planes; hence, the estimated writhing number given by

$$Wr_\perp = \left\langle \sum_{r \in \mathcal{T}_i} \epsilon_r \right\rangle_\perp \approx Wr(\mathcal{T}) \, , \qquad (8)$$

will provide a reasonable, simple measure of the average coiling of $\mathcal{T}$.

4 Algebraic Measure of Structural Complexity: Average Crossing Number

Structural complexity can be measured by counting the total number of crossings, that are present in a tangle at a given time. This quantity, which is associated with the un-oriented tangle diagram, is given by the average crossing number $\bar{C}$ by counting the total number of un-signed crossings in $\mathcal{T}_i$ and averaging over the whole domain $\mathcal{D}$ [3]. We have

$$\bar{C}(\mathcal{T}) = \left\langle \sum_{r \in \mathcal{T}_i} |\epsilon_r| \right\rangle_\mathcal{D} \, . \qquad (9)$$

Once again, it is computationally convenient to approximate this measure by the algebraic mean taken over the three principal orthogonal planes, hence

$$\bar{C}_{\perp} = \left\langle \sum_{r \in T_i} |\epsilon_r| \right\rangle_{\perp} \approx \bar{C}(\mathcal{T}) \; . \tag{10}$$

Current work done by Barenghi et al., [1] shows that $\bar{C}_{\perp}$ provides indeed a good approximation to $\bar{C}$ and it seems to be very effective to detect structural complexity.

5   Measures of Topological Entanglement: Kinetic Helicity and Directional Linking

Topological entanglement can be calculated on the basis of information about directional tropicity of $\mathcal{T}$. In this context the concept of linking number is fundamental, since it is related to the kinetic helicity H of the flow (see [7])

$$H(\mathcal{T}) = \int_{\mathcal{D}} \boldsymbol{u} \cdot \boldsymbol{\omega} \, dV \; . \tag{11}$$

For superfluid vortices helicity can be estimated through linking number measures made directly on the vortex tangle, hence avoiding the difficulties associated with the integration over the vorticity field. These are based on the concept of (Gauss) linking number $Lk_{ij} = Lk(\mathcal{L}_i, \mathcal{L}_j)$ of two loops $\mathcal{L}_i$ and $\mathcal{L}_j$, given by

$$Lk_{ij} = \frac{1}{4\pi} \oint_{\mathcal{L}_i} \oint_{\mathcal{L}_j} \frac{d\boldsymbol{X}_i \times d\boldsymbol{X}_j \cdot (\boldsymbol{X}_i - \boldsymbol{X}_j)}{|\boldsymbol{X}_i - \boldsymbol{X}_j|^3} \; , \tag{12}$$

and its limit form Lk_{ii}, given by the the Călugăreanu-White formula

$$Lk_{ii} = Wr_i + Tw_i \; , \tag{13}$$

where $Wr_i = Wr(\mathcal{L}_i)$ is the writhing number of $\mathcal{L}_i$ and $Tw_i = T_i + N_i$ is the total twist, sum of the total torsion (T_i) and intrinsic twist (N_i) of $\mathcal{L}_i$ divided by 2π (see [10] for precise definitions and physical meaning of these quantities). For a tangle of n vortex lines, each of circulation κ_i $(i \in [1, \ldots, n])$, the kinetic helicity is given by (see [11])

$$H(\mathcal{T}) = 2 \sum_{i \neq j} Lk_{ij} \, \kappa_i \kappa_j + \sum_i Lk_{ii} \, \kappa_i^2 \; . \tag{14}$$

Note that in the case of superfluid vortices, since quantized vortex lines have no internal structure we can assume $Tw_i = T_i$.

In the case of strong anisotropy of vortex flows directional tropicity provides natural reference directions to measure localized winding of vortex lines. A relative measure of winding is given by the directional linking number $\ell k_{\boldsymbol{\lambda}_i}$ of vortex lines with respect to one of the principal tropicity axes $\boldsymbol{\lambda}_i$, i.e.

$$\ell k_{\boldsymbol{\lambda}_i} = \langle \ell k(\mathcal{L}_i, \boldsymbol{\lambda}_i) \rangle_{\mathcal{D}} \; , \tag{15}$$

keeping $\boldsymbol{\lambda}_i$ fixed in the average process.

6   Relationships Between Complexity Measures and Energy Levels

It is of fundamental importance to relate measures of algebraic, geometric and topological complexity to physical properties of the system, such as kinetic helicity and energy. For this purpose it is convenient to re-write eq. (14) in a more compact form. Consider the linking numbers L_{ij} $(i, j \in [1, \ldots, n])$ as elements of a square matrix; since $L_{ij} = L_{ji}$, we can reduce the linking matrix to diagonal form, i.e.

$$
\begin{pmatrix} L_{11} & L_{12} & \ldots & L_{1n} \\ L_{21} & L_{22} & \ldots & L_{2n} \\ \ldots & \ldots & \ldots & \ldots \\ L_{n1} & L_{n2} & \ldots & L_{nn} \end{pmatrix} \rightarrow \begin{pmatrix} M_{11} & 0 & \ldots & 0 \\ 0 & M_{22} & \ldots & 0 \\ \ldots & \ldots & \ldots & \ldots \\ 0 & 0 & \ldots & M_{nn} \end{pmatrix} , \tag{16}
$$

where each element M_{ii} takes into account self- and mutual linking of the vortex lines. We can therefore re-cast eq. (14) in the form

$$
H(\mathcal{T}) = \sum_{i=1,\ldots,n} M_{ii} \, f(\kappa_i) , \tag{17}
$$

where $f(\cdot)$ is a linear function of quadratic terms in κ_i.

If the tangle is made on average of vortex filaments of same length L (obtained by an average measure over the tropicity domain $\mathcal{D}$), we can show [9] that on dimensional grounds the enstrophy Ω of the system is given by a relationship of the form

$$
\Omega(\mathcal{T}) = \int_{\mathcal{D}} |\omega|^2 \, dV = \frac{1}{L} \sum_{i=1,\ldots,n} M_{ii} \, f(\kappa_i) , \tag{18}
$$

that provides an interesting connection with helicity. Moreover, since the magnetic energy of a perfectly conducting magnetized fluid is bounded from below by the magnetic helicity H_m [8], according to the inequality

$$
E_{\min} \geq q_0 |H_m| , \tag{19}
$$

where q_0 is a positive constant, then we can expect that in steady state conditions similar bounds hold for minimum enstrophy levels or for other types of ground state energy in relation to the complexity of the physical system.

Acknowledgements

Financial support from UK PPARC (Grant GR/L63143) and EPSRC (Grant GR/K99015) is kindly acknowledged.

References

1. Barenghi, C.F., Ricca, R.L. & Samuels, D.C. (2000) How tangled is a tangle? Submitted.
2. Betchov, R. (1956) An inequality concerning the production of vorticity in isotropic turbulence. *J. Fluid Mech.* **1**, 497–504.
3. Freedman, M.H. & He, Z.-X. (1991) Divergence-free fields: energy and asymptotic crossing number. *Ann. Math.* **134**, 189–229.
4. Galanti, B., Gibbon, J.D. & Heritage, M. (1997) Vorticity alignment results for the three-dimensional Euler and Navier-Stokes equations. *Nonlinearity* **10**, 1675–1694.
5. Hussain, A.K.M.F. (1986) Coherent structures and turbulence. *J. Fluid Mech.* **173**, 303–356.
6. Kauffman, L.H. (1987) *On Knots*. Annals Study **115**, Princeton University Press.
7. Moffatt, H.K. (1969) The degree of knottedness of tangled vortex lines. *J. Fluid Mech.* **35**, 117–129.
8. Moffatt, H.K. (1991) Relaxation under topological constraints. In *Topological Aspects of the Dynamics of Fluids and Plasmas* (ed. H.K. Moffatt *et al.*), pp. 3–28. NATO ASI Series E: Applied Sciences **218**. Kluwer, Dordrecht.
9. Moffatt, H.K. & Ricca, R.L. (1991) Interpretation of invariants of the Betchov-Da Rios equations and of the Euler equations. In *The Global Geometry of Turbulence* (ed. J. Jiménez), pp. 257–264. Plenum Press, New York.
10. Moffatt, H.K. & Ricca, R.L. (1992) Helicity and the Călugăreanu invariant. *Proc. R. Soc. Lond.* A **439**, 411–429.
11. Ricca, R.L. (1998) Applications of knot theory in fluid mechanics. In *Knot Theory* (ed. V.F.R. Jones *et al.*), pp. 321–346. Banach Center Publications, Institute of Mathematics **42**, Polish Academy of Sciences, Warsaw.
12. Ricca, R.L. (2000) Towards a complexity measure theory for vortex tangles. In *Knots in Hellas '98* (Ed. McA. Gordon *et al.*), pp. 361–379. Series on Knots and Everything **24**, World Scientific, Singapore.
13. Ricca, R.L. & Berger, M.A. (1996) Topological ideas and fluid mechanics. *Phys. Today* **49** (12), 24–30.
14. Schwarz, K.W. (1988) Three-dimensional vortex dynamics in superfluid ^{4}He. *Phys. Rev.* B **38**, 2398–2417.
15. She, Z.-S., Jackson, E. & Orszag, S.A. (1990) Intermittent vortex structures in homogeneous isotropic turbulence. *Nature* **344**, 226–228.
16. Swanson, C.E. & Donnely, R.J. (1985) Vortex dynamics and scaling in turbulent counterflowing helium II. *J. Low Temp. Phys.* **61**, 363–399.
17. Vincent, A. & Meneguzzi, M. (1991) The spatial structure and statistical properties of homogeneous turbulence. *J. Fluid Mech.* **225**, 1–20.
18. Zabusky, N.J., Silver, D. & Pelz, R. (1993) Visiometrics, juxtaposition and modeling. *Phys. Today* **46**, 24–31.

The Geometry
of Magnetic and Vortex Reconnection

Gunnar Hornig

Ruhr-Universität Bochum, 44780 Bochum, Germany

Abstract. Reconnection is an important process of structure formation in fluid dynamics, occurring in the form of vortex reconnection in hydrodynamics as well as in the form of magnetic reconnection in plasmas. There is a close analogy between the quantities involved in both phenomena but, surprisingly, the process of magnetic reconnection, although complicated by the presence of a magnetic field, is geometrically simpler than vortex reconnection; it may thus serve as good starting point to understand the geometry of vortex reconnection.

A general covariant definition of reconnection is given and, starting from a simple analytic model of magnetic reconnection, the basic process of reconnection is analyzed. The model is then modified to meet the additional constraints of vortex reconnection. It is shown that, although the evolution of the vorticity near the reconnection site is stationary and two-dimensional the flow velocity is inevitably three-dimensional, and time dependent. Explicit expressions for the reconnected flux and the reconnection time are given.

1 Introduction

The notion of reconnection is found in many fields of physics: in hydrodynamics [1] for the reconnection of vortex tubes, in plasma physics for magnetic reconnection [2], in the theory of superfluids [3] for the reconnection of quantized vortex elements, as well as in cosmology for the interaction of cosmic strings [4][1]. In this contribution we will consider only reconnection as a process in the evolution of a divergence-free vector field, therefore involving the first two examples. The divergence-free field is the common ingredient to both vortex reconnection in hydrodynamics (HD) and magnetic reconnection in magnetohydrodynamics (MHD). In HD the evolution of the vorticity field is determined by the curl of the Navier-Stokes equation, while the evolution of the magnetic field in MHD is determined by the curl of Ohm's law. Both equations have the same structure:

$$\partial_t \boldsymbol{G} \ -\nabla \times (\boldsymbol{v} \times \boldsymbol{G}) = \nabla \times \boldsymbol{N} \quad \text{with:} \tag{1}$$

$\boldsymbol{v}$: the velocity of the fluid flow

$\boldsymbol{G}$: the transported field; $\boldsymbol{G} = \nabla \times \boldsymbol{v}$ in HD, $\boldsymbol{G} = \boldsymbol{B}$ in MHD;

$\boldsymbol{N}$: The non-ideal term

$\boldsymbol{N} = \partial_k \left(\nu (\partial_k v^i + \partial_i v^k) \right) \boldsymbol{e}_i$ (Viscous, incompressible HD);

$\boldsymbol{N} = \eta \nabla \times \boldsymbol{B}$ (Resistive MHD) ,

---

[1] The references given here are only examples that provide an entry to the literature

where ∂_i denotes the derivative with respect to the coordinate x^i, e_i is the corresponding unit vector and summation of repeated indices is assumed.

Because only the evolution of the divergence-free field is relevant for the definition of reconnection the other equations of hydrodynamics (HD) and magnetohydrodynamics (MHD), i.e. the equations of continuity, energy and the momentum balance in MHD are not considered here.

The main difference between HD and MHD is that in HD the transporting field, v, and the field which is transported, w, are coupled by $w = \nabla \times v$. A corresponding equation in MHD ($B = \nabla \times v$) does not exist. This extra freedom allows for more simple reconnection solutions in MHD compared to HD. It is therefore natural to start with an example of magnetic reconnection

2 Magnetic Reconnection

Note first that for vanishing resistivity equation (1) is ideal, i.e. the magnetic field is frozen into the fluid flow and the magnetic flux integrated over a comoving surface is conserved. In this case the topology of the field (B) is conserved as well and no reconnection is possible.

$$\partial_t B - \nabla \times (v \times B) = 0 \, , \tag{2}$$

$$\Rightarrow \int B \cdot n \, da = const.$$

Only if the non-idealness N is non-vanishing is reconnection possible. Consider for instance a configuration with an initially anti-parallel two-dimensional magnetic field as shown in Fig. 1. A localized non-idealness N perpendicular to the plane in which the magnetic field lies can produce a curl which reduces the initial anti-parallel field components according to (1) and adds a new perpendicular component, such that the field lines are reconnected. The non-idealness N or more exact $\nabla \times N$ plays an important role in the process. $\nabla \times N$ has to be present to allow for reconnection but it also has to be localized, that is there has to be an outside region where $\nabla \times N$ is negligible, at least on the time scales under consideration. Otherwise we would have a situation of global diffusion of the magnetic field and the term 'reconnection' would be meaningless

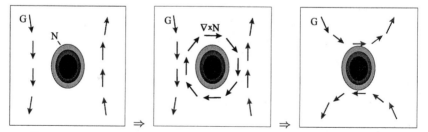

Fig. 1. Basic reconnection process with a localized non-idealness N perpendicular to the plane shown.

because there is no way of identifying field lines or flux tubes in time. Note that the localization of $\nabla \times N$ is not as restrictive as it may look at first sight. It is not necessary that N is localized, only the curl of N has to be localized. The localization is usually not a serious problem in MHD because in the most of the applications of magnetic reconnection the resistivity is very low and enhanced only in thin current sheets. In contrast in HD the viscosity is usually assumed to be constant and the localization of $\nabla \times N$ in HD is, therefore, a result of the structure of vortex sheets in most cases.

Note that 'localization' also means that there is no closed flux within the region where $\nabla \times N$ is not negligble, otherwise the same argument as above applies to this closed flux. But if all flux of the non-ideal region is connected to the 'ideal' ($\nabla \times N = 0$) surrounding, then the conservation of flux in the outside region implies a conservation of flux in the non-ideal region as well. Therefore, the effect of a localized N is only to dissipate energy and rearrange the flux but not to disspate the flux (at least on the time scale on which the reconnection acts). Hence it should be possible to describe the process as being 'ideal', i.e. satisfying (2) but with a new transport velocity u which may have a singularity. For example

$$B = [y, kx, 0], \quad u = [-x\,k\,E_z/(k^2x^2 + y^2), y\,E_z/(k^2x^2 + y^2), 0] \qquad (3)$$

is a stationary solution (Fig. 2) of (2), which has a singularity at the origin

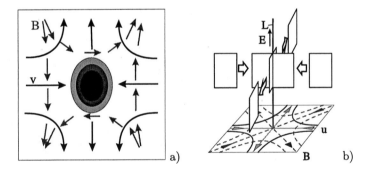

Fig. 2. (a) The flow structure and (b) the evolution of cross-sections of the magnetic flux under the singular flow u for a stationary reconnection process

of the coordinate system such that it transports the magnetic flux in a finite time along the x-z plane onto the z-axis where the flux is split and reconnected (Fig. 2b). The flux is then transported outwards along both directions of the z-y plane. The example can be derived for a magnetic field B satisfying

$$E + v \times B = \eta\,\nabla \times B \quad \text{with}$$
$$v = [-x, y, 0], \quad E = [0, 0, E_z], \quad \eta = (E_z - kx^2 - y^2)/(k - 1) \qquad (4)$$

The solution holds only in the region where $\eta > 0$. Since both $\boldsymbol{E}$ and $\boldsymbol{J}$ are perpendicular to $\boldsymbol{B}$ the transport velocity of the magnetic flux is simply

$$\boldsymbol{u} = (\boldsymbol{E} - \eta \boldsymbol{J}) \times \boldsymbol{B}/B^2 .$$

Note that $\boldsymbol{u}$ is not the fluid velocity, but where η vanishes $\boldsymbol{u}$ and $\boldsymbol{v}$ coincide.

Now (2) is only a special case of an equation which leads to reconnective solutions. Since the transport velocity has to become infinite at the reconnection line (the line where $\boldsymbol{u}$ is singular), an adequate description is a covariant one and the most general form, which leads to an equation like (2) (see [5],[6]) is

$$\epsilon^{\alpha\beta\gamma\delta} \partial_\alpha U^\nu F_{\nu\beta} = 0 \tag{5}$$

$$\Leftrightarrow \begin{cases} \partial_0 (U^0 \boldsymbol{E} + \boldsymbol{U} \times \boldsymbol{B}) + \nabla(\boldsymbol{E} \cdot \boldsymbol{U}) = 0 \\ U^0 \partial_0 \boldsymbol{B} - \nabla \times (\boldsymbol{U} \times \boldsymbol{B}) - \nabla U^0 \times \boldsymbol{E} = 0 \end{cases}$$

$$\Leftrightarrow L_U \omega_F^2 = 0 \tag{6}$$

where $F_{\alpha\beta}$ denotes the electromagnetic field tensor. The four-velocity

$$U^{(4)} = (U^0, U^1, U^2, U^3) = (U^0, \boldsymbol{U})$$

is related to the usual velocity in space by

$$\boldsymbol{u} := \frac{\boldsymbol{U}}{U^0} = \frac{\mathrm{d}\boldsymbol{X}/\mathrm{d}s}{\mathrm{d}X^0/\mathrm{d}s}, \tag{7}$$

and $L_U \omega_F^2 = 0$ is the Lie-derivative of the 2-form of the electromagnetic field with respect to the flow $U^{(4)}$. The Lie-derivative theorem then implies that the electromagnetic flux integrated over a two-dimensional comoving surface (C) is constant with respect to the parameter s which describes the $U^{(4)}$-flow in space-time.

$$(6) \Rightarrow \int_C F dA = \int_C F_{\mu\nu} dx^\mu dx^\nu = const.$$

The system (5) can be derived from

$$U^\nu F_{\nu\beta} = \partial_\beta \Phi \Leftrightarrow \begin{cases} \boldsymbol{E} \cdot \boldsymbol{U} = -\partial_0 \Phi \\ U^0 \boldsymbol{E} + \boldsymbol{U} \times \boldsymbol{B} = \nabla \Phi . \end{cases} \tag{8}$$

These equations can be considered as the most general form of an Ohm's law which lead to a conservation of electromagnetic flux in the form of (5). The potential Φ is not relevant for the two-dimensional reconnection example considered above because, for this case, $\boldsymbol{E} \cdot \boldsymbol{B}$ vanishes and $U^{(4)}$ can be chosen such that $\Phi = 0$ holds everywhere. It is, however, important for the case of reconnection in a non-vanishing magnetic field [7], which requires $\boldsymbol{E} \cdot \boldsymbol{B} \neq 0$ and hence $\Phi \neq 0$.

An additional advantage of the covariant formulation is that now the singularity in $\boldsymbol{u}$ can be represented as a null of $U^{(4)}$ and is thus open to an analysis of its structure. For instance, for the example given above the corresponding 4-velocity is given by

$$U^{(4)} = [(k^2 x^2 + y^2), -kE_z x, E_z y, 0] . \tag{9}$$

This solution is typical for reconnection because $U^0 > 0$ requires that U^0 vanishes quadratically at the null of $U^{(4)}$, while the space components are antisymmetric and thus in lowest order are linear in x and y near the null. The existence of a null of $U^{(4)}$ with an X-point structure in the space components can be used for a definition of magnetic reconnection [6]. In general the reconnection process requires a line in space along which $U^{(4)}$ vanishes, the reconnection line, denoted by L in Fig. 2 (b). For the example given this line is the z-axis, but in general the line can be curved and moving, and coincides with a field line of $\boldsymbol{B}$ as long as its velocity is small. In this case the total reconnected flux is given by the integral along the part of reconnection line where $U^{(4)}$ vanishes and over the time interval when this occurs, one has

$$\Phi_{\text{rec}} = \iint \boldsymbol{E} \cdot d\boldsymbol{l} \, dt \ .$$

Correspondingly $\int \boldsymbol{E} \cdot d\boldsymbol{l}$ is the reconnection rate for a stationary process.

It is worth mentioning that a uni-directional magnetic field ($\boldsymbol{B} = b_z(x, y)\boldsymbol{e}_z$) can also show an evolution, which requires a non-continuous $\boldsymbol{u}$. An example is where two initially isolated regions of positive b_z in an environment of negative b_z merge. We do not consider this type of 'reconnection' here since it does not require a 'cut and paste' of magnetic flux and, correspondingly, the reconnected flux as defined above vanishes in these cases.

3   Vortex Reconnection

As mentioned in the introduction there is no one-to-one correspondence between solutions of magnetic reconnection and vortex reconnection since the latter have to satisfy $\boldsymbol{w} = \nabla \times \boldsymbol{v}$. Moreover, apart from solutions with constant viscosity, the non-ideal term $\boldsymbol{N}$ is more complicated in HD.

We start by constructing a stationary two-dimensional solution for vortex reconnection, that is a solution where all quantities depend only on two space coordinates (x, y) and the vorticity vectors lie in the x-y-plane. Using the analogy with magnetic reconnection we assume that the vorticity has the form $\boldsymbol{w} = [y, kx, 0]$ as in example (3). However, the flow velocity $\boldsymbol{v} = [-x, y, 0]$, which we used in this example, does not satisfy $\boldsymbol{w} = \nabla \times \boldsymbol{v}$. It has to be modified with an additional z-component $v_z(x, y)$, such that

$$\boldsymbol{w} = \nabla v_z(x, y) \times \boldsymbol{e}_z,$$

which is inevitable if $\boldsymbol{v}$ is not to depend on z.

Moreover, from magnetic reconnection we know that an electric field along the reconnection line (the z-axis in our example) has to be present for a non-vanishing reconnection rate. The analogue of the electromagnetic field tensor in hydrodynamics is the vorticity tensor

$$W_{\mu\nu} = \partial_\mu V_\nu - \partial_\nu V_\mu \quad \mu, \nu \in \{0, 1, 2, 3\}$$

The 'electric' field component of $W_{\mu\nu}$ is therefore $-\partial_t V - \nabla V^0$. Because, $V^{(4)}$ or v, respectively, is a real fluid velocity, which we assume to be non-relativistic, we have $V^0 = 1$ and $V = v$. Thus the 'electric' field is given by $-\partial_t v$. For a stationary reconnection process this field has to be constant, hence $v_z \sim t$ in our model.

Taking these conditions into account a full solution is given by

$$v = [-x, y, y^2/2 - kx^2/2 - tE_z] \tag{10a}$$
$$\nu = E_z/(k-1) - k/(3\ k - 1)\ x^2 - 1/(k-3)\ y^2 \tag{10b}$$
$$p = p_0 + (k+1)/(6\ k - 2)\ x^2 - (k+1)/(2\ k - 6)\ y^2 \ . \tag{10c}$$

This solution satisfies (1) with a non-ideal term for viscous, incompressible HD, as given in (1), for a region near the reconnection axis where $p > 0$, $\nu > 0$ and $k > 3$.

Before discussing the properties of this solution we remark that concerning the covariant transport (5) of the vorticity tensor, the magnetic reconnection in example (3) and the above example of a vorticity reconnection are identical. They both satisfy (8) for $\Phi = 0$ and $U^{(4)}$ given by (9).

$$-U^0 \partial_t v + U \times w = 0 \tag{11}$$
$$U^0 E + U \times B = 0 \tag{12}$$

The solution (10a) is only a solution for the neighborhood of the reconnection line due to the simple polynomial assumption for the fields. Therefore, it cannot reproduce all the complex features of vortex reconnection such as 'bridging', secondary reconnection, etc. This simple system, however, already contains several basic properties of vortex reconnection. First, although we started with the most elementary reconnection process, with respect to the fluid velocity v the process is not stationary any more, since the v_z component grows linearly with time. Thus there is no true stationary vortex reconnection. Moreover, the existence of a non-vanishing v_z component breaks the two-dimensionality present in the vorticity field. This complicates the flow structure of the process enormously. But despite this complexity in the flow velocity, the reconnection process of the vorticity is comparatively simple.

Because we know the transport velocity of the vorticity we can calculate the time necessary for two flux tubes to reconnect in our example. This merging time is given by the time which is required for a point (x_0) on the x-axis to reach the z-axis under the flow u. Hence

$$\int_{x_0}^0 dt = \int_{x_0}^0 U^0/U^x \ dx = -\int_{x_0}^0 kx/E_z \ dx = x_0^2 \ k/(2E_z) \ .$$

This result is in accordance with numerical results [8], which show a quadratic dependence of the reconnection time on the distance between the flux tubes. Note that this is not an accidental feature of our model, but is due to the generic structure of reconnecting flows in space-time as explained above.

In analogy to magnetic reconnection, the reconnected flux is given by the integral along the reconnection line. The reconnected flux and the reconnection rate are therefore,

$$\Phi_{\rm rec} = -\iint \partial_t \boldsymbol{v} \cdot d\boldsymbol{l} \; dt \qquad \frac{d\Phi_{\rm rec}}{dt} = -\int \partial_t \boldsymbol{v} \cdot d\boldsymbol{l} \; . \tag{13}$$

4   Conclusions

It has been shown that the process of reconnection of a divergence-free field is basically a flux conserving process and thus can be represented by an ideal advection of the flux under a velocity field $\boldsymbol{u}$ which has a singularity. A transition to the covariant equations in space-time allows one to resolve this singularity, which was shown to be a null point of the corresponding 4-velocity $U^{(4)}$. Within this concept we strictly distinguish between the transport velocity of the flux, $\boldsymbol{u}$ or $U^{(4)}$ in space-time, respectively, and the fluid flow ($\boldsymbol{v}$). Starting from an elementary two-dimensional stationary model for magnetic reconnection, a similar solution for vortex reconnection can be constructed. This solution is, however, stationary and two-dimensional only with respect to the velocity $U^{(4)}$, whereas the construction of the corresponding flow field $\boldsymbol{v}$ requires a three-dimensional time-dependent solution. Thus, with respect to $\boldsymbol{v}$, vortex reconnection is inherently three-dimensional and time-dependent, even in its most elementary solution. It also shows that care has to be taken if passive tracers are used to follow vortex tubes because, within the region where the non-idealness is relevant, the vortex velocity $\boldsymbol{u}$ and the fluid velocity $\boldsymbol{v}$ differ significantly.

The covariant formulation also shows that the analogous quantity to the electric field in MHD is $-\partial_t \boldsymbol{v}$ in HD and thus the reconnection rate, which is known from magnetic reconnection to be the integral of the electric field parallel to the reconnection line, is given by $-\int \partial_t \boldsymbol{v} \cdot d\boldsymbol{l}$.

Acknowledgment

This work was supported by the *Volkswagen Foundation*.

References

1. S. Kida, M. Takaoka, *Vortex Reconnection*, Annu. Rev. Fluid Mech.**26**, 169 (1994)
2. D. Biskamp, *Magnetic Reconnection*, Physics Reports **237**, 179 (1994)
3. J. Koplik, H. Levine, *Vortex reconnection in superfluid helium*, Phys. Rev. Letters **71**, 1375, (1993)
4. E.P.S. Shellard, *Cosmic String interactions*, Nucl. Phys. B **282**, 624 (1987)
5. G. Hornig, *The covariant transport of electromagnetic fields and its relation to magnetohydrodynamics*, Phys. Plasmas, **4**, 646, 1997.
6. G. Hornig, 'The Evolution of Magnetic Helicity Under Reconnection'. In: *Magnetic Helicity*, ed.by A.A. Pevtsov, (American Geophysical Union, Washington 1999), pp. 157-165

7. G. Hornig, L. Rastätter, *The magnetic structure of $B \neq 0$ reconnection*, Physica Scripta, **T 74**, 34, (1997)
8. A.T.A.M. de Waele, R.G.K.M. Aarts, *Route to vortex reconnection*, Phys. Rev. Lett., **72**, 482 (1994)

Current-Sheet Formation
near a Hyperbolic Magnetic Neutral Line

Bhimsen K. Shivamoggi

Institute of Theoretical Physics, University of California,
Santa Barbara, CA 93106-4030, USA, and
University of Central Florida, Orlando, FL 32816, USA

Abstract. Two-dimensional flow of an incompressible plasma in a hyperbolic magnetic field is discussed. The effects of sweeping [7] as well as shearing [11] of the magnetic field lines by the plasma flow are considered. Exact solutions describing current-sheet evolution in this setting are given which compare very well with laboratory experiments ([3], [8]-[10]).

1   Introduction

When a plasma collapses near the neutral line of the applied magnetic field, a thin neutral current sheet is formed there. The magnetic flux continually accumulates in the region of the neutral sheet and causes the total current and the sheet width to increase. Thus, as Syrovatskii [1] pointed out, a stationary state is not really possible for a current sheet. Nonetheless, laboratory experiments (Bratenahl et al. [2], Frank [3], Gekelman et al. [4]) have shown that a quasi-stationary state is still possible for a few Alfvén times as the current sheet collapse is impeded by a pressure build-up. However, a non-stationary state ensues eventually! From a theoretical point of view, the development of a current sheet in the vicinity of a magnetic neutral line is an essentially nonlinear phenomenon. An exact solution for the MHD[1] equations for a time-dependent, two-dimensional flow of a plasma in a hyperbolic magnetic field was given by Uberoi [5] and Chapman and Kendall [6]. This solution describes the plasma carrying oppositely-directed magnetic field lines from two sides toward the neutral point, where they slip through the plasma, and are cut and reconnected, and carried outwards (see Figure 1). However, this solution had an initially current-free magnetic field, so it was not appropriate for the reconnection problem. Shivamoggi [7] modified this solution so as to remedy this defect. This modified solution predicted a sequence of events associated with the evolution of a current sheet in a hyperbolic magnetic field,

---

[1] The MHD model offers the easiest approach to describe the macroscopic interaction between plasmas and magnetic fields. Besides, many space- and laboratory-plasma phenomena can be adequately described by the MHD model. The MHD approximation is essentially valid if there is sufficient localization of the interaction of particles in physical space. This localization can be achieved either by collisions between the particles or by gyrations of the particles in a strong magnetic field. In the MHD model, the ion-dynamics play an important role and electrons with their fast-response ability provide almost instant shielding to any charge imbalances.

in agreement with laboratory experiments (Frank [3], Kirii et al. [8], [9], and Bogdanov et al. [10]) on the collapse of a plasma near the hyperbolic magnetic neutral line.

Recently, this solution was generalized (Shivamoggi [11]) to incorporate a uniform shear-strain rate in the plasma flow so that the magnetic field lines now undergo not only sweeping but also shearing by the plasma flow. The effect of the shear strain in the plasma flow was found to impede the current-sheet formation. The integrability aspects of the system of nonlinearly-coupled differential equations governing these dynamics have been investigated (Rollins and Shivamog [12]) which seem to show that the effect of the shear strain in the plasma flow is to produce chaotic evolutions in the dynamical system in question.

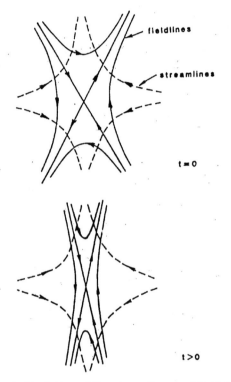

Fig. 1. Evolution of the streamlines and field lines.

2   Current-Sheet Formation at a Hyperbolic Magnetic Neutral Line in a Stagnation-Point Plasma Flow

Consider a two-dimensional problem with the flow velocity and the magnetic field confined to the xy-plane and independent of the z-coordinate.[2]

---

[2] A generalized procedure to construct solutions relaxing the latter constraint was sketched by Shivamoggi and Uberoi [13].

The MHD equations for an incompressible plasma are

$$\varrho \left(\frac{\partial}{\partial t} + \mathbf{v} \cdot \boldsymbol{\nabla} \right) \mathbf{v} = -\boldsymbol{\nabla} p - (\nabla^2 A) \boldsymbol{\nabla} A \tag{1}$$

$$\left(\frac{\partial}{\partial t} + \mathbf{v} \cdot \boldsymbol{\nabla} \right) A = \eta \nabla^2 A \tag{2}$$

where $\mathbf{v}$ is the plasma velocity, p is the plasma pressure, ϱ is the plasma mass density (which will be put equal to unity in the following) and the magnetic field $\mathbf{B}$ is given by

$$\mathbf{B} = \boldsymbol{\nabla} A \times \hat{\mathbf{i}}_z. \tag{3}$$

Let us choose the following initial conditions (Shivamoggi [7]):

$$t = 0: \quad V_x = -ax, \quad V_y = ay, \quad A = K(kx^2 - y^2), \tag{4}$$

where a, K and k are externally-determined parameters, with $a > 0$ and $k > 1$. (4) describes a stagnation-point plasma flow impinging transversely onto the $x = 0$ plane of the type involved in the laboratory experiments ([3], [8]-[10]).

The initial value of the current density $\mathbf{J}$ for the arrangement (4) is nonzero and is given by

$$t = 0 : \mathbf{J} \equiv \boldsymbol{\nabla} \times \mathbf{B} = 2K(1 - k)\mathbf{i}_z. \tag{5}$$

The Lorentz force associated with the initial current density (5) is

$$t = 0 : \mathbf{J} \times \mathbf{B} = 2K^2 k(1 - k)x\hat{\mathbf{i}}_x - 2K^2(1 - k)y\hat{\mathbf{i}}_y \tag{6}$$

which has components directed toward the origin along the x-axis and away from it along the y-axis so as to maintain the prescribed initial stagnation-point flow (4).

Let us now assume that the solution, for $t > 0$, of equations (1) and (2), commensurate with the initial conditions (4), is of the form

$$V_x(x, y, t) = -\dot{\gamma}(t)x, \quad V_y(x, y, t) = \dot{\gamma}(t)y$$

$$A(x, y, t) = K[\alpha(t)kx^2 - \beta(t)y^2], \quad p(x, y, t) = -\frac{1}{2}\nu(t)(x^2 + y^2) + p_0 \tag{7}$$

with the initial conditions:

$$t = 0 : \alpha = \beta = 1 \quad \text{and} \quad \dot{\gamma} = a. \tag{8}$$

Note that, for the solution (7), $\nabla^2 A$ is a function only of t, so that the effect of resistivity, according to equation (2), is to add a function of t to A (which leaves the magnetic field unaltered) and hence to introduce an electric field along the z-axis. We therefore neglect the effect of resistivity in the following.

Substituting (7), we obtain from equations (1) and (2),

$$\dot{\alpha} - 2\dot{\gamma}\alpha = 0 \tag{9}$$

$$\dot{\beta} + 2\dot{\gamma}\beta = 0 \tag{10}$$

$$\ddot{\gamma} = 2K^2(k^2\alpha^2 - \beta^2) \tag{11}$$

$$\nu = \dot{\gamma}^2 + \frac{K^2}{4\pi}(k\alpha - \beta)^2 \tag{12}$$

from which, we obtain, on using (8),

$$\alpha(t) = e^{2\gamma(t)} \tag{13}$$

$$\beta(t) = e^{-2\gamma(t)} \tag{14}$$

$$\dot{\gamma}^2 = K^2(k^2 e^{4\gamma} + e^{-4\gamma}) - C \tag{15}$$

with

$$t = 0 : \gamma = 0. \tag{16}$$

Here

$$C \equiv K^2(k^2 + 1) - a^2.$$

From equations (13)-(16), we obtain for small t,

$$\gamma(t) \approx at + K^2(k^2 - 1)t^2 \tag{17}$$

and for large $\gamma(t)$,

$$\gamma(t) \approx \frac{1}{2} \ln \left[\frac{1}{2Kk(t_0 - t)} \right] \tag{18}$$

where t is a constant of integration. According to (17), $\gamma(t)$ becomes infinite in a finite time (another finite-time singularity!) .[3] This is perhaps plausible because of the steady infinite energy input into the system via the stagnation-point plasma flow .

The solution (13)-(18) shows that the evolution of the streamline/field line configuration for $t > 0$ is as follows:

* the streamlines have a fixed shape (only the magnitude of the velocity changes with t),

* the magnetic separatices rotate toward each other as t increases, (see Figure 1).

Therefore, as t increases, the magnetic field becomes tangential to the y-axis, with intensity increasing indefinitely, and shows a discontinuity on the y-axis; so, a very thin neutral current sheet is formed on the y-axis. The solution (13)-(18) also shows that as the initial magnetic field gradient K increases, the magnetic separatices will rotate towards each other faster.

This evolution scenario has been confirmed by the laboratory experiments ([3], [8]-[10]) on the collapse of a plasma near the hyperbolic neutral line, which showed that:

* a thin neutral current sheet separating oppositely-directed magnetic fields formed,

---

[3] Bulanov and Sakai [14] recently demonstrated the existence of another solution of two-dimensional incompressible MHD equations which also shows a finite-time singularity. These results are compatible with a recent rigorous result of Klapper [15] that, for smooth initial conditions, no current sheet (or any other type of singularity) can form unless $\nabla \mathbf{v}$ or $\nabla \mathbf{B}$ blow up outside a finite region around the two-dimensional magnetic null point.

* during the process of sheet formation, a two-dimensional stagnation-point plasma flow in the plane normal to the current direction developed — plasma flowed toward the sheet surface at right angles on both sides and flowed out of the sheet along the sheet surface,

* an increase in the initial magnetic field gradient produced a rapid change in the structure of the magnetic field near the neutral line and hastened the current-sheet formation.

3   Effect of a Uniform Shear–Strain in the Plasma Flow

Let us now generalize the solution (13)-(18) to incorporate a uniform shear-strain in the plasma flow, and hence, investigate the effect of this shear-strain on the current-sheet formation.

Let us now choose the initial conditions to be (Shivamoggi [11]):

$$t = 0 : V_x = -ax + by, \quad V_y = ay + bx, \quad A = kx^2 - y^2 + bxy. \quad (19)$$

(19) incorporate in the plasma flow a uniform shear-strain rate characterized by the parameter b (let us take $b > 0$). Note that the initial plasma-flow is, however, irrotational because the initial value of the vorticity $\mathbf{\Omega}$, from (19), is

$$t = 0 : \mathbf{\Omega} \equiv \mathbf{\nabla} \times \mathbf{v} = \mathbf{0}. \quad (20)$$

The initial value of the current density $\mathbf{J}$ is, on the other hand, nonzero and is given by

$$t = 0 : \mathbf{J} \equiv \mathbf{\nabla} \times \mathbf{B} = 2(1 - k)\hat{\mathbf{i}}_z \quad (21)$$

and leads to the Lorentz force required to maintain the initial plasma flow (19).

Let us now assume that the solution, for $t > 0$, of equations (1) and (2), commensurate with the initial conditions (19), is of the form

$$V_x(x, y, t) = -P(t)x + S(t)y, \quad V_y(x, y, t) = P(t)y + R(t)x$$

$$A(x, y, t) = k\alpha(t)x^2 - \beta(t)y^2 + \sigma(t)xy, \quad P(x, y, t) = -\frac{1}{2}\nu(t)(x^2 + y^2) + p_0$$

$$(22)$$

with the initial conditions

$$t = 0 : \alpha = \beta = 1, \quad P = a, \quad R = S = \sigma = b. \quad (23)$$

Note that for the solution (22), $\nabla^2 A$ is a function only of t again, so that resistivity does not play an important role in the evolution of solution (22) and is therefore neglected.

The vorticity

$$\mathbf{\Gamma} = [R(t) - S(t)]\hat{\mathbf{i}}_z$$

evolves according to the equation (obtained by taking the curl of equation (1)):

$$\left(\frac{\partial}{\partial t} + \mathbf{v} \cdot \mathbf{\nabla}\right)\mathbf{\Omega} = (\mathbf{\Omega} \cdot \mathbf{\nabla})\mathbf{v} + (\mathbf{B} \cdot \mathbf{\nabla})\mathbf{J} - (\mathbf{J} \cdot \mathbf{\nabla})\mathbf{B}. \quad (24)$$

This gives

$$\frac{d}{dt}[R(t) - S(t)] = 0 \tag{25}$$

from which, on using (20), we obtain

$$R(t) = S(t), \tag{26}$$

which implies that the plasma-flow remains irrotational for all t.

Substituting (22), and using (26), we have from equations (1) and (2),

$$k\dot{\alpha} - 2kP\alpha + R^2 = 0 \tag{27}$$

$$\dot{\beta} + 2P\beta - R^2 = 0 \tag{28}$$

$$\dot{R} + 2R(k\alpha - \beta) = 0 \tag{29}$$

$$\dot{P} - 2(k^2\alpha^2 - \beta^2) = 0 \tag{30}$$

Equations (27)-(30) admit an integral:

$$I = P^2 + \frac{1}{2}R^2 - (k^2\alpha^2 + \beta^2) = \hat{C} \tag{31}$$

where

$$\hat{C} \equiv a^2 + \frac{1}{2}b^2 - (k+1). \tag{32}$$

The invariance of I is confirmed by the numerical solution of equations (27)-(30) (see Figure 2).

(31) implies that P, R, α and β may grow with t even though the quantity $(P^2 + \frac{1}{2}R^2 - k^2\alpha^2 - \beta^2)$ is bounded. Physically, this of course implies the formation of a very thin current sheet on the y-axis, as $t \Rightarrow \infty$.

(31) also shows the shear strain in the plasma flow (represented by R):

* acts as a sink for the magnetic energy,
* impedes the current-sheet formation.

This may be seen alternatively by considering the small-time behavior of the solutions of equations (27)-(30).

Putting

$$P(t) = \dot{\gamma}(t) \tag{33}$$

we obtain from equations (27) and (28), for small t,

$$\alpha(t) \approx 1 + 2\gamma(t) - \frac{1}{k}[R(t)]^2 t \tag{34}$$

$$\beta(t) \approx 1 - 2\gamma(t) + \frac{1}{k}[R(t)]^2 t. \tag{35}$$

Using (33), (34) and (35), equations (29) and (30) give

$$\ddot{\gamma} = 2(k^2 - 1) + 8(k^2 + 1)\gamma - \frac{4}{k}(k^2 + 1)R^2 t \tag{36}$$

$$\dot{R} + 2(k - 1)R = 0 \tag{37}$$

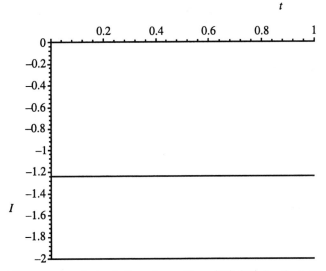

Fig. 2. I vs. t from a numerical solution of equations (27)-(30) for the initial conditions $t = 0 : P = 0.5$, $R = 1.0$, $\alpha = 1.0$, $\beta = 1.0$, and the parameter $k = 1$.

from which, we have for small t,

$$\gamma(t) \approx at + (k^2 - 1)t^2 - \frac{2b^2}{3k}(k^2 + 1)t^3. \tag{38}$$

(38) confirms that the presence of a shear strain in the plasma-flow (i.e., $b \neq 0$) tends to impede the current-sheet formation.

4   Discussion

Current-sheet formation near a hyperbolic magnetic neutral line has been investigated by including the effects of both sweeping and shearing of the magnetic field lines by the plasma flow. The field-line shearing has been found to impede the current-sheet formation. On the other hand, investigation of the integrability aspects of the system of nonlinearly-coupled differential equations governing these dynamics indicates [12] the possibility of shear-induced chaotic evolutions in the dynamical system in question.

Acknowledgements

The author is thankful to the referee for his valuable remarks and suggestions.

References

1. S.I. Syrovatskii: *Sov. Astron. J.* **10**, 270 (1966).
2. P.J. Baum, A. Bratenahl, M. Kao and R.S. White: *Phys. Fluids* **16**, 1501 (1973).
3. A.G. Frank: *Proc. Lebedev Phys. Inst.* (Moscow) **74**, 108 (1974).

4. W. Gekelman, R.L. Stenzel and N. Wild: *Phys. Sciprta* **T2/2**, 277 (1982).
5. M.S. Uberoi: *Phys. Fluids* **6**, 1379 (1963).
6. S. Chapman and P.C. Kendall: *Proc. Roy. Soc.* (London) A **271**, 435 (1963).
7. B.K. Shivamoggi: *Phys. Fluids* **29**, 769 (1986).
8. N.P. Kirii, V.S. Markov, S.O. Syrovatski, A.G. Frank and A.Z. Khodzhaev: *Proc. Lebedev Phys. Inst.* (Moscow) **110**, 121 (1979).
9. N.P. Kirii, V.S. Markov, A.G. Frank and A.Z. Khodzhaev: *Fiz. Plazmy* **3**, 538 (1977).
10. S. Yu Bogdanov, V.S. Markov, A.G. Frank and A.Z. Khodzhaev: *J. Phys. Col.* C7, **40**, Suppl. 7, C7-221-2 (1979).
11. B.K. Shivamoggi: *Phys. Lett.* A **258**, 131 (1999).
12. D.K. Rollins and B.K. Shivamoggi: *Europhys.* A, submitted for publication (2000).
13. B.K. Shivamoggi and M.S. Uberoi: *Phys. Fluids* **22**, 2247 (1979).
14. S. Bulanov and J. Sakai, *J. Phys. Soc. Japan* **66**, 3477 (1997).
15. I. Klapper, *Phys. Plasmas* **5**, 910 (1998).

Nonlocality in Turbulence

Arkady Tsinober

Faculty of Engineering, Tel Aviv University, Tel Aviv 69978

1 Introduction and Simple Examples

Nonlocality is among the three main reasons[1] why the problem of turbulence is so difficult. The term 'nonlocality' is used here in several related meanings which are clarified in the course of the discussion of the issues throughout this paper.

Taking the position that velocity fluctuations represent the large scales and the velocity derivatives represent the small scales one can state that in homogeneous (not necessarily isotropic) the large and the small scales are uncorrelated. This is expressed quantitatively via correlations between velocity and vorticity. For example, in a homogeneous turbulent flow the mean Lamb vector $\langle \omega \times \mathbf{u} \rangle = 0$, and also $\langle (\mathbf{u} \cdot \nabla)\mathbf{u} \rangle = 0$. If the flow is statistically reflexionally symmetric then the mean helicity $\langle \omega \cdot \mathbf{u} \rangle$ vanishes too. However, vanishing correlations do not necessarily mean absence of dynamically important relations. Indeed, the quantities $(\mathbf{u} \cdot \nabla)\mathbf{u} \equiv \omega \times \mathbf{u} + \nabla(u^2/2)$ and $\omega \times \mathbf{u}$, are the main 'guilty' for all we call turbulence. Both contain the large scales (velocity) and small scales (velocity derivatives, vorticity). So some kind of coupling between the two is unavoidable.

Let us begin from the *kinematic* relation between velocity and vorticity, which is just a consequence of the relation $\omega = curl\ \mathbf{u}$. Therefore any altering of ω (and strain) results in its 'reacting back' on the velocity field. This point is not that trivial as may seem. Indeed, take a Helmholz decomposition of the most significant part of the nonlinear term in NSE the Lamb vector, $\omega \times \mathbf{u} = \nabla \alpha + \nabla \times \beta$. Assuming that ω and $\mathbf{u}$ are random Gaussian and *unrelated*, i.e. $\omega \neq rot\mathbf{u}$ the result is, that $\langle (\nabla \alpha)^2 \rangle = \langle (\nabla \times \beta)^2 \rangle$. However, if $\omega = rot\mathbf{u}$ and $\mathbf{u}$ is quasi-Gaussian, i.e. obeys the zero-forth-cumulant relation, then $\langle (\nabla \alpha)^2 \rangle \sim 2 \langle (\nabla \times \beta)^2 \rangle$ i. e. in this case the *rms* of the potential part of the Lamb vector is twice larger than its solenoidal part[2]. In other words ω 'reacts back' on velocity, and consequently on $\omega \times \mathbf{u}$, even for purely kinematic reasons.

More generally, vorticity is not just small scales. It is of special importance, since together with boundary conditions the whole flow field is determined entirely by the field of vorticity. This, of course, includes the velocity field itself, and therefore the large scales are determined by the small scales and vice versa.

---

[1] The three N's: nonlinearity, nonintegrability and nonlocality.
[2] In real turbulent flows this difference is even larger.

This is the simplest indication not only for direct interaction/coupling between large and small scales but also that this interaction is bidirectional.

A natural question is what about the correlation(s) and coupling between velocity, u_i and the strain, s_{ij}. Again in a homogeneous turbulent flow they are uncorrelated $\langle u_i s_{ij} \rangle = 0$. However, velocity is correlated with small scales of 'higher order'. Namely, the correlation, $\langle u_i \nabla^2 u_i \rangle = -2 \langle s_{ij} s_{ij} \rangle = -\langle \omega^2 \rangle$, is essentially nonvanishing again for purely kinematic reasons. Nevertheless, it is of special dynamical significance as being directly related to the dissipation of turbulent energy, $\langle \epsilon \rangle = 2\nu \langle s_{ij} s_{ij} \rangle$, which for dynamical reasons remains finite for whatever small ν. Therefore the correlation $\langle u_i \nabla^2 u_i \rangle$ becomes very large at small ν. However, the corresponding correlation coefficient at large Reynolds numbers is roughly of the order $Re^{-1/4}$, i.e. becomes very small. This, of course, does not mean that the coupling between $\mathbf{u}$ and $\nabla^2 \mathbf{u}$ becomes unimportant at large Reynolds numbers. Moreover, just like in case of vorticity, strain is not just small scales, since together with boundary conditions the whole flow field is determined entirely by the field of strain [1].

One can use the above example with the Lamb vector to illustrate the dynamical aspect of this coupling. For this we retreat from homogeneous flows and consider a unidirectional *in the mean* fully developed turbulent flow such as the flow in a plane channel in which all statistical properties depend on the coordinate normal to the channel boundary, x_2, only. In such a flow a simple precise kinematic relation is valid

$$d\langle u_1 u_2 \rangle / dx_2 \equiv \langle \boldsymbol{\omega} \times \mathbf{u} \rangle_1 = \langle \omega_2 u_3 - \omega_3 u_2 \rangle \neq 0, \tag{1}$$

which is just a consequence of the vector identity $(\mathbf{u} \cdot \nabla)\mathbf{u} \equiv \boldsymbol{\omega} \times \mathbf{u} + \nabla(u^2/2)$ in which incompressibility and $d\langle \cdots \rangle / dx_{1,3} = 0$ where used, and $\langle \cdots \rangle$ means an average in some sense (e.g. time or/and over the planes $x_2 = const$, etc.). The *dynamical* aspect is that in turbulent channel flows $d\langle u_1 u_2 \rangle / dx_2 \neq 0$ is essentially different from zero at *any arbitrarily large* Reynolds number. Therefore one can see from (1) that at least some correlations between velocity and vorticity in such flows are essentially different from zero. Since vorticity is essentially a small scale quantity the relation (1) is a clear indication of a dynamically important statistical dependence between the large scales ($\mathbf{u}$) and small scales (ω). Without this dependence $d\langle u_1 u_2 \rangle / dx_2 \equiv 0$, which means that the mean flow would not 'know' about its turbulent part at all. It is noteworthy that corresponding correlation coefficients (and many other statistical characteristics) are of order 10^{-2} even at rather small Reynolds numbers. Nevertheless, as we have seen, in view of the dynamical importance of interaction between velocity and vorticity in turbulent shear flows such 'small' correlation by no means does not imply absence of dynamicaly important statistical dependence and direct interaction between large and small scales. Indeed it is this interaction that results in drastic changes of the whole mean flow.

2   Different Aspects of Nonlocality

From the formal point a process is called local if all the terms in the governing equations are differential. If the governing equations contain integral terms, then the process is nonlocal. The Navier-Stokes equations are *integro-differential* for the velocity field in both physical and Fourier space (and any other). Therefore, generally, the Navier-Stokes equations describe nonlocal processes. The problem is intimately related to the one of decompositions/representations, which is briefly mentioned below.

The property of nonlocality of Navier-Stokes equations in physical space is two-fold. On one hand, it is due to pressure ('dynamic' nonlocality), since $\rho^{-1}\nabla^2 p = \omega^2 - 2s_{ij}s_{ij}$, so that pressure is nonlocal due to nonlocality of the operator ∇^{-2}. This nonlocality is strongly associated with essentially non-Lagrangian nature of pressure. For example, replacing in the Euler equations the pressure Hessian $\frac{\partial^2 p}{\partial x_i \partial x_j} \equiv \Pi_{ij}$, which is both nonlocal and non-Lagrangian, by a local quantity $(1/3)\delta_{ij}\nabla^2 p = (1/6)\rho\{\omega^2 - 2s_{ij}s_{ij}\}$ turns the problem into a local and integrable one and allows to integrate the equations for the invariants of the tensor of velocity derivatives $\partial u_i/\partial x_j$ in terms of a Lagrangian system of coordinates moving with a particle, see references in [1]. The reason for disappearance of turbulence (and formation of singularity in finite time) in such models, called restricted Euler models, is that the eigenframe of s_{ij} in these models is *fixed* in space [2], whereas in a real turbulent flow it is oriented randomly. This means that nonlocality due to presure is essential for (self-) sustaining turbulence: no presure Hessian - no turbulence. A related aspect is that the Lagrangian acceleration Du/Dt - a kind of small scale quantity – is dominated by the pressure gradient, ∇p, [3].

Taking *curl* of the NSE and getting rid of the pressure does not remove the nonlocality. Indeed, the equations for vorticity and enstrophy are nonlocal in vorticity, ω, since they contain the rate of strain tensor, s_{ij}, due to nonlocal relation between vorticity, ω, and the rate of strain tensor, s_{ij} ('kinematic' nonlocality)[3]. The two aspects of nonlocality are related, but are not the same. For example, in compresible flows there is no such relatively simple relation between pressure and velocity gradient tensor as above, but the vorticity-strain relation remains the same.

Both aspects of nonlocality are reflected in the equations for the rate of strain tensor and total strain/dissipation, $s^2 \equiv s_{ij}s_{ij}$, and the equations for the third order quantities $\omega_i\omega_j s_{ij}$ and $s_{ij}s_{jk}s_{ki}$. An important aspect is that the latter equations contain invariant quantities $\omega_i\omega_j\Pi_{ij}$ and $s_{ik}s_{kj}\Pi_{ij}$ reflecting the nonlocal dynamical effects due to pressure and can be interpreted as interaction between vorticity and pressure and between vorticity and and strain. For

---

[3] Nonlocality of the same kind is encountered in problems dealing with the behaviour of vortex filaments in an ideal fluid. Its importance is manifested in the breakdown of the so called localized induction approximation (LIA) as compared with the full Bio-Savart induction law, see [4] and references therein.

example, the equation for $\omega_i\omega_j s_{ij}$

$$D(\omega_i\omega_j s_{ij})/Dt = \omega_i s_{ij}\omega_k s_{ki} - \omega_i\omega_j \Pi_{ij} + \nu\omega_i\nabla^2\omega_i, \tag{2}$$

shows both aspects of nonlocality of vortex stretching process. The first term in (2) (which is just the squared magnitude of the vortex stretcing vector) is strictly positive $\omega_i s_{ij}\omega_k s_{ki} \equiv W^2 > 0$. This means that the nonlinear processes involving vortex stretching (or direct interaction of vorticity and strain) always tend to increase even the *instantaneous* enstrophy production. However, the inviscid rate of change of enstrophy generation contains also a second term reflecting the interaction between vorticity and the pressure Hessian Π_{ij}. This is the term $-\omega_i\omega_j\Pi_{ij}$. Without this term the question on why $\langle\omega_i\omega_j s_{ij}\rangle > 0$ would be immeadiately answered. It appears [1] that $\langle\omega_i\omega_j\Pi_{ij}\rangle$ is positive and is about $\langle W^2\rangle/3$, i.e. in the mean the nonlinearity in (2) is reduced by this nonlocal term, since for a Gaussian velocity field $\langle\omega_i\omega_j\Pi_{ij}\rangle \equiv 0$. The nonvanishing correlation $\langle\omega_i\omega_j\Pi_{ij}\rangle$ (and also $\langle s_{ik}s_{kj}\Pi_{ij}\rangle$) is also one of the manifestations of nonlocality, the direct coupling between the large and small scales.

2.1   Direct Coupling Between Large and Small Scales

Nonlocality in the sense as discussed above is an indication of direct coupling between large and small scales. There exist massive evidence that this is really the case as there are many indications both for direct interaction/coupling between large and small scales but also that this interaction is bidirectional[4]. We mention first the well known effective use of fine honeycombs and screens in reducing large scale turbulence in various experimental facilities. The experimentally observed phenomenon of strong drag reduction in turbulent flows of dilute polymer solutions and other drag reducing additives is another example of such a 'reacting back' effect of small scales on the large scales [5]. Third, one can substantially increase the dissipation and the rate of mixing in a turbulent flow by *directly* exciting the small scales [6].

Anisotropy. One of the the manifestations of direct interaction between large and small scales is the anisotropy in the small scales. Though local isotropy is believed to be one of the universal properties of high Re turbulent flows it appears that it is not so universal: in many situations the small scales do not forget the anisotropy of the large ones. There exist considerable evidence for this which has a long history starting somewhere in the 50-ies, see references in [1], [7], [8]. Along with other manifestations of direct interaction between large and small scales the deviations from local isotropy seem to occur due to various external constraints like boundaries, initial conditions, forcing (e.g. as in DNS), mean shear/strain, centrifugal forces (rotation), buoyancy, magnetic field, etc., which usually act as an organizing factor, favoring the formation of

---

[4] Note that in case of passive objects there is no such a bidirectional relation – it is only one way.

coherent structures of different kinds (quasi-two-dimensional, helical, hairpins, etc.). These are as a rule large scale features which depend on the particularities of a given flow and thus are not universal. These structures, especially their edges seem to be responsible for the contamination of the small scales. This 'contamination' is unavoidable even in homogeneous and isotropic turbulence, since there are many ways to produce such a flow, i.e. many ways to produce the large scales. It is the difference in the mechanisms of large scales production which 'contaminates' the small scales. Hence, nonuniversality.

Let us turn first again to the 'simlpe' example above and look at the properties in the proximity of the midplane, $x_2 \approx 0$, of the turbulent channel flow. In this region $dU/dx_2 \approx 0$, but the flow is neither homogeneous nor isotropic, since though $\langle u_1 u_2 \rangle \approx 0$ in this region too, $d \langle u_1 u_2 \rangle /dx_2$ is essentially nonzero and is *finite* independently of Reynolds number as far as the data allow to make such a claim. This is also a clear indication of nonlocality, since in the bulk of the flow, i.e. *far from the boundaries*, $dU/dx_2 \sim 0$.

The first experimental evidence on anisotropy in small scales at large Reynolds numbers in the atmospheric boundary layer experiments goes back to the fact that the skewness of the derivative of temperature fluctuations is not small, as should be in locally isotropic flow, and is of order 1, whereas for a locally isotropic flow it should be close to zero, see [8] for further references. An important feature of these flows is the presence of a mean gradient of the passive scalar – the rest is not so important: the phenomenon is observed for a Gaussian and two dimensional velocity field [8].

Recently similar observations were made for the velocity increments and velocity derivatives in the direction of the mean shear both numerical and laboratory, see references in [7]. It was found that the stastistical properties of velocity increments and velocity derivatives in the direction of the mean shear do not conform with and do not confirm the hypothesis of local isotropy. *Moreover, our results imply that the large scales are directly coupled to the small scales. (The anisotropy disappears when the large scale shear is removed)*, [7]. More precisely these results imply that there is a *direct* influence of mean shear on the small scales, which is possible due to the permanent bias of the mean shear to which is exposed the field of fluctuations due to its very large *residence time* in the mean shear. One of the explanations of the results from [7] is *that the large scales are directly coupled to the small scales.* However, this does not mean that there is no such coupling *when the large scale shear is removed*. Another exlanation is that in this experiment the value of the Corrsin criterion $S_C^* = (dU/dx_2)(\nu/ \langle \epsilon \rangle)^{1/2} \approx 2.4 \cdot 10^{-2}$. This is the ratio of the Komlogorov time scale, $\tau_\eta = (\nu/ \langle \epsilon \rangle)^{1/2}$, to the time scale, $(dU/dx_2)^{-1}$, associated with the mean shear, and it should be small enough in order to have isotropy in small scales. The main problem is how small. There is no agreement on this issue, but there is evidence that in order to have one decade of isotropic inertial range in boundary layer flows (both simple and complex) at $Re_\lambda \approx 1500$ it is necessary that $S_C^* < 10^{-2}$, see [10] and references therein. This brings us to the next issue.

Other manifestations of direct coupling between large and small scales.
An important recent observation, [9] and references therein, is that conditional
statistics of small scale quantities (e.g. velocity increments, estrophy, total strain)
conditioned on large scale quantities (velocity) is not independent of the large
scale quantities as should be *if* if the large scales are *not* coupled directly to the
small scales. We stress that this observation was made in [9], in spite that in this
experiment the mean shear was rather small, less than $0.1s^{-1}$. This corresponds
to the value of the Corrsin criterion $S_C^* = (dU/dx_2)(\nu/\langle\epsilon\rangle)^{1/2} \approx 2\cdot10^{-3}$, which
is an order of magnitude smaller than in [7] and five times is lower than the value
0.01 required for local isotropy in presence of mean shear mentioned above. Two
aspects deserve special comment regarding the experiment in [9]. First, there is
a clear tendency of increase of the conditional averages of the structure functions
with the *energy* of fluctuations. Second, such a tendency, that is the direct coupling, is observed also for the smallest distance of the order of Kolmogorov scale
$\sim \eta$, which was used for estimates of the derivatives in the streamwise direction.
Similar behaviour is exhibited by the enstrophy ω^2 and the total strain $s_{ij}s_{ij}$.

The observations on the the coupling between and the 'reaction back' of the
small scales on the large by no means are not exausted by the references given
above. As an example from the atmospheric physics we bring a quotation of the
first conclusion rached at the Symposium on the nature fo so clear air turbulence
[11]: *The energy dissipated at small-scale by clear air trubulence influences the
large-scale atmospheric motion.*

Helicity. Helicity, $\int \omega \cdot u dx$, and its density, $\omega \cdot u$, deserve here also special
mentioning. The formal reason is that if $\langle u \cdot \omega \rangle \neq 0$, this is a clear indication
of direct coupling of large and small scales. So it is not surprising that in flows
with nonzero mean helicity the direct coupling between small and large scales
is stronger than otherwise. The stronger coupling between the large and small
scales in flows with nonzero mean helicity $\langle u \cdot \omega \rangle$ aids creation of large scale
structures out small scale turbuence, see [12] and references therein. As men-
tioned this does not mean that in case $\langle u \cdot \omega \rangle = 0$, or even $u \cdot \omega = 0$ as in
two-dimensional flows, such a coupling does not exist.

On closures and constitutive relations. Memory effects. The nonlocality
due to the coupling between large and small scales is also manifested in problems
related to various decompositions of turbulent flows and in the so called closure
problem. For example, in the Reynolds decomposition of the flow field into the
mean and the fluctuations and in similar decompositions associated with large
eddy simulations (LES) the relation between the fluctuations and the mean flow
(or resolved and unresolved scales in LES, etc.) is a functional. That is the field
of fluctuations at each time/space point depends on the mean (resolved) field
in the whole time/space domain. Vice versa the mean (resolved) flow at each
time/space point depends on the field of fluctuations (unresolved scales) in the
whole time/space domain. This is because the equations for the fluctuations (in-
resolved scales) contain as coefficients the mean (resolved) field. This means that

in turbulent flows point-wise flow independent 'constitutive' relations analogous to real material constitutive relations for fluids (such as stress/strain relations) can not exist, though the 'eddy viscosity' and 'eddy diffusivity' are used frequently as a crude approximation for taking into account the reaction back of fluctuations (unresolved scales) on the mean flows (resolved scales). The fact that the 'eddy viscosity' and 'eddy diffusivity' are flow (and space/time) dependent is just another expression of the strong coupling between the large and the small scales.

3   Concluding Remarks

The general conclusion is that the small scales are *not decoupled* from the large scales even at Reynolds numbers (based on the Taylor microscale) as large as 10^4. In other words the Galilean invariance is broken in the restricted sense that the properties of small scale turbulence are *not* independent of parameters characterizing the large scales, such as, e.g. the energy of velocity fluctuations. This *direct and bidirectional* interaction/coupling of large and small scales is a generic property of all turbulent flows and one of the main reasons for small scale intermittency, non-universality, and quite modest manifestations of scaling. In view of such coupling it is not clear how meaningful is the notion of inertial range and things like eddy viscosity representations of the subgrid scales.

References

1. A. Tsinober: 'Turbulence - Beyond Phenomenology', Lect. Notes Phys., **511**, (Springer, Berlin, Heidelberg 1998), pp. 85-143; A. Tsinober, An informal introduction to turbulence (Kluwer, Dordrecht, Boston 2001).
2. E.A. Novikov: *Fluid Dyn. Res.,* **6**, 79 (1990).
3. P. Vedula and P.K. Yeung , *Phys. Fluids*, **11**, 1208 (1999).
4. R.L. Ricca, D. Samuels, C.F. Barenghi: J. Fluid Mech., **391**, 29 (1999).
5. A. Gyr and H.-W. Bewersdorff: *Drag reduction of turbulent flows by additives,* (Kluwer, Dordrecht, Boston 1995).
6. J. M. Wiltse and A. Gledzer.: *Phys. Fluids*, **10**, 2026 (1998).
7. X. Shen and Z. Warhaft: Phys. Fluids, **12**, 2976 (2000).
8. Z. Warhaft: *Ann. Rev. Fluid Mech.,* **32,** 203 (2000).
9. M. Kholmyansky, and A. Tsinober: 'On the origins of intermittency in real turbulent flows', In: *Intermittency in Turbulent Flows,* ed. by J.C. Vassilicos, (Cambridge, Cambridge University Press 2000), pp. 183-192.
10. S.G. Saddoughi: J. Fluid Mech., **348**, 201 (2000).
11. Y.-H. Pao, A. Goldburg: *Clear air turbulence and its detection,* (Plenum, New York 1969).
12. Droegemeier, K.K., Lazarus, S.M. and Davies-Jones, R.: *Month. Weather Rev.,* **121**, 2006 (1993).

Part VIII

Helium 3 and Other Systems

Quantized Vorticity in Superfluid ^{3}He-A: Structure and Dynamics

R. Blaauwgeers[1,2], V.B. Eltsov[1,3], M. Krusius[1], J. Ruohio[1], and R. Schanen[1,4]

[1] Low Temperature Lab, Helsinki University of Technology, FIN-02015 HUT, Finland
[2] Kamerlingh Onnes Lab, Leiden University, 2300 RA Leiden, The Netherlands
[3] Kapitza Institute, Kosygina 2, Moscow 117334, Russia
[4] CRTBT-CNRS, BP 166, F-38042 Grenoble Cedex 09 FRA, France

Abstract. Superfluid ^{3}He-A displays the largest variety in vortex structure among the presently known coherent quantum systems. The experimentally verified information comes mostly from NMR measurements on the rotating fluid, from which the order-parameter texture can often be worked out. The various vortex structures differ in the topology of their order-parameter field, in energy, critical velocity, and in their response to temporal variations in the externally applied flow. They require different experimental conditions for their creation. When the flow is applied in the superfluid state, the structure with the lowest critical velocity is formed. In ^{3}He-A this leads to the various forms of continuous (or singularity-free) vorticity. Which particular structure is created depends on the externally applied conditions and on the global order-parameter texture.

1 Superfluid ^{3}He

The accepted textbook example of a superfluid has traditionally been ^{4}He-II. However, in many respects the ^{3}He superfluids display more ideal behaviour, both in their theoretical description and their macroscopic properties. One remarkable difference is the absence of remanent vorticity in most experimental setups for superfluid ^{3}He. In ^{4}He-II, the formation of vortex lines via mechanisms intrinsic to the superfluid itself is observed only in rare cases, most notably in the superflow through a sub-micron-size orifice where the vortex lines are blown out of the immediate vicinity of the aperture and are trapped on surface sites far away where the flow is small. In ^{3}He superfluids, where the vortex core radii are at least 100 times larger, intrinsic critical velocities can be measured simply with bulk superfluid flowing past a flat wall. For this reason rotating measurements have proven very efficient in their study, quite unlike in ^{4}He-II.

The hydrodynamics of the ^{3}He superfluids abound with new features which have only marginally been investigated. Some experimental work has been performed on vortex tangles in the context of quench-cooled non-equilibrium transitions from the normal liquid to the quasi-isotropic ^{3}He-B [1] and also around vibrating wire resonators in the zero temperature limit [2]. Little if any work has been reported on turbulent flow or on vortex networks in the highly anisotropic ^{3}He-A phase [3]. Nevertheless, this is the phase with the much richer variety of response to externally applied flow. Here we describe preliminary studies in this

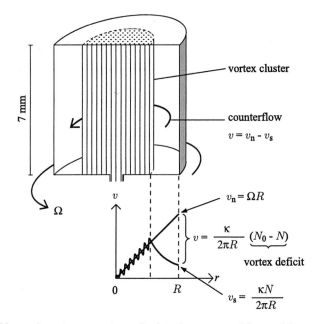

Fig. 1. Vortex lines in a rotating cylindrical container. Metastable states, which include fewer vortex lines than the equilibrium state, consist of a central cluster with rectilinear vortex lines and a surrounding annular region of vortex-free counterflow. The Magnus force from the circulating counterflow confines the lines in the cluster to the density $n_v = 2\Omega/(\nu\kappa)$, when the cluster rotates like a solid body with the container. If Ω is increased, the cluster contracts, the counterflow velocity increases and ultimately reaches the critical velocity limit v_c, where a new vortex is created. If Ω is reduced, the counterflow velocity diminishes, the cluster expands and eventually reaches the annihilation threshold where vortices in the outermost circle of lines are pushed to the cylindrical wall for annihilation. The graph on the bottom illustrates how the counterflow velocity changes as a function of the radial position in the container. Also shown how its value $v(R)$ at the cylinder wall depends on the vortex-line number N, when $N_0 \approx \pi R^2 n_v$ denotes their number in the equilibrium state at the rotation velocity Ω.

direction, which explain the critical flow velocity [4] and the dynamic response to rotation in the A phase [5]. Unusual new features are observed, which all can be explained by the structural properties of the anisotropic order-parameter field, but which also suggest that more surprises can be expected. The critical velocity [6] and the response of vortex lines to a dynamic rotation drive [7] in ^{3}He-B have been measured earlier. These measurements attest to more traditional behaviour, as can be expected for vortex lines with singular cores in an order-parameter field with many orders of magnitude weaker anisotropy.

Much of the existing information on new vortex structures in the ^{3}He superfluids has been derived from NMR measurements on bulk liquid samples contained in a rotating cylinder with smooth walls. The principle of this "rotating bucket" method is outlined in Fig. 1. Here all sensors can be placed outside the cylinder and one can investigate the undisturbed order-parameter field in the bulk superfluid, in the presence of different types of quantized vorticity as well as the conditions in which they are created.

2   Order-Parameter Texture and Superflow in ³He-A

The beauty of ³He-A lies in its anisotropy: Although the underlying material, liquid ³He, is isotropic, an all-pervading anisotropy arises from the condensation into a coherent p-wave paired state. Here the two fermion quasiparticles forming a Cooper pair have relative angular momentum $L = 1$ and total spin $S = 1$ [8]. The spin structure of the condensate is characterized by the formation of spin up-up and down-down pairs so that the total spin $\mathbf{S}$ has zero projection on an axis, which traditionally is denoted by the unit vector $\hat{\mathbf{d}}$ ($\perp \mathbf{S}$). The Cooper pairs also have a preferred direction for their orbital momentum, denoted by the unit vector $\hat{\ell}$, so that the projection of the net orbital momentum $\mathbf{L}$ on the $\hat{\ell}$ axis is positive. Finally, also the quasiparticle excitation spectrum is anisotropic: In momentum space the energy gap vanishes at two opposite poles of the Fermi surface, located on the $\hat{\ell}$ axis.

A number of competing interactions act in unison to produce a smooth variation of the $\hat{\ell}$ and $\hat{\mathbf{d}}$ vector fields over the container volume. These vector fields, which in the absence of singularities are smoothly continuous, are called the orbital and spin textures. The dipole (or spin-orbit) coupling connects the orbital and spin textures via a free-energy density $f_D = -g_D (\hat{\mathbf{d}} \cdot \hat{\ell})^2$. The magnetic anisotropy energy in an externally applied magnetic field $\mathbf{H}$ is written as $f_H = g_H (\hat{\mathbf{d}} \cdot \mathbf{H})^2$. These two orientational forces on $\hat{\mathbf{d}}$ balance each other at a characteristic value of the field $\mathbf{H}$. This is called the dipole field, $H_D = \sqrt{g_D/g_H} \sim 1$ mT, below which the dipole coupling wins and the texture becomes *dipole locked*, with $\hat{\mathbf{d}}$ and $\hat{\ell}$ either parallel or anti-parallel. At higher fields, $H \gg H_D$, $\hat{\mathbf{d}}$ is forced to lie perpendicular to $\mathbf{H}$. This is the case in NMR measurements: $\hat{\mathbf{d}}$ is everywhere contained in the plane perpendicular to $\mathbf{H}$ and varies smoothly in this plane. This texture is coupled to the orbital alignment such that $\hat{\ell}$ is also for the most part forced parallel to $\hat{\mathbf{d}}$.

Important exceptions, when $\hat{\ell}$ becomes decoupled from the planar $\hat{\mathbf{d}}$ orientation (in the presence of a magnetic field $H \gg H_D$), are topological defects like the *dipole-unlocked* central part of a vortex line, which is also called the "soft vortex core" (Fig. 2). To minimize the loss in dipole energy, the soft core has a radius on the order of the dipolar healing length $\xi_D = \hbar/(2m_3 v_D) \sim 10$ μm, where $v_D \approx \sqrt{g_D/\rho_{s\parallel}} \sim 1$ mm/s is the order of magnitude of the so-called dipolar velocity, ie. the flow velocity at which the orienting force on $\hat{\ell}$ from an externally applied superflow matches the bending or gradient energy of $\hat{\ell}$, which maintains the spatial coherence of the order parameter.

The rigidity of the continuous $\hat{\ell}$ texture, which arises from the anisotropy forces, from superfluid coherence, and from the boundary conditions, explains why fluid flow in the A phase is not always dissipative, in spite of the gap nodes in the quasiparticle spectrum. For most $\hat{\ell}$ textures the velocity of the superfluid

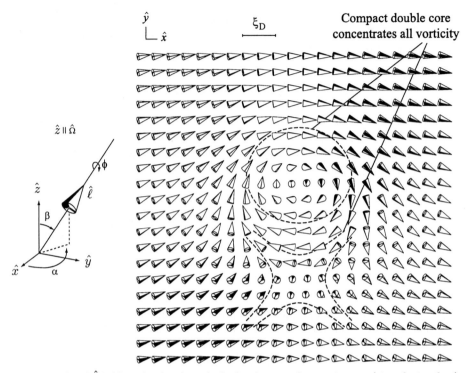

Fig. 2. Orbital $\hat{\ell}$ field in the dipole-unlocked soft core of a continuous (singularity-free) vortex. The figure shows the orientational distribution of $\hat{\ell}$ in the plane perpendicular to the vortex axis. It extends over a solid angle of 4π, which corresponds to 2 quanta of circulation. Equivalently, by following one full circle around the outer periphery of the figure it is seen that $\hat{\ell}$ rotates twice around its own axis, ie. the phase angle ϕ changes by 4π on circling the soft core once in the dipole-locked region far away, where $\hat{\ell}$ is uniformly oriented. This is the most common structure of quantized vorticity which evolves in ^{3}He-A in a magnetic field, when rotation is started in the superfluid state. We call it the double-quantum vortex or continuous unlocked vortex (CUV). (From Ref. [9])

fraction v_s can be nonzero, as seen from the experimental fact that their critical velocities of vortex formation have a finite, albeit small value. Conceptually the single most dramatic difference from the isotropic case is the feature that superflow does not need to remain curl-free in ^{3}He-A with gap nodes [10]. The superflow velocity is written as

$$\mathbf{v}_s = \frac{\hbar}{2m_3}(\nabla\phi - \cos\beta\,\nabla\alpha)\,, \tag{1}$$

where α, β, and ϕ are the local azimuthal, polar, and phase angles of $\hat{\ell}$. This means that the vorticity

$$\nabla\times\mathbf{v}_s = \frac{\hbar}{2m_3}\sin\beta\,(\nabla\beta\times\nabla\alpha) \tag{2}$$

becomes nonvanishing in those regions of the orbital texture where the $\hat{\ell}$ orientation is not contained within one single plane. On forming the circulation of $\mathbf{v}_s$ along a closed path, which encircles such a region with inhomogeneous $\hat{\ell}$ orientations,

$$\nu\kappa = \oint \mathbf{v}_s \cdot d\mathbf{r} = \frac{\hbar}{2m_3}\mathcal{S}(\hat{\ell}) , \tag{3}$$

one finds that the number (ν) of circulation quanta ($\kappa = h/(2m_3)$) is given by the solid angle $\mathcal{S}(\hat{\ell})$ over which the $\hat{\ell}$ orientations extend within the encircled region. In other words, the circulation is related to the topological charge of the $\hat{\ell}$ field.

In the soft core of the dipole-unlocked singularity-free vortex (Fig. 2), the orientational distribution of the $\hat{\ell}$ field covers a solid angle of 4π and $\nu = 2$. This configuration is an example of a *skyrmion*, which consists of two halves, a circular and a hyperbolic Mermin-Ho vortex, which also are known as *merons*. The circular half covers the 2π orientations in the positive half sphere and the hyperbolic those in the negative half.

The dipole coupling exerts an extra torque on spin precession and gives rise to frequency shifts in NMR. Experimentally a most valuable consequence is the fact that dipole-unlocked regions experience a frequency shift which is different from that of the locked bulk liquid and moreover a characteristic of the $\hat{\ell}$ texture within the soft core of the defect. Different structures of topological defects give rise to absorption in satellite peaks where both the frequency shift of the peak and its intensity are a characteristic of the defect structure [11]. This property provides a measuring tool which differentiates between defects and where the absorption intensity can be calibrated to give the number of defects (Fig. 3).

3   Double-Quantum Vortex Line

The generic rotating experiment consists of an acceleration – deceleration cycle in the rotation drive, as shown in Fig. 4. This experiment gives reproducible results in ³He superfluids where remanent trapped vortex filaments can be avoided in a container with smooth walls. In Fig. 4 the NMR spectrometer has been tuned to the frequency of the satellite from the double-quantum vortex line and its peak height is recorded as a function of the rotation velocity Ω. The rate of change $|\,d\Omega/dt\,| \sim 10^{-4}$ rad/s² is kept as slow as possible so that dynamic effects do not influence the result, but such that long-term drifts neither become important. During increasing Ω it is possible to discern from the measuring noise a staircase-like pattern in the satellite peak height when $\Omega \geq \Omega_c$. The periodicity in this signal as a function of Ω calibrates the circulation associated with one vortex line: $\Delta\Omega = \nu\kappa/(2\pi R^2)$. This provides the proof for $\nu = 2$, namely that the vortex line created in the externally applied flow is doubly quantized [13].

The intercept of the acceleration record with the horizontal axis, when $\Omega \geq \Omega_c$, determines the critical velocity v_c in Fig. 4. The fact that this section is linear proves that v_c remains constant during all of the acceleration. On plotting the corresponding counterflow velocity at the cylinder wall, $v(\Omega) =$

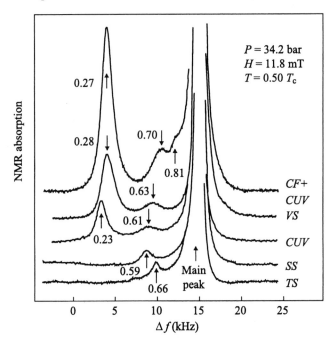

Fig. 3. NMR spectroscopy of topologically stable defects of the order-parameter field in ^{3}He-A. The measured NMR absorption is plotted as a function of the frequency shift $f - f_0$ from the Larmor value $f_0 = \gamma H_0$, where the resonance takes place in the normal phase. The large truncated peak on the right represents the resonance absorption of the dipole-locked bulk superfluid, with a relative frequency shift $R_\perp^2 = 1$. The different satellite peaks represent: **CUV** — double-quantum vortex lines (Fig. 2) in the equilibrium rotating state at $\Omega = 0.58$ rad/s, which in a cylinder of radius $R = 2.5$ mm corresponds to 150 vortex lines. At a temperature as low as $0.49\,T_c$, the satellite spectrum includes both a large primary and a small secondary peak. **VS** — equilibrium state of the vortex sheet in the same conditions. **SS** — soliton sheet with splay structure and with the sheet oriented vertically parallel to the cylinder axis ($\Omega = 0$). **TS** — soliton sheet with twist structure and oriented transverse to the cylinder axis ($\Omega = 0.11$ rad/s, which is below but close to Ω_c). **CUV + CF** — double-quantum vortes lines in a cluster surrounded by vortex-free counterflow close to the critical velocity threshold ($\Omega = 2.0$ rad/s). The two satellites with the normalized frequency shifts $R_\perp^2 = 0.27$ and 0.70 are the primary and secondary double-quantum vortex peaks while that with $R_\perp^2 = 0.81$ is caused by the counterflow. (In this measurement the temperature is higher ($T = 0.54\,T_c$) which explains the larger $R_\perp^2$ values of the vortex satellites than in the CUV spectrum.) The NMR field **H** is oriented parallel to the rotation axis. (From Ref. [11])

$\Omega R - \nu \kappa N/(2\pi R)$, it is seen that the noise in v_c can be explained to arise from experimental sources. Thus vortex formation proceeds here in the form of a regular periodic process and displays no measurable stochastic behaviour, which could be associated with nucleation across an energy barrier. Indeed, a simple argument, which is presented below, shows that the nucleation energy barrier is

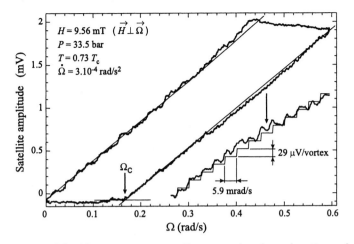

Fig. 4. Response of double-quantum vortex lines to a closed acceleration – deceleration cycle in the rotation drive Ω: The peak height of the primary NMR satellite plotted as a function of the externally applied rotation. Initially on increasing Ω from zero, vortex-free counterflow is created. This is known as the Landau state in superfluids and as the Meissner state (with complete flux expulsion) in superconductors. At Ω_c the critical counterflow velocity, $v_c = \Omega_c R$, is reached at the cylinder wall and the first vortex line is created. This brings about a reduction in the flow velocity at the wall by $\Delta v = \nu\kappa/(2\pi R)$. When Ω is increased further, the process is repeated periodically and the slanting section with the slope $dN/d\Omega = 2\pi R^2/(\nu\kappa)$ is recorded. This means that here the critical velocity remains constant at $v_c = \Omega_c R = \Omega R - \nu\kappa N/(2\pi R)$. At maximum amplification this section can be seen to mimic a periodic signal of staircase pattern, as shown in the insert. The upper branch of the acceleration – deceleration cycle is measured during decreasing Ω. Here the excess vortex-free counterflow is first reduced, until the cluster reaches the annihilation threshold, upon which vortex lines start to annihilate during further deceleration. At the annihilation threshold the cluster is separated from the cylinder wall by a counterflow annulus of minimum width: $d \approx \beta R/(2\sqrt{\Omega})$. This is comparable to the inter-vortex distance within the cluster: $d \approx [\nu\kappa/(2\pi\Omega)]^{1/2}$. In practice the number of vortex lines at the annihilation threshold [12] is equal to that in the equilibrium state: $N_0 = \pi R^2 n_v (1 - \beta/\sqrt{\Omega})$, where $\beta \approx 0.09$ for double-quantum vortex lines in the present experimental container. The true equilibrium state is obtained, in principle, by cooling through T_c in rotation at constant Ω. Annihilation in the upper branch can be seen to be a more random process, in which a larger number of lines may be removed approximately simultaneously from the outermost circle of vortex lines. (From Ref. [13])

so large compared to thermal energy that it cannot be overcome by any usual nucleation mechanisms. Instead, the applied counterflow has to be increased to the point where the barrier height goes to zero [6]. Thereby vortex formation becomes essentially an instability.

The reason for the high nucleation barrier at low applied flow velocities is the large length scale ξ_D on which the vortex has to be formed. The energy stored per unit length in the superflow around the vortex core is of order $\sim \rho_s v^2 \kappa^2$. This

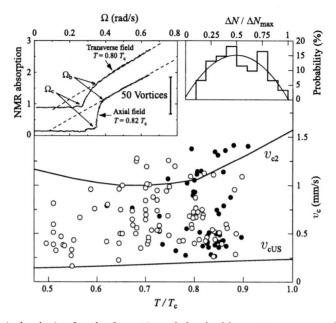

Fig. 5. Critical velocity for the formation of the double-quantum vortex. Vortex formation can proceed as a regular periodic process, as in Fig. 4, or it may start as a
burst-like event which then goes over into the periodic process, as shown in the top-left
insert. In the burst-like event a large number of vortex lines is formed simultaneously
at a bulk-liquid texture instability. Here a first order transition occurs in the order-
parameter texture and it is suddenly transformed to a new configuration in which the
critical velocity is generally lower than in the original texture. In the transformed texture the periodic process turns on during a further increase of Ω and the corresponding
Ω_c is obtained by extrapolating back to zero peak height of the vortex satellite. Vortex
formation at a texture instability is possible only at high temperatures ($T \gtrsim 0.7\,T_c$),
where the energy barriers separating different textures can still be overcome. The inset
on the top right shows as a histogramme the number of vortex lines ΔN which are
produced in the burst, normalized to the equilibrium number ΔN_{max} at the rotation
velocity Ω_b after the burst: $\Delta N_{max} = [2\pi R^2 \Omega_b/(\nu\kappa)]\,(1 - \beta/\sqrt{\Omega_b})$. The data for the
periodic process are marked as (o) while burst-like events are denoted as (•). Measuring
conditions: $\mathbf{H} \parallel \mathbf{\Omega}$, $H =9.9 - 15.8$ mT, $P =29.3 - 34.2$ bar. (From Ref. [4])

has to be compared to the kinetic energy of the applied superflow at the cylinder
wall in a volume comparable to that where the instability occurs, $\sim \rho_s v^2 \xi_D^2$. Thus
the flow velocity for creating the vortex has to be of order $v_c \sim \nu\kappa/(2\pi\xi_D)$. This
shows that the velocity for reaching the instability decreases with length scale
and is in ^{3}He-A comparable to the dipole velocity $v_D \propto 1/\xi_D$, ie. the velocity
required to break dipole locking. The barrier height, in contrast, increases $\propto \xi_D^2$.
In ^{4}He-II the appropriate length scale is the superfluid coherence length $\xi \sim 0.1$
nm, which is of atomic size, and gives a barrier height of order 1 K. In ^{3}He
superfluids the barrier is higher and the temperature lower, both by at least three
orders of magnitude. Therefore vortex-formation takes place at an instability.

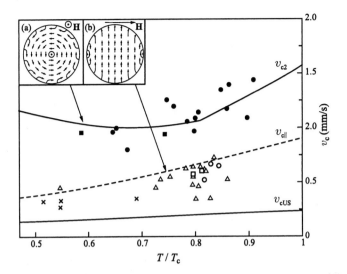

Fig. 6. Critical velocity of vortex formation for selected order-parameter textures. The curves represent the respective calculated bulk superfluid flow instability [14]. The absolute instability limit at v_{c2} applies to the originally homogeneous ($\hat{\ell} \parallel \mathbf{v}$) orbital texture in inset (a). Here $\hat{\ell}$ is confined to the transverse plane and is dipole-locked everywhere, when $\Omega < \Omega_c$, except within the surface layer and in the very center (where there is either a singular disclination line or a dipole-unlocked radial 2π Mermin-Ho vortex). This limit can be compared to measurements ($\bullet$, $\blacksquare$) in which the sample is cooled in rotation through T_c, to obtain the global equilibrium texture. The lowest v_c is measured ($\times$) when the NMR spectrum shows the signature from the transverse twist soliton (cf. Fig. 3). These data points can be compared to the calculated v_c of a dipole-unlocked soliton v_{cUS}. In transverse magnetic field, $\mathbf{H} \perp \mathbf{\Omega}$, the equilibrium global texture in inset (b) is also obtained by cooling through T_c in rotation and gives data points ($\circ$, $\square$) which should be compared to the calculated instability at $v_{c\parallel}$. This texture includes two singular disclination lines, located diametrically opposite each other at the cylinder wall (marked with a black dot on the horizontal diameter in inset (b)). The transverse field measurements seem to be less sensitive to the annealing requirement since other data ($\triangle$) with varied prehistories are not very different. (From Ref. [4])

As outlined above, one might think that the instability-determined critical velocity in ^{3}He-A is a well-defined quantity. It should depend only on the externally applied conditions, such as temperature (T) and pressure (P), which determine the ^{3}He-A properties ρ_s and ξ_D. In a smooth-walled container it should not depend on the wall properties, since in the A phase vortex formation has to occur essentially within the bulk liquid: At the cylinder wall $\hat{\ell}$ is oriented perpendicular within a surface layer of width comparable to ξ_D, owing to a rigid boundary condition. However, experimentally it is immediately concluded that the formation process has a lot of variability: It can have the regular appearance shown in Fig. 4, or it can take the burst-like form shown in Fig. 5 (inset at top left). Also v_c depends on the previous experimental history which the sample has been subjected to in the superfluid state [4].

All this variation in the characteristics of the critical velocity can be measured in one and the same smooth-walled sample container. This is quite different from ^{3}He-B, where vortex formation in the same cylinder proceeds as a rule as a reproducible and well-behaved regular process, similar to that in Fig. 4 [6]. In ^{3}He-B the critical velocity is not history dependent and typically at least an order of magnitude larger, as can be expected when a singular-core vortex has to be formed on the length scale of the superfluid coherence length $\xi(T, P) \sim 10$ – 100 nm.

The seemingly unruly behaviour of ^{3}He-A can be explained by the dependence of the instability velocity on the global order parameter texture, which is still a poorly controlled and understood feature of the experiments. In Fig. 5 various measurements on v_c have been collected, regardless of the earlier history of the samples. The data points, accumulated from three different sample cylinders, seem to fall between a maximum and a minimum limit. Note that there are no measurements with zero or very small v_c, proving that ^{3}He-A is a true superfluid. The global order parameter texture in the cylinder depends on the history of sample preparation. In Fig. 6 the measurements are grouped according to what type of global texture is expected on the basis of the procedure which was used to prepare the sample. This figure now provides some credibility to the notion that the global texture can be influenced by the sample preparation method, that the critical velocity indeed depends on the texture, and that theoretically calculated estimates [14] of the bulk-liquid flow instability for the different textures provide reasonable upper or lower bounds for the measured data. The measurements on v_c seem to provide the first experimental tool for characterizing the global texture, however indirectly. Also the independence of the observed features on the container exemplifies the fact that in the A phase vortex formation is a truly intrinsic bulk-liquid process, well separated from the container wall (unlike even ^{3}He-B).

The lowest critical velocities in Fig. 6 are recorded when a dipole-unlocked soliton (cf. Fig. 3) is present in the container and is oriented perpendicular to the rotation axis. The soliton is a planar wall of width $\sim \xi_D$ which separates bulk liquid in two different, but degenerate minima of the dipole energy, $\hat{d} \uparrow\uparrow \hat{\ell}$ and $\hat{d} \uparrow\downarrow \hat{\ell}$, while within the wall the dipole energy is not minimized. This explains the reduced critical velocity since in this case a dipole-unlocked region exists at the cylinder wall where the counterflow velocity is maximized and which can seed the formation of a dipole-unlocked vortex. The resulting structure, the intersection of a double-quantum vortex line with a transverse soliton sheet, is an example of a metastable $\hat{\ell}$ field with complicated knot-like continuous topology. This unstable configuration can be maintained in a long cylinder to moderately high rotation ($\Omega \lesssim 0.5$ rad/s) [15].

However, the most surprising case is that when a vertical dipole-unlocked soliton is present. Here the soliton sheet is oriented approximately parallel to the rotation axis and connected to the cylinder wall along two dipole-unlocked

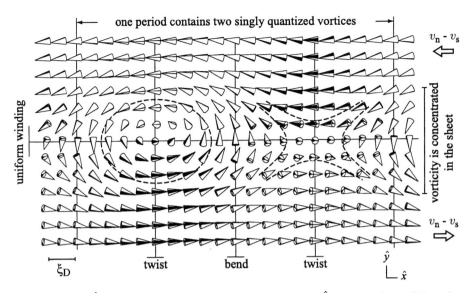

Fig. 7. Orbital $\hat{\ell}$ field of the vortex sheet. Outside the sheet $\hat{\ell}$ is oriented parallel to the sheet, but in opposite directions on the two sides of the sheet. The sheet itself consists of a periodic linear chain of alternating circular and hyperbolic Mermin-Ho vortices, the two constituents of the double-quantum vortex in Fig. 2. (From Ref. [9])

lines over the whole length of the container. As a function of Ω the critical velocity is then initially close to zero and increases in a nonlinear fashion. In rotation this situation leads to a new stable order-parameter structure which is known as the *vortex sheet*, a combined topologically stable object with planar structure into which linear quantized circulation has been integrated.

4   Vortex Sheet

In classical turbulence a vortex sheet is known as a thin interface across which the tangential component of the flow velocity is discontinuous and within which the vorticity approaches infinity [16]. In superfluids vortex sheets with quantized circulation, which separate two regions with irrotational flow, were briefly discussed in the late 1940's and early 1950's, to explain the observations from rotating experiments [17]. However, in an isotropic superfluid like ^{4}He-II, the vortex sheet was not found to be stable with respect to break up into isolated vortex lines. In quantum systems the vortex sheet was first experimentally identified in the anisotropic ^{3}He-A phase in 1994 [18]. The existence of vortex sheets in the form of domain walls which incorporate half-integer magnetic flux quantization is also discussed in unconventional superconductors with ^{3}He-A-like structure such as Sr_2RuO_4, UPt_3 or $U_{1-x}Th_xBe_{13}$ [19].

In ^{3}He-A the vortex sheet is formed from a soliton wall into which Mermin-Ho vortices with continuous structure have been incorporated. As shown in Fig. 7,

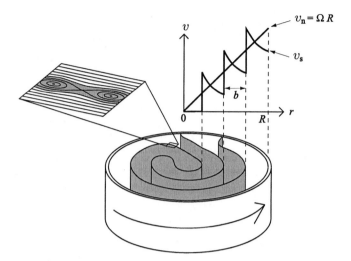

Fig. 8. Large-scale structure of the vortex sheet in its equilibrium configuration in the rotating container. The vortex sheet is formed when ^{3}He-A is accelerated to rotation in an applied magnetic field ($H > H_D$) and a soliton wall parallel to the rotation axis exists in the container. If the magnetic field is oriented along $\mathbf{\Omega}$, the equilibrium configuration of the folding is a double spiral. The graph on the top right illustrates the radial distributions of the normal and superfluid velocities. The latter is discontinuous across the sheet and differs from that of individual vortex lines in Fig. 1.

the sheet is made up of a alternating linear chain of singly quantized vortices with circular and hyperbolic $\hat{\ell}$ winding such that a fully periodic structure results. Like in the case of vortex lines, the vortex sheet is translationally invariant in the direction parallel to the rotation axis $\mathbf{\Omega}$. The large-scale structure in the equilibrium configuration is a continuously meandering foil which is attached at both ends at two vertical connection lines to the cylindrical wall (Fig. 8). It is these connection lines where circulation quanta are added to or removed from the sheet.

The hydrodynamic stability of the vortex sheet was calculated by Landau and Lifshitz [20]. By considering the equilibrium state with the kinetic energy from the flow between the folds and the surface tension σ from the soliton sheet, it is concluded that the distance between the parallel folds has to be $b = (3\sigma/\rho_{s\parallel})^{1/3} \, \Omega^{-2/3}$. This is somewhat larger than the inter-vortex distance in a cluster of vortex lines. The areal density of circulation quanta has approximately the solid-body value $n_v = 2\Omega/\kappa$. This means that the length of the vortex sheet per two circulation quanta is $p = \kappa/(b\Omega)$, which is the periodicity of the order-parameter structure in Fig. 7. The NMR absorption in the vortex-sheet satellite measures the total volume of the sheet which is proportional to $1/b \propto \Omega^{2/3}$. The nonlinear dependence of absorption on rotation is the experimental signature of the vortex sheet, in addition to its low value of critical velocity.

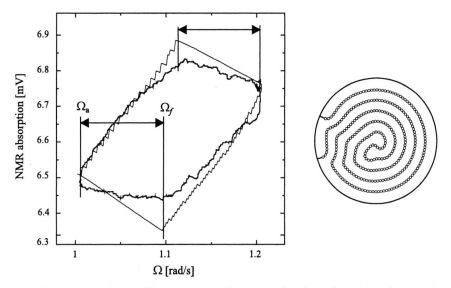

Fig. 9. Response of the equilibrium vortex sheet to a closed acceleration – deceleration cycle in the rotation drive. This measurement is used to define the critical velocity of the vortex sheet as a function of Ω. During increasing Ω the threshold, where more circulation is added, is denoted as Ω_f, while the threshold of annihilation, where circulation is removed, is Ω_a. The experimental result (thick line) is compared to a simulation calculation [21] (thin line) on the vortex-sheet configuration which is shown on the right at $\Omega_a = 1.0$ rad/s. In the vortex-sheet meander the circles denote the actual center positions of each circulation quantum. The measurement has been performed in a smooth-walled fused quartz cylinder ($2R = 3.9$ mm). The measuring conditions are the same as in Figs. 13 and 14.

The vortex sheet is formed whenever a vertical dipole-unlocked soliton sheet is present in the container and rotation is started. The reason for this is the low critical velocity at the dipole-unlocked connection lines between the sheet and the cylinder wall. This facilitates the creation of new circulation quanta, which is then immediately incorpotated in the structure of the sheet. The critical velocity of the vortex sheet is well below that of isolated vortex lines. Nevertheless, when a new circulation quantum is added to the sheet, it experiences repulsion from the circulation which already resides in the sheet close to the connection line. Owing to this small Ω-dependent barrier, the critical velocity becomes Ω-dependent. Experimentally the critical velocity can be defined from Fig. 9 as $v_c^* = (\Omega_f - \Omega_a)R$, the separation in rotation velocities between the thresholds where new circulation is formed at (Ω_f) and existing circulation is annihilated at (Ω_a).

The measured critical velocity in Fig. 10 follows qualitatively the dependence

$v_c^*(\Omega) \propto \sqrt{\Omega}$. This approximate relation illustrates the characteristics of the vortex sheet: The Magnus force $F_M = \rho_s \kappa v$ from the vortex-free counterflow, with the velocity $v = (\Omega - \Omega_a)R \approx 2d\Omega$ at the cylinder wall, attempts to pull

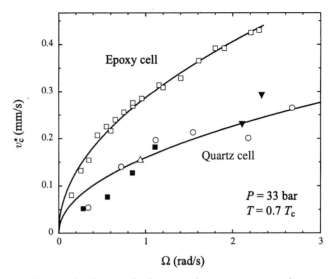

Fig. 10. Measured critical velocity v_c^* of vortex-sheet expansion, when new circulation is added as a function of the applied rotation Ω. In contrast to the regular periodic process in vortex-line formation, where v_c is a constant as a function of Ω, here $v_c^*(\Omega) \propto \sqrt{\Omega}$. The measurements represent the situation when a vertical soliton wall is already present in the cylinder and the vortex sheet has been adiabatically grown in an axially oriented magnetic field to its equilibrium configuration. The solid curves are fits with $\sqrt{\Omega}$ dependence. The rougher epoxy wall has a larger magnitude of critical velocity, perhaps because of pinning of the connection lines which resists smooth readjustment in the folding of the vortex sheet as a function of Ω. The different symbols of data points (quartz cell) illustrate the reproducibility of the results from one adiabatically grown vortex sheet to another.

additional circulation into the sheet. It is opposed by the repulsion from the circulation, which already resides in the sheet closest to the connection line, at a distance equal to the width d of the circulation-free annulus. The balance between the repulsion $F_r = (\rho_s \kappa^2/d) \ln [d/(\alpha \xi_D)]$ and F_M gives a critical counterflow velocity of order $v_c \sim \sqrt{2\kappa\Omega}$.

Actually, because of the presence of the two connection lines, a vortex-sheet state does not have full axial symmetry in the distribution of its vortex-free counterflow even in the double-spiral configuration of Fig. 9 (ie. at the connection line the distance of the first circulation quantum from the wall is different from that in the outermost circular spiral part). This causes a small difference in the measured $v_c^* = (\Omega_f - \Omega_a)R$ from the real value of v_c at the connection lines. The two can be connected if the sheet configuration is known. A qualitative difference between these two critical velocities is that the true value of v_c does not approach zero when $\Omega \to 0$, but remains finite. This is expected since some counterflow velocity is required when new circulation is formed even at a dipole unlocked connection line of a circulation-free soliton sheet, since also the attractive interaction of the emerging vortex line with its image forces within the wall have to be overcome.

5   Dynamic Response

So far we have looked at the response of the ^{3}He-A order-parameter field to what are essentially adiabatic changes in the rotation velocity. But what happens if the rotation is changing rapidly with time? Because of its lower critical velocity, it is found that the vortex sheet becomes the preferred structure in rapidly changing rotation, rather than individual double-quantum vortex lines. Consequently in ^{3}He-A, the response to high-frequency perturbations in the rotation drive is explained by the dynamic properties of the vortex sheet. The central feature becomes the interplay between the large-scale configuration of the sheet and its confined circulation. The first illustration of these considerations is provided by the measurements in Fig. 11.

In the experiment of Fig. 11 the complete NMR absorption spectrum is slowly recorded while the rotation is harmonically swinging back and forth around zero with a short period of 50 s. At low amplitude of back-and-forth rotation the circulation enters in the form of double-quantum vortex lines (upper spectrum). The lines are periodically formed and then annihilated during both the positive and negative half cycles of rotation. During the positive and negative half cycles the vorticity has opposite signs of circulation: All vortex lines are annihilated approximately by the time when rotation goes through zero. Therefore the envelope of the vortex-line satellite is modulated at twice the frequency of the rotation drive. To record the spectrum the rate of the NMR field sweep has to be much slower than the period of the rotation drive: The horizontal scale, which here is plotted in terms of the linearly changing NMR field sweep, is actually the common time axis for all variables.

A rarely seen feature of the upper spectrum is the modulation in the absorption on the high-field flank of the large bulk-liquid peak. This absorption component is created by the slight dipole unlocking in the vicinity of the surface layer on the cylinder wall when the counterflow velocity gets sufficiently large. The modulation of this absorption is approximately in anti-phase with that in the vortex-line satellite, ie. minimum absorption in the vortex-satellite is reached approximately at the same time when absorption in the counterflow signal is at maximum. Since the total area in the absorption spectrum must be constant as a function of time, this means that absorption is transferred periodically between the different parts of the spectrum.

The lower spectrum in Fig. 11 is recorded with a larger amplitude of back-and-forth rotation. In this case the circulation enters in the form of a vortex sheet, which now provides a more stable response than double-quantum vortex lines. (During the first few half cycles the response may be in the form of lines, but it soon goes spontaneously over into the vortex sheet.) The modulation of the absorption in the vortex-sheet satellite is similar to that of the double-quantum vortex peak. The major change is the absence of the counterflow signal, since

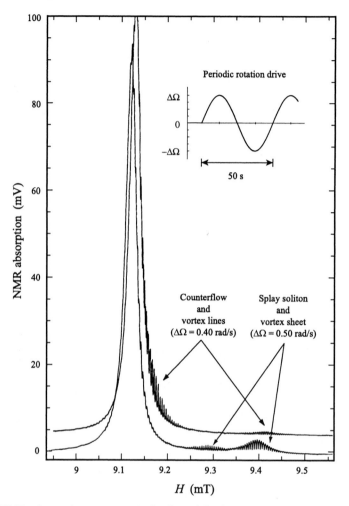

Fig. 11. NMR absorption spectra in back-and-forth rotation. The measurement is performed by sweeping slowly the polarizing magnetic field H at constant rf excitation frequency ($f = 310$ kHz) across the NMR region while the sample is in sinusoidal back-and-forth rotation: $\Omega(t) = \Delta\Omega \sin\omega t$. The field sweep is much slower than the period of the rotation drive ($2\pi/\omega = 50$ s). Thus the rotation appears as a modulation envelope on the field-dependent NMR absorption. 1) When the amplitude $\Delta\Omega$ of the sinusoidal rotation is sufficiently small ($v_D/R \lesssim \Delta\Omega = 0.40$ rad/s) the circulation enters in the form of double-quantum vortex lines (*upper spectrum*). Here the absorption is transferred periodically between the vortex and counterflow satellites (spectrum **CUV + CF** in Fig. 3). 2) When the rotation amplitude is increased ($\Delta\Omega = 0.50$ rad/s) the circulation goes into the vortex sheet (*lower spectrum*). Now the absorption is shifted between the vortex-sheet and soliton satellites (spectra labeled as **VS** and **SS** in Fig. 3). (Measuring conditions: $2R = 3.87$ mm, $\mathbf{H} \parallel \mathbf{\Omega}$, $P = 34$ bar, $T = 0.7\,T_c$)

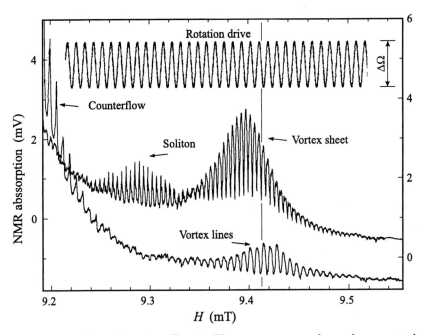

Fig. 12. Satellite absorptions from Fig. 11. The topmost trace shows the correct phase of the rotation drive with respect to that of the NMR absorptions in the different satellites. The horizontal axis is actually time t since all variables, the NMR field sweep $H(t)$, rotation $\Omega(t)$, and the two absorption signals, have been plotted on a common time axis.

now the critical velocity is lower and the counterflow velocity is limited below a smaller value. The second major difference is the presence of the absorption in the splay soliton satellite. This absorption component appears when the circulation is absent around $\Omega \approx 0$ and the vortex-sheet satellite is at minimum. Thus in the lower spectrum the absorption is transferred between the vortex-sheet and soliton satellites and their modulation signals are in anti-phase.

The high-field tails of the spectra are repeated in larger scale in Fig. 12, to display the relative phases of the different variables. A comparison of these satellite signals demonstrates concretely the differences between the responses in terms of linear and planar states of vorticity. The vortex sheet satellite is larger in amplitude because the critical velocity is lower and the amplitude $\Delta\Omega$ of the rotation drive is larger; therefore the peak represents more circulation quanta than that of the double-quantum vortex lines. Also the vortex sheet absorption tracks more closely the phase of the rotation drive (topmost signal), again because of the smaller critical velocity. This is also the explanation for the pointed and narrow minima in the modulation envelope of the vortex-sheet satellite and the corresponding maxima in the soliton satellite. The good reproducibility of the modulated soliton absorption shows that the soliton is metastable in the container at $\Omega = 0$. Actually, in this measurement the configuration of the vortex

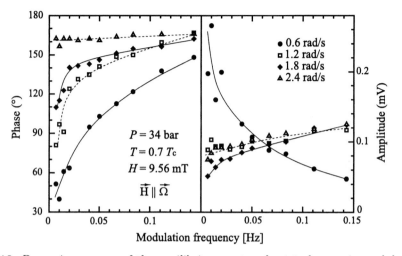

Fig. 13. Dynamic response of the equilibrium vortex sheet to harmonic modulation of the rotation drive: $\Omega(t) = \Omega_0 + \Delta\Omega \sin \omega t$. The left panel shows the phase of the modulated absorption component in the peak height of the vortex-sheet satellite, relative to the rotation drive (in degrees). The right panel gives the amplitude of the modulated absorption (peak to peak), referred to in mV at the output of the cryogenic preamplifier operated at LHe temperature [6]. The solid curves are guides to the eye. The measuring conditions are the same as in Fig. 11 with $\Delta\Omega = 0.050$ rad/s. (From Ref. [24])

sheet is not that of the equilibrium double spiral but one where multiple pieces of sheet exist which all are attached separately along two contact lines to the vertical cylinder wall [5].

A straightforward continuation of the measurements in Fig. 12 is to monitor the satellite peak height as a function of the modulation frequency, to find the dynamic response to changes in the rotation drive. In the case of double-quantum vortex lines, the satellite intensity (or peak height at constant line width) monitors in first approximation the number of lines. In this situation only the fast radial motion of vortex lines can be measured. This is performed by stopping rotation very abruptly and by measuring the time dependence of vortex-line annihilation. The tail of this signal is given by $N(t) = N(0)(1 + t/\tau_F)^{-1}$, where $N(0)$ is the number of vortex lines in the container at the moment $t = 0$, when rotation has come to a stop ($\Omega(t) = 0$). Such measurements give the characteristic decay time $\tau_F = (\nu\kappa n_v(0)\, \rho_n B/\rho)^{-1}$, where $\nu\kappa n_v(0)$ is the density of circulation quanta at $t = 0$ and $\rho_n B/\rho$ is the dissipative mutual friction. This mode of vortex-line motion has been monitored in the ^{3}He superfluids for double-quantum vortices in A phase [22] and for singular-core vortices in B phase [7]. Both measurements are consistent with other determinations of mutual friction [23]. Other resonance techniques can be used to record also the slow approach of vortex lines to equilibrium at constant Ω, after an initial disturbance has been switched off. This motion is predominantly azimuthal in character for an

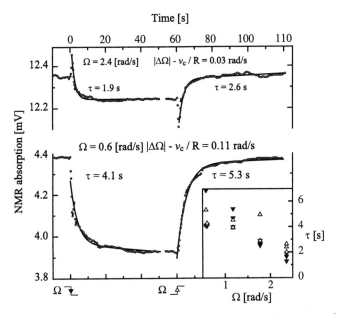

Fig. 14. Response of the equilibrium vortex sheet to a step change in the rotation drive. On the left Ω is reduced by $\Delta\Omega = 0.15$ rad/s and on the right it is increased back to the original value. The solid curves are exponential fits. The corresponding time constants are plotted in the inset, with ($\blacktriangledown$) for downward and ($\triangle$) for upward step changes. (From Ref. [24])

isolated vortex cluster and in the ^{3}He superfluids it is over damped, ie. has pure exponential time dependence [7].

In the vortex sheet the circulation is topologically confined within the sheet where it can move in two ways: either along the sheet or by forcing the entire sheet to contract or expand. Both types of motion proceed only via changes in the length and folding of the sheet. A slow final readjustment in the configuration of the sheet follows these faster initial responses, if the texture is left to anneal at constant Ω. The central feature is thus the interplay between the large-scale configuration of the sheet and its confined circulation. This aspect is illustrated by the measurements in Figs. 13 and 14.

The satellite intensity of the vortex sheet is proportional to the total volume of the sheet. The circulation is distributed as a continuous periodic chain along the sheet where the total number of quanta N is given by the length L of the meander in the transverse plane: $N = L/(p/2) = 2bL\Omega/\kappa$. Thus the satellite signal has to be proportional to

$$L = \frac{1}{2}\kappa N \left(\frac{\rho_{s\parallel}}{3\sigma\Omega}\right)^{1/3}, \tag{4}$$

which means that it displays a dependence on Ω even at constant N, as evident in Fig. 9. Therefore, also the readjustment to small disturbances in Ω at constant circulation can be measured for the vortex sheet with this technique [24].

The response of the peak height of the vortex-sheet satellite is shown in Fig. 13 for harmonic modulation of the rotation drive and in Fig. 14 for a step change. In both cases the change by $\Delta\Omega$ about the average rotation Ω_0 is kept as small as possible, to maintain N constant. However, because of finite measuring resolution the condition $2\Delta\Omega < \Omega_f - \Omega_a$ (cf. Fig. 9) is valid only at large Ω_0 in the measurement of the harmonic response. In the step-response measurement it is not satisfied even at the highest Ω_0 value.

Let us first look at the characteristics in the limit of large Ω_0. The left panel in Fig. 13 shows that the phase shift between the response and the drive is then approaching 180°. The right panel shows that this response occurs at small amplitude. The out-of-phase behaviour at constant N is expected from Eq. (4): To maintain solid-body-rotation, the vortex sheet contracts during increasing rotation and expands during decreasing Ω.

This out-of-phase signal at constant N identifies the origin of the sharp out-of-phase spike in the step response at large Ω_0 (upper trace in Fig. 14). The spike is then followed by a slow exponential in-phase response which has to be associated with the change in N. The amplitude of this component is small at large Ω_0 (upper trace) but grows much larger at low Ω_0 where the change in N becomes the dominant effect. The time constant of the exponential signal is plotted as a function of Ω_0 in the inset at bottom right in Fig. 14. In the harmonic response this means that with decreasing Ω_0 the phase shift is reduced and the amplitude grows.

The time sequence of different processes in the dynamic response of the equilibrium vortex sheet with double-spiral configuration is thus the following: (i) The fastest response with a time constant of order 1 s occurs in the density of the circulation quanta. In the axial field with circularly spiraling folding this can happen only perpendicular to the sheet, ie. the sheet either contracts or expands at constant N. (ii) The adjustment of the number of circulation quanta to the equilibrium value occurs slower on a time scale of order 10 s. This requires the motion of the circulation along the sheet in unison and the corresponding readjustment in the length of the sheet. (iii) The slowest component is the annealing of the vortex sheet at constant Ω which takes place on a time scale of minutes. In the present range of high and intermediate temperatures all motions are exponentially over damped – a generic property of ^{3}He superfluids where the kinematic viscosity of the normal component is so large that it can be considered to be clamped to corotation with the container.

More recent measurements demonstrate that the response of ^{3}He-A to a rotation drive, which includes components at high frequency and large amplitude, is dominated by the vortex sheet and its properties. Again the basic features are here derived from structural considerations: The large-scale configuration of the vortex sheet is altered substantially from the equilibrium state such that the dynamic response becomes greatly enhanced [5]. This leads to considerable gains in the kinetic energy under rapidly changing flow conditions which readily compensate the small increases in textural energies.

6   Summary and Future Work

The hydrodynamics of ^{3}He-A will undoubtedly be a source for more surprises. It is well known that with heat flow superfluid – normal fluid thermal counter currents can be produced which drive dissipative time-dependent $\hat{\ell}$ textures [25]. In rotational flow a large number of different structures of quantized vorticity have been identified. These make it possible for the superfluid to mimic optimally solid-body rotation, depending on temperature, magnetic field, the oder-parameter texture, and the properties of the rotation drive.

NMR has been the most efficient method for distinguishing between different vortex textures, by providing the possibility to probe the order-parameter texture from outside the rotating container. The method is well suited for the measurement of a vortex cluster which is translationally invariant in the direction parallel to the rotation axis. However, both the vortex texture and the NMR spectrum change if the orientation of the polarization field is rotated. Therefore the NMR method is less suited for precise measurements of vortex states which are less regular in configuration than rotating vortex clusters. An exception are random tangles of vorticity where the *average density* of dipole-unlocked vorticity can be continuously monitored.

Uniform rotation is not generally suited for the study of turbulence. Sufficiently long-lived turbulent states, which could efficiently be recorded with present continuous-wave NMR techniques, have not been produced by rotating ^{3}He-A at temperatures above $0.5\,T_c$. At lower temperatures the normal fluid fraction is rapidly depleted and the hydrodynamic response time of the superfluid fraction increases, owing to the reduced coupling with the walls of the rotating container. If the A phase can be maintained in metastable state in low magnetic fields to these low temperatures [26], then it might become possible to generate transient, but long-lived turbulent states by rotation techniques. For such studies present measuring techniques would be adequate. We might thus expect to see new development in the hydrodynamics of ^{3}He-A which would illuminate the properties of an anisotropic superfluid in the zero temperature limit.

Acknowledgements

This work was funded in part by the EU – Improving Human Potential – Access to Research Infrastructures programme under contract EC HPRI-CT-1999-50.

References

1. V.B. Eltsov, M. Krusius, G.E. Volovik in: *Proc. Low Temp. Phys.*, Vol. XV, ed. W.P. Halperin (Elsevier Sciece B.V. Amsterdam, 2001); *preprint* (http://xxx.lanl.gov/abs/cond-mat/9809125)
2. S.N. Fisher, A.J. Hale, A.M. Guénault, G.R. Pickett: Phys. Rev. Lett. **86**, 244 (2001)
3. H.M. Bozler in: *Helium three*, eds. W.P. Halperin, L.P. Pitaevskii (Elsevier Sciece B.V. Amsterdam, 1990)

4. V.M. Ruutu, J. Kopu, M. Krusius, Ü. Parts, B. Plaçais, E.V. Thuneberg, W. Xu: Phys. Rev. Lett. **79**, 5058 (1997)

5. V.B. Eltsov, R. Blaauwgeers, N.B. Kopnin, M. Krusius, J.J. Ruohio, R. Schanen, E.V. Thuneberg: to be published

6. Ü. Parts, V.M. Ruutu, J.H. Koivuniemi, Yu.M. Bunkov, V.V. Dmitriev, M. Fogelström, M. Huebner, Y. Kondo, N.B. Kopnin, J.S. Korhonen, M. Krusius, O.V. Lounasmaa, P.I. Soininen, G.E. Volovik: Europhys. Lett. **31**, 449 (1995); V.M. Ruutu, Ü. Parts, J.H. Koivuniemi, N.B. Kopnin, M. Krusius, J. Low Temp. Phys. **107**, 93 (1997)

7. E.B. Sonin, Y. Kondo, J.S. Korhonen, M. Krusius: Europhys. Lett. **22**, 125 (1993); Phys. Rev. B **47**, 15113 (1993-II)

8. D. Vollhardt, P. Wölfle: *The superfluid phases of helium 3* (Taylor & Francis, London, 1990)

9. V.B. Eltsov, M. Krusius in: *Topological defects in ^{3}He superfluids and the non-equilibrium dynamics of symmetry-breaking phase transitions*, eds. Yu. Bunkov and H. Godfrin (Kluwer Academic Publ., Dordrecht, 2000), p. 325

10. N.D. Mermin, T.L. Ho: Phys. Rev. Lett. **36**, 594 (1976); N.D. Mermin: Phys. Today **34**, 46 (1981)

11. V.M. Ruutu, Ü. Parts, M. Krusius: J. Low Temp. Phys. **103**, 331 (1996)

12. V.M. Ruutu, J.J. Ruohio, M. Krusius, B. Plaçais, E.B. Sonin, Wen Xu: Phys. Rev. B **56**, 14089 (1997-I); Physica B **255**, 27 (1998)

13. R. Blaauwgeers, V.B. Eltsov, M. Krusius, J.J. Ruohio, R. Schanen, G.E. Volovik: Nature **404**, 471 (2000)

14. R. Kopu, R. Hänninen, E.V. Thuneberg: Phys. Rev. B **62**, 1 Nov (2000)

15. V.M.H. Ruutu, Ü. Parts, J.H. Koivuniemi, M. Krusius, E.V. Thuneberg, G.E. Volovik: Pis'ma v ZhETF **60**, 659 (1994) [JETP. Lett. **60**, (1994)

16. P.G. Saffman: *Vortex dynamics* (Cambridge Univ. Press, Cambridge UK, 1992)

17. For the early history on quantized vorticity in ^{4}He-II see R.J. Donnelly: *Quantized vortices in Helium II* (Cambridge Univ. Press, Cambridge UK, 1991)

18. Ü. Parts, E.V. Thuneberg, G.E. Volovik, J.H. Koivuniemi, V.H. Ruutu, M. Heinilä, J.M. Karimäki, M. Krusius: Phys. Rev. Lett **72**, 3839 (1994); E.V. Thuneberg: Physica B **210**, 287 (1995); M.T. Heinilä, G.E. Volovik: Physica B **210**, 300 (1995); Ü. Parts, V.M. Ruutu, J.H. Koivuniemi, M. Krusius, E.V. Thuneberg, G.E. Volovik: Physica B **210**, 311 (1995)

19. M. Sigrist: Physica B, **280**, 154 (2000) and references therein; T. Kita: Phys. Rev. Lett. **83**, 1846 (1999)

20. L. Landau, E. Lifshitz: Dokl. Akad. Nauk. **100**, 669 (1955)

21. V.B. Eltsov: J. Low Temp. Phys. **121**, 387 (2000)

22. P.J. Hakonen, V.P. Mineev: J. Low Temp. Phys. **67**, 313 (1987)

23. T.D.C. Bevan, A.J. Manninen, J.B. Cook, H. Alles, J.R. Hook, H.E. Hall: J. Low Temp. Phys. **109**, 423 (1997); Phys. Rev. Lett. **77**, 5086 (1996); *ibid.* **74**, 750 (1996)

24. V.B. Eltsov, R. Blaauwgeers, M. Krusius, J.J. Ruohio, R. Schanen: Physica B **284–288**, 252 (2000)

25. D.N. Paulson, M. Krusius, J.C. Wheatley: Phys. Rev. Lett. **37**, 599 (1976); H.E. Hall, J.R. Hook: The hydrodynamics of superfluid ^{3}He, in: *Prog. Low Temp. Phys.*, Vol. IX, ed. D.F. Brewer (Elsevier Sciece B.V. Amsterdam, 1985), p. 143

26. P. Schiffer, M.T. O'Keefe, M.D. Hildreth, H. Fukuyama, D.D. Osheroff: Phys. Rev. Lett. **69**, 120 (1992); *ibid.* **69**, 3096 (1992); *Prog. Low Temp. Phys.*, Vol. XIV, ed. W.P. Halperin (Elsevier Sciece B.V. Amsterdam, 1995), p. 159

Vortices in Metastable ^{4}He Films

Ralf Blossey

Universität Essen, Fachbereich Physik, D-45117 Essen, Germany

1 Wetting Properties of ^{4}He on Weak-Binding Alkali Metals

1.1 Wetting Transitions of Liquid Helium

Liquid helium wets almost all surfaces. This behaviour results from its small dielectric permittivity, $\varepsilon = 1.057$, which leads to attractive forces between the helium atoms weaker than those between helium and other atoms [1]. In 1991 Cheng et al., however, predicted that ^{4}He does *not* wet certain alkali metals. Instead, for some of these metallic substrates, a first-order wetting transition was foreseen [2]. In the alkali metals, weakly bound outer electrons cause a short-range repulsion for the helium atoms. The resulting temperature-dependent balance between repulsive short-range and attractive long-range forces is at the origin of the wetting phenomenon in liquid helium.

Shortly after the theoretical prediction the non-wetting behaviour of ^{4}He on Cs at ultralow temperatures was experimentally confirmed for ^{4}He/Cs ([3], for a review [4]). Both the wetting transition at $T_w \approx 1.9$ K, and first-order prewetting transitions were observed, accompanied by strong hysteresis effects upon undercooling [5] (termed 'anomalous nucleation', [6]). Subsequently, the wetting behavior of ^{4}He on Rb has been investigated ([7], [8] and references therein). The findings are still somewhat controversial; e.g., it is not entirely clear whether there is a wetting phase transition of ^{4}He on (nominally pure) Rb at coexistence; if it exists, the wetting temperature is very low ($T_w \leq 300$ mK). Further, the superfluid transition in the ^{4}He-film on Rb appears to deviate in several respects from the standard Kosterlitz-Thouless transition scenario [8].

Figure 1 shows a schematic phase diagram of the system ^{4}He/Cs in temperature T and chemical potential difference $\Delta\mu$. The first-order wetting transition occurs at coexistence of bulk liquid and gas phases, $\Delta\mu = 0$, at $T = T_w$. At $T = T_\lambda$ occurs the superfluid transition. The off-coexistence continuation of the first-order transition is the prewetting line, $\Delta\mu_p(T)$, ending in the prewetting critical point at $\Delta\mu_p(T_p)$. The superfluid transition in a film of finite thickness occurs at the Kosterlitz-Thouless line $\Delta\mu_\lambda(T)$.

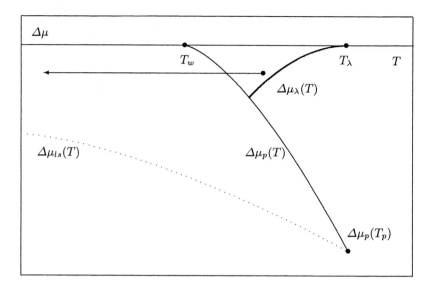

Fig. 1. Equilibrium wetting phase diagram of ^{4}He/Cs in the parameters temperature T and chemical potential difference $\Delta\mu$ between the bulk liquid and gas phase. Further: lower surface spinodal $\Delta\mu_{ls}$ (dotted); quench line (arrow) (see Secs.2/3)

1.2   Interface Model Description of Wetting Transitions

A convenient description of both equilibrium and non-equilibrium aspects of wetting transitions is based on the effective interface Hamiltonian [9]

$$\mathcal{H}_0[f] = \int d^2x \left[\gamma\sqrt{g} + V(f) - \Delta\mu f\right] \tag{1}$$

where $f(x)$ is the local height above the (planar) substrate $x \in \mathbf{R}^2$. The first term $\gamma\sqrt{g} = \gamma(1 + (\nabla f)^2)^{1/2}$ is the capillary contribution of the liquid-vapor interface with surface tension $\gamma_{LV} = \gamma$. Typically, a squared-gradient approximation to $\sqrt{g} \approx 1 + (1/2)(\nabla f)^2 + \dots$ suffices. The second term in (1) is an effective interface potential $V(f)$ resulting from effective, i.e. temperature-dependent, interactions between the liquid-vapor interface and the solid. For the case of a first-order wetting transition between a microscopic film of thickness $f = f_0$ and a macroscopic film with $f = \infty$, $V(f)$ generally has a strongly asymmetric double-well structure, as shown in Figure 2. For long-ranged molecular interactions the asymptotic decay of $V(f)$ for $f \to \infty$ is given by $V(f) = A/f^2$ where A is the Hamaker constant, determined by the dielectric properties of the three media involved (solid, liquid, gas) [1]. Finally, the last term in eq.(1) is the chemical potential contribution with $\Delta\mu \equiv \mu - \mu_c$, where μ_c is the chemical potential at coexistence of the bulk gas and liquid phases.

The average thickness $\bar{f}$ of a wetting layer on a wall is a convenient order parameter for a discussion of equilibrium wetting states and the transitions between

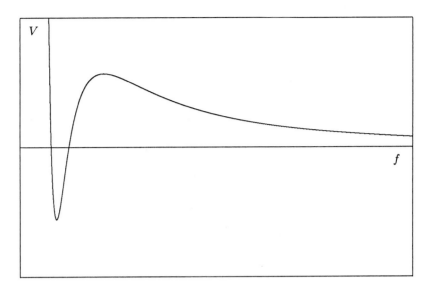

Fig. 2. Effective interface potential $V(f)$ for a first-order wetting transition

them. It fulfills the equation of state obtained from the variation $\delta\mathcal{H}/\delta f|_{f=\bar{f}} = 0$, $V'(f)|_{f=\bar{f}} = \Delta\mu$, where $V' \equiv dV/df$. For homogeneous film states, the discussion of the minima of V allows the construction of the wetting part of the phase diagram of Figure 1.

2   Lifetime of an Undercooled Film

A metastable thick wetting layer of ⁴He on Cs can be created by undercooling a film from above the prewetting line into the nucleation region located between $\mu_p(T)$ and a lower spinodal line $\mu_{ls}(T)$, whereby the location of the lower spinodal is determined by the conditions $V'(f) = \Delta\mu$ and $V''(f) = 0$, where $f = \bar{f}$ is the thickness of the undercooled thick film. The undercooled layer resides on the substrate until, by thermal fluctuations, a hole of critical size (or excess free energy) has formed. The shape of the critical nucleus is determined by the variational equation to (1), $\delta\mathcal{H}_0[f]/\delta f = 0$, assuming cylindrical symmetry, i.e. $f = f(r)$,

$$\gamma\left(f''(r) + \frac{1}{r}f'(r)\right) = V'(f) - \Delta\mu \tag{2}$$

where the prime denotes differentiation with respect to the argument. The term on the left-hand side of the equation corresponds to the curvature-contribution of the hole profile. A critical hole obeys the boundary conditions

$$f(\infty) = \bar{f} \equiv F \;, \;\; f'(0) = 0\,. \tag{3}$$

The solution to eqs.(2) and (3) determines the excess free energy of the critical hole via

$$\mathcal{E}_c(T, \Delta\mu) \equiv \mathcal{H}_0[f(r)] - \mathcal{H}_0[F].\qquad(4)$$

Within a Fokker-Planck description of the nucleation process [10] the excess free energy determines the probability for the formation of a critical hole in a wetting film via the Arrhenius-type factor

$$\Gamma \propto \exp -[\mathcal{E}_c/(k_B T)]\qquad(5)$$

where k_B is Boltzmann's constant, and T is the temperature. The lifetime τ of the film is then given by $\tau \sim \Gamma^{-1}$.

The experimentally observed anomalously long lifetime of an undercooled helium film can be explained by the properties of critical holes at deep temperature quenches close to coexistence. When $\Delta\mu \to -0$ for $T \ll T_w$, F and consequently the critical depth $F_c \equiv F - f_0$ diverges. In this regime the critical hole profile at distances far from the wall, $f \gg f_0$, is determined by the asymptotic behaviour of $V(f)$ for $f \to \infty$. The critical hole then has a funnel-like shape with a finite inner radius R_c [11]. With $\Delta\mu \to -0$ one has $R_c = const.$, while the central depth F_c of the funnel-shaped hole diverges according to

$$F_c \sim |\Delta\mu|^{-\frac{1}{3}},\qquad(6)$$

and the excess free energy scales as

$$\mathcal{E}_c \sim F_c^2 \sim F^2.\qquad(7)$$

If the quench is performed at a finite chemical potential difference and approaches the lower spinodal, the excess free energy of the critical hole vanishes according to [12]

$$\mathcal{E}_c \sim T - T_{ls}.\qquad(8)$$

3   Application to ^{4}He/Cs

The exponential dependence of the nucleation rate Γ on $\mathcal{E}_c$ stands in the way of an experimental verification of the scaling results: the observation times $t > \tau$ are very long close to coexistence $\Delta\mu \to -0$, while upon the approach to the spinodal, hole nucleation becomes an exponentially fast process. Recently, it has been proposed that for a superfluid helium film a controlled deformation of the liquid-vapor interface ('surface dimple') can be a suitable candidate for a reduction of the nucleation barrier [13]. The presence of a surface deformation affects the nucleation kinetics: the nucleation of a hole can now occur from an already existing 'hole'-like excitation of the interface. This can lead to a reduction of the excess free energy barrier of the critical nucleus so that, to a first approximation, the excess free energy is modified by replacing $\mathcal{E}_c$ by $\mathcal{E}_c - \mathcal{E}_d$, where $\mathcal{E}_d$ denotes the excess surface free energy of the surface dimple.

For concreteness, we compare two temperature quenches from high to low temperatures. Consider $\Delta\mu < 0$ fixed. Quench 1 starts at a temperature slightly above the KT-line, i.e. from a normal ^{4}He film. Quench 2 starts from a temperature slightly below the KT-line, hence from a superfluid film. Both quenches end at the same temperature $T_f \ll T_w$. Since both initial states can be very close to each other, the probability for homogeneous nucleation of a critical hole differs only little.

When the hole is nucleated from a superfluid film, however, the excess free energy required for the formation of a critical hole will be different from that of a normal fluid due to the presence of vortices with the associated Bernoulli pressure [14]

$$P(r) = \begin{cases} -R_0/r^2 & \text{for } r > \xi \\ -(R_0/\xi^2)(2-(r/\xi)^2) & \text{for } r < \xi \end{cases} \tag{9}$$

where $R_0 = \varrho_s \kappa_c^2/8\pi^2$ with the mass density ϱ_s of the superfluid and κ_c as the circulation quantum. The Bernoulli pressure modifies the effective interface Hamiltonian $\mathcal{H}[f] = \mathcal{H}_0[f] + 2\pi \int dr\, r P(r) f$. By comparison of terms with $\mathcal{H}_0$ we see that the Bernoulli pressure effectively acts as a local change in chemical potential.

We can now determine the shape of the vortex-caused surface deformation in the same manner as that of the critical hole. We first focus on a 'thick' but finite film. The interface potential $V(f) - \Delta\mu f$ for the thick film can be approximated by a parabola centered at the equilibrium thickness $f = F$. Using F as a reference level by introducing $\zeta \equiv f - F$, the surface deformation of the interface is determined by the harmonic functional

$$\mathcal{H}[\zeta] = 2\pi \int_0^\infty dr\, r[(\gamma/2)(\nabla\zeta)^2 + \phi(\zeta) - P(r)\zeta] \tag{10}$$

with $\phi(\zeta) \equiv (V''(F)/2)\zeta^2 = (3A/F^4)\zeta^2$. The depth of the dimple is found to behave according to [13,14]

$$F_d \equiv |\zeta(0)| \approx \frac{\nu}{2}[ln(2a/\xi)]^2\,, \tag{11}$$

with the capillary length $a = (\gamma/V''(F))^{1/2}$. For the dimple radius one has

$$R_d \equiv |r(\zeta = \zeta(0)/2)| \approx \xi \exp[(F_d/\nu)^{1/2}]\,, \tag{12}$$

while for the dimple free energy one has from eq.(11)

$$\mathcal{E}_d \equiv -\mathcal{H}[\zeta] = \gamma\nu^2 I_0 \sim F_d^2\,, \tag{13}$$

where I_0 is of the order $\mathcal{O}(ln(a/\xi))$. $\mathcal{H}[\zeta]$ is *negative*, expressing the fact that the thick film lowers its free energy by forming the dimple. Table I lists the results for the capillary length, the dimple depth, radius and free energy for ^{4}He/Cs at two film thicknesses, $F = 100$ Å and $F = 200$ Å with $A = 10K/k_B$. The magnitude of F_d for ^{4}He is fairly small, while the radius of the dimple is already

Table 1. Dimple properties

F (Å)	a (Å)	F_d (Å)	R_d (Å)	$\mathcal{E}_d(K)$
100	670	12.28	149.98	24.46
200	2680	17.67	405.49	41.59

considerable: for $F = 200$ Å, $R_d \approx 2F$. These results indicate that the vortex dimple is, at least in the ^{4}He/Cs system, indeed a weak perturbation for a thick film $F \to \infty$ for $\Delta\mu \to -0$. However, when the size of the dimple matches that of the critical hole, which occurs upon the approach to the spinodal, the difference in surface free energies $\Delta\mathcal{E} \equiv \mathcal{E}_c - \mathcal{E}_d$ shrinks. It vanishes according to the power law [15]

$$\Delta\mathcal{E} \sim (T_f - T_*)^{1/2} \tag{14}$$

where $T_*(\Delta\mu)$ is the temperature, defined as the point beyond which surface deformations have become unstable. At a given value of $\Delta\mu$, $T_* > T$, so that T_* corresponds to a *shifted spinodal temperature* [13].

4   Conclusions

The consideration of a nucleation process in the presence of localized surface deformations might open up the possibility of performing quantitative experiments on the nucleation of holes in undercooled wetting layers. A comparison of the decay of metastable normal liquid and superfluid layers could therefore allow a test of the predictions of the scaling theory of critical holes. The scenario described here is in addition affected by the number n of circulation quanta present in the vortex, since $\mathcal{E}_d \sim R_0^2 \sim (n\kappa_c)^4$. Further modifications of the dimple-assisted dewetting scenario are: 1) the thinning of the superfluid film due to the Kontorovich effect [16]. Superfluid flow of the film modifies the asymptotics of the effective interface potential via $V(f) = A/f^2 - \varrho_s j^2/(2f)$ where $j = v_c f$ is the critical current. For typical critical velocities $v_c \approx 0.6$ m/s the Kontorovich effect will be appreciable for film thicknesses $F > 200$ Å[15]. 2) P.G. de Gennes has recently pointed out that on an extremely weak-binding surface such as Rb the balance between Bernoulli pressure and the boundary tension of ^{4}He 'pancakes' may give rise to unusually large vortex cores [17]. 3) The disordered surfaces that are common to Cs and Rb substrates might affect the presented scenario by introducing disorder-dominated hysteresis effects [18]. They are however less relevant for dewetting than for wetting processes, since in the former case their difference is clearly discernable [18].

Acknowledgments

Support by the DFG under the Leibniz grant (Di387/2-1) and the Schwerpunktsprogramm 'Wetting and Structure formation at Interfaces' (Bl356/2-1) is gratefully acknowledged.

References

1. Israelachvili J. (1992) Intermolecular and surface forces. 2nd ed., Academic Press, London
2. Cheng E., Cole M.W. et al. (1991) Helium prewetting and nonwetting on weak-binding substrates. Phys. Rev. Lett. **67**, 1007-1010
3. Nacher P.-J., Dupont-Roc J. (1991) Exprimental evidence for nonwetting with superfluid helium. Phys. Rev. Lett. **67**, 2966-2969
4. Cheng E., Cole M.W. et al. (1993) Novel wetting behaviour in quantum films. Rev. Mod. Phys. **65**, 557-567
5. Rutledge J.E., Taborek P. (1992) Prewetting phase diagram of ^{4}He on cesium. Phys. Rev. Lett. **69**, 937-940
6. Schick M., Taborek P. (1992) Anomalous nucleation at first-order wetting transitions. Phys. Rev. B **46**, 7312-7314
7. Wyatt A.F.G., Klier J., Stefanyi P. (1995) Prewetting of ^{4}He on Rb: coexistence of two superfluid films? Phys. Rev. Lett. **74**, 1151-1154
8. Philips J.A., Ross D. et al. (1998) Superfluid onset and prewetting of ^{4}He on rubidium. Phys. Rev. B **58**, 3361-3370
9. Brezin E., Halperin B.I., Leibler S. (1983) Critical wetting: the domain of validity of mean field theory. J. Phys. (Paris) **44**, 775-783
10. Blossey R. (1995) Nucleation at first-order wetting transitions. Int. J. Mod. Phys. **9**, 3489-3525
11. Bausch R., Blossey R. (1994) Lifetime of undercooled wetting layers. Phys. Rev. E **50**, R1759-R1761
12. Bonn D., Indekeu J.O. (1995) Nucleation and wetting near surface spinodals. Phys. Rev. Lett. **74**, 3844-3847
13. Blossey R. (1998) Dimple-assisted dewetting in rotating superfluid films. Phys. Rev. B **57**, R14048-R14051
14. Harvey K.C., Fetter A.L. (1973) Free surface of a rotating superfluid. J. Low. Temp. Phys. **11**, 473-481
15. Blossey R. (2000) Dimples and Dents. Heterogeneous Nucleation in Undercooled Wetting Layers. Habilitation Thesis Heinrich-Heine-Universität Düsseldorf
16. Kontorovich V.M. (1956) Effect of the rate of flow of a He II film on its thickness. Sov. Phys. JETP **3**, 770-771
17. de Gennes P.G. (1999) Dry vortices in thin helium films. C.R. Acad. Sci. Paris t.**327 II**, 1337-1343
18. Blossey R., Kinoshita T., Dupont-Roc J. (1998) Random-field Ising model for the hysteresis of the prewetting transition on a disordered substrate. Physica A **248**, 247-272

Quantum Hall Effect Breakdown Steps and Possible Analogies with Classical and Superfluid Hydrodynamics

L. Eaves

School of Physics & Astronomy, University of Nottingham,
Nottingham NG7 2RD, UK

Abstract. The breakdown of the integer quantum Hall effect at high currents sometimes occurs as a series of regular steps in the dissipative voltage drop measured along the Hall bar. The steps were first seen clearly in two Hall bars used to maintain the US Resistance Standard, but have also been reported in other devices. This paper discusses the origin of the steps in terms of an instability in the dissipationless flow induced by inter-Landau level tunnelling processes in local microscopic regions of the Hall bar. It is proposed that electron-hole pairs are generated in the quantum Hall fluid in these regions and that the electronic motion can be envisaged as a quantum analogue of the von Karman vortex street which forms when a classical fluid flows past an obstacle. A possible analogue with vortex formation in superfluids is also discussed briefly.

Keywords: Quantum Hall effect, Tunnelling, Hydrodynamics, Quantum Vortices

1 Introduction

The relation between the quantum Hall effect (QHE) [1,2] and superfluid hydrodynamics appears a tenuous one at first sight. Yet in 1985 Muirhead, Vinen and Donnelly[3] pointed out an interesting analogy between the motion of a vortex in a superfluid film and electron motion in two dimensions under the action of crossed electric and magnetic fields. In the analogy, the vortex circulation corresponds to the electron charge, the Magnus force to the Lorentz force and the superfluid density to the strength of the magnetic field. These authors then went on to consider the quantum description of the vortex motion and how vortices can be generated by tunnelling from the edge of the film.

Another possible analogy for the two systems is the small value of kinematic viscosity of He II and the low level of resistive dissipation when current flows in a semiconducting Hall bar sample under ideal quantum Hall effect conditions.

The purpose of this article is two-fold: first, to propose a model based on inter-Landau level tunnelling to explain an unusual step-like breakdown of the dissipationless state in the quantum Hall effect and, second, to draw attention to possibly interesting parallels between this model and well-established ideas of vortex pair formation in superfluids. When considering the analogies we should be aware that the electrons which make up the quantum Hall fluid are moving in an electrostatic potential which we can consider as comprising two parts: the first part arises from the presence of the Hall electric field due to the edge charges, and the second part arises from the randomly-located remote donor ions. It is well-established that this random potential plays an important role in the QHE [2]. The parallel situation for He II wou ld be flow in the presence of small obstacles, e.g. charged ions or possibly ^{3}He atoms.

In the integer QHE [1], a two-dimensional electron fluid carries an almost dissipationless current and the ratio of the current to the Hall voltage is quantised in units of e^2/h. This quantisation is so precise that it is used to define the unit of electrical resistance. Above a critical value of current, the dissipative voltage, V_x, measured along the direction of current flow, increases rapidly, leading to QHE breakdown. The microscopic mechanism, or mechanisms, responsible for breakdown have been a topic of interest and controversy for almost twenty years [2].

In two samples used to maintain the US Resistance Standard [4,5], Cage and co-workers observed a staircase of ten or more steps in the longitudinal voltage, V_x, at breakdown as shown in Fig. 1.

The steps have a regular height, $\Delta V_x \approx 5mV$ and are accompanied by hysteresis and intermittent noise. The data were obtained by measuring V_x with a large current I flowing along the length of a bar-shaped sample (see Fig. 1(a) - inset) made from a GaAs/(AlGa)As semiconductor heterostructure. The figure shows sweeps of the applied magnetic field B over an extended range around $12T$ at which the Landau level filling factor $\nu = 2$. Despite the extensive literature on QHE breakdown, reviewed in ref. [2], the precise origin of the steps has remained uncertain. In particular, they are not readily explicable in terms of the properties of the magneto-conductivity tensor components, σ_{xx} and σ_{xy}, which describe successfully most of the macroscopic properties of the quantum Hall fluid (QHF).

QHE breakdown corresponds to the disruption of dissipationless current flow (I) along the voltage equipotentials and to an abrupt increase of V_x from a very low value. The dissipative current, i, flowing perpendicular to the equipotentials is given by $2e^2V_x/h$. Prior to breakdown I/i can be as large as $\sim 10^6$-10^7, values approaching the Reynolds numbers achievable with cryogenic fluids. The dissipative current i can arise from inter-Landau level tunnelling [6–11]. In this process, an electron in the lower filled Landau level scatters elastically (or quasi-elastically with the emission of a low energy acoustic phonon) across the Landau level energy gap into the empty state of the upper unfilled Landau level. Since the wavefunctions of the Landau states are strongly localised along the direction perpendicular to the equipotentials and have strongly decaying gaussian tails, a large electric field is required for significant spatial overlap between the initial and final states of the same energy, i.e. the electric field E must be $\sim \hbar\omega_c/e\lambda_B$, where $\omega_c = eB/m$ is the cyclotron frequency, m is the electron effective mass ($= 0.07m_e$ for GaAs) and $\lambda_B = \sqrt{\hbar/eB}$ is the magnetic length. Alternatively, this condition can be expressed in the form $mv_d^2 \sim \hbar\omega_c$, where $v_d = E/B$ is the drift speed of the cyclotron orbit centre. For the experimental conditions of Fig. 1, the average value of the component of the electric field E along y (the average Hall field) is relatively small, $E_y = V_H/w \sim 7 \times 10^3 Vm^{-1}$, almost two orders of magnitude too small to produce significant wavefunction overlap.

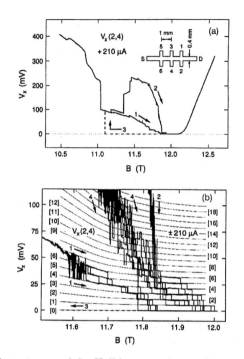

Fig. 1. QHE breakdown in one of the Hall bar samples used to maintain the US Resistance Standard at the National Institute of Standards and Technology, Gaithersburg. (a) Plots of longitudinal voltage V_x versus magnetic field B at $T = 1.3K$, in the region of the $\nu = 2$ filling factor, $I = 210\mu A$. The arrows show the hysteresis for various sweeps of B. (b) A detail of the breakdown curve, showing the large number of steps in V_x. (Both figures from ref. [4], courtesy of NIST)

The essence of the model proposed here is that E and hence v_d can be much larger in *localised* microscopic regions of the Hall bar due to the proximity of the edge of the Hall bar and the presence of charged impurities. It is proposed that in these regions the QHF is unstable against inter-Landau level tunnelling. This effect can be envisaged as a quantum analogue of the generation of vortex-pairs in a von Karman vortex street, when a classical fluid flows past an obstacle, and is also related to vortex pair formation in superfluids.

2   Model and Comparison with Experiment

For simplicity, let us assume a quadratic dependence of the electron potential energy ϕ in the local region, as shown schematically in Fig. 2(a). Outside this region, the Hall field is smaller. Over the distance $\sim s$ along the y-direction, we write $e\phi \approx m\omega_0^2 y^2/2$. This model potential is a function of only one unknown parameter ω_0, and corresponds to a large local velocity gradient $\Omega = dv_x/dy = \left(d^2\phi/dy^2\right)/B = \omega_0^2/\omega_c$. The equipotentials shown in Fig. 2(b) give rise to this

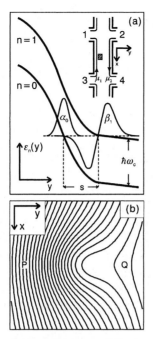

Fig. 2. a Model for spatial variation of energy of the two lowest Landau levels, $n = 0$ and $n = 1$ in a localised region of QHE breakdown. The eigenfunctions $|\alpha_0\rangle$ and $|\beta_1\rangle$ are in the $n = 0$ (filled) and $n = 1$ (empty) Landau levels respectively, at filling factor $\nu = 2$. They have the same energy and spatially overlap each other. The length s is the distance over which the cyclotron orbit centre jumps in an inter-Landau level tunnelling transition. The inset shows schematically the geometry of the Hall bar sample. V_x is measured between probes 1 and 3. (The Hall voltage V_H is measured across 1 and 2 or 3 and 4). The small cross-hatched region indicates schematically a breakdown region near the edge of the sample. The difference between chemical potentials at the two edges, $\mu_1 - \mu_2 = eV_H$. **b** Plot of the voltage equipotentials in the plane of the quantum Hall fluid for a region of large velocity gradient in which breakdown occurs. The energy variation of the Landau levels shown in box (a) above corresponds to that along the line PQ.

type of velocity gradient. Due to the character of electron motion in crossed electric and magnetic fields, they also correspond to the directions of dissipationless flow in this region. A potential distribution of this kind, where E is locally much larger than E_y, can arise from the combined effect of a charged donor impurity located current close to the QHF, and the large Hall field due to the charge distribution near the edge of the sample [10,12–14]. These charges are essentially unscreened by the presence of the dissipationless QHF [14,15].

Fig. 2(a) also shows the wavefunction of a state $|\alpha_0\rangle$ in the lowest filled Landau level ($n = 0$) and of a state $|\beta_1\rangle$ in the $n = 1$ level. The two states have the same energy (the dispersion of the Landau level energies is due, of course, to the spatial variation of ϕ). For simplicity, we neglect spin. The orbit centres of $|\alpha_0\rangle$ and $|\beta_1\rangle$ are separated by a distance $\Delta y = s$, given by

$$mw_0^2 s^2/2 \approx \hbar\omega_c \tag{1}$$

For small ω_0, spatial overlap between the wavefunctions is negligible. However, if ω_0 is sufficiently large, overlap occurs (see Fig. 2a) and an electron in state $|\alpha_0\rangle$ can undergo a quantum transition to the unfilled state $|\beta_1\rangle$. An inter-Landau level transition of this kind leads to the dissipative breakdown current, i. It requires a jump of orbit centre position which can be induced by the presence of the charged impurity which destroys translational symmetry. To compare our model with the data, in particular with the measured value of ΔV_x, we need to determine the size of dissipative current i generated by inter-Landau level tunnelling. In principle, the tunnelling rate could be calculated using Fermi's Golden Rule. The problem is that inter-Landau level tunnelling creates an electron in the upper, previously empty, Landau level and an empty state, or hole, in the previously filled lower Landau level. The two particles experience a Coulombic attractive interaction and also tend to screen the local potential which generated them, thereby perturbing the drift motion of other electrons in adjacent states of the lowest Landau level. Since the interaction between the electron-hole pair (or magneto-exciton) and the QHF is quite complicated, even in the absence of impurity or edge charges [16], we cannot make a calculation based on a simple independent particle picture. To avoid this difficulty, let us consider the following semiclassical description, which seems to account for the essential features of the breakdown mechanism.

At the critical breakdown values of current and magnetic field, we assume that the electric field in the local breakdown region is just large enough to induce inter-Landau level tunnelling. This process can be envisaged as electron-hole pair formation, the electron occupying state $|\beta_1\rangle$ and the hole corresponding to the electron missing from state $|\alpha_0\rangle$ in the otherwise filled lower Landau level. The presence of the electron-hole pair acts to screen the locally strong electric field, thus temporarily inhibiting further generation of pairs by tunnelling. However, due to the presence of the local electric field, the electron-hole pair drift away from the region in which they are formed. For the potential distribution shown in Fig. 1, assuming an independent particle description, the hole left in state $|\alpha_0\rangle$ has a high $\mathbf{E} \times \mathbf{B}$ drift velocity, whereas the electron in $|\beta_1\rangle$ has a lower drift velocity, each given by the local value of the electric field. We account qualitatively for the effect of the Coulombic attraction between the electron and hole by assuming that the pair drift at a speed determined by the mean electric field, which is $mw_0^2 s/2e$. The mean drift speed is therefore $\langle v_d \rangle \approx \Omega s/2$ and the pair move a distance $\sim s$ from the formation region in a time $\tau = 2/\Omega\gamma$, where γ is a numerical factor ~ 1. At this separation, they no longer screen effectively the strong local electric field around the impurity, so this is restored to its original strength, leading to the creation of another electron-hole pair. Assuming that the time τ is the controlling factor in the formation of electron-hole pairs, we can take the rate of generation to be $\approx \gamma\Omega/2$. On a longer timescale, given by the intra-Landau level energy relaxation time $\tau_e \approx 10^{-10}s$, we assume that the electron-hole pair breaks up, the two particles diffusing away from each other with velocity components perpendicular to the equipotentials [17]. This process can occur for each particle by emission of low energy acoustic phonons through intra-Landau level scattering processes. The dissipative backscattering current $i \approx e\gamma\Omega/2$ causes an incremental increase in V_x given by

$$\Delta V_x = hi/2e^2 = h\gamma\Omega/4e \qquad . \qquad (2)$$

The steps shown in Fig. 1 would then correspond to successive contributions to the dissipative current of other breakdown regions with a similar local potential distribution due to the presence of a charged impurity, each adding $\sim e\gamma\Omega/2$ to the dissipative current as B is swept away from the $\nu = 2$ value. To test the model, we now compare the observed value of ΔV_x with the value of ω_0 required to generate inter-Landau level tunnelling. Using equation 1 and recalling that $\Omega = \omega_0^2/\omega_c$, we write equation 2 in the form

$$e\Delta V_x = \pi\gamma\hbar^2/ms^2 \qquad . \qquad (3)$$

This equation allows us to relate the observed value of $\Delta V_x = 5mV$ to the orbit centre separation, s, rather than to ω_0. Setting $\gamma \approx 1$, we then obtain $s^2/\lambda_B^2 \approx \pi\hbar\omega_c/e\Delta V_x$, which gives $s \approx 3.5\lambda_B$, where $\lambda_B = \sqrt{\hbar/eB}$. This value of s agrees quite well with the condition for alignment of the classical turning points of the wavefunctions $|\alpha_0\rangle$ and $|\beta_1\rangle$, namely when $s_{TP} = (\sqrt{3}+1)\lambda_B$. This condition corresponds to the threshold of the strong increase in the matrix element for inter-Landau level transitions [11].

The model can also be tested against the recently-reported voltage steps for QH breakdown of hole gases [18]. Here the typical step height is $\Delta V_x = 1mV$ for breakdown at $\nu = 2$. Using equation 3 and the measured value of the hole effective mass, $m_h = 0.15m_e$ [19], we obtain $s = 2.3\lambda_B$, which is also quite close to the value of s_{TP}. The hole gas data therefore provide further support for the model.

Recalling that $\Omega = \omega_0^2/\omega_c$ we can use equation 2 to confirm that the size of ω_0 required for breakdown is consistent with that produced by an unscreened charged impurity. For such an impurity located at a distance Δz from an idealised strictly 2-dimensional QHF, it is easily shown that the step height of $\Delta V_x = 5mV$ observed at $B \approx 12T$ by Cage et al corresponds to a value of Δz given by $(h/16\pi\varepsilon_r\varepsilon_0 B\Delta V_x)^{1/3} \approx 13nm$, where $\varepsilon_r \approx 12$ is the relative permittivity of GaAs. This value corresponds very closely to the mean distance of the QHF from the GaAs/(AlGa)As interface [20] and indicates that charged interface impurities can generate the local velocity required for our breakdown model.

To conclude this section, we note that the model may also be relevant to the results of experiments by Bliek et al [21] in which the QHE breakdown was induced by passing current through a narrow constriction, $1\mu m$ wide. The critical current for breakdown in the constriction corresponds to the typical value of electric field required for inter-Landau level scattering [9]. These measurements are also of interest because steps appear as regular values of resistance, $R_x = V_x/I$, rather than of voltage, V_x. Briefly, the model proposed here can account for this difference using the following argument (fuller details will be published elsewhere). In a very narrow constriction, we can assume that the overall form of the Hall voltage profile $V(y)$ across the constriction, i.e. across the entire region of current flow, is mainly determined by the edge charges. The form of $V(y)$ is a logarithmic function, with a prefactor determined by, and proportional to, the total current I flowing through the constriction [10,12,14]. The local velocity gradient, $\Omega(y) = \left(d^2\phi/dy^2\right)/B$, is then also proportional to I. From equation 2, it can be seen that this would then lead to increments, ΔV_x, which are proportional to I, i.e. to constant increments of ΔR_x. Further analysis based on this model and on the observed size of the steps, $\Delta R_x \approx 200\Omega$, indicates that the breakdown region in the channel is confined to a thin ($\sim 0.1\mu m$) layer close to one edge of the constriction, in reasonable agreement with model calculations [12].

3   Analogies with Classical and Quantum Fluids

Note that the proposed mechanism for the dissipative current i is closely analogous to the formation of vortex-antivortex pairs which are created when a classical fluid flows past an obstacle, e.g. a cylinder, above a critical Reynolds number, $R_e \approx 60$. In particular, the value of i/e, the electron-hole pair generation rate, corresponds to the rate of vortex shedding in a classical von Karman vortex street [22]. An insight into the analogy can be gained by applying Madelung's hydrodynamic formulation of the Schrödinger wave equation [23] to the QHF. We can envisage the electron motion as comprising an $\mathbf{E} \times \mathbf{B}$ drift motion along the equipotentials combined with an internal cyclotron motion with a vorticity equal to ω_c [24]. The formation of an electron-hole pair in an almost dissipationless QHF due to inter-Landau level tunnelling in the vicinity of a charged impurity would then correspond to the creation of an electron (vortex) in the previously empty $n = 1$ Landau level and the loss of an electron (or, creation of hole = antivortex) in the previously filled lower Landau level. In this picture, the creation of vortices of opposite sign, but equal magnitude, in the upper and lower Landau levels can be compared with the conservation of circulation (vorticity) in almost inviscid classical fluids.

Interestingly, using the classical concept of eddy viscosity for 2d flow [25] with the tunnelling distance s playing the role of the mixing length, it is easily shown that each step in V_x in Fig. 1 corresponds to an eddy viscosity increment of $\sim s^2\Omega/2 \sim \hbar/m \sim 10^{-3}m^2s^{-1}$. This value is $\sim 10^5$ times larger than the corresponding value of the circulation quantum $(\hbar/M_{He})$ for superfluid He; hence a QHF is relatively stable against electron-hole pair (=vortex pair) generation at high flow speeds, compared to the analogous case of an ideal superfluid He film [3,26].

It is also interesting to note that the hysteresis and intermittent noise which frequently accompany QHE breakdown (see Fig. 1) are features which are also characteristic of the transition from laminar to turbulent flow in classical fluids through pipes at high Reynolds numbers [22].

The hydrodynamic description proposed here may also be relevant to another recent experiment on QHE breakdown [27], which suggests an analogy with experiments on grid turbulence. Also, the recently reported non-local (delayed) nature of breakdown in narrow Hall bars [28] is somewhat analogous to the breakdown of laminar pipe flow in Reynolds's classic experiments [22].

4 The Breakdown Steps and Their Relation to Other Types of QHE Breakdown

Komiyama and Kawaguchi [17] have proposed a thermal instability or bootstrap electron heating model to account for the type of QHE breakdown observed in many samples. This model accounts successfully for the size of critical breakdown current, I_c, of a wide variety of Hall bars of differing widths, studied by Kawaji and co-workers [29], and by others. The model also successfully predicts the $B^{3/2}$ dependence of I_c, as observed by Kawaji and co-workers [29], [30]. However, the type of breakdown described by the Komiyama-Kawaguchi model is probably of a different type from that observed by Cage and co-workers. In particular, the occurrence of bootstrap heating may lead to such a strong dissipation that it masks the small breakdown steps which are best observed in high resolution B-sweeps.

Finally, we note the rich, but seemingly diverse, variety of breakdown behaviour reported in recent publications: (i) the type described by the Komiyama-Kawaguchi model [17,28]; (ii) the step- breakdown of the type discussed here [4,5,9,18,21]; (iii) the breakdown observed at relatively low current density in narrow, high mobility Hall bars, also probably due to inter-Landau level scattering close to the sample edge [10]; (iv) the breakdown measurements which show a relatively well-behaved dependence on magnetic field (filling factor), current and temperature, which can be used to construct a phase diagram with a phenomenologically similar form to that found for Type II superconductors [31].

5 Summary

A model has been proposed to explain the voltage steps observed in the breakdown of the QHE in two samples used to maintain the US Resistance Standard. The model suggests that we can view this type of breakdown as an instability in the flow of the QHF. This instability can arise from the generation of electron-hole pairs by inter-Landau level tunnelling due to the presence of a charged impurity in local regions of the Hall bar, where the electric field is large. This type of process has analogies with vortex pair formation in classical fluids and superfluids.

Acknowledgement

I am grateful to Dr. M. Cage and NIST for permission to use Fig. 1, and to my colleagues F. W. Sheard and K. A. Benedict for helpful discussions. This work was partly supported by the Engineering and Physical Sciences Research Council (UK).

References

1. K. von Klitzing, G. Dorda and M. Pepper, Phys. Rev. Lett. 45, 494-497 (1980).
2. For a comprehensive review of QHE breakdown, see G. Nachtwei, Physica E 4 79-101 (1991).
3. C. M. Muirhead, W. F. Vinen and R. J. Donnelly, Phil. Trans. R. Soc. Lond. A 311, 433-467 (1984).
4. M. E. Cage, J. Res. Natl. Inst. Stand. Technol. 98, 361-373 (1993). This reference, and reference 4 below, describe the voltage steps observed in the two US Resistance Standard samples with reference numbers GaAs(7) and GaAs(8).
5. C. F. Lavine, M. E. Cage, and R. E. Elmquist, J. Res. Natl. Inst. Stand. Technol. 99, 757-764 (1994).
6. H. L. Störmer, et al., in: J. D. Chadhi, W. A. Harrison (Eds.), Proc. of the 17th Int. Conf. on the Physics of Semiconductors (17th ICPS 1984), Springer, Berlin, 1985 p.267.
7. O. Heinonen, P. L. Taylor and S. M. Girvin, Phys. Rev. 30, 3016-3019 (1984).
8. L. Eaves, P. S. S. Guimares and J.-C. Portal, J. Phys. C: Solid State Phys. 17, 6177-6190 (1984).
9. L. Eaves and F. W. Sheard, Semicond. Sci. Technol. 1, 346-349 (1986).
10. N. Q. Balaban, U. Meirav, H. Shtrikman and Y. Levinson, Phys. Rev. Lett. 71, 1443-1446 (1993).
11. C. Chaubet, A. Raymond and D. Dur, Phys. Rev. B 52, 11178-11192 (1995).
12. A. H. MacDonald, T. M. Rice and W. F. Brinkman, Phys. Rev. 28, 3648-3650 (1983).
13. D. J. Thouless, J. Phys. C 18, 6211-6218 (1985).
14. C. W. J. Beenakker and H. van Houten, in Solid State Physics: Semiconductor Heterostructures and Nanostructures, pp.1-228, eds. H. Ehrenreich and D. Turnbull, Academic Press, 1991.
15. D. B. Chklovskii, B. I. Shklovskii and L. I. Glazman, Phys. Rev. B 46, 4026-4034 (1992).
16. C. Kallin and B. I. Halperin, Phys. Rev. B 30, 5655-5668 (1984).
17. S. Komiyama and Y. Kawaguchi, Phys. Rev. B 61, 2014-2027 (2000).
18. L. Eaves et al., Physica E 6,136-139 (2000).
19. B. E. Cole et al., Phys. Rev. B 55, 2503-2511 (1997).
20. F. Fang and W. E. Howard, Phys. Rev. Lett. 16, 797 (1966).
21. L. Bliek et al., Surf. Sci. 196, 156 (1988).
22. See for example Fluid Dynamics for Physicists, T. E. Faber (Cambridge University Press, 1995).
23. E. Madelung, Z. Phys. 40, 322 (1926).
24. L. Eaves, Physica B 272, 130-132 (1999).
25. G. I. Taylor, Proc. Roy. Soc. London Series A, 85 685-705 (1932).
26. R. P. Feynman, Application of quantum mechanics to liquid helium, in Progress in Low Temperature Physics I (ed. C. J. Gorter), North-Holland. Chapter 2 (1955).
27. I. I. Kaya, G. Nachtwei, K. von Klitzing and K. Eberl, Phys. Rev. B 58, R7536-R7539 (1998).

28. S. Komiyama, Y. Kawaguchi, T. Osada and Y. Shiraki, Phys. Rev. Lett. 77, 558-561 (1996).
29. S. Kawaji, Semicond. Sci. Technol. 11, 1546-1551 (1996).
30. S. Kawaji et al., J. Phys. Soc. Japan 63, 2303-2313 (1994).
31. L. B. Rigal et al, Phys. Rev. Lett. 82, 1249-1251 (1999).

Atomic Bose Condensate with a Spin Structure: The Use of Bloch State

Hiroshi Kuratsuji

Department of Physics, Ritsumeikan University, Kusatsu City, 525-8577, Japan

Abstract. We study a novel aspect of the anisotropic bose condensates that is inspired by the recently discovered alkari atom bose condensates. To formulate the theory, we use the Bloch state which provides a suitable tool for describing the order parameter possessing with the spin components. The Landau-Ginzburg type Lagrangian is reduced to that of ferromagnet fluid which contains the kinectic energy fluid as well as the Heisenberg Hamiltonian. Specifically we scrutinize a possible vortex state for this condensate possessing with spin structure; the profile for a single vortex as well as the equation of motion.

1 Introduction

The discovery of the Bose-Einstein condensates (BEC) in alkari atomic vapours has opened a new direction of condensed matter physics [1–3] besides the conventional quantum condensates such as superfluid helium or quantum liquid. The purpose of this report is to address a specific aspect of this newly established BEC; namely, we want to explore the physical effect that is inherent in the spin structure of the order parameter for the atomic condensate. The spin nature of the order parameter is caused by the "hyperfine spin" of constituent alkari atoms. Some aspects of the spin degree that arises from hyperfine spin was first suggested by using the spinor field [4]. From experimental point of view, the condensate incorporating spin degree can be realized in the optical trap [5] that has been invented as a complement of the magnetic trap, since the spin degree survives as an active degree, whereas in the magnetic trap the spin degree is frozen along the magnetic field line. In the sequel to this experiment, the term of the "spinor condensate"was introduced in [6], which indeed predicts the polar state as well as the ferromagnetic state for the special case of the spin magnitude 1.

The bose condensates possessing with spin structure can be regarded as an "anisotropic quantum fluid". Apart from the atomic BEC, the superfluid He3-A provides a typical anisotropic quantum fluid, which is indeed quite different from the conventional superfluid He4 governed by the scalar order parameter. The order parameter for He3A is written in terms of the vector order parameter causing the so-called l-texture [7] that resembles the director in liquid crystals.

In what follows, we shall give a general formalism of the atomic bose condensate accomodating spin texture and subsequently address some specific aspects inherent in spin structure especially, vortex state. In order to carry out this, we adopt the Landau-Ginzburg (LG) Lagrangian which has similar form with the one developed for the dynamics of He3A [8]. Here the essential point is to use the Bloch state [9] in order for describing the spin structure of the order parameter. The use of the Bloch state can incorporate any magnitude of spin that includes the condensate with spin 1 as a special case that has been previously given. The LG Lagrangian is reduced to that of the ferromagnet fluid for which the Lagrangian consists of two terms; the fluid kinetic energy as well as the Heisenberg ferromagnet. We consider the profile of a single vortex as well as its motion and the resultant equation of motion gives the geometrical force that is caused by the topological invariant known as the Mermin-Ho relation.

2   Order Parameter and Lagrangian

To describe the spin nature of the order parameter, it is natural to adopt the spin states which is denoted by $|J, M\rangle$ with $M = -J \sim +J$. Specifically, we are concerned with the lowest state $|J, -J\rangle$. Using the usual vector notation, it is written as $(0, \cdots, 1)^T \equiv \Psi_0$. From this vector, the general vector which directs $\mathbf{n} = (\sin\theta\cos\phi, \sin\theta\sin\phi, \cos\theta)$ is constructed by rotating the Ψ_0 by the rotation operator; $R(\theta, \phi) = \exp[i\theta(\mathbf{k} \cdot \mathbf{J})]$ where $\mathbf{k} = (\cos\phi, \sin\theta, 0)$. Thus, following the general procedure, we have the vector

$$\Psi_\xi = \exp[\eta\hat{J}_+ - \eta^*\hat{J}_-]\Psi_0 = (1 + |\xi|^2)^{-J}\exp[\xi\hat{J}_+]\Psi_0 \tag{1}$$

with $\xi = \tan\frac{\theta}{2}\exp[-i\phi]$ and $\hat{J}_\pm$ the raising and lowering operator for spin. This state vector is known to be the Bloch states (alias spin coherent state). For the case of $J = 1$, we have

$$\Psi = (1 + \tan^2\frac{\theta}{2})(\xi^2, \sqrt{2}\xi, 1)^\dagger$$

$$= e^{i\phi}(\sin^2\frac{\theta}{2}e^{i\phi}, \sin\frac{\theta}{2}\cos\frac{\theta}{2}, \cos^2\frac{\theta}{2}e^{-i\phi}) \tag{2}$$

and for $J = \frac{1}{2}$ it follows that $\Psi^\dagger = (\cos\frac{\theta}{2}, \sin\frac{\theta}{2}e^{-i\phi})^\dagger$. Thus we regard equation (1) as the general order parameter carrying the spin of arbitrary magnitude:

$$\Psi = \Psi_\xi\Delta_0\exp[i\alpha] \tag{3}$$

Δ_0 represents the magnitude of the condensate which is fixed to be constant though it spacially varies in general. Actually, the order paramter is determined up to a phase factor, $\exp[i\alpha]$, but it can be chosen arbitrarily so we set $\alpha = 0$ in the following argument. Having defined the vector order parameter realized by the Bloch state, we give the Lagrangian that governs the dynamics of this order parameter. Here it is natural to set it in the similar form with the superfluid He4:

$$L = \int(\frac{i\hbar}{2}(\Psi^\dagger\dot{\Psi} - c.c) - H(\Psi^\dagger, \Psi))dx \tag{4}$$

and the Hamiltonian density

$$H = \frac{\hbar^2}{2m}\nabla\Psi^\dagger\nabla\Psi + V(\Psi^\dagger, \Psi) \tag{5}$$

Here $V(\Psi^\dagger, \Psi)$ represents the interaction potential that comes from the atomic two-body interactions as well as the trapping potential which is chosen such that the stability of the condensate Δ_0 is maintained. The concrete form of this will be given later.

3   Hydrodynamical Equation

Now we use the the Bloch form to reduce the Lagrangian. In order to carry out this, we need to have some elementary but crucial calculations concerning the derivative of the Bloch state, namely, for the time derivative

$$\frac{\partial \Psi_\xi}{\partial t} = \dot{N} \exp[\xi \hat{J}_+]\Psi_0 + N\dot{\xi}\hat{J}_+ \exp[\xi \hat{J}_+]\Psi_0 \tag{6}$$

with $N = (1+|\xi|^2)^{-J}$. The gradient is also obtained by replacing the dot by ∇. Then after some tedious manipulation, we get the first term of the Lagrangian, L_C (called the canonical term):

$$i\hbar \Psi^\dagger \dot{\Psi} - c.c. = J\hbar\Delta_0^2 \frac{(\xi^*\dot{\xi} - c.c)}{1+|\xi|^2} \tag{7}$$

Hence the Hamiltonian term becomes

$$H = \frac{1}{2}m\Delta_0^2 \mathbf{v}^2 + \frac{\hbar^2 \Delta_0^2}{2Jm}\frac{\nabla\xi^* \nabla\xi}{(1+|\xi|^2)^2} + V(|\xi|^2) \tag{8}$$

where

$$\mathbf{v} = \frac{J\hbar}{m}\frac{i(\xi^*\nabla\xi - c.c)}{1+|\xi|^2} \tag{9}$$

which reresents the velocity field (see below). In the above derivation, use is made of the relation

$$N^2 \Psi_0^* \exp[\xi^*\hat{J}_-]\hat{J}_+ \exp[\xi\hat{J}_+]\Psi_0 = \frac{2J\xi}{1+|\xi|^2} \tag{10}$$

which is an expectation value of the spin operator with respect to the Bloch state. This written in terms of the angular variable;

$$L_x = J\sin\theta\cos\phi, \; L_y = J\sin\theta\sin\phi, \; L_z = J\cos\theta, \tag{11}$$

which defines the effective spin carried by the order parameter. If using the angle polar coordinate, the hydrodynamical form of the Lagrangian becomes

$$L = \int [J\Delta_0^2(1 - \cos\theta)\dot{\phi} - H(\theta, \phi)]d\mathbf{r} \tag{12}$$

The first term of the reduced Hamiltonian is regarded as a kinectic energy where the velocity field $\mathbf{v}$ is written in terms of the angular variable:

$$\mathbf{v} = \frac{J\hbar}{m}(1 - \cos\theta)\nabla\phi \tag{13}$$

The second term of the Hamiltonian has the same form as the continuous Heisenberg model of ferromagnet [10], namely, it is rewritten as

$$H_S \simeq \int [(\nabla\theta)^2 + \sin^2\theta(\nabla\phi)^2]d\mathbf{r} \tag{14}$$

From the formal aspect, the above Lagrangian has the same form as the superfluid He4 described by the scalar order parameter. The action principle $\delta \int Ldt = 0$ leads to the canonical equation of motion

$$J\Delta_0^2 \frac{d\theta}{dt} = -\frac{1}{\sin\theta}\frac{\delta H}{\delta\phi}, \quad J\Delta_0^2\frac{d\phi}{dt} = \frac{1}{\sin\theta}\frac{\delta H}{\delta\theta}, \tag{15}$$

In terms of the spin variable, it turns out to be

$$\dot{\mathbf{L}} = \mathbf{L} \times \frac{\delta H}{\delta\mathbf{L}} \tag{16}$$

The stationary solutions are obtained by putting as $\dot\theta = 0, \dot\phi = 0$ and around the stationary solutions one can get the small oscillation that is just realized by spin wave. Besides this, we may have the case of "no flow", namely, $\mathbf{v} = 0$ leading to $\cos\theta = 1$ which gives the aligned spin $L_3 = J$. Indeed the aligned state is expected from the energy consideration if we take into account of the proper form of the interaction potential.

4   Vortex State

We now consider the vortex state that is inherent in the spin texture. In the following argument, it is assumed that the system is two-dimensional plane (x, y).

4.1   The Profile of a Single Vortex

We consider the following nonlinear potential which comes from the spin-spin interaction;

$$V(\Psi^*, \Psi) = g \int \Psi^*\hat{S}_3\Psi\dot\Psi^*\hat{S}_3\Psi dx \tag{17}$$

with the positive coupling constant g. In terms of the angular variable, this turns out to be

$$\tilde{V} = g\Delta_0^2 J^2 \int \cos^2\theta d^2x \tag{18}$$

The configuration of the single vortex is controlled by the angle variables (θ, ϕ). A static solution for the one vortex is obtained by choosing the phase function $\phi = n\tan^{-1}(\frac{y}{x})$, with $n = 1, 2, \cdots$. being the winding number, together with the profile function θ that is given as a function of the radial variable r. Thus the static Hamiltonian is written in terms of the field $\theta(r)$:

$$H = \frac{\hbar^2\Delta_0^2 J}{4m}\int[(\frac{d\theta}{dr})^2 + \frac{n^2}{r^2}(2J + \cos^2\frac{\theta}{2})4\sin^2\frac{\theta}{2} + g'\cos^2\theta]rdr \tag{19}$$

where $g' = \frac{4gJm}{\hbar^2}$. The profile function $\theta(r)$ may be derived from the extremum of H, namely, the Euler-Lagrange equation leads to

$$\frac{d^2\theta}{d\xi^2} + \frac{1}{\xi}\frac{d\theta}{d\xi} - \frac{n^2}{2\xi^2}(4J\sin\theta + \sin 2\theta) + \sin 2\theta = 0 \qquad (20)$$

where we adopt the scaling of the variable: $\xi = \sqrt{g'/2}\,r$. In order to examine the behavior of $\theta(\xi)$, we need a specific boundary condition at $\xi = 0$ and $\xi = \infty$. We impose at the origin such that $\theta(0) = 0$, whereas at $\xi = \infty$, we adopt two typical possibilities: A) $\theta(\infty) = \pi$ and B) $\theta(\infty) = \pi/2$. If we define the unit vector as $\mathbf{l} = \mathbf{L}/J$ we have $l_3(0) = 1$ for both caseses A), B) and we have $l_3(\infty) = -1$ for case A) and $l_3(\infty) = 0$ for case B). This feature indicates that the spin field which directs *upward* gradually converts into the state of *downward* or *outward*. First we consider the behavior near the origin $\xi = 0$, for which the differential equation behaves like the Bessel equation, so we see the power law

$$\theta(\xi) \simeq \xi^k \qquad (21)$$

with $k^2 = n^2(2J+1)$. Next we examine the behavior at $\xi = \infty$. This is simply performed by checking the stability for two cases mentioned above: (A) and (B). Now for the case (B), if putting $\theta(\xi) = \frac{\pi}{2} + \alpha$, with α the infinitesimal deviation, so the linearized equation near $\xi = \infty$, $\alpha'' - \alpha \simeq 0$, which results in $\alpha \simeq \exp[-\xi]$. This suggests that the solution behaves as at $\xi = \infty$,

$$\theta(\xi) = \frac{\pi}{2}(1 - \exp[-\xi]) \qquad (22)$$

On the other hand, for the case (A) we have $\alpha'' + \alpha \simeq 0$, which gives $\alpha \simeq \exp[i\xi]$ meaning the oscillatory behavior. This simply implies that the solution with $\theta(\infty) = \pi$ does not converge to the stable solution and this feature means that the case (A) is not relevant. Keeping mind of the above general feature, we here give a numerical behavior of the function $\theta(\xi)$ (Fig.1).

4.2 Vortex Dynamics

We now consider a possible motion of a single vortex. The dynamics of a vortex is described by the collective coordinates $\mathbf{X}(t) = (X, Y)$ that represent the center of the moving vortex, and the space coordinates in the angle variables are replaced as $\mathbf{x} \to \mathbf{x} - \mathbf{X}(t)$. Using the chain rule, $\frac{\partial\phi}{\partial t} = \frac{\partial\phi}{\partial\mathbf{X}}\dot{\mathbf{X}}$, the canonical term L_C becomes

$$L_C = J\hbar\Delta_0^2 \int (1 - \cos\theta)\nabla\phi \cdot \dot{\mathbf{X}} d^2x$$

$$= m\Delta_0^2 \int \mathbf{v} \cdot \dot{\mathbf{X}} d^2x \qquad (23)$$

where $\mathbf{v}$ means the velocity field given above, in other words, the momentum density $\mathbf{p}$ is defined to be canonically conjugate with $\mathbf{X}$ such that $\mathbf{p} = M\Delta_0^2\mathbf{v}$

with $\rho = M\Delta_0^2$ (the mass density). The velocity field $\mathbf{v}$ does not bear any singularity near the origin, and such vortices are called "soft-core" as in the case of the superfluid He3-A. The equation of motion for a vortex is given by the Euler- Lagrange equation

$$\frac{d}{dt}\frac{\partial L}{\partial \dot{\mathbf{X}}} - \frac{\partial L}{\partial \mathbf{X}} = \mathbf{F}_{ex} \qquad (24)$$

This is rewritten as the balance of two types of forces; $\mathbf{F}_C + \mathbf{F}_T = \mathbf{F}_{ex}$ where $\mathbf{F}_T$ is given by $\mathbf{F}_T \equiv -\frac{\partial T}{\partial \mathbf{X}}$, where T is the kinetic energy term of spin fluid (i.e. the first term of eq.(8)). The RHS represents the force for balance that may come from the other types of non-dissipative force as well as dissipative force. We get the force from the canonical term:

$$\mathbf{F}_C = 2m\Delta_0^2(\mathbf{k} \times \dot{\mathbf{X}})[\int (\nabla \times \mathbf{v})_z d^2 x] \qquad (25)$$

where the vector $\mathbf{k}$ is the z-directed unit vector. There is no contribution coming from the kinetic energy term as is intuitively guessed. The force thus derived is considered to be a counterpart of the Magnus-type force for the anistropic bose condensates, that is the same as the force for a vortex in ferromagnet. We consider the meaning of the integral in the (25). The integrand, which is nothing but the continuous vorticity $\omega = \nabla \times \mathbf{v}$

$$\omega = J\hbar/m \sin\theta \nabla\theta \times \nabla\phi$$
$$= \frac{J\hbar}{m}\mathbf{1} \cdot \left(\frac{\partial \mathbf{l}}{\partial x} \times \frac{\partial \mathbf{l}}{\partial y}\right) \qquad (26)$$

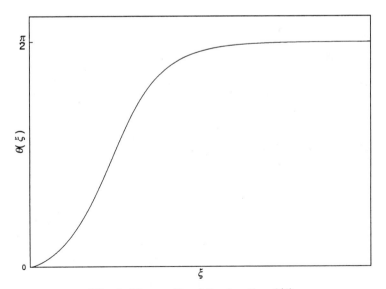

Fig. 1. The profile of the function $\theta(\xi)$

This relation is a counterpart of the Mermin-Ho relation for the case of the superfluid He3A. The integral over the plane is convered into the one over the field space

$$\sigma = J \int \sin\theta d\theta d\phi \tag{27}$$

so finally we get

$$\mathbf{F} = 2m\Delta_0^2\sigma(\mathbf{k} \times \dot{\mathbf{X}}) \tag{28}$$

This is just the same form as the expression given for the vortex occuring in the superfluid He3A [8].

5   Summary

In this report we have presented a hydrodynamical theory for the atomic bose condensate accomodating the spin texture. This is naturally carried our by adopting the Bloch state (or spin coherent state), which enables us to formulate the anisotropic nature of the bose condensate. This is closely connected with the superfluid He3A. The LG Lagrangian is brought to the fluid dynamical form for the spin liquid. The spin vortex has been explicated from this fluid dynamical Lagrangian in both static and dynamical aspects.

References

1. M.H.Anderson, J.R.Ensher, M.R.Matthews, C.E.Weiman and E.A.Cornell, Science **269**, 198(1995).
2. C.C.Bradley, C.A.Schakett, J.J. Tollett, and R.G.Hulet, Phys.Rev.Lett.**75**, 1687(1995).
3. K.B.Davis, M.O.Mewes, M.R.Andrews, N.J.Van Druten, D.S.Durfree, D.M.Kurn, and W.Ketterle, Phys.Rev.Lett.**75**, 3969(1995).
4. T.L.Ho and V.B.Shenoy, Phys.Rev.Lett.**77**,2595(1996).
5. D.M.Stamper-Kurn, M.R.Andrews, A.P.Chikatur, S.Inoue, H.H.Miesner, J.Stenger, and W.Ketterle, Phys.Rev.Lett.**80**,2027(1998).
6. T.L.Ho,Phys.Rev.Lett.**81**,742(1998).
7. G.Volovik and M.M.Salomaa, Rev.Mod.Phys.**59**,533(1987).
8. H.Kuratsuji and H.Yabu, Phys.Rev.**B59**, 11175(1999).
9. F.T.Arecchi, E.Courtens, R.Gilmore and H.Thomas, Phys.Rev.**A6**,2211(1972).
10. H.Ono and H.Kuratsuji, Phys.Lett.**186A**,255(1994).

Quantum Dynamics of Vortex–Antivortex Pairs in a Circular Box

Vittorio Penna

Dipartimento di Fisica and Unità INFM, Politecnico di Torino,
C.so Duca degli Abruzzi 24, I-10129 Torino, Italy

Abstract. After reviewing the quantum dynamics of unbounded vortex pairs, the case of a vortex-antivortex pair in a box with a circular boundary is investigated. The appropriate algebraic framework in which performing the quantization is reconstructed and the structure of the vortex pair spectrum is discussed.

1 Introduction

One of the early attempts to study the quantum dynamics of two-dimensional (2d) pointlike massless vortices was performed in Ref. [1] within the canonical quantization for an almost 2d system formed by an array of parallel vortex lines. While the generalization of such a model to a gas of vortex lines that are closed (and thus endowed with an arbitrarily complex 3d topology) has raised dramatic, still open, formal problems [2], the simpler structure of 2d vortices has permitted to investigate several aspects of vortex quantum dynamics (VQD), from the possible fractional character of their statistics [3], [4] and the influence of boundary effects on VQD [5], to the emergence of point vortices, dynamically quantized, within both a second quantized $|\Psi|^4$- field theory [6], and the quantum-phase model for Josephson junctions arrays [7].

A renewed interest for 2d vortex dynamics (VD) has been prompted by the recent, experimental as well as theoretical, developments in the context of both superfluidity and superconductivity [8], [9]. The great improvements concerning both the measurement techniques and the observations of microscopic processes [10], [11] should render quantum aspects of VD accessible to experimental detection. This is certainly true for the Bose-Einstein condensate (BEC) physics which represents a rich source of new experimental scenarios in which vortex creation is effected and VD can be studied [12].

This paper is devoted both to investigate the VQD of a vortex-antivortex (VA) pair in the presence of a circular reflecting wall, and to illustrate the formal tools one employs to study the spectral properties of vortex pair models. The construction of the model Hamiltonian is based on the virtual charge method which accounts for the boundary confining the fluid through an appropriate configuration of virtual vortices [13], [5]. Although the quantization of the unbounded VA pair is apparently trivial (its spectrum essentially coincides with that of a free particle), the interactions between the virtual vortices and the pair vortices has the effect to confine the latter on closed trajectories that are thus liable to be quantized. This aspect is particularly investigated in the sequel.

The VQD originated by the VA pair Hamiltonian is studied by means of the spectrum generating algebra method [14] applied in [15] to the vortex pair scattering from a disk. This consists in identifying first a complete set of dynamical degrees of freedom forming a Lie algebra (the *dynamical algebra*) and in deriving then the Hilbert space of the system from the unitary irreducible representations (UIR) of the associated Lie group. For the VA pair [15] such a group is SU(1,1) whose UIR's are completely classified. Here, we exploit repeatedly this formal set-up to characterize the weakly excited quantum states of the VA dynamics.

2   Canonical Quantization of Planar Vortices

The Hamiltonian of N pointlike vortices in a frictionless fluid is given by [16]

$$H(z_1, ..., z_n) = -(\rho/4\pi) \, \Sigma_{i \neq j} \, k_i k_j \, \ln[|z_i - z_j|/a] \, , \qquad (1)$$

where ρ is the fluid planar density, $z_j := x_j + iy_j \in \mathbf{C}$ describes the jth vortex position on the ambient plane, and the vorticity of the jth vortex $k_j = hn_j/m_H$ (m_H is the fluid atom mass) is quantized ($n_j \in \mathbf{Z}$) due to the Feynman-Onsager condition [17]. The length a represents the vortex core size stating the minimum distance between a vortex ($k_j > 0$) and an antivortex ($k_j < 0$) before coalescence processes occur. The Hamiltonian equations are derived from Eq. (1) via the Poisson brackets [16]

$$\{F, G\} = \Sigma_j [F_{x_j} G_{y_j} - G_{x_j} F_{y_j}]/\rho k_j \quad , \qquad (2)$$

($E_z := \partial E/\partial z$, $E = F, G$) involving, in turn, the ρk_j-dependent canonical brackets $\{x_i, y_j\} = \delta_{ij}/(\rho k_j)$. Then vortex coordinates can be regarded as a complete set of canonically conjugate variables whose *momenta* are defined as $p_j = \rho k_j y_j$. Also, one can easily check that $J_z = -\rho \Sigma_j k_j (x_j^2 + y_j^2)/2$ and $J_\pm = \rho \Sigma_j k_j (x_j \pm iy_j)$ ($J_\pm := J_x \pm iJ_y$) are constants of motion namely $\{F, H\} = 0$ with $F = J_a, J_*$, ($a = x, y, z$), and fulfil the equation

$$J_* = -(2CJ_z + J_x^2 + J_y^2)/\rho := \Sigma_{i \neq j} (k_i k_j/2) \left[(x_i - x_j)^2 + (y_i - y_j)^2 \right] , \qquad (3)$$

where $C = \rho \, \Sigma_j k_j$ is related to the total vorticity. J_* identifies with the Casimir function of the e(2)-like algebraic structure $\{J_z, J_\pm\} = \mp iJ_\pm$, $\{J_x, J_y\} = C$ (hereafter we denote it by $e_*(2)$) which fully emerges when $C = 0$. Physically, the conserved quantities J_x (J_y), J_z represent, up to a constant factor, the x (y) component of the fluid total momentum and the fluid total angular momentum, respectively. The quantization of the VD through the *canonical rule* $[x_i, p_j] = \delta_{ij} i\hbar$, ($p_j := \rho k_j \, y_j$ ensures the canonical form) entails that the symmetry operators J_a form the algebra $e_*(2)$

$$[J_x, J_y] = i\hbar C \, , [J_z, J_x] = i\hbar J_y \, , [J_z, J_y] = i\hbar J_x \, . \qquad (4)$$

In view of the fact that the constants of motion can be used to integrate the dynamical equation if they mutually commute, Eqs. (4) imply that the many-body wave function for the 2d vortex gas is characterized by three macroscopic quantum numbers two of which are, of course, H and J_*, while the third one can be chosen as an arbitrary linear combination $I = aJ_x + bJ_y + cJ_z$ since $[H, I] = 0$.

2.1   The Spectrum of Unbounded Vortex Pairs

The spectrum of the vortex-vortex pair can be reconstructed within the quantization procedure of the harmonic oscillator (HO) [3], [5]. Its generalization to a pair with two arbitrary vortex charges k_1 and k_2 has been discussed in [15], where a thorough analysis of the energy spectrum degeneracy is also given and related to the choice of I. A special case (that relevant for the VA model we are presenting) occurs when $k_1 = -k_2$ (pure VA case) which leads to the free particle spectrum instead of the HO spectrum.

To explain the source of such a basic difference as well as for illustrative purposes we first review the general case $k_1 \neq -k_2$. The two-vortex Hamiltonian given by Eq. (1) is $H(z_1, z_2) = -(\rho k_1 k_2 / 4\pi) \ln[J_*/(k_1 k_2 \rho a^2)]$, where $|z_1 - z_2|^2$ has been replaced by $\rho k_1 k_2 J_*$ due to Eq. (3). The set of energy eigenstates thus coincides with the J_* spectrum. Upon assuming $0 \leq |k_2| \leq k_1$ (such range suffice to describe any possible pair), let us express J_* through the canonical conjugate variables $P = J_y$, $X = J_x/C$, $p = [k_1 k_2 \rho/(k_1 + k_2)](y_1 - y_2)$, $x = x_1 - x_2$, satisfying the commutators $[X, p] = [x, P] = 0$ and $[x, p] = [X, P] = i\hbar$. Then the logarithm argument can be written in the HO form

$$J_*/(k_1 k_2 \rho) = |z_1 - z_2|^2 = (p^2 + w^2 x^2)/w^2 , \tag{5}$$

where the frequency w reads $w = k_1 k_2 \rho/(k_1 + k_2)$. This implies that the states

$$\Psi_n(x; |w|) = C_{\ell,n} \exp[-x^2/2\ell^2] H_n(x/\ell) , \tag{6}$$

where $C_{\ell,n} := \sqrt{2^n \pi \ell\, n!}$, $\ell^2 = \hbar/|w|$, that satisfies the secular equation $(p^2 + w^2 x^2)\, \Psi_n = \hbar|w|(2n+1)\Psi_n$, represent the eigenstates of $|z_1 - z_2|^2$ with eigenvalues

$$S_n(w) = (\hbar/|w|)(2n + 1). \tag{7}$$

A possible way to describe the energy-level degeneracy is that supplied by the quantum number of the conserved quantity $I \equiv J_z$ (this generates the rigid rotations of vortices around the ambient plane origin which leave H unchanged). The complete set of eigenstates $\Psi_{nm}(x, X) = \Psi_n(x; |w|) \Psi_m(X; |C|)$ is readily obtained by rewriting Eq. (3) as $J_z = -(\rho J_* + J_x^2 + J_y^2)/2C$. This makes evident that the HO-like wave function $\Psi_m(X; |C|) \equiv f_{L,m} \exp[-X^2/2L^2] H_m(X/L)$, where $f_{L,m} := \sqrt{2^m \pi L\, m!}$, $L^2 = \hbar/|C|$, diagonalizes $J_x^2 + J_y^2 = P^2 + C^2 X^2$. Hence the eigenvalues associated with J_z have the form

$$\Lambda_m(n; w) = -\hbar(sg(w)n + m + 1), \quad (sg(w) := w/|w|) \tag{8}$$

where we have exploited the fact that $\rho J_*/C = \hbar(2n + 1)w/|w|$ and $C = \rho(k_1 + k_2) > 0$, due to our initial assumptions. Such a description of the energy degeneracy entails a significant geometric-quantum picture. One can show [15] that the integers (n, m) state that the two vortices are confined along two circles of radii $\mathcal{R}_j = |k_1 k_2|\sqrt{S_n(w)}/|k_j||C|$, $j = 1, 2$, while the *locus* allowed for the vorticity center is the circle of radius $X^2 + Y^2 = [\hbar(2m + 1)/|C|]^{1/2}$, labeled by

m. The uncertainty principle prevents one from getting any further information on the angular position both of the vorticity center and of the vortex position on their own circles. The effects of other choices of I are illustrated in Ref. [15].

For $k_2 \to -k_1$ (pure VA case) the canonical scheme based on x, X, p, P breaks down in that $1/(k_1+k_2) \to \infty$. In particular, J_x and J_y tends to x and $y = y_1 - y_2$, respectively, which now commute since $\rho k_1 [x, y] \equiv [J_x, J_y] = i\hbar C \equiv 0$. Such a situation imposes the use of the new (canonically conjugate) variables

$$x\,,\ y\,,\ P_x = (y_1 + y_2)/2\sigma\,,\ P_y = -(x_1 + x_2)/2\sigma, \quad (\sigma := 1/\rho k) \qquad (9)$$

where $-k_2 = k_1 = k$, which turns out to be completely disjoint from those employed in the case when $k_2 \neq -k_1$, and obeys the standard relations $[x, P_x] = [y, P_y] = i\hbar$, $[x, P_y] = [y, P_x] = 0$. Actually P_x, P_y (the vorticity center coordinates, up to the factor $k\rho$) and x, y do not play prefixed roles so that, depending on the interactions involved by the dynamical problem, they can be regarded either as *momenta* or as position variables. If H does not contain further interactions, the simplest choice is to consider vortex wave functions in the *momentum* space picture, namely to treat P_x, P_y as coordinates. Then plane waves

$$\Phi_{\mathbf{K}}(P_x, P_y) = e^{i(Q_x P_x + Q_y P_y)}/2\pi \qquad (10)$$

diagonalize $x^2 + y^2$ namely H, $\hbar Q_x$ and $\hbar Q_y$ being the eigenvalues of $x = i\hbar\partial_{P_x}$ and $y = i\hbar\partial_{P_y}$, respectively. Information on P_x and P_y is, of course, completely missing since the pair cannot localize anywhere in the ambient plane.

3   Pair Quantum Dynamics in a Circular Box

The potential flow function associated to a single vortex placed at z_j inside a circular box is given by $V(z, z_j) = -(k_j/2\pi)\,\mathrm{Re}\ln\left[R(z_j - z)/(R^2 - z\bar{z}_j)\right]$, where R is the box radius. From the kinetic energy functional $H = (\rho/2)\int d^2r\, v^2(r)$, where $v(r) = \nabla V \wedge e_z$ is the velocity field, r is the planar vector version of z, and $e_z \cdot r = 0$ ($|e_z| = 1$), one derives [5], [13] the vortex-pair Hamiltonian

$$H_* = (4\pi)^{-1}[\rho k_1 k_2 \ln|(R^2 - z_1\bar{z}_2)/R(z_1 - z_2)|^2 + \Sigma_j\, \rho k_j^2 \ln\left(1 - |z_j|^2/R^2\right)]. \qquad (11)$$

In the VA case ($k_1 = k = -k_2$) H_* can be written as $H_* = (\rho k^2/4\pi)\ln\mathcal{A}$, where

$$\mathcal{A} = |z_1 - z_2|^2 S_1 S_2/[R^2 S_1 S_2 + R^4|z_1 - z_2|^2]. \quad (S_j := R^2 - |z_j|^2) \qquad (12)$$

This is also available in the form $\mathcal{A} = 1/\{\nu/[2(A_3 - A_1)] + \nu^2/[(A_3 - \nu)^2 - A_4^2]\}$ with $\nu := R^2/2\hbar\sigma$, whose new variables

$$A_\pm = [(2\sigma P_y \pm ix)^2 + (2\sigma P_x \pm iy)^2]/8\hbar\sigma, \ A_3 = \sigma[P_x^2 + P_y^2 + (x^2 + y^2)/4\sigma^2]/4\hbar$$

($A_\pm := A_1 \pm iA_2$ furnish A_1) are recognized to be the generators of the algebra $su(1,1)$ since they satisfy the equations $[A_3, A_\pm] = \pm A_\pm$, $[A_-, A_+] = 2A_3$ when using the commutators pertaining variables (9).

Such a su(1,1) formulation is important for two reasons. First, it allows one to recognize explicitly the constant of motion A_4 ensuring the model integrability. A simple calculation shows, in fact, that the so-called Casimir operator $Q = A_3^2 - A_1^2 - A_2^2$ can be written as $Q = A_4^2 - 1/4$, where $A_4 = (xP_y - yP_x)/2\hbar = (x_1^2 + \sigma_1^2 p_1^2 - x_2^2 - \sigma_2^2 p_2^2)/4\hbar$ with $\sigma_j = 1/\rho k_j$. Consistently, one can check that $[A_s, Q] = 0$, $(s = 1, 2, 3)$ namely $[A_s, A_4] = 0$ so that $[H_*, A_4] = 0$. Second, this fact suffices to identify the Hilbert space basis relying on which the VQD of the bounded pair can be developed. Observing that $\Psi_m(x_j; \sigma)$, defined as in Eq. (6), fulfils the secular equation $[x_j^2 + \sigma_j^2 p_j^2]\Psi_m = \hbar\sigma(2m + 1)\Psi_m$ $(|\sigma_j| = \sigma)$, one finds

$$A_4 \Psi_\mathbf{m}(x_1, x_2) = 2\sigma\hbar(m_1 - m_2) \Psi_\mathbf{m}(x_1, x_2) \quad (\mathbf{m} := (m_1, m_2)),$$

where $\Psi_\mathbf{m}(x_1, x_2) = \Psi_{m_1}(x_1; \sigma)\Psi_{m_2}(x_2; \sigma)$ with $m_j \in \mathbf{N}$ is an element of the Hilbert space basis. The ensuing HO picture strongly differs from the plane wave basis (10) of the unbounded case. Another macroscopic quantum effect is that $4\hbar\sigma A_4 \equiv |z_1|^2 - |z_2|^2$ (the difference, up to a factor π, of the disk areas relative to the radii $|z_1|, |z_2|$) quantally ranges in $\mathbf{Z}$ because of the HO form of $|z_j|^2 = x_j^2 + \sigma_j^2 p_j^2$ with $\hbar\sigma(2m_j + 1) \leq R^2$ owing to the boundary confinement.

3.1   Spectral Structure of Low Energy States

We concentrate now on two special cases related to the system ground-states and their weakly excited states. In second case, we show how the macroscopic exchange symmetry (ES) characterizing H_* ($\mathcal{A}$ is symmetric under $z_i \leftrightarrow z_j$, $i \neq j$) seems to prelude a level splitting. A complete description of the spectrum properties which necessarily requires a preliminary analysis of the phase space structure will be developed elsewhere. From Eq. (12) one recognizes the two (main) ground-state configurations $z_1 \equiv z_2$ ($\rightarrow A_4 = 0$) and $|z_i| \equiv R$, $|z_j| < R$ ($\rightarrow A_4 \neq 0$) (the latter can be realized in two independent ways linked via the ES), we will denote with $\mathcal{C}_1$ and $\mathcal{C}_2$, respectively. Real vortices endowed with finite cores cannot satisfy such conditions in that, below the core size, pair annihilations take place (notice how, close to the boundary, one of the two vortices is virtual: actually this process might/should be modeled in a different way to account for a possible boundary microscopic structure). Nevertheless, $\mathcal{C}_1$, $\mathcal{C}_2$ can be used as reference configurations to describe weakly excited states exempt from VA annihilations. The VA dynamics is governed by the equations

$$i\dot{z}_j = k_j \left[1/(\bar{z}_j - \bar{z}_k) + z_k/(R^2 - z_k z_j^*) - z_j/(R^2 - |z_j|^2)\right]/(2\pi), \quad (k \neq j) \quad (13)$$

derived from H_* by using Eq. (2). Based on them, one can show that when $A_4 = 0$ ($\rightarrow |z_1| \equiv |z_2|$) any solution reduces (up a suitable rotation around the origin) to a pair of curves $z_1(t)$, $z_2(t)$ such that $z_2 = -z_1^*$ for each time t. Such curves are closed since the boundary deviates vortices approaching it. When $A_4 = 0$, low energy states related to $\mathcal{C}_1$ correspond to vortices that are very close initially and placed, e.g., at the box center ($|x_j(0)| \ll R$, $y_j(0) = 0$). After travelling towards the boundary with an almost constant VA distance, vortices scatter (at $y_1 = y_2 \simeq R$) in opposite directions "captured" by their virtual companion, and start to move close to the boundary; finally the pair is reconstituted when they reach the position $y_1 = y_2 \simeq -R$ thus realizing a periodic motion (exchanging vortex initial positions inverts their initial velocities and the motion directions but the curves remain the same).

The double nature affecting this case is confirmed by the fact that $\mathcal{A} \simeq |z_1 - z_2|^2[1 - R^2|z_1 - z_2|^2/S_1 S_2]$ for $R \gg |z_j|$, [since $S_j \simeq R^2$, dynamics depends on $|z_1 - z_2|^2$ which involves a free particle motion as that accounted by Eq. (10)], whereas $\mathcal{A} \simeq S_1 S_2[1 - S_1 S_2/R^2|z_1 - z_2|^2]$ for $|z_j| \simeq R$, $z_2 \equiv -z_1^*$ ($S_j \simeq 0$, $|z_1 - z_2|^2 \approx 4R^2$ entails an HO-like motion of vortices along the boundary). For deciding which behavior dominates quantally, let us introduce $q^2 = (x^2 + y^2)/R^2$, $p^2 = \sigma^2(P_x^2 + P_y^2)/R^2$ to rewrite $\mathcal{A}$ as

$$\mathcal{A} \equiv (1 - q^2 - p^2)^2 - \delta^2/4 + \mathcal{A}^2/(4q^2 - \mathcal{A}). \quad (\delta/2 := 2\hbar\sigma A_4/R^2) \qquad (14)$$

For small value of $\mathcal{A}$ the implicit function $p(q; \mathcal{A})$ defined on a suitable q interval $I_{\mathcal{A}}$ practically coincides with the circle-arc $p = [1 - q^2 - \sqrt{\delta^2/4 + \mathcal{A}}]^{1/2}$ since the term with $\mathcal{A}^2$ is negligible. The latter contains a singularity which drastically makes $p(q; \mathcal{A})$ go to zero, for $q \to \sqrt{\mathcal{A}}/2$. Physically, this represents the excaped annihilation of the two vortices when they reconstitute the pair, after travelling along the boundary. A simple approximation of this (classical) behavior is given by $p_*(q; \mathcal{A}) \simeq [e - q^2 - d/4q^2]^{1/2}$ which gives $e = q_1^2 + q_2^2$, $d = 4q_1^2 q_2^2$ by imposing $p_*(q_i) = 0$ ($I_{\mathcal{A}} = [q_1, q_2]$ where $q_1^2 = \mathcal{A}/4$, $q_2^2 \simeq 1 - (\mathcal{A} + \delta^2/4)^{1/2}$).

The fact that the resulting classical orbits well mimic the exact ones suggests to approximate the Schrödinger problem (SP) (14) through the *effective* one $[p^2 + q^2 + d/4q^2]\Psi(x,y) = \Lambda\Psi(x,y)$ [$\Lambda = e(\mathcal{A})$ must be imposed]. Solving exactly SP (14) requires a separate work due to its complexity; a possible approach based on deriving the dynamical algebra of the SP with eigenvalue-dependent generators is provided in Ref. [18]. The same approach is easily applicable to the effective SP thanks to the further $su(1,1)$ realization [18] $K_\pm = (\mathbf{r} \mp ic\mathbf{P})^2/4c - cg/4r^2$, $K_3 = c(\mathbf{P}^2 + g/r^2)/4 + r^2/4c$, where $\mathbf{P} = (P_x, P_y)$, $\mathbf{r} = (x, y)$. The basic presence of the Coulomb term $1/r^2$, allows one to write the effective SP as $K_3\Psi(x,y) = (R^4\Lambda/\sigma)\Psi(x,y)$ provided $g \equiv dR^4/\sigma^2$, $c \equiv \sigma$. The reduction of this secular equation to a 1d radial problem by inserting $r^2 := x^2 + y^2$, $\theta := \arctan(y/x)$ shows how it is equivalent to the SP $[(d^2/dz^2) - g/z^2 - z^2/c^2 + 4(n - J)/c]\psi_{nJ}(z) = 0$ whose solutions are

$$\psi_{nJ}(z) = (-1)^n D_{nJ}\sqrt{2/z}(z^2/c)^{(\alpha+1)/2}\exp[-z^2/2c]L_n^\alpha(z^2/c), \quad \alpha := -2J - 1.$$

The eigenstates of the effective SP are easily written in terms of such solutions, up to a factor $e^{in\theta}$ diagonalizing the angular part of $\mathbf{P}^2$. This proves how configurations $\mathcal{C}_1$ are expected to have a discrete spectrum whose states may have a free particle behavior just locally.

We conclude showing that $\mathcal{C}_2$ excited states have a strongly quantized spectrum. Also, we consider the circumstances that allows the level splitting (LS) suggested by the ES. Classically, $\mathcal{C}_2$ states are characterized by a vortex running counter-clockwise along the boundary while the other weakly oscillates at the box center; the vortex exchange caused by ES places an antivortex moving clockwise at the boundary but leaves $\mathcal{A}$ unchanged. The fact that the ES action connects distinct orbits close to a pair of isoenergetic minima in the phase space suggests a possible LS. For a particle with energy E_0 in a two-well potential field [19] the LS is given by $\Delta E = E_- - E_+ \propto \exp[-\int_0^d dx\, p(x, E_0)/\hbar]$, where $E_\pm = E_0 \mp \Delta E/2$ are the energies of the symmetrized well states $\psi_\pm$ and $x = d$ is the inversion point related to E_0. Setting $\mathcal{A} \simeq S_1 S_2[1 - S_1 S_2/R^2|z_i|^2]$ if $z_i \gg z_j$, and using the A_4 eigenstates $\Psi_{\mathbf{m}}(x_1, x_2)$ constrained by $2m_r + 1 \leq R^2/\hbar\sigma, r = 1, 2$, the eigenvalues of the $\mathcal{A}$ operator are found to be

$$\lambda(\ell, k) = s_1 s_2 \{1 - s_1 s_2/[R^2\hbar\sigma(2\ell + 1)]\},$$

where $s_r = 1 - \hbar\sigma(2m_r + 1)/R^2$, and $m_i = \ell$ $(m_j = k \ll \ell)$ for the vortex close to (far from) the boundary. Even if the same eigenvalue belongs to the state $\Psi_{\mathbf{m}}(x_2, x_1)$, obtained by the ES action, symmetrized states such as $\Psi_\pm = \Psi_{\mathbf{m}}(x_1, x_2) \pm \Psi_{\mathbf{m}}(x_2, x_1)$ are physically prohibited since an infinite potential barrier is generated by $\mathcal{A}^2/(4q^2 - \mathcal{A})$ in the momentum $p(q, \mathcal{A})$ derived from Eq. (14). Such a problem could be skipped if the A_4 symmetry is lost. Changes of the constant of motion A_4 during the system evolution (caused, e.g., by external interactions as well as by a boundary with $R \neq const$) allow, in fact, to connect the two energy minima through state paths avoiding the potential barrier.

References

1. A. L. Fetter: Phys. Rev. **162**, 143 (1967).
2. M. Rasetti, T. Regge: Physica A **80**, 217, (1975); G.A. Goldin, et al.: Phys. Rev. Lett. **67**, 3499 (1991); V. Penna, M. Spera: J. Math. Phys. **33**, 901 (1992).
3. R. Y. Chiao, A. Hansen et al.: Phys. Rev. Lett. **55**, 2887 (1985); F.D.M. Haldane, Y.-S. Wu: *ibid.* **55**, 1431 (1985); G.A. Goldin et al.: *ibid.* **58**, 174 (1987).
4. J.M. Leinaas: Ann. Phys. (N.Y.) **198**, 24 (1990).
5. V. Penna: Physica A **152**, 400 (1988);
6. A.M. Thompson, J.M.F. Gunn: Physica C **235**, 2953 (1994).
7. R. Fazio, G. Schön: Phys. Rev. B **43**, 5307 (1991); see also references therein.
8. *Proceedings of the 20th International Conference on Low Temperature Physics*, ed. by R.J. Donnelly, [Physica B **194-196**, (1994)].
9. R. Iengo, G. Jug, Phys. Rev. B **54**, 13207 (1996); H. H. Lee, J. M. F. Gunn: *ibid* **57**, 7892 (1998); D.J. Thouless et al: Int. J. Mod. Phys. B **13**, 675 (1999);
10. J.C. Davis et al.: Phys. Rev. Lett. **69**, 323 (1992); G.G. Ihas et al.: ibid. **69**, 327 (1992); see also the articles by E. Varaquaux and R. Packard in the present volume.
11. *Proceedings of the ICTP Workshop on Josephson Junction Arrays*, ed. by H. A. Cerdeira, S. R. Shenoy [Physica B **222**, 336-370 (1996)].
12. F. Dalfovo et. al.: Rev. Mod. Phys. **71**, 463 (1999); M. R. Matthews et. al.: Phys. Rev. Lett. **83**, 2498 (1999); K. W. Madison et. al.: Phys. Rev. Lett. **84**, 806 (2000);
13. R. Lupini, S. Siboni: Nuovo Cimento A **106**, 957 (1991).
14. W.M. Zhang, D.H. Feng and R. Gilmore: Rev. Mod. Phys. **62**, 761 (1990).
15. V. Penna: Phys. Rev. B **59**, 7127 (1999).
16. P. G. Saffman: *Vortex Dynamics* (Cambridge University Press, Cambridge, 1992).

17. L. Onsager: Nuovo Cimento Suppl. **6**, 249 (1949).
18. A. Inomata, H. Kuratsuji and C. C. Gerry, *Path Integral and Coherent States of SU(2) and SU(1,1)*, (World Scientific, Singapore, 1992).
19. L.D. Landau, E.M. Lifsits *Quantum Mechanics* (Pergamon, Oxford, 1957).

Index of Topics

The reference to each item is an article (indicate by a numeral) in a chapter (indicated by a roman numeral). For example V.7 means article 7 in Section 5.

Druck: Strauss Offsetdruck, Mörlenbach
Verarbeitung: Schäffer, Grünstadt

Lecture Notes in Physics

For information about Vols. 1–539
please contact your bookseller or Springer-Verlag

Monographs
For information about Vols. 1–26
please contact your bookseller or Springer-Verlag